Good Cascade Impactor Practices, AIM and EDA for Orally Inhaled Products

Terrence P. Tougas • Jolyon P. Mitchell
Svetlana A. Lyapustina
Editors

Good Cascade Impactor Practices, AIM and EDA for Orally Inhaled Products

Springer

Editors
Terrence P. Tougas
Analytical Development
Boehringer Ingelheim Pharmaceuticals Inc.
Ridgefield, CT, USA

Jolyon P. Mitchell
Trudell Medical International
London, ON, Canada

Svetlana A. Lyapustina
Drinker Biddle & Reath LLP
Washington, DC, USA

ISBN 978-1-4614-6295-8 ISBN 978-1-4614-6296-5 (eBook)
DOI 10.1007/978-1-4614-6296-5
Springer New York Heidelberg Dordrecht London

Library of Congress Control Number: 2013932873

Printed on acid-free paper

Springer is part of Springer Science+Business Media (www.springer.com)

Preface

The measurement of the aerodynamic size properties of aerosols from orally inhaled products has been a long-standing topic of interest to stakeholders because of the need to provide measurements that can ultimately be related to the likely deposition in the human respiratory tract and therefore to subsequent clinical response to the active pharmaceutical ingredient(s). The challenge has been the limited choice of measurement equipment available that meets all of the requirements demanded of it. Historically, the full-resolution multistage cascade impactor has been the workhorse in the laboratory undertaking inhaler performance testing, but its successful operation requires a high degree of operator skill and know-how, and the measurement process is both laborious and time consuming.

This book has been the outcome of several years work within two cross-industry organizations: the Cascade Impaction Working Group of the International Pharmaceutical Consortium on Regulation and Science (IPAC-RS) and the Cascade Impactor Sub-Team of the European Pharmaceutical Aerosol Group (EPAG). The abbreviated impactor measurement (AIM) concept was developed in the middle of 2000 out of the realization that there is a need to offer users of the cascade impaction technique, the opportunity to make measurements of the important aerodynamic size-based metrics, without the need to determine the aerodynamic particle size distribution. From these developments came the realization that in the product quality environment, the minimum number of metrics that is needed relating to the sum and ratio of the large and small particle fractions, are all that is needed to define the APSD. If the boundary dividing them was chosen appropriately, could result in better decision-making than the currently accepted practice in the USA of accumulating the mass of active pharmaceutical ingredient collected on each stage of a full-resolution impactor into groupings that broadly reflected coarse, fine, and extra-fine particulates. This new approach was termed efficient data analysis (EDA) and is the companion concept to AIM, although EDA may be applied to data from either full or abbreviated impactor measurements.

As this volume was in the process of early development, it became evident that in order to describe the AIM and EDA concepts meaningfully, it would be necessary to place these approaches into the context of the existing apparatuses that are defined in the pharmacopeial literature. This material is currently in disparate locations and is therefore not always easy to find quickly. A chapter has therefore been devoted to the description of the underlying theory, including key descriptive information concerning the compendia apparatuses. This chapter also addresses the potential impact that assumptions concerning the properties of the collection efficiency profiles of cascade impactor stages may have both on measurements by full and abbreviated systems, to provide assurance to the reader that such fundamental concerns have been addressed.

In addition to a chapter that reviews in detail and updates the so-called good cascade impactor practice (GCIP) concept developed in February 2002 through the US-based Product Quality Research Institute, this book also contains information on how AIM and EDA could be applied at various stages in the orally inhaled drug product life cycle. There are also several case studies that illustrate how these concepts have already been applied to the assessment of currently marketed products. The intriguing question 'When could AIM fail…?' has also been addressed through scenarios in which the underlying physical processes that influence the size distribution of inhaler aerosols are considered as well as by failure modes analyses related to the cases of pressurized metered dose inhaler and dry-powder-inhaler categories of drug products.

An extensive chapter contains a compilation of the large number of experimental validations of the AIM concept that have been undertaken since 2008. Its purpose is again to provide confidence in the robustness of the concept and, at the same time, to highlight the precautions that should be considered when beginning the process of implementing an AIM-based regimen with or without EDA.

A chapter is included that addresses the regulatory and compendia pathways that will likely need to be followed before either or both concepts become fully accepted as routine approaches by all stakeholders involved in the process of inhaler performance evaluation.

Towards the end of this book, a chapter has been included that considers how the AIM concept might be adapted in the future to explore the possibilities for greater correlation than has been possible hitherto, between impactor-generated measures of particle size and likely particle deposition in the human respiratory tract. This is a highly active topic of current research as the development of robust in vitro–in vivo relationships for the class of orally inhaled products as a whole continues to pose severe challenges.

This volume concludes with a chapter that attempts to look forward and present ideas to encourage further research into the application of both AIM and EDA concepts.

This compilation of knowledge concerning the cascade impaction technique will be of interest to all those who are involved with the day-to-day management of orally inhaled product quality testing, as well as to the researcher seeking to know how AIM and EDA might be applied in novel ways. The authors have provided new insights that will help both the novice and experienced user of the impaction method come to grips with either concept.

At the same time, the authors and editors acknowledge that this field is evolving rapidly, and no single published work could capture all of the possible angles. This book is but a stepstone on the way towards much deeper understanding of the various aspects of cascade impactor testing, and particle sizing in general, of the pharmaceutical aerosols. Future investigations and publications will undoubtedly elaborate on concepts presented here as well as introduce new data and considerations. One of the goals *Good Cascade Impactor Practices, AIM and EDA for Orally Inhaled Products* could serve, therefore, is to provide a helpful backdrop and to strengthen a foundation for further work in this important and actively debated area.

The editors pay tribute to the effort that has gone into development of each contributed chapter, making it both information rich and authoritative and therefore a valuable resource.

Ridgefield, CT, USA — Terrence P. Tougas
London, ON, Canada — Jolyon P. Mitchell
Washington, DC, USA — Svetlana A. Lyapustina

Acknowledgments

The coeditors wish to acknowledge the input from many organizations and individual contributors who are not identified as coauthors of chapters for this book. In this context, they acknowledge the support of the Cascade Impaction Working Group of IPAC-RS and the Impactor Sub-Team of EPAG. The following individuals are also acknowledged:

William Doub, FDA, USA
Prasad Peri, FDA, USA
Marjolein Weda, RIVM, Netherlands
Anthony J. Hickey, University of North Carolina at Chapel Hill and Chair of Aerosol Expert Sub-committee of the USP General Chapters Dosage Forms Committee, USA
Paul Curry, Abbott Laboratories and Member of Aerosol Expert Sub-committee of the USP General Chapters Dosage Forms Committee, USA
Mike Smurthwaite, Westech Instrument Services, UK
Frank Chambers, Covance, UK
David A. Lewis, Vectura, UK
Adam Watkins, Vectura, UK
Andrew D. Cooper, Mylan, UK, formerly Pfizer, UK
Dave Russell-Graham, Mylan, UK, formerly Pfizer, UK
Geoffrey E. Daniels, GSK, UK
Melanie Hamilton, GSK, UK
Fabienne Després-Gnis, Aptar Pharma, France
Gerallt Williams, Aptar Pharma, France
Philippe G.A. Rogueda, Watson Pharmaceuticals, UK, formerly Novartis Pharma, UK
Gracie Sheng, MAP Pharmaceuticals, USA
Elna Berg, Emmace Consulting AB, Sweden, formerly AstraZeneca, Sweden
Yvonne Sizer, Melbourn Scientific, UK
Teresa Russell, Melbourn Scientific, UK
Mark W. Nagel, Trudell Medical International, Canada

Valentina A. Avvakoumova, Trudell Medical International, Canada
Cathy C. Doyle, Trudell Medical International, Canada
Rubina S. Ali, Trudell Medical International, Canada
Heather Schneider, Trudell Medical International, Canada
Mary Devlin Capizzi, Drinker Biddle & Reath, USA
Rebekah A. Grabowski, Drinker Biddle & Reath, USA

Contents

Chapter 1
Introduction

Terrence P. Tougas, Jolyon P. Mitchell, Beth Morgan, and Helen Strickland

Abstract The abbreviated impactor measurement (AIM) and efficient data analysis (EDA) concepts for the in vitro assessment of orally inhaled drug products (OIPs), comprising pressurized metered-dose inhalers (MDIs), soft mist inhalers (SMIs), dry powder inhalers (DPIs), and nebulizing systems, are the topics covered by this book. The chief aims are to provide to potential adoptees of these new methods answers to "how to" questions, as well as to those charged with regulatory oversight assurance that both concepts are fully founded on valid scientific principles. A chapter reviewing the cascade impaction method applied to the assessment of OIPs is also included in order to provide the reader with the necessary background material before exploring the extensions to current methodologies associated with AIM and EDA. Both concepts are related, but do not necessarily need to be applied together, nor is AIM an all-purpose replacement for OIP aerodynamic particle size distribution (APSD) measurements by full-resolution cascade impactor (CI).

T.P. Tougas (✉)
Boehringer Ingelheim Pharmaceuticals Inc., 900 Ridgebury Road,
Ridgefield, CT 06877-0368, USA
e-mail: terrence.tougas@boehringer-ingelheim.com

J.P. Mitchell
Trudell Medical International, 725 Third Street, London,
ON N5V 5G4, Canada
e-mail: jmitchell@trudellmed.com

B. Morgan • H. Strickland
GlaxoSmithKline, Zebulon Manufacturing and Supply,
1011 N. Arendell Avenue, Zebulon, NC 27597, USA
e-mail: beth.e.morgan@gsk.com; helen.n.strickland@gsk.com

T.P. Tougas et al. (eds.), *Good Cascade Impactor Practices, AIM and EDA for Orally Inhaled Products*, DOI 10.1007/978-1-4614-6296-5_1,

1.1 Purpose

It is well understood and accepted that the aerodynamic particle size distribution (APSD) of the pharmaceutical aerosol emitted upon actuation of an OIP is a critical in vitro quality attribute [1]. This is because the OIP has to produce an aerosol with appropriate aerodynamic properties to deposit in a reproducible manner the inhaled medication beyond the upper airway comprising the oropharynx into the airways of the lungs. Ultimately, the size properties of OIP aerosols that are measured in the laboratory should be described in ways that are meaningful in the clinical context, as well as suitable for the purpose of product quality control (QC) during development and production. In practice, meeting the former need has proven a difficult task, mainly because apart from that natural variability in breathing behavior associated with the same patient at different occasions, factors such as patient-to-patient differences in airway caliber as well as disease modality play an important part in the eventual clinical outcomes. Furthermore, clinical measures, such as forced expiratory volume in one second (FEV_1), are intrinsically less sensitive to discern small changes in aerosol particle size distribution, compared with currently available laboratory methods. Given this framework, the purpose of this book is to present cascade impaction as the most suitable method for the laboratory determination of OIP aerosol size properties, both in the context of product QC and in the still developing field of linking these in vitro measures of size to clinical performance of the product.

The multistage CI, originally developed as an air sampling device for use in occupational hygiene, has in the past 30-plus years been adopted and adapted by the community involved with the laboratory measurement of aerosols emitted from OIPs. This outcome is primarily because the impactor-based size-separation process is based on the aerodynamic diameter-related size, rather than on physical (i.e., microscopic) size of the airborne particles [2]. This aerodynamic size scale is more appropriate to describe the motion and ultimate deposition of aerosol particles in the airways of the human respiratory tract (HRT) [3, 4]. The ability to recover quantitatively the drug substance(s), also known as active pharmaceutical ingredient(s) (APIs), of the aerosolized formulation from the components of the CI is also regarded as being of crucial importance by regulatory agencies [5]. This situation precludes the adoption of more rapid and less labor-intensive techniques that do not have the API-detection capability (Table 1.1).

The development of ways in which the cascade impaction method might be improved grew out of increasing awareness by stakeholders in the early to mid 2000s that the existing full-resolution CI-based procedures were not only labor intensive, but prone to a multiplicity of errors arising from the large number of individual (and still mostly manual) operations that have to be performed correctly [1].

The AIM concept was the first of these ideas that, although identified conceptually in the mid-1990s [6], did not gain traction until the false but pervasive paradigm that a multistage impactor somehow is a simulator of the HRT was dispelled by

Table 1.1 Classification of size-analysis methods for OIPs

Aerodynamic methods	Non-aerodynamic methods
Multistage CI and multistage liquid impinger: • Full-resolution CIs provide complete APSDs but are slow and labor intensive • Abbreviated impactors increase speed and throughput and decrease labor and time requirements. They provide simplified APSD metrics such as coarse, fine, and extra-fine particles (potentially relevant for relating to HRT deposition) or based on large and small particles (for product QC)	Laser diffractometry (LD): • Rapid with high size resolution • Can be made noninvasive (no need to extract a sample of the aerosol from that created by the OIP for measurement) • No API specificity, hence inapplicable to mixtures of APIs or API + excipient(s) • The most useful technique for size-analyzing the large droplets from nasal sprays
Time-of-flight (TOF) methods: • Measure APSD directly with high size resolution, but weighting in the distribution is particle count rather than mass-based • Rapid compared with cascade impaction • No API specificity—hence may be inapplicable to mixtures of APIs or API + excipient(s)	Phase (laser) Doppler particle size analysis (PDPA): • Similar to LD in terms of resolution and rapidity • No API specificity, hence may be inapplicable to mixtures of APIs or API + excipient(s) • Complex signal rejection criteria can make representative sampling difficult
–	Single particle light scattering [optical particle counting (OPC)]: • Similar to LD in terms of resolution and rapidity • No API specificity, hence may be inapplicable to mixtures of APIs or API + excipient(s) • A sampling system is needed to allow the aerosol to pass through the measurement zone at a defined velocity
–	Microscopy-automated image analysis: • Moderately fast with automated image analysis, but care is needed to define particle boundaries • When combined with Raman Chemical Imaging may provide specificity to API content

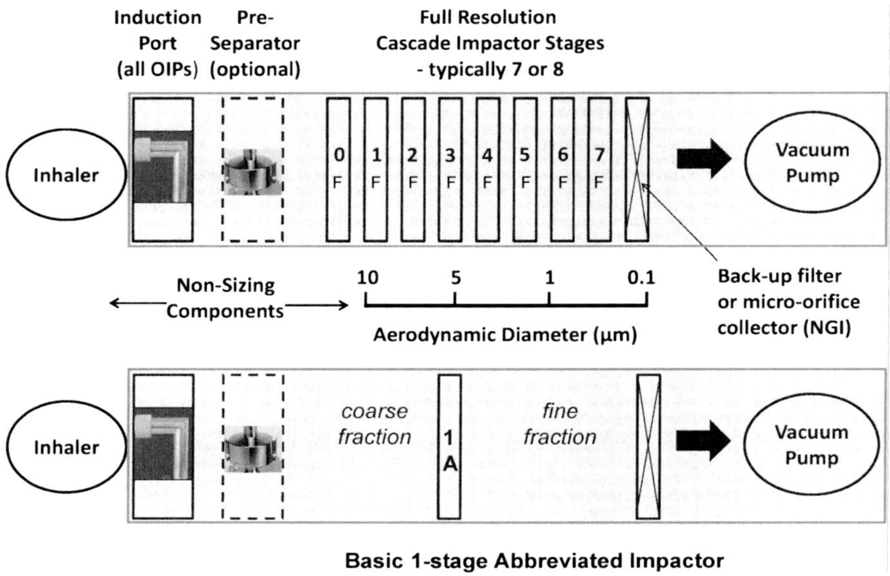

Fig. 1.1 Hypothetical full-resolution CI measurement system (*top*) and a basic 1-stage AIM configuration having its cut-point close to 5.0 µm d_{ae} (*bottom*); the symbols "F" and "A" on the stages identify full-resolution and abbreviated CI configurations, respectively

comparison of the relative size selectivity of individual stages (high) with that of particle deposition in the various compartments of the HRT (low) [7]. In its simplest form, AIM seeks to reduce the components of the cascade impaction system to the minimum number required to be able to determine meaningful metrics related to the cumulative mass of API both finer than a certain aerodynamic diameter (d_{ae}) and larger than this size boundary. Typically, this size boundary is chosen at or close to 5 µm to define so-called fine and coarse mass fractions in accordance with guidance provided in the European (Ph. Eur.) [8] or the United States (USP) [9] Pharmacopeias (Fig. 1.1). Although this figure depicts the abbreviated and full-resolution CI stage cut-point (d_{50}) sizes to match, this agreement may not always be necessary (i.e., it may be appropriate to interpolate data from the full-resolution CI to the stage d_{50} size(s) of the abbreviated system).

AIM-based systems can also include an additional stage with its cut-point size chosen to be close to 1.0 µm d_{ae} for the separate determination of extra-fine mass fraction (Fig. 1.2).

The coarse mass fraction will likely include the additional API recovered from the non-sizing components of the impactor system, such as the induction port and pre-separator (if used). This refinement in the assessment of the contribution to the emitted aerosol by coarse particles is essential when evaluating the behavior of

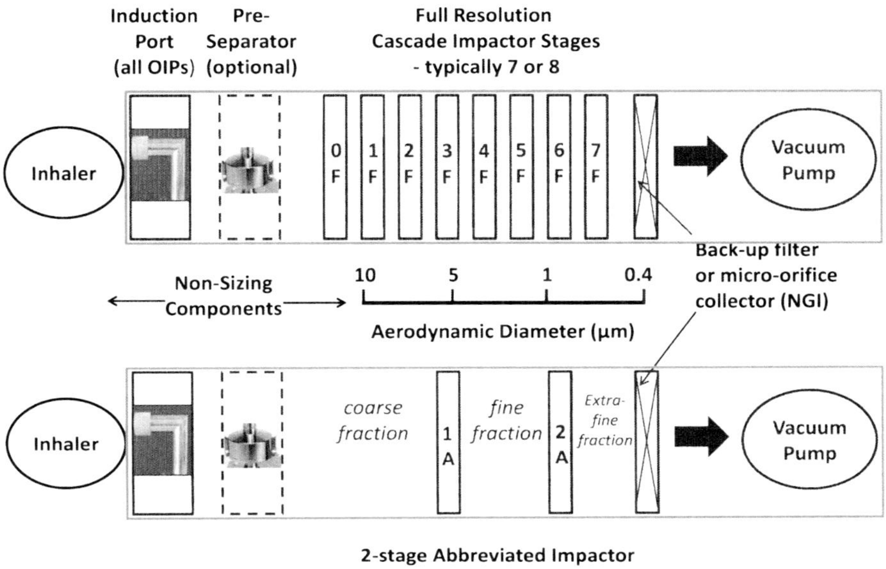

Fig. 1.2 Hypothetical full-resolution CI measurement system (*top*) compared with a 2-stage AIM configuration having cut-points with d_{ae} values close to 1 μm to provide discrimination of extra-fine from fine particles and 5 μm to separate fine from coarse particles (*bottom*); the symbols "F" and "A" on the stages identify full-resolution and abbreviated CI configurations, respectively

add-on devices, in particular spacers and valved holding chambers (VHCs) that are frequently prescribed for use with MDIs [10].

Other enhancements include incorporation of one or more "dummy" stages not containing a collection surface before the first size-separating stage in order to increase the internal dead space within the abbreviated impactor to be closer to that of the full-resolution system. This measure can be important if low-volatile substances, such as ethanol, are present in the formulation being aerosolized. The various AIM-based options that have been tried with different OIPs are examined in Chap. 10.

The EDA concept was developed in parallel with the AIM concept, with the purpose of providing a more discriminating set of APSD-related metrics than the so-called stage groupings, which relate to the mass of API recovered from all components of a full-resolution CI and grouped, typically into three or four adjacent size ranges (Fig. 1.3) [11]. EDA is not intended to be a replacement for existing CI data analysis procedures, such as the determination of fine particle mass (*FPM*) <5 μm in aerodynamic diameter, or the grouping of adjacent stages in a given CI to form a reduced number of measures that can be related to likely deposition locations in the respiratory tract. However, it will be shown explicitly in Chaps. 7 and 8 of the book that EDA does have the capability to be more discriminating of movements of the aerosol APSD than these other techniques of data analysis.

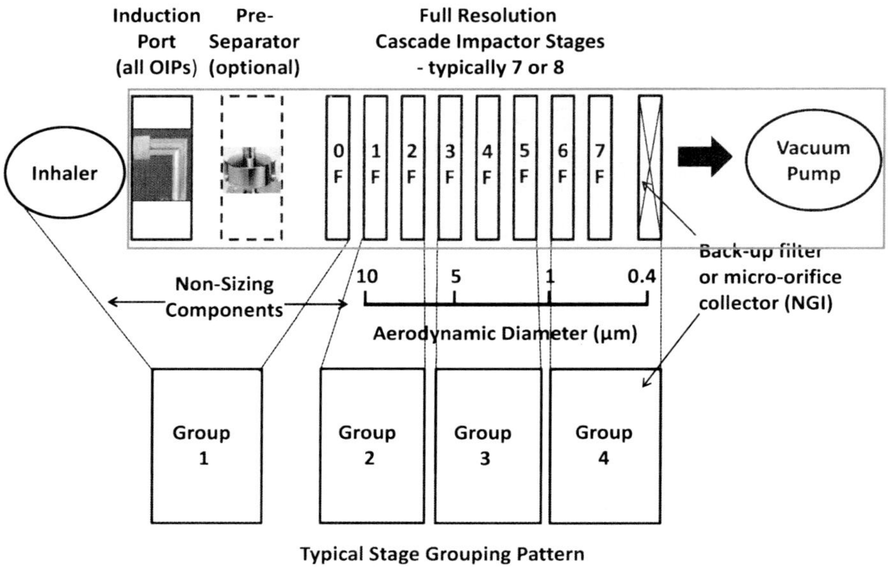

Fig. 1.3 Typical grouping of API mass collected in full-resolution CI; group 1 is the aggregated mass collected by the non-sizing components; groups 2, 3, and 4 define the coarse, fine, and extra-fine mass components of the CI size-fractionated aerosol, respectively; the symbol "F" on the stages identifies full resolution CI configuration

In EDA, consideration of the API deposition data from the CI is limited to the total API mass that is size-fractionated, defined as the impactor-sized mass (*ISM*). Mathematically, *ISM* represents the area under the curve (*AUC*) of the differential mass-weighted APSD. However, it should be noted that in the traditional definition of *AUC*, this quantity is estimated from the best fit of a model describing a process that results in a continuous curved relationship between two independent variables. In contrast, individual values of API mass are assigned to separate stages of the CI, each possessing a small number (no more than eight with the compendial apparatuses) of different and discrete size bounds that are contiguous in the represented differential mass-weighted APSD. In this book, *AUC* will, however, be used as an approximation to the quantity. *ISM*, in the context of EDA, can be conveniently thought of as defining the total mass of API from the OIP aerosol that is sized-fractionated and represented by the continuous form of the APSD.

ISM is further subdivided into large particle mass (*LPM*) and small particle mass (*SPM*) with the boundary ideally fixed at a size close to or identical with the mass median aerodynamic diameter (*MMAD*), representing the central moment of the APSD (Fig. 1.4). The *MMAD* is therefore a function of the distribution of API mass across all stages of the CI. The ratio metric *LPM*/*SPM* is independent of *ISM*, conferring a powerful advantage for EDA for the detection of small shifts in APSD either to finer or coarser sizes.

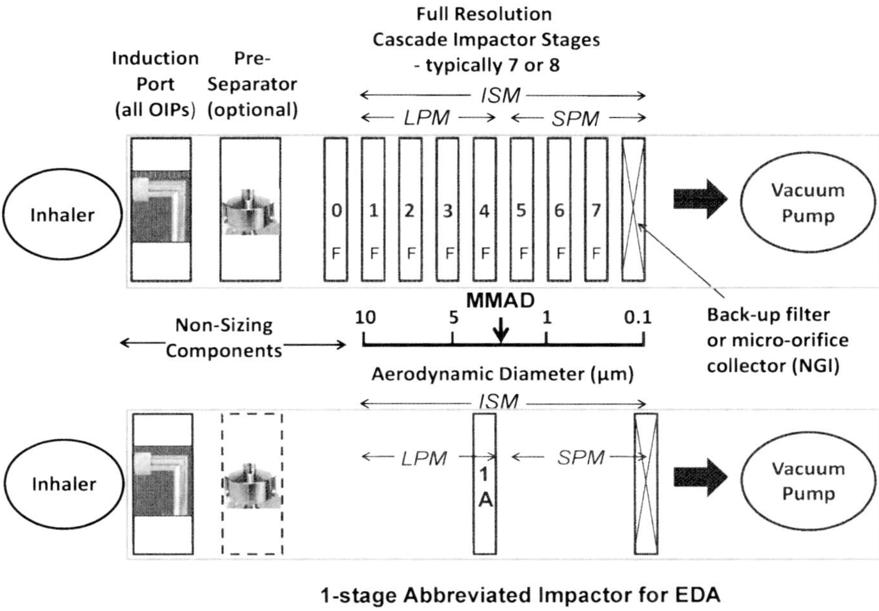

Fig. 1.4 Application of EDA to measurements made by a hypothetical full-resolution CI compared with a single-stage AIM-based system as required for the use of EDA-based metrics; the symbols "F" and "A" on the stages identify full-resolution and abbreviated CI configurations, respectively

The relationships between all the metrics associated with EDA of an OIP-emitted aerosol and their linkage with the various determination methods available either by full-resolution or abbreviated CI apparatuses are summarized in Table 1.2, and a summary as to their derivation is given in Fig. 1.5. Note that EDA can be applied *either* to full-resolution CI data *or* to measurements made by an abbreviated system (Fig. 1.4). It can therefore be linked with the AIM concept, but it is not exclusively restricted to abbreviated CIs.

Table 1.3 provides a "road map" for readers to familiarize themselves with the primary purposes for the acquisition of each of the metrics associated with the CI method, as well as the nature of the metric itself (i.e., whether the outcome from a direct measurement or by calculation based on other direct measurements).

1.2 Scope

This book begins with a review of all aspects of the pharmacopeial methods (US and European) for APSD assessment of OIPs (in Chap. 2). The following chapter describes the physical processes that underlie aerosol formation from OIPs and the

Table 1.2 Relationships between EDA metrics and determination methods

Metric	Metric determination method when fractionation measurements are obtained by		
	Full-resolution CI and assay of each individual deposition sites	Full-resolution CI and assay of combined selected deposition sites	AIM apparatus and assay of combined selected deposition sites
API mass deposited on a site	Actual mass measurement for each stage deposition site selected for assay of API	Actual mass measurement of stage deposition sites selected for combined assay and/or individual assay of API	Actual mass measurement of abbreviated impactor stage deposition sites selected for individual assay of API
Grouped stages	*Mathematical*: direct sum of API mass for selected non-sized and sized CI measurements	Actual API mass measurement of stage deposition sites selected for combined assay for API	N/A
Fine particle mass (*FPM*)	*Mathematical*: direct sum of API mass on deposition sites in CI where stage cutoff sizes (d_{50}) are below ca. 5 µm in aerodynamic diameter (boundary diameter)	Actual API mass measurement of stage deposition sites selected for combined assay for API where cutoff sizes (d_{50}) are below ca. 5 µm in aerodynamic diameter	N/A
Large particle mass (*LPM*)	*Mathematical*: direct sum of API mass measurements for deposition sites with d_{50} values larger than boundary diameter	Actual API mass measurement for deposition sites with d_{50} values larger than boundary diameter for combined assay	Actual API mass measurement of abbreviated impactor deposition site with d_{50} values larger than boundary diameter for assay
Small particle mass (*SPM*)	*Mathematical*: direct sum of API mass measurements for deposition sites with d_{50} values smaller than boundary diameter	Actual API mass measurement for deposition sites with d_{50} values smaller than boundary diameter for combined assay	Actual API mass measurement of abbreviated impactor deposition site with d_{50} values smaller than boundary diameter for assay
Impactor-sized mass (*ISM*)	*Mathematical*: direct sum of API mass measurements for deposition sites with d_{50} values larger and smaller than boundary diameter	*Mathematical*: direct sum of API mass measurements from deposition sites larger and smaller than boundary diameter	Direct sum of API mass measurements of abbreviated impactor deposition sites larger and smaller than boundary diameter for assay

Ratio of LPM to SPM	*Mathematical*: ratio of direct sum of API mass measurements for deposition sites with d_{50} values larger to direct sum of API mass measurements for deposition sites smaller than boundary diameter		
AUC	Integration of continuous form of the distribution of API mass as function of aerodynamic diameter derived from CI calibration	N/A	N/A
MMAD	*Mathematical*: estimation of aerodynamic diameter for 50th percentile of cumulative APSD of API mass deposited in the CI	N/A	N/A
Total emitted mass (*TM*) ex OIP	*Mathematical*: direct sum of API mass associated with non-sizing and sizing components of the CI system		

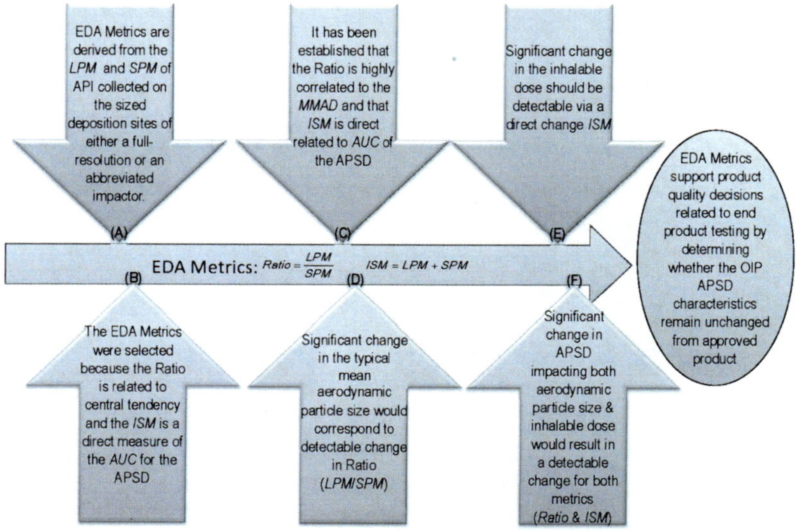

Fig. 1.5 Derivation of EDA metrics

mechanisms that might result in changes to the APSD of the sampled aerosol from these products. Chapter 4 reviews good cascade impactor practice (GCIP) from the standpoint of the laboratory manager or technical staff implementing either full-resolution or abbreviated impactor measurements. The twin concepts of AIM and EDA are introduced in Chap. 5, where a detailed explanation is given as to why these new approaches are needed and how they fit together. Chapter 6 presents the typical lifecycle of an OIP and develops an approach that can be used to help decide when it is more appropriate to use AIM or full-resolution measurements. Chapter 7 sets out the theoretical basis for EDA, based on a measurement systems analysis (MSA) approach. Chapter 8 elaborates on how EDA can improve the decision-making process in the context of OIP quality control, based on statistical arguments. Chapter 9 introduces a series of case studies, in which EDA has been applied, to guide the reader in best practices; this chapter also takes a look at ways in which EDA might theoretically fail to detect shifts in APSD. Chapter 10 presents the outcomes from a large number of experimental studies undertaken to validate the AIM concept, largely by member organizations of the European Pharmaceutical Aerosol Group (EPAG) and the Cascade Impaction Working Group of the International Pharmaceutical Aerosol Consortium on Regulation and Science (IPAC-RS). Chapter 11 examines the regulatory and compendial pathways that will likely need to be followed as AIM and EDA concepts mature into the mainstream of OIP performance testing. Chapter 12 suggests how the AIM concept might be developed to provide measures of OIP performance that are more closely linked with particle deposition behavior in the human respiratory tract. Chapter 13 is a summary of

Table 1.3 "Road map" for EDA-related metrics

Measurement scenario	Metric	Primary purpose	Determination method
Fractionation by Full-resolution Impaction and Measurement by Assay of each Individual Deposition sites Units API mass deposited	API mass deposited on an individual CI stage (M_i)	Product characterization in vitro comparison	Direct quantification after recovery and API assay
	Area under the curve (*AUC*) of the mass-weighted APSD		Integration of continuous form of mass-weighted APSD as function of aerodynamic diameter
	Mass median aerodynamic diameter (*MMAD*)		*Statistical, graphical, or mathematical* estimation of the central tendency of the mass-weighted APSD as the size that corresponds to the 50th mass percentile
	Total (emitted) mass from the OIP (*TM*)	Product characterization Product acceptance in vitro comparison	*Mathematical*: direct sum of API mass recovered from both sizing and non-sizing components of CI system
	Grouped stages	Product acceptance	*Mathematical*: direct sum of API mass recovered from selected collections of adjacent stages and external components of CI system
Fractionation by Full-resolution or abbreviated impaction and Measurement by Assay of each Individual Deposition sites Units API mass deposited	Large particle mass (*LPM*)	Product acceptance	*Mathematical*: direct sum of API mass recovered from stages with d_{50} sizes larger than boundary diameter
	Small particle mass (*SPM*)		*Mathematical*: direct sum of API mass recovered from stages with d_{50} sizes smaller than boundary diameter
	Impactor-sized mass (ISM) [ISM = LPM + SPM]		*Mathematical*: direct sum of API mass recovered from stages with d_{50} sizes both larger and smaller than boundary diameter
	Ratio: LPM/SPM		*Mathematical*: ratio of direct sum of API mass recovered from stages with d_{50} sizes larger than boundary diameter to mass recovered from stages with d_{50} sizes smaller than boundary diameter

(continued)

Table 1.3 (continued)

Measurement scenario	Metric	Primary purpose	Determination method
Fractionation by Full-resolution impaction and Measurement by Assay of each Individual Deposition sites	API mass deposited on an individual CI stage (M_i)	Product acceptance	Direct quantification after recovery and API assay
	Area under the curve (*AUC*) of the mass-weighted APSD	N/A	N/A
	Mass median aerodynamic diameter (*MMAD*)	N/A	N/A
Units API mass deposited	Total (emitted) mass from the OIP (*TM*)	Product acceptance	*Mathematical*: direct sum of API mass recovered from both sizing and non-sizing components of CI system
	Grouped stages		*Mathematical*: direct sum of API mass recovered from selected collections of adjacent stages and external components of CI system
	Large particle mass (*LPM*)		*Mathematical*: direct sum of API mass recovered from stages with d_{50} sizes larger than boundary diameter
	Small particle mass (*SPM*)		*Mathematical*: direct sum of API mass recovered from stages with d_{50} sizes smaller than boundary diameter

current thinking about the future of both concepts in the context of OIP testing, and Chap. 14 provides a series of concluding statements that reflect the current position concerning both concepts. The book also contains a glossary of abbreviations and acronyms in common usage in connection with OIP testing, together with an index of keywords to aid in searching the text.

References

1. Bonam M, Christopher D, Cipolla D, Donovan B, Goodwin D, Holmes S, Lyapustina S, Mitchell J, Nichols S, Petterson G, Quale C, Rao N, Singh D, Tougas T, Van Oort M, Walther B, Wyka B (2008) Minimizing variability of cascade impaction measurements in inhalers and nebulizers. AAPS PharmSciTech 9(2):404–413
2. Mitchell JP, Nagel MW (2003) Cascade impactors for the size characterization of aerosols from medical inhalers: their uses and limitations. J Aerosol Med 16(4):341–377
3. Rudolph G, Köbrich R, Stahlhofen W (1990) Modeling and algebraic formulation of regional aerosol deposition in man. J Aerosol Sci 21(S1):S403–S406
4. Heyder J, Svartengren MU (2002) Basic principles of particle behavior in the human respiratory tract. In: Bisgaard H, O'Callaghan C, Smaldone GC (eds) Drug delivery to the lung. Dekker, New York, NY
5. Christopher D, Curry P, Doub B, Furnkranz K, Lavery M, Lin K, Lyapustina S, Mitchell J, Rogers B, Strickland H, Tougas T, Tsong Y, Wyka B (2003) Considerations for the development and practice of cascade impaction testing including a mass balance failure investigation tree. J Aerosol Med 16:235–247
6. Van Oort M, Roberts W (1996) Variable stage-variable volume strategy for cascade impaction. In: Dalby RN, Byron PR, Farr EJ (eds) Respiratory drug delivery-V. Interpharm Press, Buffalo Grove, IL, pp 418–421
7. Dunbar C, Mitchell JP (2005) Analysis of cascade impactor mass distributions. J Aerosol Med 18(4):439–451
8. European Directorate for the Quality of Medicines and Healthcare (EDQM) (2012) Preparations for inhalation: aerodynamic assessment of fine particles, Chapter 2.9.18. In: European Pharmacopeia, 7th edn. Strasbourg, France
9. United States Pharmacopeial Convention (USP) (2012) Aerosols, metered-dose inhalers, and dry powder inhalers USP35-NF30, Chapter 601. Rockville, MD
10. Mitchell JP, Dolovich MB (2012) Clinically relevant test methods to establish *in vitro* equivalence for spacers and valved holding chambers used with pressurized metered dose inhalers (pMDIs). J Aerosol Med Pulm Deliv 25(4):217–242
11. Tougas TP, Christopher D, Mitchell JP, Strickland H, Wyka B, Van Oort M, Lyapustina S (2009) Improved quality control metrics for cascade impaction measurements of orally inhaled drug products (OIPs). AAPS PharmSciTech 10(4):1276–1285

Chapter 2
Current Approaches to APSD Measurements of OIPs Based on Inertial Impaction

Jolyon P. Mitchell and Daryl L. Roberts

Abstract The AIM and EDA concepts are founded on the principles of inertial impaction of aerosol particles under laminar flow conditions. This chapter examines the current CI systems that are recognized by the pharmaceutical compendia, providing a summary of the key parameters that affect the size-resolving capability of each system. The potential for bias introduced through the assumption that individual stage collection efficiency curves are step functions at the calibration size is explored, with attention given to the effect of removing stages in order to achieve an AIM-based configuration. Non-sizing accessories, such as the induction port (IP) entry and preseparator (PS), are discussed and the chapter concludes with consideration of how add-on devices, such as spacers and VHCs that are commonly used in conjunction with MDIs, should be evaluated.

2.1 Introduction

Aerosols produced by all types of OIP comprise either solid particles or liquid droplets suspended in air that the patient receiving therapy is intended to inhale. If suspended particles are present, each droplet itself will strictly be inhomogeneous, if one or more particles are incorporated within each droplet. For simplicity, the term "particles" will be used from now onwards to include droplets as well as solid particles. Aerosols are by definition semistable phenomena, in that the size distribution of particles continually changes with time due to several physical processes happening simultaneously [1]. The most important of these processes affecting the

J.P. Mitchell (✉)
Trudell Medical International, 725 Third Street, London, ON N5V 5G4, Canada
e-mail: jmitchell@trudellmed.com

D.L. Roberts
MSP Corporation, 5910 Rice Creek Pkwy, Ste 300, Shoreview, MN 55126, USA
e-mail: droberts@mspcorp.com

T.P. Tougas et al. (eds.), *Good Cascade Impactor Practices, AIM and EDA for Orally Inhaled Products*, DOI 10.1007/978-1-4614-6296-5_2,

size range of OIP-produced aerosols (0.5- to 10-μm aerodynamic diameter) are as follows (in no particular order of priority) [2]:

1. Gravitational sedimentation
2. Turbulent deposition to adjacent surfaces
3. Particle–particle agglomeration (if the particle concentration is sufficiently high)
4. Flash evaporation of highly volatile propellants associated with the aerosol formation process with MDIs
5. Evaporation/condensation of associated low-volatility substances (e.g., ethanol cosolvent incorporated with some MDI formulations as well as ambient moisture, if present)
6. Molecular (Brownian) diffusion

The latter process is only significant with the finest particles < ca. 0.5-μm physical (geometric) diameter. The influence of these processes on CI-measured APSDs is reviewed in Chap. 9 (Sect. 9.4) in the context of evaluating how resulting changes in APSD may or may not be detected by the efficient data analysis (EDA) metrics.

This book is concerned with the CI method that determines the aerodynamic rather than physical (e.g., geometric) size of such particles, as would be measured by means of a microscopy-based technique [1]. As a general rule, particles in the size range from about 0.5 to 10 μm in aerodynamic diameter will deposit somewhere in the human respiratory tract (HRT) and larger particles in the upper airways (oropharyngeal region), with finer particles penetrating progressively further into the 23 generations of the airways of the lungs before the finest ultimately reach the alveolar sacs in which gas exchange takes place.

Detailed descriptions of aerosol mechanics associated with transport through the HRT can be found in the books by Hinds [1] and Finlay [2]. The explanation given here is intended to provide the basics in order to understand the capability and limitations of the inertial impaction method for size-characterizing aerosols emitted from OIPs.

Particle motion throughout the airways of the respiratory tract is assumed to take place following Stokes's law, in that the relative velocity of the gas at the surface of the particle is zero. Under Stokesian motion in a stagnant support gas, the force (F_d) acting on a particle in a fluid (air) comprises both form and frictional components, such that

$$F_d = 3\pi\eta_a v_t d_p \tag{2.1}$$

where η_a is the air viscosity which is a function of the temperature of the air, v_t is particle terminal velocity, and d_p is the particle volume-equivalent diameter [i.e., the diameter of a spherical particle of the same volume and the same as the physical (geometric) diameter for particles possessing spherical geometry]. Particles rapidly reach their setting velocity (v_t) when released from rest, and the drag force (F_d) is balanced by the gravitational force F_g, where $F_g = mg$, so that

$$v_t = \frac{\rho_p d_p^2 g}{18\eta} \tag{2.2}$$

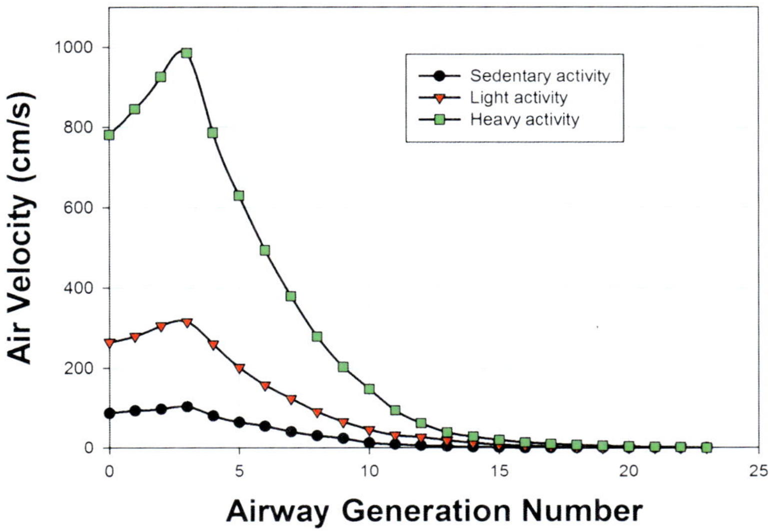

Fig. 2.1 Particle velocity-airway generation profiles in the HRT for three levels of adult physical activity; sedentary is defined as $V_t = 500$ mL, 14 breaths/min, and PIFR = 14 L/min; light activity is $V_t = 1{,}291$ mL, 15.5 breaths/min, and PIFR = 40 L/min; heavy activity is $V_t = 2{,}449$ mL, 40 breaths/min, and PIFR = 120 L/min (*adapted from* [3])

Moving from the stagnant environment in which Stokes's law is defined to consider particle motion in the flow through the respiratory tract, an interesting feature is the steady decrease in velocity with increasing airway generation number beyond generation 3 (Fig. 2.1) [3]. In contrast, CIs size particles by increasing the velocity of air from stage to stage, as the aerosol passes through the apparatus. Furthermore, their diameters may decrease slightly during passage, due to the continuous evaporation of any volatile liquid species present, whereas in the HRT, some drug particles may grow due to the presence of near-saturation relative humidity conditions in their immediate surroundings. Hence, although the physical principles of impaction apply equally to the CI as to the lung, these aerosol measurement devices do not in any sense "simulate" the lung, as will be discussed further below.

Aerodynamic diameter (d_{ae}) represents the size scale that takes into account the effect of both particle density and shape on mobility in any flow field, such as in the respiratory tract, and is related to d_p, which is the diameter of a spherical particle (dynamic shape factor, χ, of unity and density of 10^3 kg/m^3) through the expression

$$d_{ae} = \left[\frac{\rho_p}{\rho_0 \chi} \right]^{1/2} d_p C_c \tag{2.3}$$

where ρ_p is the particle density (kg/m^3). C_c is the Cunningham slip correction factor, which reflects reduction in F_d due to "slip" of the particle by adjacent gas molecules. For particles with $d_{ae} < 1.0$ µm, C_c can be described in terms of the particle Knudsen number (Kn_p) in which the size of the particle in terms of d_p is related to the mean free path length of a molecule of the surrounding air (λ), where [1]

$$Kn_p = 2\lambda / d_p \tag{2.4}$$

and

$$C_c = 1 + 0.5Kn_p\left[2.34 + 1.05\exp\left(-0.195Kn_p\right)\right] \tag{2.5}$$

The value of C_c asymptotically approaches unity for particle sizes larger than a few micrometers and can generally be ignored unless the mass-weighted APSD (either differential or cumulative plot of API mass versus d_{ae} [1]) contains an appreciable proportion of particles < 0.5 μm.

The multistage CI links the determination of mass of API in the OIP aerosol with d_{ae} and has been therefore accepted by both the compendial [4, 5] and regulatory [6, 7] authorities as the apparatus of choice for sizing these aerosols. Alternative aerodynamic particle size characterization techniques, such as particle time of flight (TOF) in accelerating motion, can size aerosols based on the d_{ae} scale more rapidly than the CI method [8]. However, such near real-time methods do not, at the present time, provide traceability to the mass of API(s) in the formulation through the use of well-defined and widely accepted analytical assay principles (i.e., high-performance liquid chromatography combined with a quantitative spectroscopic detection method, usually UV/visible light absorption or fluorescence that is API specific) [9]. This latter attribute is regarded as being of critical importance by the major regulatory agencies [6, 7]. The foci of this chapter, and the book as a whole, are therefore on OIP aerosol particle sizing by CI-based methods.

2.2 The CI Is Not an In Vitro Analog of the Human Respiratory Tract

It has often, but erroneously, been promoted that multistage CI-derived APSD data can directly indicate the fate of the particles in the human respiratory tract (Fig. 2.2).

If this situation was totally true, such data would be ideal for estimating the likelihood of clinical response in studies of the efficacy and safety of OIPs. However, it is important at the outset to realize that the CI is fundamentally not a lung simulator, as more mechanisms governing particle deposition are at play in the HRT. The difference between the two becomes readily apparent when overlaying the collection efficiency curves of the ACI, chosen as a representative multistage CI, with the corresponding empirical/theoretical particle collection efficiency curves associated with the morphological regions within the respiratory tract (Fig. 2.3).

The relatively poor size selectivity of the HRT has been recognized for many years by the community of aerosol scientists involved with environmental protection [11], where three subfractions covering inhalable particles are currently

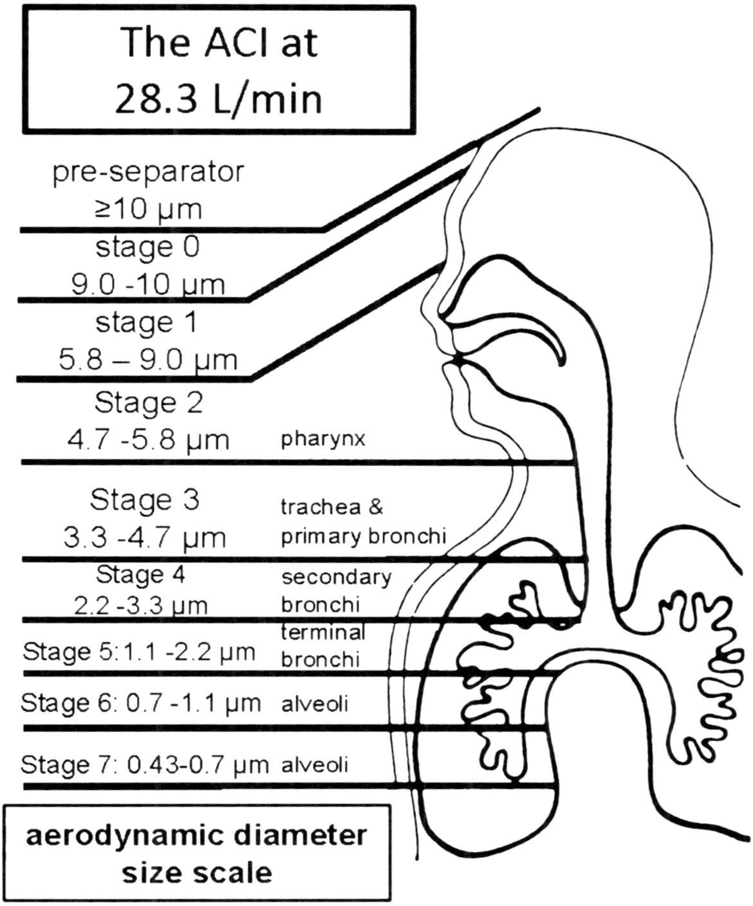

Fig. 2.2 A postulated, but unfounded, relationship of the ACI to particle deposition in the human respiratory tract (*adapted from the operators' manual for the ACI*)

recognized: (1) PM_{10}, suspended coarse particles with $d_{ae} < 10$ µm; (2) $PM_{2.5}$, suspended fine particles with $d_{ae} < 2.5$ µm, analogous to the majority of the fine particles from OIPs; and (3) $PM_{1.0}$, suspended ultrafine particles with $d_{ae} < 1.0$ µm (approximately equivalent to extrafines from OIPs).

The US FDA, in its 1998 draft Guidance for Industry document relating to Chemistry, Manufacturing and Controls for OIPs [7], recommended treating cascade impaction data as follows: "Data from full resolution CI measurements may also be presented in terms of the percentage of the mass found on the various stages and accessories relative to the label claim. Acceptance criteria may be proposed in terms of appropriate groupings of stages and/or accessories. However, if this approach is

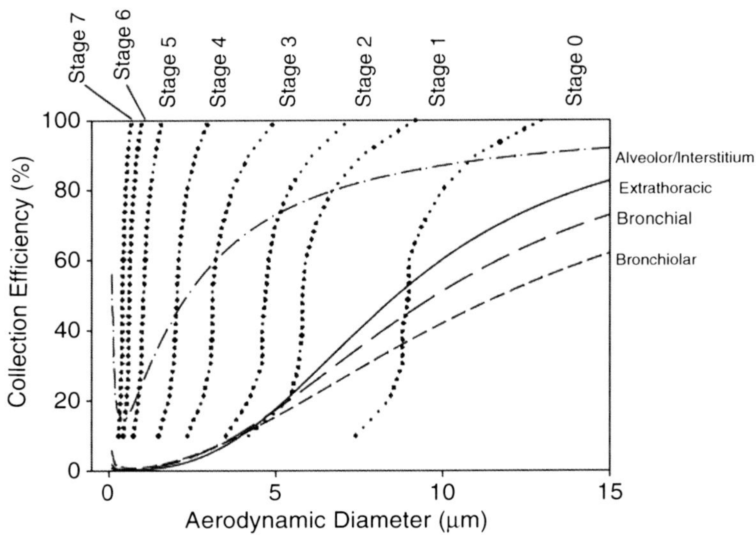

Fig. 2.3 Collection efficiencies of the ACI stages ($Q = 28.3$ LPM) and of morphological regions of the lung (PIFR = 28.3 LPM; $V = 2$ L; healthy male) (*from* [10]—*used with permission*)

used, at a minimum there should be three to four groupings to ensure future batch-to-batch consistency of the particle size distribution." Currently, these specifications often require that the ratio of the upper to lower allowed mass of API on stage groupings be 1.5:1 or 2:1. However, these ratios are based neither on drug class specific nor on mechanism of action, or on clinical factors, such as patient age and disease modality.

In relation to single clinical dose testing, the corresponding European regulatory guidance [6], which is harmonized with equivalent Health Canada (HC) guidance [12], focuses on the determination of fine particle mass with $d_{ae} < 5$ μm and only refers to pooling of mass on stages in the context of limited analytical sensitivity for the API. For APSD determination, the EMA and HC guidance documents state: "individual stage particle size distribution data should be provided for the batches used in these studies, as well as data on batches representative of the commercial process."

From the perspective of product QC, however, OIP performance metrics based on stage groupings or even stage-by-stage data are confounded with respect to changes in APSD, for example, because mass transferred from one group to an adjacent group in direction of either increasing or decreasing aerodynamic size is inevitably associated with a compensating decline in the mass associated with the group from which the mass was transferred (Fig. 2.4). In Chaps. 7 and 8, it will be demonstrated that a suboptimal performance of these metrics for product quality control (QC) purposes exists compared to the capability of the EDA method.

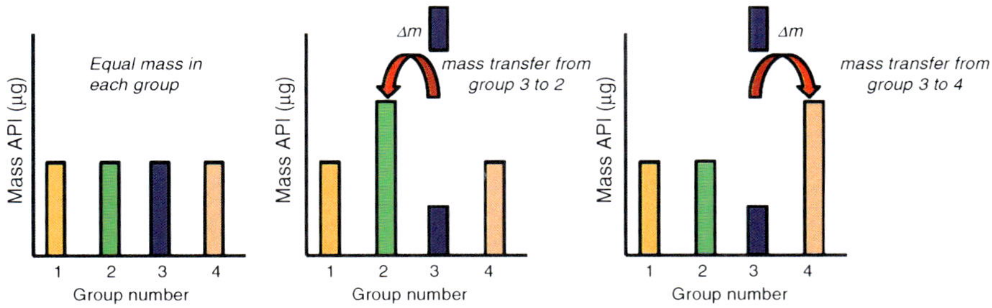

Fig. 2.4 API mass transfer from one stage grouping to its neighbor results in an equal and opposite change to both groupings

2.3 Capability of the Cascade Impaction Method

The multistage CI is fundamentally a particle size-fractionating instrument, making use of differences in particle inertia in (ideally) a laminar flow field to result in either collection on a solid substrate or impingement into a liquid. CIs are designed to operate at a constant flow rate. This limitation therefore becomes an issue in connection with DPI testing, in which it is necessary to approximate an idealized inspiratory maneuver in order to aerosolize and disperse the aliquot of dry powder either from a reservoir or from a punctured retaining capsule/blister. A compromise is reached in the compendial DPI testing procedures, in that although the CI is operated from zero flow at start of sample, sufficient volume of air is drawn through the system (typically 4 L), so that the finest suspended particles have time to pass through the entire apparatus, enabling the CI to properly size fractionate the emitted dose. This requirement places a limit on the lowest sample volume it is reasonable to use, because below this volume there will be insufficient time for the aerosol to pass entirely through the CI system to be fractionated. This minimum volume is based on the magnitude of the internal dead volume of the CI together with the preseparator (PS), if used, and induction port (IP). Recent work undertaken by the European Pharmaceutical Aerosol Group (EPAG), in which this sample volume has been intentionally reduced to be close to the dead volume of the system (ca. 2 L), has confirmed that the APSD is significantly affected in the case of the NGI under such circumstances [13]. In the NGI, operated at 60 L/min, $FPF_{0.94\text{–}4.46\mu m}$ was shown to be systematically reduced with decreasing sample volume, reflecting the fact that the finer particles have had insufficient time to move through the CI to the stages at which they would normally be collected. In contrast, the ACI appeared not to be as sensitive as the NGI, with $FPF_{0.76\text{–}6.18\mu m}$ remaining nearly independent of sample volume. The underlying reason for this anomalous behavior is not obvious, but it is believed to be likely related to flow maldistribution in the ACI. Given this situation, it is good sampling practice to ensure that the sample volume exceeds the dead volume by at least a factor of two.

The use of CIs in conjunction with breathing simulators is at first sight at odds with the constant flow rate limitation. However, as will be discussed in Chap. 12, various groups have found ways to effect such an interface, enabling the OIP to be operated either through a half cycle (i.e., inspiration only) for DPIs or in conjunction with a continuously varying flow rate cycle associated with mimicking tidal respiration. At the same time, the CI is simultaneously sampling the resulting aerosol at a fixed flow rate. In general, these arrangements are complex [9] and have therefore thus far not been incorporated into the compendial methods. There are also limited validation data available, so that caution is urged before implementing such an approach.

A further limitation of the CI technique is its relatively limited size-resolving capability compared with that obtainable by optical particle detection methods, such as TOF, OPC, LD, PDPA, and even microscopy image analysis [8]. The resolution of a CI is restricted by the finite size range within which a given stage operates, as will be seen in the next section of this chapter. The NGI, designed to optimize resolution, is capable of providing an APSD with at most five measurements of mass within the important range from 0.5- to 5.0-μm aerodynamic diameter [14]. Attempts to develop a CI with greater than this size resolution will fail because the resulting APSDs will inevitably contain intrinsic bias brought about through substantial overlap of the collection efficiency-size profiles of adjacent stages.

2.4 Fundamentals of Inertial Size Fractionation Affecting Multistage CI Performance

The principles of multistage CIs in the context of in vitro testing of OINDPs were reviewed in 2003 by Mitchell and Nagel [9]. Little of a fundamental nature in terms of the underlying theory has changed since this date, although there has been a concerted effort to acquire a more robust understanding of the processes associated with calibration and validation of full-resolution cascade impaction systems [14, 15].

Key aspects of cascade impaction theory are presented in this chapter to familiarize the reader with sufficient background information to understand the advantages and, perhaps more importantly, the limitations of the AIM-based techniques that are introduced in Chap. 5.

A typical CI used for OIP testing comprises several stages (Fig. 2.5), each of which functions as a size fractionator of the incoming aerosol in a gas stream moving at constant velocity [proportional to volumetric flow rate (Q)]. In concept, a single-stage impactor comprises a jet or nozzle plate containing one or more circular or slot-shaped orifices located a fixed distance from a collection surface that is usually horizontal. The stage functions by classifying incoming particles of various sizes (d_{ae}) on the basis of their differing inertia, the magnitude of which reflects the resistance to a change in direction of the laminar flow streamlines.

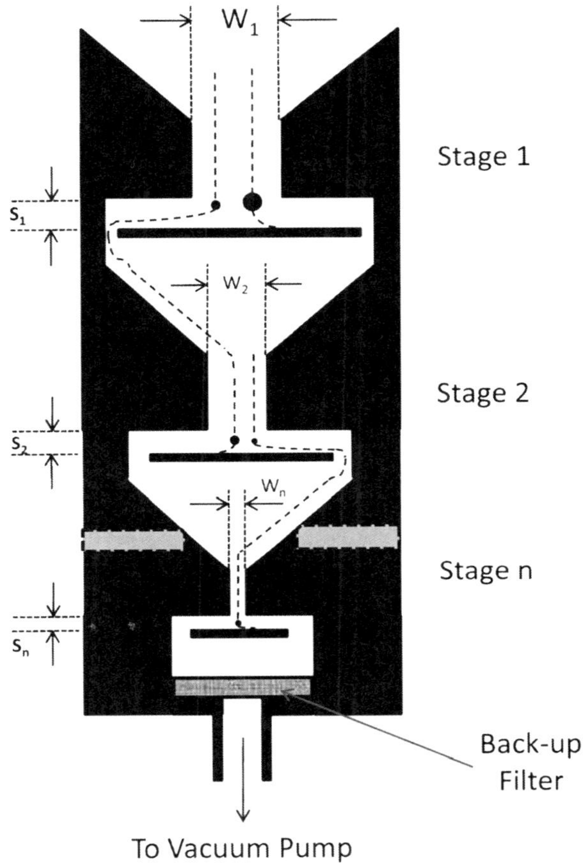

Fig. 2.5 Idealized *n*-stage cascade impactor

As the incoming flow passes through the nozzle plate, the streamlines diverge on approach to the collection surface, whereas the finite inertia of the particles causes them to cross the streamlines, if sufficiently large. The dimensionless Stokes number (*St*), which is the ratio of the stopping distance of a particle to a characteristic dimension, in this case the nozzle diameter, *W* (or average diameter, for a multiorifice stage), describes the process. It will be seen later that the square root of *St* defines a critical particle size in terms of aerodynamic diameter that will succeed in leaving the airstream and reaching the collection surface for a particular stage geometry (Fig. 2.6).

The modern theory underlying impactor function has been developed over the past 35 years based firstly upon solving the Navier–Stokes equations for steady, incompressible, and isothermal conditions that define the gas flow field having viscosity, η, and density, ρ_g, modeling the simplified geometry of a single-stage "ideal" impactor. This operation is undertaken in the absence of particles.

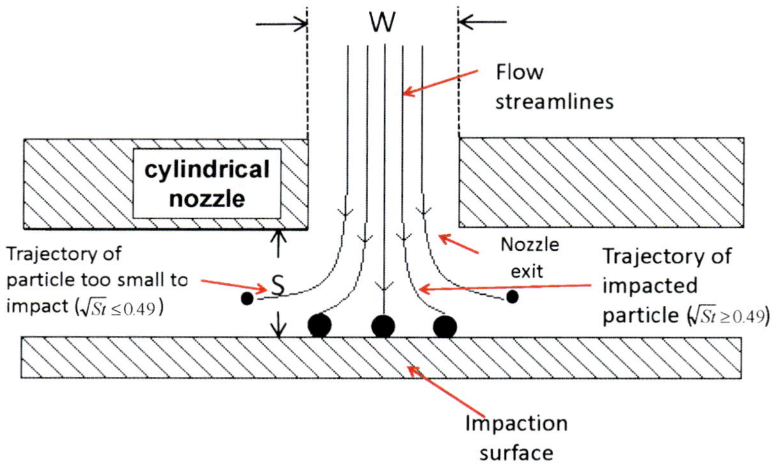

Fig. 2.6 Hypothetical single-stage impactor showing trajectories of particles that are collected and those that miss the collection substrate

Subsequently, Newton's equation of motion is used to model the passage of different-sized particles through the ideal impactor. The process may ultimately be extended to evaluate the particle transport and size fractionation through impactor stages having various nozzle dimensions (diameter, *W*, and length, *L*) as well as several nozzle-to-collection surface distances (*S*) [16]. Flow is normally assumed to be laminar within the stages of the CI. However, if gas flow and particle transport through the induction port inlet are also being modeled, an appropriate model of turbulence such as the low Reynolds number (LRN) κ-ω approach can be introduced, based on its ability to accurately predict pressure drop, velocity profiles, and shear stress for transitional and turbulent flows [17].

Returning to the idealized single-nozzle (jet) impactor (Fig. 2.5), *St* is related to *W* through the expression

$$St = \frac{\rho_p C_c d_p^2 U}{18 \eta W} \tag{2.6}$$

in a self-consistent set of units, based on d_p. *U* is the linear velocity of the particle, which can be considered the same as the surrounding local air velocity when the flow rate through the impactor is constant. The Cunningham slip correction factor, C_c, that takes into account the faster settling of particles whose size is close to that of the mean free path of the surrounding gas is defined by (2.4) and (2.5).

The particle collection efficiency (*E*) of an ideal impactor stage, expressed as a percentage, will increase in a stepwise manner between limits of 0–100%. In practice, for a well-designed stage, *E* is a monotonic sigmoidal function of either *St* or d_p that increases steeply from $E \approx 0\%$ to >95%, reaching its maximum steepness

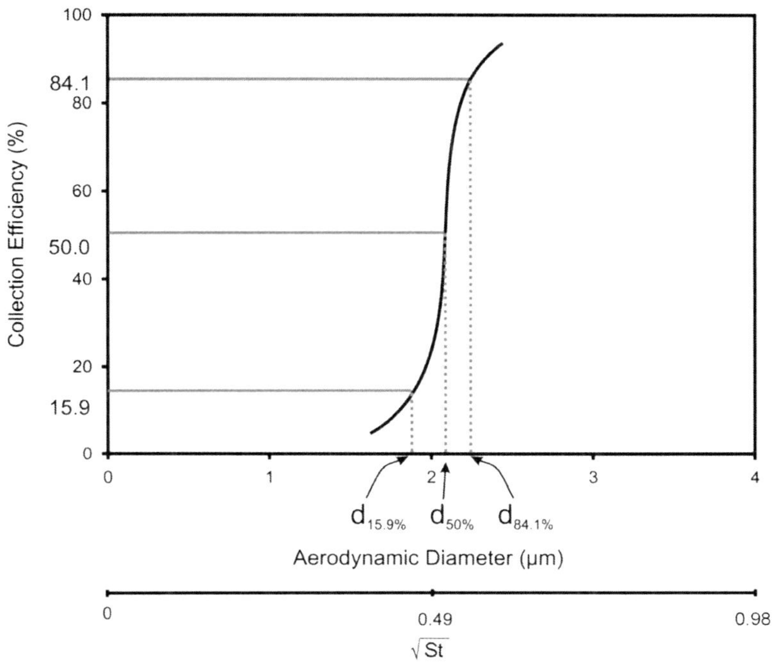

Fig. 2.7 Hypothetic single stage impactor collection efficiency curve

when E is 50% (Fig. 2.7). At this location, defined as the cut size ($d_{p,50}$) or effective cutoff diameter,

$$\sqrt{C_{c,50}}\, d_{p,50} = \left[\frac{9\eta n W}{\rho_0 U}\right]^{1/2} \sqrt{St_{50}} \tag{2.7}$$

or in terms of volumetric flow rate (Q),

$$\sqrt{C_{c,50}}\, d_{p,50} = \left[\frac{9\pi\mu n W^3}{4\rho_0 Q}\right]^{1/2} \sqrt{St_{50}} \tag{2.8}$$

for a multiorifice stage comprising n circular nozzles. Equations (2.7) and (2.8) can be written in terms of the more usually encountered aerodynamic diameter scale ($d_{ae,50}$), substituting (2.3) for d_p.

The assumption is often made that the mass of particles larger than d_{50} (the size corresponding to E_{50}) that penetrate the stage is exactly compensated by the mass associated with particles finer than this size that are collected. Thus, the cut size can be defined as a single-valued constant for a given stage at a fixed flow rate. Particles with $d_{ae} \geq d_{50}$ are assumed to be fully collected, whereas all particles finer than d_{50} are

Table 2.1 Values of d_{50} (μm) and $\mathrm{GSD}_{\mathrm{stage}}$ for the NGI size-fractionating stages at flow rates associated with its archival calibration

Q=	15 L/min		30 L/min		60 L/min		100 L/min	
Stage	d_{50}	$\mathrm{GSD}_{\mathrm{stage}}$	d_{50}	$\mathrm{GSD}_{\mathrm{stage}}$	d_{50}	$\mathrm{GSD}_{\mathrm{stage}}$	d_{50}	$\mathrm{GSD}_{\mathrm{stage}}$
1[a]	14.1	1.39	11.7	1.34	8.06	1.33	6.12	1.35
2	8.61	1.16	6.40	1.19	4.46	1.21	3.42	1.26
3	5.39	1.15	3.99	1.21	2.82	1.24	2.18	1.27
4	3.30	1.12	2.30	1.11	1.66	1.17	1.31	1.22
5	2.08	1.14	1.36	1.11	0.94	1.17	0.72	1.20
6	1.36	1.13	0.83	1.14	0.55	1.15	0.40	1.28
7	0.98	1.12	0.54	1.17	0.34	1.20	0.24	1.38

[a]Preceded by USP/Ph.Eur. induction port—values are slightly different if a preseparator is also used

deemed to have penetrated the stage. The sharpness of a given CI stage efficiency curve is defined in terms of the geometric standard deviation (GSD_{stage}) by analogy with the properties of the lognormal distribution function:

$$GSD_{\mathrm{stage}} = \sqrt{\frac{d_{84.1}}{d_{15.9}}} \tag{2.9}$$

A stage having its GSD_{stage} at unity corresponds to the ideal size separator. In practice, however, GSD_{stage} is ideally <1.2. GSD_{stage} values in excess of this limit do occur with compendial CIs, particularly for those stages that size fractionate particles larger than about 5 μm aerodynamic diameter where gravitational settling contributes significantly to the size-separation process. Values of d_{50} and $\mathrm{GSD}_{\mathrm{stage}}$ can be determined by calibration of the impactor with uniform-sized particles of known mean values of d_{ae}; however, the process is both time-consuming and highly exacting [18–20]. A further refinement of the calibration process is to undertake a so-called archival calibration of a single CI whose stage dimensions (principally nozzle diameters) have been intentionally manufactured to be as close as possible to their nominal values. This procedure avoids the need to calibrate each and every CI at the time of manufacture and also after use for validation purposes. Such a calibration was undertaken for the first time early in the development for the NGI at volumetric flow rates of 100, 60, and 30 L/min [21] and later extended to 15 L/min to enable this CI to be used for the measurement of nebulizing systems [22]. Values of d_{50} and $\mathrm{GSD}_{\mathrm{stage}}$ from these archival calibrations are summarized in Table 2.1.

Equivalent information is given in Table 2.2 for the ACI at 28.3 L/min that is widely used for MDI-based evaluations. However, to date, an archival calibration of the ACI has not been undertaken, so the actual values of d_{50} and $\mathrm{GSD}_{\mathrm{stage}}$ are presented based on the historic data provided by Vaughan in 1989 [19]. These calibration-produced d_{50} values are compared with the corresponding nominal values that have been adopted by manufacturers of this impactor since that time. It is important to note that it is common practice to work with the nominal values of d_{50} for the stages of ACI. However, Stein and Olson, in an extensive investigation of impactors manufactured before the year 1995, reported that significant deviations in

Table 2.2 Values of d_{50} (μm) and GSD_{stage} for the size-fractionating stages of the ACI at Q=28.3 L/min

Stage	Nominal d_{50}	Calibration d_{50}	Calibration GSD_{stage}
0	9.0	9.0	1.15
1	5.8	6.0	1.17
2	4.7	5.7	1.20
3	3.3	3.1	1.22
4	2.1	2.1	1.20
5	1.1	0.9	1.23
6	0.7	0.6	1.21
7	0.4	ND	ND

ND not determined

nozzle diameters from nominal were possible as the result of imperfections during manufacture as well as the expected result of wear (corrosion/plugging) in normal use [23]. Later in this chapter, the importance of having an accurate understanding of each stage collection efficiency curve profile will be explained, since an understanding of the issues involved is essential in the context of abbreviating full-resolution impactors by the process of removing intermediate stages that is at the heart of the AIM concept.

Returning to the basic impactor theory, Marple and Liu [16] and Rader and Marple [24] identified that the value of $\sqrt{St}$ at E_{50}, defined as $\sqrt{St_{50}}$, should be close to 0.49 for well-designed circular-profiled nozzles, where differences in particle inertia dominate the size-separation process. The compendial CIs include only round nozzles, although other geometries, such as a slit, are possible. However, at least two other parameters have a secondary influence on the particle size-separation performance of a CI. The ratio of nozzle-to-collection surface distance/nozzle diameter (S/W) describes the geometry of the stage. The value of $\sqrt{St_{50}}$ is unaffected by small variations in S/W if S/W is between 1.0 and 10.0. At the same time, the dimensionless flow Reynolds number (Re_f), defined as

$$\mathrm{Re_f} = \frac{\rho_g UW}{\eta} \tag{2.10}$$

should ideally be in the range 500–3,000 to minimize the value of GSD_{stage}. However, this criterion may be overstringent [25], as indicated by the summary of the ACI design information presented in 1998 by Marple et al. for the ACI [26], where several stages do not meet this criterion yet function effectively as size fractionators at 28.3 L/min (Table 2.3).

The ratio of nozzle throat length (T) to W can also influence size-separation efficiency, decreasing $\sqrt{St_{50}}$ with increasing T/W [16]. However, the effect is likely to be small with commercially available impactors, where T/W is typically <10 [25].

Cross-flow, induced by air exiting the nozzles near the center of the nozzle plate and flowing outwards past other air jets located near the periphery of the nozzle cluster, can prevent the air jets near the edge of the cluster from reaching

Table 2.3 Original design characteristics of the ACI at 28.3 L/min (*From* [26]—*used with permission*)

Stage	0	1	2	3	4	5	6	7
d_{50}[a] (μm)	9.0	5.8	4.7	3.3	2.1	1.1	0.65	0.43
Re_f	163	221	110	141	188	292	394	782
S/W[b]	0.4	0.54	2.37	3.05	4.07	6.32	8.54	8.54
X_c	–	–	1.2	0.93	0.69	0.44	0.33	0.16

[a]Nominal cut-point sizes
[b]The nominal value of S is 1.6 mm (see Table 2.5)

the impaction plate with multinozzle designs [28]. Under these circumstances, particles that would otherwise be collected are instead captured by the cross-flow and transferred beyond the collection surface. Increased interstage losses as well as bias towards smaller sizes can therefore result with multistage impactors. The dimensionless cross-flow parameter (X_c) is defined as

$$X_c = \frac{nW}{4D_c} \tag{2.11}$$

where n is the number of nozzles per stage having nominal nozzle diameter, W, and D_c is the diameter of the overall cluster of nozzles in a multinozzle impactor stage. Values of X_c are listed in Table 2.3 for the ACI calibration by Vaughan [19].

It follows from its definition that X_c for a given stage remains unaltered if the design of the CI is fixed irrespective of design flow rate, as will be seen later is the case with any of the Marple-Miller model 150/160 variants or the NGI. This parameter should be <1.2 to avoid cross-flow-related problems that can lead to unacceptably high internal wall losses. This criterion is achieved for all of the commonly encountered CIs used with inhaler testing, with the notable exception of stage 2 of the ACI, where X_c is 1.2.

In 2000, Nichols and Smurthwaite [29] and Nichols et al. [30] extended the use of the ACI to higher flow rates primarily for the purpose of performance-testing DPIs. At 60 L/min, stage 7 is removed from the bottom of the stack and replaced by a new stage "-1" that is located immediately above stage "0." If the ACI is to be used at 90 L/min, stage 6 is also removed and replaced by a further new stage "-2" (Table 2.4).

Stage-1 comprises 96 nozzles, each with 4.5-mm diameter, whereas stage-2 contains 95 nozzles, each with 5.5-mm diameter. Importantly, in 2004, Byron et al. published definitive data for current production ACIs, superseding performance data developed from previous versions of this CI as the result of drawing changes to make manufacture of this impactor a more robust process [31]. These changes, together with the alternative high flow rate configurations, modify the original ACI design characteristics given in Table 2.3 to those summarized in Table 2.5.

Small differences and discrepancies remain between various reports of the values of parameters in the ACI. For example, one manufacturer's user guide establishes the d_{50} of stage 5 at 90 L/min to be 0.44 μm (BGI, Waltham, MA, USA). Although this value appears to be unsupported by a technical reference, it is rather

Table 2.4 Stage d_{50} values for the ACI with standard configuration for operation at 28.3 L/min compared with specialized configurations for use at 60 and 90 L/min (*From* [27]—*used with permission*)

Stage	28.3 L/min	60 L/min	90 L/min
−2	–	–	8.0
−1	–	8.6	6.5
0[a]	9.0	6.5	5.2
1	5.8	4.4	3.5
2	4.7	3.2	2.6
3	3.3	1.9	1.7
4	2.1	1.2	1.0
5	1.1	0.55	0.22
6	0.7	0.26	–
7	0.4	–	–

[a]The version of stage 0 used at 60 and 90 L/min has external modification permitting another stage, rather than the inlet adapter cone to be fitted above it. Its internal characteristics and performance are unaltered

Table 2.5 Design characteristics of the current ACI for $Q = 28.3$, 60, and 90 L/min

Stage	−2[a]	−1[a]	0	1	2	3	4	5	6	7
W (mm)	5.50	4.50	2.55	1.89	0.914	0.711	0.533	0.343	0.254	0.254
N	95	96	96	96	400	400	400	400	400	201
Re_f at 28.3 L/min	–	–	163	221	110	141	188	292	394	782
Re_f at 60 L/min	–	194	346	469	233	299	399	619	835	–
Re_f at 90 L/min	241	291	518	703	350	448	598	929	–	–
S/W[b] at all Q	0.29	0.36	0.63	0.85	1.75	2.25	3.00	4.66	6.30	6.30
X_c at all Q	N/A	N/A	N/A	N/A	1.2	0.93	0.69	0.44	0.33	0.16

[a]Stage −2 and −1 data provided by Mike Smurthwaite (Westech Instrument Services, UK), 2011—other data from Byron et al. [27] (Used with permission)
[b]The nominal value of S used in the calculations of S/W is 1.6 mm, but values between 1.5 and 1.7 mm have been observed with individual CIs

close to the value predicted by extrapolation from the 60 L/min value of 0.55 μm. Indeed the values of d_{50} of stage 6 at 60 L/min and stage 5 at 90 L/min recorded in Table 2.4 are much smaller than the values predicted by extrapolating from the previous lower flow rate figures. And although the user guides of manufacturers of the ACI may cite the Vaughan calibration of the ACI at 28.3 L/min [19] as definitive, Vaughan actually reported the d_{50} for stage 2 to be 5.7 μm, rather than the generally accepted value of 4.7 μm for this particular stage.

Note also that the value of 4.7 μm for stage 2 at 28.3 L/min has recently been confirmed by Roberts, undertaking a more rigorous calibration with monodisperse standard particles akin to that used for the archival NGI calibration measurements [32]. It should also be noted that the jet-to-plate distance, S, is 1.6 mm, as reported by Roberts [14].

Table 2.6 Flow Reynolds numbers (Re_f dimensionless) associated with the four-stage MSLI at different flow rates (Q)

	Stage	1	2	3	4
	S/W	0.38	0.39	0.50	2.22
Q (L/min)	X_c	N/A	N/A	N/A	0.31
20		1,132	2,018	3,552	1,485
30		1,701	3,030	5,334	2,250
40		2,267	4,043	7,115	3,002
50		2,835	5,049	8,886	3,749
60		3,401	6,061	10,667	4,500
70		3,962	7,073	12,449	5,252

Table 2.7 Stage d_{50} values (μm) for the four-stage MSLI at $Q = 60$ L/min (*From* [26]—*used with permission*)

Stage	1	2	3	4
d_{50}	13.0	6.8	3.1	1.7

Overall, the available literature indicates that some historical differences exist between commercially available ACIs and some uncertainties exist in the performance characteristics of the basic ACI design.

In the case of the multistage liquid impinger (MSLI), cited as one of the standard apparatuses for aerodynamic assessment of OIP aerosols in the European Pharmacopoeia the values of *S*/*W* are lower than the recommended range for all except stage 4 and values of Re_f are in general much higher than their equivalents with the ACI at ca. 30 L/min. Table 2.6 summarizes the design properties of the MSLI at different flow rates [33].

Particle bounce and re-entrainment are avoided altogether because particles are captured in the impingement fluid ready for recovery and subsequent assay. However, this design results in shallower collection efficiency curves that are indicative of poorer size selectivity, and the d_{50} sizes (Table 2.7) are therefore likely to be more dependent upon small changes in either *S* or *W* compared with either the ACI or NGI systems.

The Marple-Miller series of five-stage impactors (MMI), cited in the US Pharmacopeia, are currently available in three sizes, all based on the original work of Miller [34] and Marple et al. [35]. A low-flow rate version (model 150P) was developed in the late 1990s to permit MDIs with add-on devices intended for low-flow patients to be tested at more appropriate conditions of use (4.9 and 12 L/min) [36]. This addition enables measurements to be made by MMIs at flow rates ranging from 4.9 to 90 L/min with d_{50} values that are all located within the useful range for all types of OIP testing (Table 2.8).

Stage collection efficiency curves for all versions of this CI expressed either in terms of d_{ae} or $\sqrt{St}$ are steep and associated with GSD_{stage} values that are close to or below 1.2 (Table 2.9).

Table 2.8 Stage d_{50} values (μm) for the Marple-Miller five-stage CIs

	Model and design flow rate (Q)				
	150P		150	160	
Stage	4.9 (L/min)	12.0 (L/min)	30 (L/min)	60 (L/min)	90 (L/min)
1	10.0	10.0	10.0	10.0	8.1
2	7.2	4.7	5.0	5.0	4.0
3	4.7	3.1	2.5	2.5	2.0
4	3.1	2.0	1.25	1.25	1.0
5	0.77	0.44	0.63	0.63	0.5

Table 2.9 Design characteristics of the Marple-Miller CI variants at Q=4.9, 12.0, 30, 60, and 90 L/min

Model	Parameter	Q (L/min)	Stage 1	Stage 2	Stage 3	Stage 4	Stage 5
150/160	Re_f	30, 60	3,160	1,240	1,240	1,260	1,260
		90	4,740	1,860	1,860	1,890	1,890
	S/W	30, 60, 90	1.0	1.0	1.0	1.2	2.5
	X_c	30, 60, 90	N/A	0.37	0.37	0.37	0.36
150P	Re_f	4.9	918	539	453	386	548
		12.0	1,765	1,320	1,109	945	1,342
	S/W	4.9	1.71	5.36	4.43	3.84	5.03
		12.0	1.31				
	X_c	4.9, 12.0	N/A	0.18	0.21	0.25	0.17

These CIs were first designed for OIP testing to make use of collection cups as particle collectors, with the purpose of improving productivity. However, API recovery can be more difficult than with the simpler geometry of collection plates with recovery procedures requiring more than contact with solvent to dissolve the collected particles. The model 160 (high flow) MMI is the standard configuration intended for use at 60–90 L/min. The model 150 MMI has half the number of nozzles per stage compared with the model 160 MMI, for use as an alternative to the ACI for MDI characterization at 30–60 L/min.

Although these CIs have not been as widely adopted as the ACI, NGI, or MSLI, internal losses for models 150 and 160 MMIs reported by Marple et al. were no more than 5% of the incoming aerosol at worst case (4 μm < d_{ae} < 6 μm), decreasing to <1% for finer particle sizes and <2% for larger particles [35]. These measurements were based on calibration with monodisperse droplets.

The model 160 MMI has subsequently been reported as having internal losses at 60 L/min with at least one DPI (Bricanyl Turbuhaler®, AstraZeneca, Sweden) that were comparable with those indicated by Marple et al. [35], provided that precautions were taken to eliminate particle bounce and re-entrainment by coating the collection surfaces with a tacky surface (silicone oil) [37]. By comparison, losses within the low-flow MMI have also been reported as being <5% of the material balance from two types of MDI-generated formulations [36].

Table 2.10 Values of X_c, S/W, Re_f, and d_{50} (μm) for the NGI size-fractionating stages at flow rates (Q) associated with its archival calibration

All flow rates								
Stage	1	2	3	4	5	6	7	MOC[a]
X_c	N/A[b]	N/A[b]	0.35	0.41	0.61	0.84	0.85	0.94
S/W	1.0	2.0	3.0	3.0	3.0	3.1	4.9	7.1
$Q=15$ L/min								
d_{50}	14.1	8.61	5.39	3.30	2.08	1.36	0.98	Do not use
Re_f	1,482	724	404	338	230	167	166	
$Q=30$ L/min								
d_{50}	11.8	6.40	3.97	2.30	1.36	0.83	0.54	0.36
Re_f	2,938	1,435	801	669	455	328	324	149
$Q=60$ L/min								
d_{50}	8.06	4.46	2.82	1.66	0.94	0.55	0.34	0.14
Re_f	5,876	2,870	1,602	1,339	909	757	647	298
$Q=100$ L/min								
d_{50}	6.12	3.42	2.18	1.31	0.72	0.40	0.24	0.07
Re_f	9,793	4,783	2,671	2,231	1,515	1,095	1,079	496

[a]*MOC* micro-orifice collector—used as an alternative to a backup filter; note that the size corresponding to 80% collection efficiency (d_{80}) rather than d_{50} values are quoted for this equivalent to a backup filter, rather than an impaction stage

[b]Single-nozzle geometry; therefore, cross-flow parameter does not apply

In summary, there is some freedom in establishing the precise location of the collection surface beneath the nozzle plate with many impactor designs except perhaps the MSLI. This flexibility is available for the ACI and NGI, which are the apparatuses most likely these days to be used in OIP aerosol characterization. More importantly from a user's standpoint, small variations in the precise thickness (depth) of a tacky coating that could be applied under normal circumstances to the collection surfaces, to improve particle adhesion, are unlikely to affect stage performance. As will be seen in Chap. 10, the ability to mitigate bias due to particle bounce and blowoff from the collection surface is particularly important with abbreviated systems when used to characterize both MDI- and DPI-generated aerosols.

The NGI represents the state of the art for CI design. Based on their experience with the Marple-Miller series of CIs, the developers of the NGI took the opportunity to ensure that the criterion for *S/W* was met for all stages (Table 2.10). Likewise, the multinozzle designs of stages 3–7 and the MOC were arranged such that values of X_c were significantly <1.2. Some values of Re_f are <500 for certain stages, particularly those that size fractionate the finer particles and when operating this CI at the lower flow rates within its range [38]. Nevertheless, all stage collection efficiency curves for this impactor throughout the original design range of operation from 30 to 100 L/min were shown to be sharply defined with minimal overlap [21]. Even at the lower flow rate extension at 15 L/min for nebulizer testing, these curves are still acceptable for size-fractionating purposes [22]. Note, however, that the MOC should be followed by a final filter or the internal filter option be used to replace the MOC when used at 15 L/min [22].

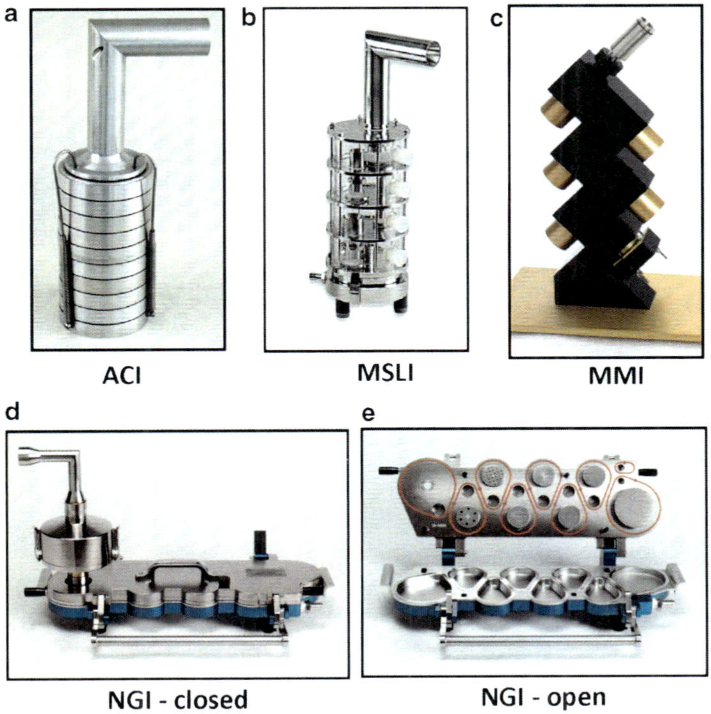

Fig. 2.8 External appearance of the (**a**) ACI, (**b**) MSLI, (**c**) MMI, and (**d**) and (**e**) NGI

In the designs of the ACI (Fig. 2.8a), MSLI (Fig. 2.8b), and MMI (Fig. 2.8c) systems, the individual impaction stages are connected together in a more-or-less vertical alignment. It is notable, however, that the NGI has all of its stages located adjacent to each other in the horizontal plane (Fig. 2.8d, e). This configuration was chosen primarily for ease of use for semi- or fully automated operation.

The function of a multistage CI is primarily to fractionate the incoming aerosol into progressively finer particle sizes, beginning with the coarsest particles. The linear air velocity, U, is therefore increased in a series of well-defined steps as the aerosol particles move from one stage to the next throughout the sequence. These increases are put into effect primarily by reductions in the nozzle diameter (W) from one stage to the next in the series. However, the number of nozzles per stage (n) as well as the number of stages within the CI can also be adjusted to optimize size resolution and minimize pressure drop across especially the stages that size separate the finest subfractions. It is important to note that inserting more than five stages per decade of particle size is counterproductive with this technology, because the stage(s) immediately before a given stage will interfere with efficient particle collection, due to the nonideal nature of their collection efficiency curves. This practical restriction therefore limited the design of the NGI such that a maximum of five stages with d_{50} sizes in the range from 0.5 to 5 µm aerodynamic diameter was possible at most flow rates within the range 30–100 L/min [38].

2.5 Potential Bias Arising from the Assumption That Stage Collection Efficiency Profiles Are Each Step Functions at the d_{50} Value Appropriate to the Stage Under Consideration

In the previous section, the performance characteristics for the most widely used CIs were described on the basis of an assumption that each stage d_{50} size is truly representative of the particle aerodynamic diameter at which size fractionation takes place. Hence, APSDs are usually reported with data points for the stages on the size axis based on their d_{50} values so that in a typical data analysis method, the size of particles on each CI stage is considered to be larger than its d_{50}. With this assumption, it is possible to calculate the values of *MMAD* and GSD representing measures of central tendency and spread respectively of an APSD. Note that the determination of GSD is calculated on the basis that the APSD is both unimodal and lognormal, assumptions that are in most cases valid for OIP-generated aerosols.

In atmospheric aerosol research where CIs are also widely used, the nonideal nature of stage collection efficiency profiles has been tackled by many groups, with well-accepted methods in place, based on MPS data inversion that is predicated on knowing the functional form of the stage collection efficiency curves [39–41]. However, this approach had not been brought into the assessment of OIP aerosols until recently. It is highly germane to the abbreviation of CIs, since any overlap that might be significant in the collection efficiency profiles of adjacent stages of the full-resolution CI configuration is likely no longer present.

Roberts and Mitchell have therefore evaluated theoretically the potential impact of assuming a step function change in stage collection efficiency in both the full-resolution ACI and NGI systems [42] and have also extended the analysis to include abbreviated versions of these CIs [43]. In the analysis of the full-resolution CIs, they postulated an aerosol entering the NGI or ACI to have a lognormal distribution of particles with known MMAD and GSD, an assumption not far from the true condition for most OIPs. On this basis, the fraction of the mass that deposits on stage "*N*" is given by

$$f_{\mathrm{N}} = \frac{1}{\sqrt{2\pi}}\frac{1}{\ln\sigma_{\mathrm{g}}}\bullet\int_{0}^{\infty}\frac{1}{x}exp\left[-\frac{1}{2}\frac{(\ln x-\ln[MMAD])^{2}}{(\ln\sigma_{\mathrm{g}})^{2}}\right]\bullet E_{\mathrm{N}}(x)\left[\left(1-E_{0}(x)\right)\left(1-E_{1}(x)\right)\ldots\left(1-E_{\mathrm{N-1}}(x)\right)\right]\mathrm{d}x \qquad (2.12)$$

Table 2.11 Parameters for ACI stage collection efficiency curves at $Q=28.3$ L/min

Stage	A (μm^{-1})	B (dimensionless)
0	−1.324	12.51
1	−1.874	11.39
2	−1.928	9.604
3	−2.808	9.329
4	−4.494	9.668
5	−8.258	8.787
6	−14.82	9.483
7	−17.76	7.725

Here, σ_g is the geometric standard deviation (GSD). The quantities $E_0 \ldots E_N$ are the fractional efficiency curves of stages 0–N of the CI.

They then developed analytical forms of the real stage efficiency curves of both CIs. For the NGI, these were based on the hyperbolic tangent functional form first described by Rader et al. [40] to compute how this model aerosol distributed itself on each stage of the CI under consideration:

$$E_i(d_{pc}) = \tanh\left[\left(\frac{d_{pc}}{Y_i}\right)^{Z_i}\right] \tag{2.13}$$

in which d_{pc} is the modified particle diameter corrected for slip in accordance with the expression

$$d_{pc} = d_p\sqrt{\frac{\rho_p}{\rho_0}}\sqrt{C_c} = d_{ae}\sqrt{C_c\left(d_{ae}\right)} \tag{2.14}$$

and Y_i and Z_i are best fit parameters for each impaction stage. The values of the parameters in (2.13) that fit the archival calibration data for the NGI are in the supplementary information associated with Roberts [14] and available on-line at http://www.tandfonline.com/doi/suppl/10.1080/02786820903204060

For the ACI at 28.3 L/min, Roberts and Mitchell used the exponential function developed by Gulak et al. [44] for the stage efficiency curves:

$$E_i(d_{ae}) = \frac{1}{1+\exp(A_i d_{ae} + B_i)} \tag{2.15}$$

Table 2.11 lists the values of the parameters A_i and B_i that best fitted the adjusted calibration data of Vaughan for this CI.

When calculated MMAD and GSD parameters for selected APSDs were compared with those of hypothetical aerosols likely to be encountered with OIPs that entered the NGI or ACI (using CITDAS® software; version 3.00, Copley Scientific, Nottingham, UK), the differences were at most 5% for the MMAD and 11% for the

Table 2.12 Comparison of input MMAD and GSD with CITDAS®-generated values for both ACI and NGI (*From* [42], *courtesy of DL Roberts and JP Mitchell*)

Input APSD		ACI: Q = 28.3 L/min		NGI: Q = 30.0 L/min	
MMAD (μm)	GSD	MMAD (μm)	GSD	MMAD (μm)	GSD
1.10	2.0	1.16	1.98	1.10	2.00
2.50	2.0	2.53	1.99	2.57	2.08
4.00	2.0	3.96	1.93	4.12	2.07
5.00	2.0	4.77	1.96	5.02	2.06
4.00	1.5	4.00	1.50	4.17	1.57
4.00	1.2	3.98	1.26	4.13	1.33

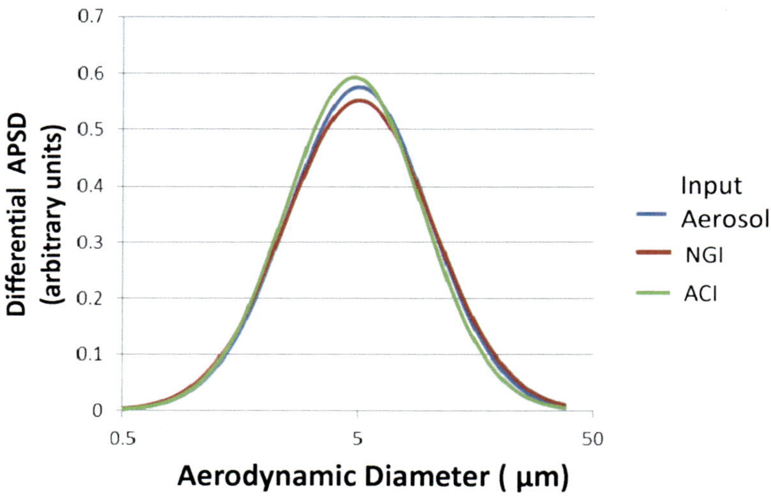

Fig. 2.9 Input and calculated APSDs from a hypothetical OIP-generated unimodal, lognormal distributed aerosol having MMAD of 5 μm and GSD of 2.0 (*From* [42], *courtesy of D L Roberts and J P Mitchell*)

GSD values (Table 2.12). The larger discrepancies arose as the result not so much because of the nonideal nature of stage collection efficiency curves but because of the inaccuracies of the two-point linearization of the lognormal curve fit. These divergences were more evident when the MMAD of the input aerosol was nearly identical with one of the stage d_{50} values.

The difference between the input and the calculated aerosol APSDs is illustrated in the example shown in Fig. 2.9 where the input MMAD and GSD were chosen to be 5 μm and 2.0, respectively. This example was chosen because the differences between input and calculated APSDs arose mostly as the result of overlap between neighboring CI stages. The calculated APSD from the NGI data was very close to that of the input aerosol, whereas the calculated APSD for the ACI was shifted only slightly to finer sizes, mostly as a result of the significant overlap between stage 2 and stages 1 and 3 with this CI. This finding most probably arises because of the

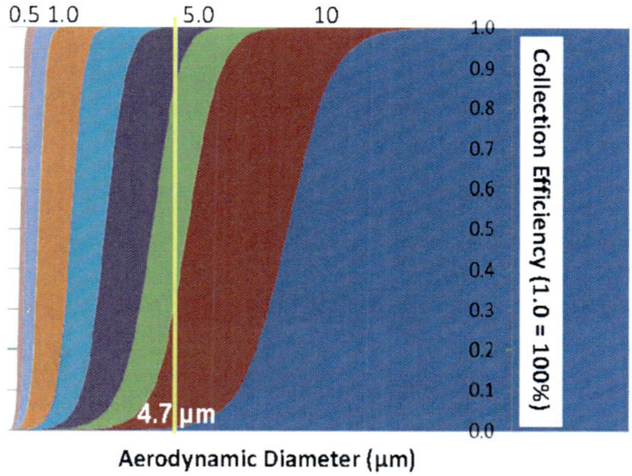

Fig. 2.10 Stage collection efficiency curves for the full-resolution ACI at 28.3 L/min, based on the adjusted calibration data of Vaughan [19] (*From* [42], *courtesy of D L Roberts and J P Mitchell*)

symmetry of each collection efficiency curve about the d_{50} value. Under such a condition, the mass of API that is associated with particles large enough that they should have been captured is almost counterbalanced by the mass of API contained in finer particles that should have penetrated the stage in question.

Roberts and Mitchell concluded that these differences would likely be unimportant in understanding the aerosol coming from an OIP [42]. Their assessment was based in part on recent industry guidance from the EMA, in which bounds of ±15% are deemed equivalent from in vitro measurements in regulatory submissions (i.e., for target dose and for comparison of CI stage deposition between reference and test OIP [44]).

In their follow-on assessment of abbreviated ACI and NGI systems, Roberts and Mitchell adapted the same approach by comparing abbreviated with full-resolution CI results calculated from hypothetical aerosols with unimodal and lognormal APSDs [43]. It is important to note that the amount of overlap from stage to neighboring stage in an abbreviated CI is greatly reduced or even abolished compared with the situation that exists in a full-resolution CI of the same design. As a result, however, even for the same input aerosol, the APSD of the aerosol approaching a given stage of an abbreviated impactor is different to that which exists when approaching the sequence of individual stages of the full-resolution system. This situation exists even if some of the stages of the abbreviated CI are identical to those of the full-resolution CI.

In the case of the ACI operated at 28.3 L/min, the problem was summarized by comparing the stage collection efficiency curves for the full-resolution CI operated at 28.3 L/min to the corresponding profiles of the remaining stages of the particular abbreviated ACI configuration investigated, in which measures of fine and extrafine particle fractions are sought by retaining just stages 2 and 5 and the final filter from the full-resolution system (the stage numbering in the abbreviated CI corresponds with that of the full-resolution impactor for simplicity). Hence, although stage 2 in the abbreviated ACI was physically the first impaction stage, its construction is that of stage 2 of a full-resolution ACI. When all the stages were present (Fig. 2.10),

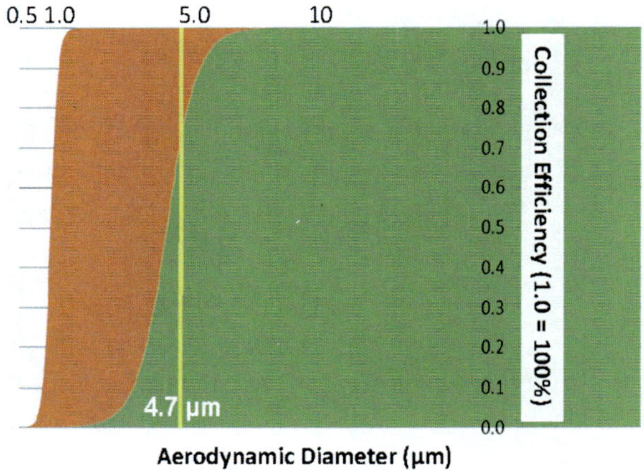

Fig. 2.11 Collection efficiency curves for the abbreviated ACI at 28.3 L/min, based on the adjusted calibration data of Vaughan [19] for stages 2 and 5 of the ACI (*From* [43], *courtesy of D L Roberts and J P Mitchell*)

overlap of the collection efficiency curves for stages 1 and 3 with that from stage 2 is prominent. Hence, a 4.7-µm aerodynamic diameter particle could potentially be captured on stage 1 or 2 or 3.

However, when only stages 2 and 5 are present, the overlap region between these two stages is absent, and a 4.7-µm-diameter particle would either be captured or pass the first stage (Fig. 2.11). Consequently, of the particles that would have a chance of collecting on stages 1 or 2 or 3 with the full-resolution ACI, a larger mass fraction is available to approach stage 2 of the abbreviated ACI configuration. In consequence, more particles can penetrate stage 2 of the abbreviated apparatus than would be the case if stage 1 is present.

It follows that summing up the masses collected on stages 0, 1, and 2 of the full-resolution system will not give the identical result as the mass collected by stage 2, in the absence of stages 0 and 1, with the abbreviated CI configuration.

The approach to develop a theoretical interpretation of the behavior above was similar to that described already for the full-resolution CIs [42]. On this occasion, though, an aerosol that is both unimodal and lognormal in terms of its APSD distribution was considered as it enters and passes through the abbreviated impactor.

To calculate how a lognormal aerosol distributes itself throughout an impactor with some stages removed, the stage efficiency functions, $E_i(x)$, in (2.15) were set equal to zero for stages of the full-resolution ACI that are removed and (2.12) applied to the remaining stages of the abbreviated ACI configuration.

Roberts and Mitchell also considered an abbreviated NGI likewise, by postulating that the user removes the airflow after stage 3 and reintroduces it just upstream of stage 6 with special cups for the NGI, as described by Svensson and Berg [45]. In this arrangement, the abbreviated NGI is much like a full-resolution NGI but with fewer stages. It is important to note that Mitchell and Roberts did not consider in this analysis the Fast Screening Impactor (FSI, MSP Corp., St Paul, MN, USA), as this abbreviated system is based on the NGI preseparator [46] and is therefore not simply a full-resolution NGI comprising fewer stages.

Roberts and Mitchell first calculated how the lognormally distributed aerosol described above would distribute in all stages of a full-resolution ACI or NGI, accounting for the actual stage efficiency curves. They then calculated where in the appropriate abbreviated apparatus the mass actually distributes, according to the stage efficiency curves for this reduced system. Next, they compared these values to the relevant values of mass from summed stages of the full-resolution parent CI. Finally, to complete the validation, they compared the mass on the summed stages from the parent CI and the mass on the abbreviated impactor stages to the *actual* mass in the size range present in the input aerosol.

Equation (2.12) was evaluated numerically using Simpson's rule to calculate the fraction of the input aerosol that goes to a given stage of either a full-resolution or an abbreviated CI configuration. Note that numerical integration was not necessary for calculating the inlet aerosol mass in a given size range, because an analytical expression could be derived by setting the $E_i(x)$ functions in (2.13) equal to step functions at the impactor d_{50} sizes:

$$f_N = \frac{1}{2}\left\{ erf\left[\frac{\ln\left(\frac{D_{50,N-1}}{\bar{D}}\right)}{\sqrt{2}\ln(\sigma_g)}\right] - erf\left[\frac{\ln\left(\frac{D_{50,N}}{\bar{D}}\right)}{\sqrt{2}\ln(\sigma_g)}\right]\right\} \tag{2.16}$$

The error function (erf) given by the expression from Matthews and Walker [47] was used:

$$erf(x) = \frac{2}{\sqrt{\pi}}\int_0^x exp(-h^2)dh \tag{2.17}$$

For the incoming aerosol that is in the size range larger than the d_{50} size for the first stage, the value of $d_{50,N-1}$ is set to infinity, and it is known that erf(∞) = 1.

As in the investigation of the full-resolution CIs, a series of model aerosols were evaluated, having MMAD values in the range 1.0–5.0 μm and GSD values selected between 1.2 and 2.0. There was always less mass found on the first stage of the abbreviated ACI than on the corresponding sum of the stages on the full-resolution CI (see the two examples shown in Figs. 2.12 and 2.13). This bias was ≤5% for the broadest (GSD = 2.0) APSDs comprising the largest particles (MMAD = 5.0 μm). However, it increased to as much as 11% for narrow (GSD = 1.2) and finer (MMAD = 1.0 μm) aerosols. In the context of Figs. 2.12 and 2.13, it should be noted that a ratio of mass on "AIM" to "summed stage" of unity would represent an abbreviated system without bias arising from the nonideal nature of the stage collection efficiency curves.

Similar trends were also detected with the NGI system analysis (Figs. 2.14 and 2.15). However, the bias associated with the abbreviated NGI was ≤2.5% regardless

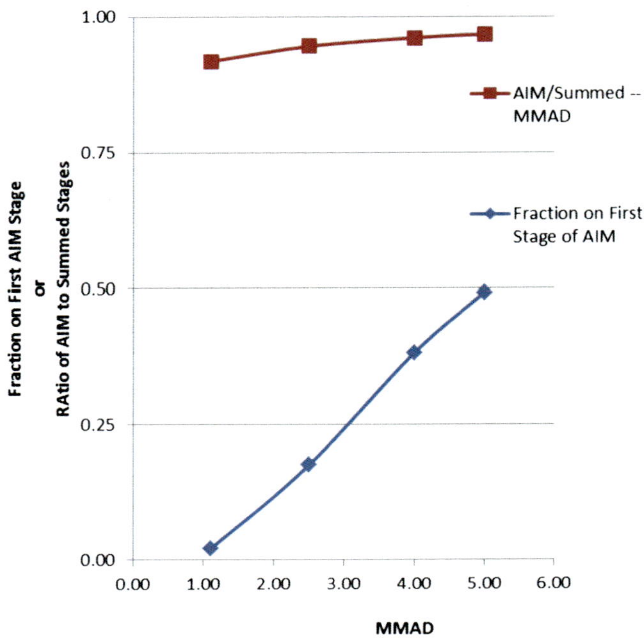

Fig. 2.12 ACI system: mass fraction on first AIM collection stage or ratio of AIM system compared with corresponding summed stages—MMAD varied in the range 1–5 μm with GSD fixed at 2.0 (*From* [43], *courtesy of D L Roberts and J P Mitchell*)

of APSD profile. Roberts and Mitchell concluded that such a small discrepancy would likely not be observable experimentally. In addition, for an aerosol with MMAD of 4.0 μm, the fraction on the first stage of the abbreviated NGI was close to 0.5 and nearly independent of the dispersity of the input aerosol (Fig. 2.15). This was not the case for the corresponding ACI data where the mass fraction on the first stage (stage 2) reduced as the dispersity of the incoming aerosol decreased (Fig. 2.12). The improved robustness of the NGI system derives from the relative sharpness of the individual NGI stage efficiency curves compared with those of the ACI and the consequent minimal overlap between neighboring stages [21]. The increase in the mass fraction collected by the first stage (stage 2) of the abbreviated ACI with the model aerosols having larger MMAD values (Fig. 2.12) and also for the corresponding situation with the NGI system (Fig. 2.14) would be a linear function of MMAD if there is no bias associated with the nonideal nature of the stage collection efficiency curves.

From this analysis of both CI systems, Roberts and Mitchell observed that bias in measures of the ratio of "AIM" to "summed stages" is systematically in one direction but can sometimes expand and on other occasions reduce the difference between the input aerosol and the abbreviated CI results in terms of mass fractions. Importantly, no cases were observed in which the abbreviated ACI deviated from

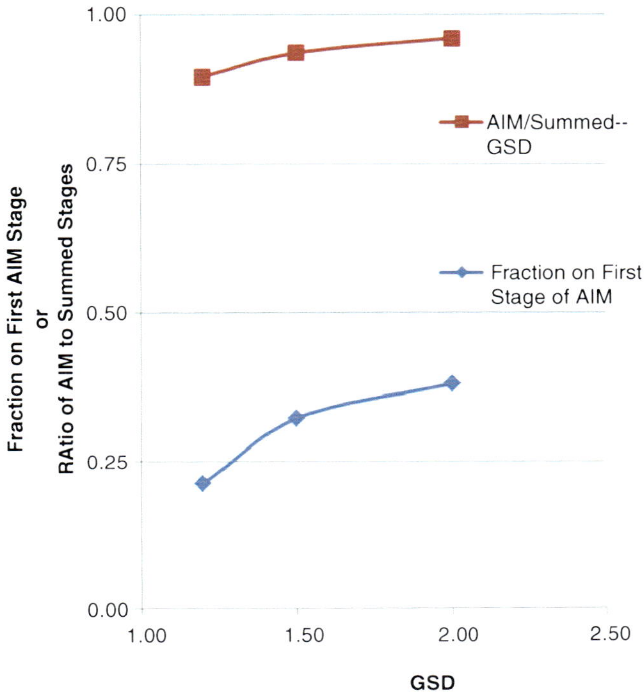

Fig. 2.13 ACI system: mass fraction on first AIM collection stage or ratio of AIM system compared with corresponding summed stages—GSD varied with MMAD fixed at 4.0 μm (*From* [43], *courtesy of D L Roberts and J P Mitchell*)

the input aerosol by more than about 10%. However, it is believed that the potential exists for larger discrepancies to arise if the incoming aerosol is more monodisperse than those considered and also if it contains finer particles in the range of interest.

In summary, although bias from the assumption of ideal stage collection efficiency behavior appears small with both systems, it can likely be ignored for the NGI. However, account may need to be taken of this effect in the most accurate work with the ACI and particularly when comparing data from abbreviated systems developed from this CI design.

2.6 Overview of the Compendial Full-Resolution CI-Based Methods for OIP Aerosol Characterization

Currently, the requirements for OIP in vitro performance assessment throughout the product life cycle dictate the need for measurements by full-resolution multistage CI, since this type of apparatus has good compatibility for the assessment of all formats currently manufactured (Table 2.13).

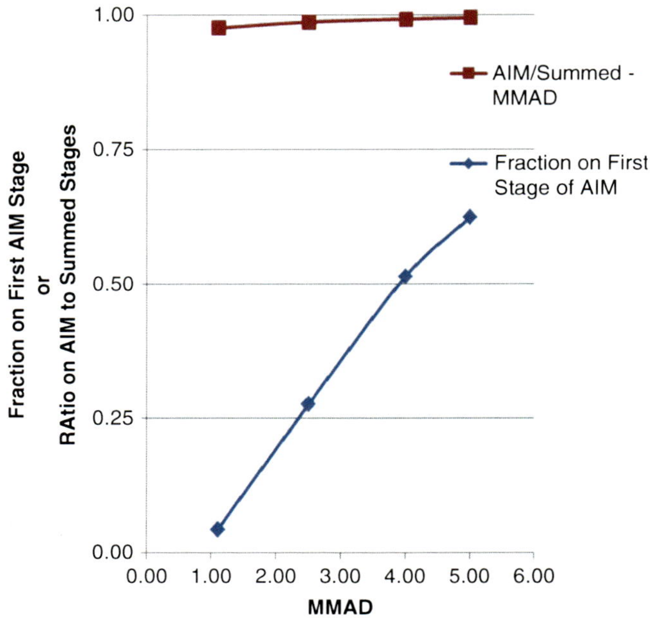

Fig. 2.14 NGI system: mass fraction on first AIM collection stage or ratio of AIM system to corresponding summed stages—MMAD varied with GSD fixed at 2.0 (*From* [43], *courtesy of D L Roberts and J P Mitchell*)

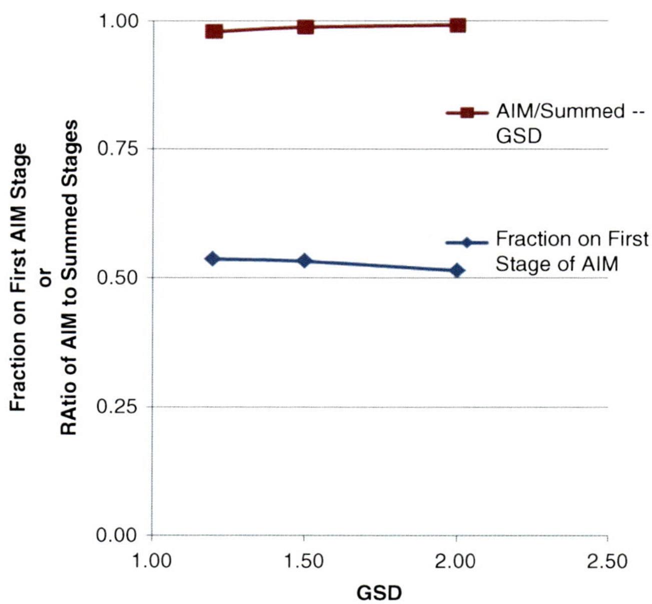

Fig. 2.15 NGI system: mass fraction on first AIM collection stage or ratio of AIM system to corresponding summed stages—GSD varied with MMAD fixed at 4.0 μm (*From* [43], *courtesy of D L Roberts and J P Mitchell*)

Table 2.13 Suitability of the cascade impactor method for the various inhaler classes

Inhaler class	Inhalation route	Suitability
MDI	Oral	Good
DPI		Good
SMI		Good
Nebulizer		Good
Nasal MDI	Nasal[a]	Fair
Aqueous nasal spray		Poor

[a]Nasal inhaled products are shown to illustrate relatively poor compatibility with the CI method but are outside the scope of the book

Table 2.14 Multistage CI apparatuses recognized in the European and United States Pharmacopeias as of 2011

Impactor or impinger	US Pharmacopeia	European Pharmacopoeia
Glass twin impinger (TI)	Not referenced	Apparatus A
Andersen eight-stage (ACI) nonviable impactor without preseparator	Apparatus 1 for MDIs	Apparatus D
Marple-Miller impactor model 160	Apparatus 2 for DPIs	Not referenced
ACI with preseparator	Apparatus 3 for DPIs	Apparatus D
Multistage liquid impinger (MSLI)	Apparatus 4 for DPIs	Apparatus C
Next generation pharmaceutical impactor (NGI)	Apparatus 5 for DPIs Apparatus 6 for MDIs	Apparatus E

The CI method is most suitable for the assessment of aerosols from MDI, DPI, and SMI inhaler forms, although in the case of DPI testing, the compendial methods require the impactor to be started from zero flow at the beginning of the sample, simulating an inhalation maneuver, whereas it is evident from the previous sections that CIs ideally operate at a fixed flow rate. Their applicability to the assessment of nebulizing systems is slightly more problematic, in that precautions need to be taken to prevent heat transfer from the CI to the aqueous aerosols that are produced by these systems [48].

Likewise, precautions need to be taken with sprays from nasal MDIs, as these frequently comprise aqueous solutions. Aqueous nasal spray characterization is normally undertaken by laser diffractometry (LD), as the size range of the droplets, which typically lies between 20 and 300 μm diameters, is well outside the range of operation of a CI [49]. This is the primary reason why this class of inhalers is largely outside the scope of this publication. However, single-stage impactors with d_{50} close to 10 μm aerodynamic diameter have been used to identify the small subfraction of ultrafine droplets potentially capable of penetrating the nasopharynx and entering the airways of the lungs [50].

Chapter 601 of the US Pharmacopeia and the corresponding monograph 2.9.18 in the European Pharmacopoeia permit several different types of CI to be used for OIP evaluations (Table 2.14).

In a survey undertaken in 2006 by EPAG in connection with CI qualification testing, the ACI was identified as the most widely used multistage CI, with the NGI following closely in second place [51]. The MSLI came third, but this apparatus is not in use outside of European regulatory submissions. Other systems were seldom used. It should be noted that the glass TI (Apparatus A of the European Pharmacopoeia) would qualify as an abbreviated impactor method, since it size classifies the incoming aerosol into only two subfractions with its d_{50} size at 6.4 µm aerodynamic diameter, when operated at its design flow rate of 60 L/min.

2.7 Preseparators and Induction Ports

Some form of inlet to the CI is required to ensure that the aerosol produced by the inhaler enters the impactor in a reproducible manner. Most OIPs are designed such that they emit their medication in the horizontal plane and entrances to impactors are typically oriented in the vertical downward direction. The solution to the problem is the induction port, which also mimics to a greater or lesser extent depending on its design, the human oropharynx. There are many designs of induction port currently in use, with their shape and capacity reflecting differing viewpoints on how inhaler aerosols should be sampled. Seven examples are illustrated in Fig. 2.16 [52]. By far the most commonly encountered is the relatively

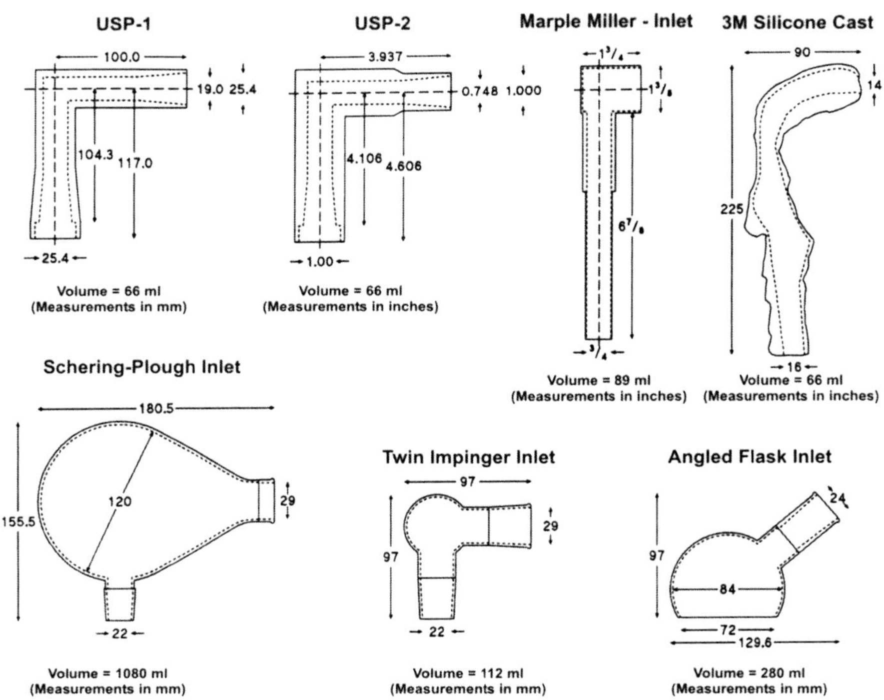

Fig. 2.16 Assortment of induction ports used in OIP testing (*From* [52]—*used with permission*)

straightforward metal right-angle bend (so-called USP/Ph.Eur. design) described in both regional pharmacopeias.

It is important to note that this inlet is not intended to be an accurate representation of the upper airway but rather a simplified arrangement that reproducibly collects incoming aerosol for the purpose of product quality assessment. Chapter 12 contains information about more anatomically appropriate inlets, including the new generation of "idealized" geometries that have been developed at the University of Alberta, Canada.

In its role as a model of the entrance to the respiratory tract, the induction port collects almost all the fast moving "ballistic" component of MDI-produced aerosols that is formed by flash evaporation of the propellant and therefore likely to deposit in the oropharynx [53]. It also serves to remove larger and often aggregated particles generated by most DPIs. The USP/Ph.Eur. design is intended to provide a common benchmark to compare different formulations by standardizing its critical dimensions (internal diameter and unobstructed path length). Unfortunately, unlike a CI stage, its collection efficiency is not easily determined, since there is significant turbulence at flow rates encountered typically for inhaler testing [54], and particles in the ballistic component from MDIs have velocities greater than that of the surrounding airflow when entering the induction port. Nevertheless, recently, Zhou et al. reported calibration data with monodisperse particles for this inlet (Fig. 2.17), from which they estimated d_{50} sizes to be 14.4- and 20.2-μm aerodynamic diameters for flow rates of 60 and 30 L/min, respectively, by fitting their experimental data to flow rate-dependent empirical relationships linking deposition fraction and d_{ae} [55].

Their term "deposition fraction" can be considered equivalent to collection efficiency. They found that d_{50} at the lowest flow rate investigated (15 L/min) could not be obtained, as the collection efficiency was found to be <50% regardless of particle size to an upper limit close to 30 μm aerodynamic diameter (Fig. 2.17). It should be noted that Zhou et al. grease coated the interior surfaces to mitigate particle bounce and re-entrainment, a practice that is possible but rarely undertaken with OIP testing.

In summary, although several attempts have been made to develop an understanding of how this particular induction port performs, given the complexity of the fluid dynamics and particle motion through its flow path, particularly when capturing the ballistic fraction emitted by MDIs, it should not be regarded as an additional impactor stage in routine work.

A preseparator is often required to be located immediately after the induction port when sampling aerosols produced from DPIs, since these formulations in many instances contain the API attached to the surface of much larger (lactose) carrier particles. The shear forces generated by inhalation from the DPI detach some, but not all, of the API particles from the carrier material, but the detached particles are sufficiently fine to penetrate beyond the oropharyngeal region into the lungs [56]. Both pertinent USP and Ph.Eur. compendial monographs refer to the use of a preseparator for DPI-based particle size distribution measurements with the ACI, recommending that its interior surfaces be coated either with a tacky agent in the same

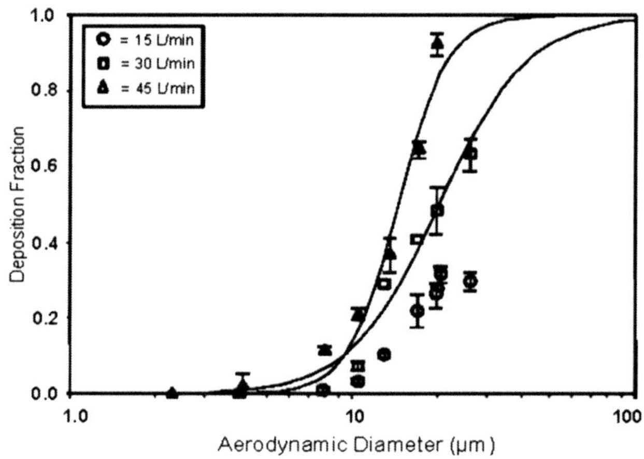

Fig. 2.17 Calibration data for the USP/Ph.Eur. induction port at 15, 30, and 60 L/min (*From* [55]—*used with permission*)

way as the collection surfaces of the CI stages or with up to 10 mL of a suitable solvent to eliminate particle bounce and re-entrainment. A similar requirement exists also for use of the NGI under these circumstances.

The preseparators for the ACI and NGI are illustrated in Figs. 2.18 and 2.19, respectively. Ideally, the preseparator should not starve the first stage of the impactor of particles. However, the preseparator used with the ACI has been shown by calibration at 28.3 L/min to have its cut size close to 9 μm in aerodynamic diameter, almost identical with the d_{50} size of the first stage (stage 0 in Table 2.2) at this flow rate [58]. Its sharpness of cut is relatively poor by comparison to equivalent values for the CI stages ($GSD_{pre\text{-}sep}$ is about 1.55), due primarily to the strong influence of gravity. The preseparator therefore starves the second as well as the first stage of this impactor.

Recognizing the limitations of the ACI preseparator, the consortium developing the NGI designed its preseparator to function as a single component [57] but with two distinct steps in the size-separation process (Fig. 2.19). The incoming aerosol is first passed through a so-called scalper impingement stage that removes the coarsest particles. The remaining aerosol then passes immediately through a more conventional impaction stage before leaving the preseparator. Calibration with monodisperse particles has demonstrated that its $GSD_{pre\text{-}sep}$ is close to 1.3 at 30, 60, and 100 L/min. This value is only slightly greater than estimates of GSD_{stage} for well-designed CI stages in which inertial rather than gravitational or turbulent size-separation predominates. Its measured d_{50} values (10.0, 12.7, and 14.9 μm at 100, 60, and 30 L/min, respectively) are sufficiently separated from the corresponding values for stage 1 (6.07, 8.29, and 11.4 μm) that starvation of the first stage of the CI does not occur to a significant extent.

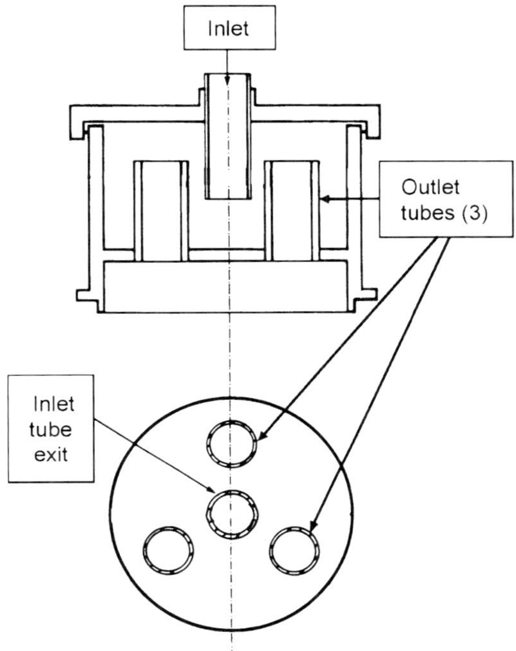

Fig. 2.18 Side cross section and top view of the preseparator for the ACI

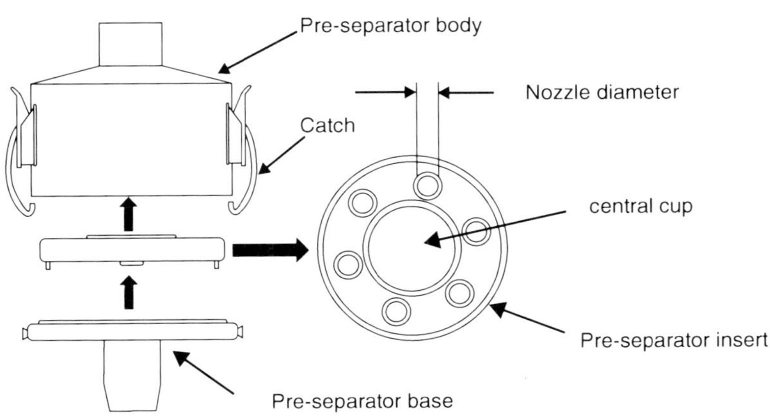

Fig. 2.19 Preseparator for the NGI (*From* [57]—*used with permission*)

2.8 Add-On Devices Used with MDIs

The evaluation of add-on devices that are widely prescribed for use with MDIs [59–61] poses some additional considerations beyond those that exist for the testing of the inhaler alone. Spacers are open extensions to the MDI actuator mouthpiece whose primary purpose is to place distance between the inhaler and the patient, such that plume development and impaction of the ballistic fraction takes place within the spacer itself (Fig. 2.20).

As such, they do not retain the medication should the patient fail to either inhale at the time of MDI actuation or exhale into the add-on device. Their evaluation is therefore probably best accomplished following the compendial methods for MDI performance evaluation, without any attempt to simulate delayed inhalation.

The valved holding chamber (VHC), on the other hand, has a valve at its exit that enables the aerosol to be retained if the patient is poorly coordinated. These devices are equipped with either a mouthpiece (Fig. 2.21) or face mask (Fig. 2.22) as the patient interface, as they are used by patients of all ages from infants to geriatrics.

As such, it follows that it is logical that these devices be evaluated simulating delayed inhalation, rather than by the methods in the pharmacopeias that currently require the tester to sample the aerosol immediately upon MDI actuation.

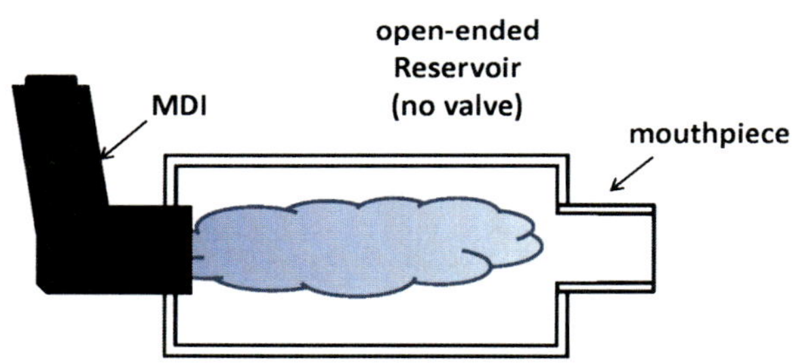

Fig. 2.20 Open-tube spacer for use with MDIs

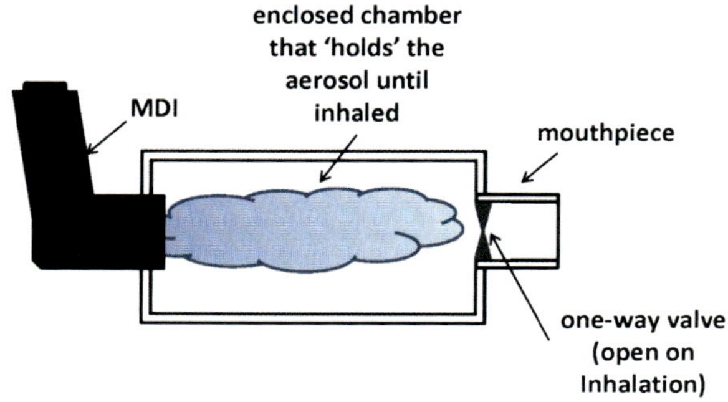

Fig. 2.21 VHC mouthpiece as patient interface for use with MDIs

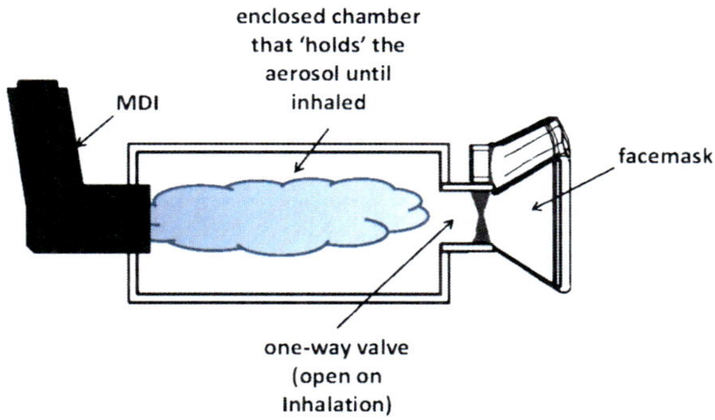

Fig. 2.22 Valved holding chamber with face mask as patient interface for use with pressurized metered-dose inhaler

Given the lack of guidance in the (then) compendial methods, about 10 years ago, a group of Canadian stakeholders met to develop a national standard that for the first time set out detailed methods for the evaluation of these add-on devices, including sections on the simulation of delayed inhalation following inhaler actuation [62]. This standard retains the CI as the apparatus of choice for the evaluation of emitted aerosol APSD but contains the following extensions to existing methodology [63]:

(a) Measurements are made initially with the MDI alone following the compendial method to provide benchmark data.
(b) The VHC is then evaluated simulating a 2 s delay following inhaler actuation, chosen to mimic the behavior of a typical uncoordinated user; testing at other delay intervals is also possible.
(c) The flow rate chosen should be appropriate for the intended age range of the VHC; for example, an infant VHC might be evaluated at close to 5 L/min, an add-on for a small child at 12 L/min and a product for adult use at ca. 30 L/min.

This approach has recently been proposed for adoption as a future informative chapter in the USP through the publication of a Stimulus to Revision article in *Pharmacopeial Forum*, in which the pharmacopeial organization publishes proposed revisions to USP-NF for public review and comment. [64]. This proposal has been partly driven by the recommendation from at least one regulatory agency that new MDI applications should contain in vitro data with a nominated spacer/VHC [65].

Although the Canadian Standard describes the framework for testing by CI, it does not contain any detail concerning suitable apparatus(es) to undertake measurements simulating delayed inhalation. At the present time, there is therefore no standardized "delay" apparatus. However, work undertaken at Trudell Medical International shortly after the standard was developed resulted in the apparatus shown in Fig. 2.23, seen in conjunction with a model 150P MMI [66].

The use of the NGI at 15 L/min, the lowest flow rate for which calibration data are available, would, in principle, also be suitable at least for testing VHCs intended for use by small children. In the so-called delay apparatus [67], simulation of the

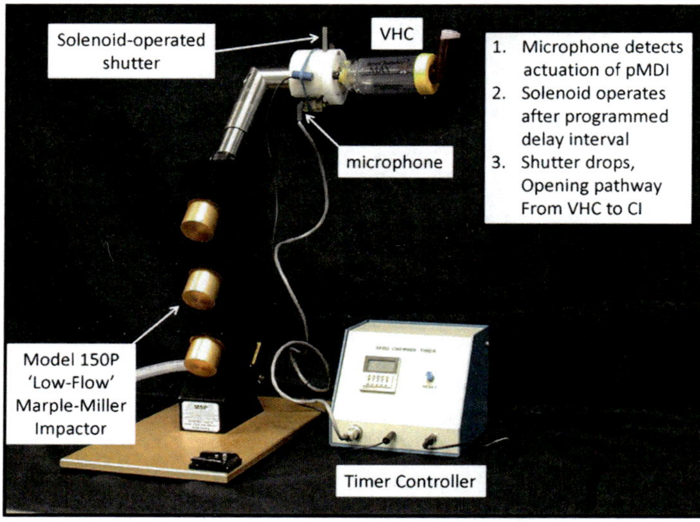

Fig. 2.23 Trudell Medical International "delay" apparatus coupled to USP/Ph.Eur. induction port with low-flow five-stage MMI developed for testing inhalation devices intended for use by infants or small children

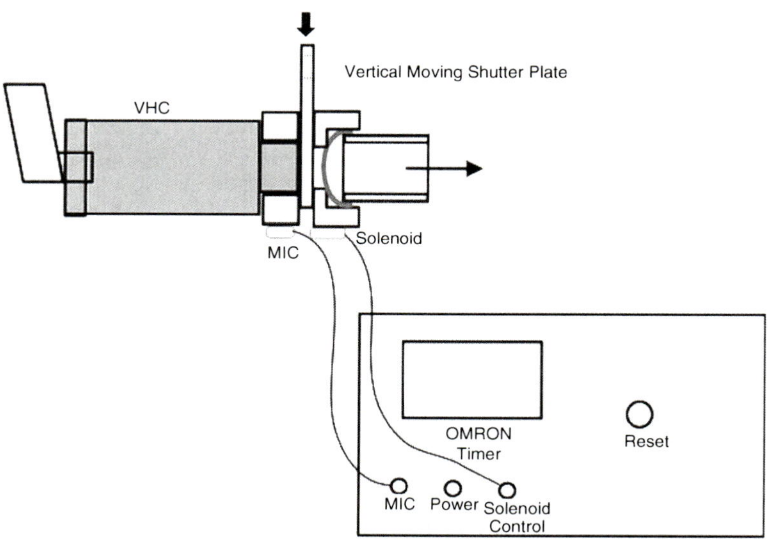

Fig. 2.24 Schematic diagram of Trudell Medical International "delay" apparatus showing working components (*From* [67]—*used with permission*)

elapsed time between MDI actuation and the onset of sampling is achieved by interposing a mechanical shutter housed in a purpose-built adapter that locates on axis between the VHC and induction port leading to a CI that is operated at a flow rate appropriate for the intended patient category for the VHC (Fig. 2.24).

The shutter is set in the elevated (closed) position during connection of the VHC, MDI actuation, and the subsequent delay period, but air can still be sampled by the impactor at the desired flow rate via a bypass channel located on the side of the shutter facing the induction port. A microphone (MIC) located on the adapter detects the sound emitted at actuation of the MDI, starting a timer that operates a solenoid valve immediately after the expiry of the selected delay period. This movement retracts the shutter support pin, thereby permitting the shutter to drop, opening a direct flow path from the VHC to the CI. The adapter introduces <5 ml additional volume to the aerosol pathway from the VHC to the CI system. The "delay" apparatus avoids the need to start the vacuum pump for the CI after the elapsed delay, a situation that would result in undefined stage d_{50} sizes, while the flow rate is increasing to its final stable value.

Although systems have been developed to enable a CI to be used in conjunction with a breathing simulator [68–70], which would be the obvious extension of the delay technique to mimic more closely tidal breathing, so far a standardized arrangement has not yet been developed to the point at which it could be incorporated as part of a compendial procedure for the evaluation of VHCs.

References

1. Hinds WC (1999) Aerosol technology: properties, behavior, and measurement of airborne particles, 2nd edn. Wiley, New York, NY
2. Finlay WH (2001) The mechanics of inhaled pharmaceutical aerosols: an introduction. Academic, London
3. Sbirlea-Apiou G, Katz I, Caillibotte G, Martonen T, Yang Y (2007) Deposition mechanics of pharmaceutical particles in human airways. In: Hickey AJ (ed) Inhalation aerosols: physical and biological basis for therapy. Informa Healthcare, New York, NY, pp 1–30
4. European Directorate for the Quality of Medicines and Healthcare (EDQM) (2012) Preparations for inhalation: aerodynamic assessment of fine particles. Section 2.9.18—European Pharmacopoeia [—Apparatus B in versions up to 4th edn. 2002] Council of Europe, 67075 Strasbourg, France
5. United States Pharmacopeial Convention (2012) USP 35-NF 30 Chapter 601: aerosols, nasal sprays, metered-dose inhalers and dry powder inhalers. United States Pharmacopeial Convention, Rockville, MD
6. European Medicines Agency (EMA) (2006) Guideline on the pharmaceutical quality of inhalation and nasal products (2006) EMEA/CHMP/QWP/49313/2005 Corr. 2006. London. http://www.ema.europa.eu/docs/en_GB/document_library/Scientific_guideline/2009/09/WC500003568.pdf. Visited 22 Aug 2011
7. United States Food and Drug Administration (FDA) (1998) Draft Guidance: Metered Dose Inhaler (MDI) and Dry Powder Inhaler (DPI) Drug Products Chemistry, Manufacturing and Controls Documentation. United States Federal Drug Administration, Rockville, MD. Docket 98D-0997 http://www.fda.gov/downloads/Drugs/GuidanceComplianceRegulatoryInformation/Guidances/ucm070573.pdf. Visited 22 Aug 2011
8. Mitchell JP, Bauer R, Lyapustina S, Tougas T, Glaab V (2011) Non-impactor-based methods for sizing of aerosols emitted from orally inhaled and nasal drug products (OINDPs). AAPS PharmSciTech 12(3):965–988

9. Mitchell JP, Nagel MW (2003) Cascade impactors for the size characterization of aerosols from medical inhalers: their use and limitations. J Aerosol Med 16:341–377
10. Dunbar C, Mitchell JP (2005) Analysis of cascade impactor mass distributions. J Aerosol Med 18(4):439–451
11. World Bank Group (1999) Pollution prevention and abatement handbook. The World Bank Group, Washington, DC
12. Health Canada (2006) Guidance for industry on the pharmaceutical quality of inhaled and nasal products. File Number: 06-106624-547, Ottawa, ON. Available at: http://www.hc-sc.gc.ca/dhp-mps/alt_formats/hpfb-dgpsa/pdf/prodpharma/inhalationnas-eng.pdf
13. Mohammed H, Roberts D, Copley M, Hammond M, Mitchell J (2012) Effect of sampling volume on dry powder inhaler (DPI)-emitted aerosol aerodynamic particle size distributions (APSDs) measured by the Next Generation pharmaceutical Impactor (NGI) and the Andersen eight-stage cascade impactor (ACI). AAPS PharmSciTech 13(3):875–882
14. Roberts DL (2009) Theory of multi-nozzle impactor stages and the interpretation of stage mensuration data. Aerosol Sci Technol 43(11):1119–1129
15. Chambers F, Ali A, Mitchell J, Shelton C, Nichols S (2010) Cascade impactor (CI) mensuration—An assessment of the accuracy and precision of commercially available optical measurement systems. AAPS PharmSciTech 11(1):472–484
16. Marple VA, Liu BYH (1974) Characteristics of laminar jet impactors. Environ Sci Technol 8:648–654
17. Vinchurkar S, Longest PW, Peart JP (2009) CFD simulations of the Andersen cascade impactor: model development and effects of aerosol charge. J Aerosol Sci 40:807–822
18. Mitchell JP, Costa PA, Waters S (1986) An assessment of an Andersen Mark-II cascade impactor. J Aerosol Sci 19(2):213–221
19. Vaughan NP (1989) The Andersen cascade impactor: calibration, wall losses, and numerical simulation. J Aerosol Sci 20(1):213–221
20. Asking L, Olsson B (1997) Calibration at different flow rates of a multistage liquid impinger. Aerosol Sci Technol 27(1):39–49
21. Marple VA, Olson BA, Santhanakrishnan K, Roberts DL, Mitchell JP, Hudson-Curtis BL (2003) Next generation pharmaceutical impactor. Part II: archival calibration. J Aerosol Med 16(3):301–324
22. Marple VA, Olson BA, Santhanakrishnan K, Roberts DL, Mitchell JP, Hudson-Curtis BL (2004) Next generation pharmaceutical impactor. Part III: extension of archival calibration to 15 L/min. J Aerosol Med 17(4):335–343
23. Stein SW, Olson BA (1997) Variability in size distribution measurements obtained using multiple Andersen Mark II cascade impactors. Pharm Res 14(12):1718–1725
24. Rader DJ, Marple VA (1985) Effect of ultra-Stokesian drag and particle interception on impactor characteristics. Aerosol Sci Technol 4(2):141–156
25. Marple VA (2002) Private communication. University of Minnesota, Minneapolis, MN
26. Marple VA, Olson BA, Miller NC (1998) The role of inertial particle collectors in evaluating pharmaceutical aerosol systems. J Aerosol Med 11S1:S139–S153
27. Nichols SC (2000) Calibration and mensuration issues for the standard and modified Andersen cascade impactor. Pharm Forum 26(5):1466–1469
28. Fang CP, Marple VA, Rubow KL (1991) Influence of cross-flow on particle collection characteristics of multi-nozzle impactors. J Aerosol Sci 22:403–415
29. Nichols SC, Smurthwaite M (1998) The Andersen cascade impactor: calibration data, operation at various airflow rates and modified for use with DPIs at various airflow rates. In: Byron PR, Dalby RN, Farr SJ (eds) Respiratory drug delivery-VI. Interpharm Press, Buffalo Grove, IL, pp 393–396
30. Nichols SC, Brown DR, Smurthwaite M (1998) New concept for the variable flow rate Andersen cascade impactor and calibration data. J Aerosol Med 11S1:S133–S138
31. Byron PR, Cummings RH, Nichols SC, Poochikian G, Smurthwaite MJ, Stein SW, Truman KG (2004) Selection validation of cascade impactor methods. In: Dalby RN, Byron PR, Peart J,

Suman JD, Farr SJ (eds) Respiratory drug delivery-IX. Davis HealthCare International Publishing, River Grove, IL, pp 485–487
32. Roberts DL (2012) Private communication. MSP Corporation, St Paul, MN
33. De Boer AH, Gjaltema D, Hagerdoorn P, Frijlink HW (2002) Characterization of inhalation aerosols: a critical evaluation of cascade impactor analysis and laser diffraction technique. Int J Pharm 249:219–231
34. Miller NC (1994) A cascade impactor for aerodynamic size measurement for MDIs and powder inhalers. In: Byron PR, Dalby RN, Farr SJ (eds) Respiratory drug delivery IV. Interpharm Press, Buffalo Grove, IL, pp 342–343
35. Marple VA, Olson BA, Miller NC (1995) A low-loss cascade impactor with stage collection cups: calibration and pharmaceutical inhaler applications. Aerosol Sci Technol 22:124–134
36. Olson BA, Marple VA, Mitchell JP, Nagel MW (1998) Development and calibration of a low-flow version of the Marple-Miller impactor. Aerosol Sci Technol 29:307–314
37. Hindle M, Byron PR, Miller NC (1996) Cascade impaction methods for dry powder inhalers using the high flow rate Marple-Miller impactor. Int J Pharm 134:137–146
38. Marple VA, Roberts DL, Romay FJ, Miller NC, Truman KG, Van Oort M, Olsson B, Holroyd MJ, Mitchell JP, Hochrainer D (2003) Next generation pharmaceutical impactor. Part 1: design. J Aerosol Med 16(3):283–299
39. Crump JG, Seinfeld JH (1981) A new algorithm for inversion of aerosol size distribution data. Aerosol Sci Technol 1(1):15–34
40. Rader DJ, Mondy LA, Brockmann JE, Lucero DA, Rubow KL (1991) Stage response calibration of the Mark III and Marple personal cascade impactors. Aerosol Sci Technol 14(3): 365–379
41. O'Shaughnessy PT, Raabe OG (2003) A comparison of cascade impactor data reduction methods. Aerosol Sci Technol 37(2):187–200
42. Roberts DL, Mitchell JP (2011) Influence of stage efficiency curves on full-resolution impactor data interpretation. Drug Delivery to the Lungs-22, The Aerosol Society, Edinburgh, pp 181–184. Available at: http://ddl-conference.org.uk/index.php?q=previous_conferences. Visited 4 Aug 2012
43. Roberts DL, Mitchell JP (2011) Influence of stage efficiency curves on interpretation of abbreviated impactor data. Drug Delivery to the Lungs-22, The Aerosol Society, Edinburgh, pp 177–180. Available at: http://ddl-conference.org.uk/index.php?q=previous_conferences. Visited 4 Aug 2012
44. Gulak Y, Jayjock E, Muzzio F, Bauer A, McGlynn P (2010) Inversion of Andersen cascade impactor data using the maximum entropy method. Aerosol Sci Technol 44(1):29–37
45. Svensson M, Berg E (2010) Measuring the fine particle dose using inter-stage filters in the NGI: an overview of two methods, AIM Specialty Workshop, Drug Delivery to the Lung 21, The Aerosol Society, Edinburgh, pp 382–385. Available at: http://ddl-conference.org.uk/index.php?q=previous_conferences. Visited 4 Aug 2012
46. Roberts DL, Romay FJ (2009) Design of the fast screening impactor based on the NGI pre-separator, Drug Delivery to the Lung 20, The Aerosol Society, Edinburgh, pp 206–210. Available at: http://ddl-conference.org.uk/index.php?q=previous_conferences. Visited 4 Aug 2012
47. Mathews J, Walker RL (1970) Mathematical methods of physics. W.A Benjamin, Menlo Park, CA
48. Finlay WH, Stapleton KW (1999) Undersizing of droplets from a vented nebulizer caused by aerosol heating during transit through an Andersen impactor. J Aerosol Sci 30(1):105–109
49. Mitchell JP, Nagel MW, Nichols SC, Nerbrink O (2006) Laser diffractometry as a technique for the rapid assessment of aerosol particle size from inhalers. J Aerosol Med 19(4):409–433
50. Doub WH, Adams WP (2002) Measurement of drug in fine particles from aqueous nasal spray by cascade impactor. Presented at the annual meeting of American Association of Pharmaceutical Scientists, Toronto. Poster T3033 (abstract)

51. Mitchell JP (2006) Good practices of qualifying cascade impactors (CIs): a survey of members of the European Pharmaceutical Aerosol Group (EPAG). Drug Delivery to the Lungs-16, The Aerosol Society, London, UK, pp. 189–192, abstract in. J Aerosol Med 19(2):232
52. Dolovich M, Rhem R (1998) Impact of oropharyngeal deposition on inhaled dose. J Aerosol Med 11(S1):S112–S115
53. Hickey AJ, Evans RM (1996) Aerosol generation for propellant-driven metered dose inhalers. In: Hickey AJ (ed) Inhalation aerosols: physical and biological basis for therapy. Dekker, New York, NY, pp 417–439
54. Stein SW, Gabrio BJ (2000) Understanding throat deposition during cascade impactor testing. In: Dalby RN, Byron PR, Farr SJ, Peart J (eds) Respiratory drug delivery VII. Serentec Press, Raleigh, NC, pp 573–576
55. Zhou Y, Sun J, Cheng Y-S (2011) Comparison of deposition in the USP and physical mouth-throat models with solid and liquid particles. J Aerosol Med Pulm Drug Deliv 24(6):277–284
56. Dunbar CA, Hickey AJ (2000) Evaluation of probability density functions to approximate particle size distributions of representative pharmaceutical inhalers. J Aerosol Sci 31:813–831
57. Roberts DL, Romay FA, Marple VA, Miller NC (2000) A high-capacity pre-separator for cascade impactors. In: Dalby RN, Byron PR, Farr SJ, Peart J (eds) Respiratory drug delivery VII. Serentec Press, Raleigh, NC, pp 443–445
58. Mitchell JP, Costa PA, Waters S (1988) An assessment of an Andersen Mark-II cascade impactor. J Aerosol Sci 19(2):213–221
59. Mitchell JP, Nagel MW (2007) Valved holding chambers (VHCs) for use with pressurized metered-dose inhalers (MDIs): a review of causes of inconsistent medication delivery. Prim Care Respir J 16(4):207–214
60. Dolovich MB, Ahrens RC, Hess DR, Anderson P, Dhand R, Rau JL, Smaldone GC, Guyatt G (2005) Device selection and outcomes of aerosol therapy: evidence-based guidelines. Chest 127(1):335–371
61. Rubin BK, Fink JB (2005) Optimizing aerosol delivery by pressurized metered-dose inhaler. Respir Care 50(9):1191–1200
62. Canadian Standards Association (CSA) (2008) Spacers and holding chambers for use with metered-dose inhalers. Mississauga, Ontario. CAN/CSA/Z264.1-02:2008 (revised). Available at: http://www.shopcsa.ca/onlinestore/GetCatalogItemDetails.asp?mat=2013981
63. Dolovich MB, Mitchell JP (2004) Canadian Standards Association standard CAN/CSA/Z264.1-02:2002: a new voluntary standard for spacers and holding chambers used with pressurized metered-dose inhalers. Can Respir J 11(7):489–495
64. Mitchell JP, Poochikian G, Hickey AJ, Suggett J, Curry P, Tougas T (2011) In vitro assessment of spacers and valved holding chambers used with pressurized metered-dose inhalers: the need for a USP chapter with clinically relevant test methods. Pharm Forum 37(4). On-line at: http://www.usppf.com/pf/pub/index.html. Visited 3 Jan 2012
65. European Medicines Agency (EMA) (2009) Guideline on the requirements for clinical documentation for orally inhaled products (OIP) including the requirements for demonstration of therapeutic equivalence between two inhaled products for use in the treatment of asthma and chronic obstructive pulmonary disease (COPD) in adults and for use in the treatment of asthma in children and adolescents. CPMP/EWP/4151/00 Rev. 1. London
66. Nagel MW, Schmidt JN, Doyle CC, Varallo VM, Mitchell JP (2003) Delay testing of valved holding chambers (VHCs) with a new apparatus. Drug Delivery to the Lungs-14, The Aerosol Society, London, pp 79–82
67. Rau JL, Coppolo DP, Nagel MW, Avvakoumova VA, Doyle CC, Wiersema KJ, Mitchell JP (2006) The importance of nonelectrostatic materials in holding chambers for delivery of hydrofluoroalkane albuterol. Respir Care 51(5):503–510
68. Foss SA, Keppel JW (1999) In vitro testing of MDI spacers: a technique for measuring respirable dose output with actuation in-phase or out-of-phase with inhalation. Respir Care 44(12):1474–1485

69. Finlay WH, Zuberbuhler P (1999) In vitro comparison of salbutamol hydrofluoroalkane (Airomir) metered dose inhaler aerosols inhaled during pediatric tidal breathing from five valved holding chambers. J Aerosol Med 12(4):285–291
70. Janssens HM, De Jongste JC, Fokkens WJ, Robben SGF, Wouters K, Tiddens HAWM (2001) The Sophia anatomical infant nose-throat (Saint) model: a valuable tool to study aerosol deposition in infants. J Aerosol Med 14(4):433–441

Chapter 3
Physical Causes of APSD Changes in Aerosols from OIPs and Their Impact on CI Measurements

Helen Strickland, Beth Morgan, and Jolyon P. Mitchell

Abstract The successful implementation of AIM and/or EDA principles to the in vitro assessment of inhalable aerosols emitted from OIPs requires the user of such methods to have a basic understanding of how these particles and/or droplets interact with the human respiratory tract (HRT) upon inhalation. Such processes are inextricably governed by the underlying physical processes associated with these semi-stable systems, and all of the changes influencing particle size affect the entire APSD. This chapter looks at both aspects in some detail, in particular paying attention to how small changes in APSD might be detected by full-resolution CI systems. The information presented herein is a prelude to Chap. 9, in which case studies are presented to demonstrate the sensitivity of EDA metrics to such changes.

3.1 Introduction

It is worthwhile briefly reviewing how aerosols are formed in the first place from the different inhaler classes, before exploring the ways in which aerosols emitted by OIPs are currently measured in the laboratory, and later in the book, how the AIM and related EDA concepts may make the process more efficient and effective in the

H. Strickland (✉)
GlaxoSmithKline, N226, 1011 North Arendell, Zebulon, NC 27597, USA
e-mail: helen.n.strickland@gsk.com

B. Morgan
GlaxoSmithKline, Zebulon Manufacturing and Supply,
1011 N. Arendell Avenue, Zebulon, NC 27597, USA
e-mail: beth.e.morgan@gsk.com

J.P. Mitchell
Trudell Medical International, 725 Third Street, London, ON N5V 5G4, Canada
e-mail: jmitchell@trudellmed.com

T.P. Tougas et al. (eds.), *Good Cascade Impactor Practices, AIM and EDA for Orally Inhaled Products*, DOI 10.1007/978-1-4614-6296-5_3,

quality decision-making process. There are several textbooks available that address the fundamentals of the various aerosol generation processes, and the reader is referred to these for more detailed information [1–5]. Although somewhat older than the other texts, the book edited by Morén et al. is particularly significant because it addresses how inhaler aerosols interact with the respiratory tract in the context of the diagnosis and treatment of lung diseases.

The creation of aerosols containing medication for inhalation in general involves one of the following basic processes [6]:

(a) Rapid flash evaporation of a metered dose of formulation containing the API(s) in either a high-volatile hydrofluoroalkane (HFA) or an older similarly high-volatile chlorofluorohydrocarbon (CFC) propellant by means of a metered-dose inhaler (MDI).
(b) Dispersion of a dry powder containing the API(s); in so-called passive DPIs, dispersion takes place by the energy associated with the vacuum created during inhalation, and in the newer so-called active systems, the energy comes from an external source.
(c) Atomization of bulk liquid containing the API(s) by various means, including the mechanical breakup of liquid forced through one or more fine orifices (soft mist inhalers—SMIs), the expansion of air from a narrow orifice entraining the liquid stream by the Bernoulli principle (jet nebulizer), the application of ultrasonic energy (ultrasonic nebulizer), the electromechanical vibration of one or more orifices or capillaries (vibrating mesh/membrane nebulizer), and the application of high electrostatic charge to the liquid stream (electro-spray), all of which create instability in the emanating liquid stream(s).
(d) A few inhalers (currently in development) make use of controlled evaporation of a low-volatile liquid containing the thermally stable medication by means of an integral heat source, followed by condensation as the vapor moves away from the heat source to exit the inhaler.

There are thus many variants associated with each form of inhaler. As well as hand-operated MDIs, where the patient has to take care of coordinating inhaler actuation, in recent years there has been increasing interest in the development of breath-actuated MDIs [7] because such devices avoid the coordination issue altogether. Nebulizers, by virtue of their external gas source (either from a fixed compressed gas supply or portable compressor), also avoid the need for patient coordination, although breath-enhanced and breath-actuated devices provide improved drug output during patient inspiration [8]. Alternatively, tube spacers and VHCs have been developed as add-on devices for MDIs during the past 30 years to assist patients using MDIs, as they also eliminate the need for perfect coordination with inhaler actuation [9]. Passive DPIs, by virtue of the fact that they make use of the inspiratory effort from the patient to create the aerosol, ensure coordination, but their effective delivery of medication may be impeded if the energy transfer to the dry powder is reduced because the user cannot generate sufficient effort [10]. However, relatively high-resistance DPIs have been shown to provide better medication delivery to the lower respiratory tract than do lower resistance devices [10].

The provision of a patient-independent source of energy to disperse the powder, as takes place within an active DPI, is one way to address this limitation [6]. There is more variety in the design of different DPIs than associated with the other OIP classes, and this diversity may have implications when attempting to connect the patient interface of the inhaler to the CI system.

At the outset, it is important to realize that all of the changes influencing particle size of OIP-generated aerosols that have just been outlined affect the entire APSD. That is, they are not particularly size-selective. It follows that there are no processes which would selectively make the mass of API collecting on one stage of a CI grow and that on its neighboring stage decrease, without a concomitant change in the rest of the APSD profile.

In 2000, a consensus statement was developed by a group of experts assembled to evaluate critically each of these forms of aerosol generator [9]. This statement includes much useful information concerning the advantages and limitations of each OIP-based aerosol generator from the perspective of the practicing clinician and based on currently available medications at the time of publication. Individual papers from these authors, describing each type of inhaler class can be found in the same issue (June 2000) of the peer-reviewed journal, *Respiratory Care*. Although CFC propellants for MDIs have almost entirely been withdrawn from most first-world markets since that statement was published, the descriptions of the principles of aerosol generation provided at this conference still apply to HFA-propellant operated MDIs today.

3.2 Deposition of Aerosols in the Human Respiratory Tract

The human respiratory tract (HRT) has evolved to act as an airborne particle size classifier to prevent the ingress of inhaled aerosols to the gas-exchange region in the distal lung [11]. Thus, moving from the proximal carina to the distal alveolar sacs, the airways in successive generations steadily decrease in diameter, such that particles from an incoming aerosol are deposited in size order, beginning with the largest [5]. The dimensions of the airways enlarge as the result of growth from the fetal to adult stage [12]. API delivery by pulmonary inhalation exploits the use of aerosolized particles which are least efficiently trapped in the upper airways and best able to reach the conducting airways (for local drug action) or the respiratory airways (for either local action or systemic absorption) [13–15].

The current International Commission on Radiological Protection (ICRP) human respiratory tract model for radiological protection provides a profile of the likely deposition fate of incoming aerosol particles based on their aerodynamic size (Fig. 3.1).

These curves are based on a large repository of data associated with radiological protection and therefore relate to the prevention rather than delivery of particulates to the HRT.

The aerodynamic diameter size scale (d_{ae}), introduced in Chap. 2, defines most closely the transport and deposition of particles in the critical size range from about

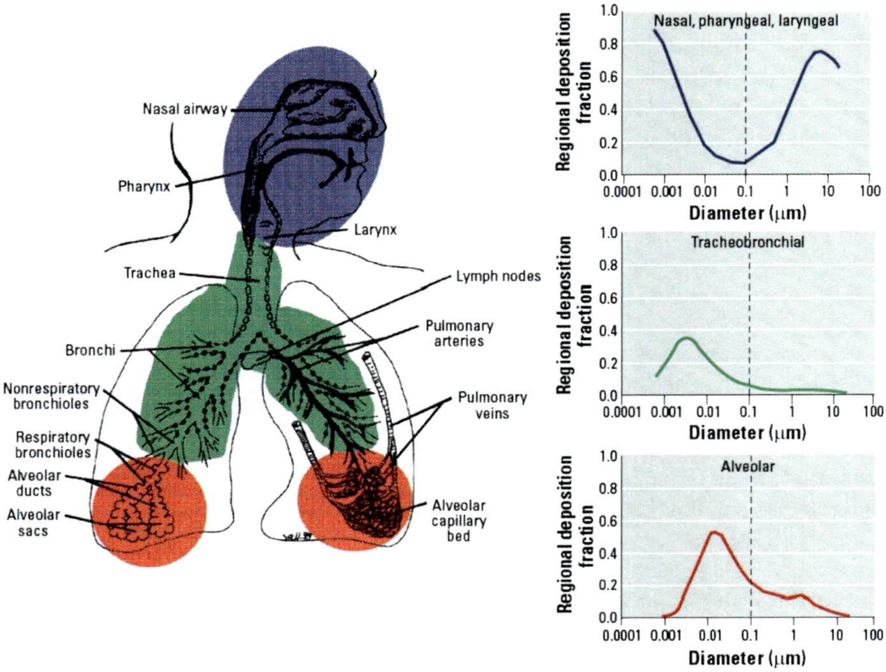

Fig. 3.1 Regional deposition of particles in the adult respiratory tract, based on the ICRP human respiratory tract model for radiological protection (*From* [15]—*used with permission*)

0.5 to 15 μm. In the case of an oral tidal-breathing adult, particles with d_{ae} larger than approximately 5 μm deposit mainly in the oropharyngeal region, central (bronchial) airway deposition peaks with particles with d_{ae} between 7 and 9 μm, and peripheral (alveolar) deposition in the lung reaches a maximum with particles having d_{ae} between 2 and 4 μm (Fig. 3.2). Particle aerodynamic size is therefore a sensitive measure of likely HRT deposition location. This fact has formed the basis of several attempts to develop either mathematically rigorous in vitro-in vivo correlations (IVIVCs) or somewhat less predictive relationships (IVIVRs) from the standpoint of proportionality, thus far with limited success [16]. This situation is due to a variety of reasons, not least the relative insensitivity of clinical metrics based on lung mechanics to the action of APIs whose purpose and function are other than dilatation of the airways (e.g., inhaled corticosteroids that are used to reduce underlying inflammatory disease).

In summary, the creation and delivery of an aerosol containing micron-sized particles by any of the OIP forms mentioned previously is favorable for the efficient delivery of medication past the oropharynx (upper respiratory tract) to the airways of the lungs where the receptors are located for the treatment of the various diseases associated with this organ, commonly but not restricted to asthma, COPD, and

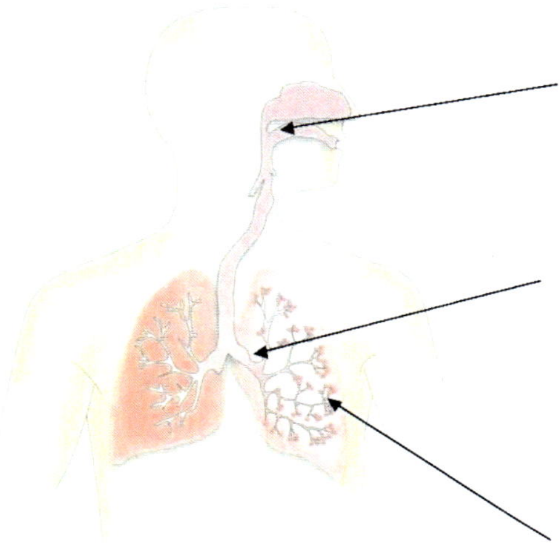

Fig. 3.2 Regional particle deposition in an oral tidal-breathing adult (*Courtesy of Trudell Medical International*)

cystic fibrosis. There is also the potential for delivery of medication for systemic rather than topical application via the lungs [17] because of the large surface area for gas exchange to the blood circulation in the alveolar region.

3.3 Linking Aerosol Behavior to APSD Changes

3.3.1 Aerosol Formation from OIPs

This chapter has introduced the concepts that inhaled aerosols are useful for both topical and systemic delivery of medication to the patient, using the HRT as the portal of entry. However, aerosols are by their nature quasi-stable two-phase systems comprising either solid particles or liquid droplets (from now onwards referred to collectively as particles) separated from each other by a supporting gas [18].

3.3.1.1 Dry Powder Inhalers

The aerosols produced by dry powder inhalers (DPIs) are the result of a combination of several factors that result in the de-aggregation of the bulk powder that is stored either as a single-dose pre-metered unit (e.g., blister or capsule) until use or as bulk powder within the inhaler. The physicochemistry of DPI formulations is complex and beyond the scope of this book; however, there are comprehensive reviews on the topic for readers interested in more detail [19–22].

In the context of aerosol formation, the use of large lactose carrier particles that are an order of magnitude or more larger in physical diameter than the primary API-containing particles is one way to avoid problems with cohesiveness of a formulation, if it contained just the latter [23]. Alternatively, API-containing particles may be made less cohesive by preprocessing [24] or forming them to have a nearly spherical shape [25]. As a general rule, the formulators seek properties that optimize the efficiency with which single aerosol particles are formed in the airflow generated by the patient upon inhalation. The vast majority of DPIs currently on the market are passive, in that they rely on the inspiratory effort of the patient to provide the energy to de-aggregate and aerosolize the API-containing particles [26]. However, a few newer devices are considered "active," in that powder dispersion is accomplished by a suitable component of the inhaler, such as an electronic vibrator or impeller. Whichever option is under consideration, ultimately the APSD of the aerosol that is inhaled therefore depends upon a combination of the formulator's ability to optimize the powder disaggregation process and the device developer's ability to adjust DPI resistance to focus the available energy where it is needed. In terms of making measurements of DPI aerosol APSDs, the compendia have standardized the process, whereby sampling a fixed 4 L volume of air from the inhaler occurs at the designated flow rate, using a critical orifice at the flow adjustment valve to achieve sonic flow, rather than simulating the profile of an individual inhalation [27, 28]. Under these circumstances, small changes in APSD are likely to originate as much from the result of the environmental conditions under which the measurements are conducted (in particular high relative humidity that may affect powder cohesiveness [29] and the presence or absence of electrostatic charge [30]), as they are from the inhaler itself.

3.3.1.2 Pressurized Metered-Dose Inhalers

MDI-based aerosol formation is also closely linked to the physicochemical characteristics of the formulation contained in the pressurized canister, in particular if the API is in solution or suspension [31]. However, the presence of the liquid propellant under pressure introduces the ability for these inhalers to self-deliver the aerosol to the patient when actuated, as the result of flash evaporation that takes place upon actuation as the metered dose of liquid is exposed to ambient pressure and temperature [32]. Device design factors, in particular the construction of the metering valve and orifice, play an important role in determining the APSD of the aerosol that is ultimately released [33] for the patient to inhale. The APSD that is measured by CI during MDI testing can be influenced by the temperature and relative humidity of the testing environment, especially at high temperatures and relative humidity levels [34]. For example, Lange and Finlay observed APSD changes when making measurements at temperatures above about 35°C with the relative humidity above about 95% [35]. They attributed their findings as being likely due to changes in the droplet

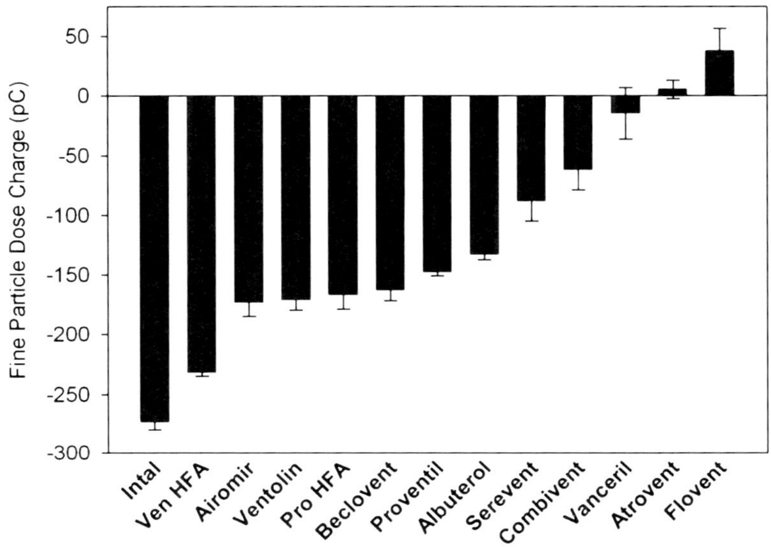

Fig. 3.3 Intrinsic electrostatic charge series for different MDI products (*From* [36]—*used with permission*)

evaporation rate caused by interaction of water vapor with the surfaces of the evaporating droplets. However, such extreme conditions are unlikely to be encountered in testing laboratories with basic heating ventilation and air conditioning controls in place for the building.

Electrostatic charge associated with triboelectrification of the metered-dose contents as they emerge from the metering valve and are atomized results in intrinsic charge associated with the aerosol that is different from one product to another (Fig. 3.3) [36]. The charge associated with fine particles (fine particle dose in Fig. 3.3) depends on many factors including the formulation itself, as well as the materials used for the metering valve system. This intrinsic charge combined with the presence of surface charge associated with the inhaler mouthpiece or add-on device that may be present is a major source of APSD variability [37] and must be mitigated or better eliminated altogether, for reproducible results to be obtained in cascade impactor-based measurements [38].

3.3.1.3 Nebulizing Systems and Soft Mist Inhalers

Nebulizing systems and SMIs operate on the basis of atomization of a liquid (usually an aqueous solution or suspension of the API in physiologically normal saline) by various methods, including pneumatic pressure, ultrasonics, and mechanical pressure through ultrafine orifices (SMIs), or by applying vibration of the bulk liquid against a mesh or membrane containing a multitude of similar-sized orifices [39–42].

The pneumatic atomization process generally produces bimodal droplet distributions [43, 44]. Baffles and other obstructions are therefore located close to the orifice(s) through which the liquid stream emerges as the result of the Bernoulli effect, when the driving gas pressure is applied to operate the nebulizer. The aerosol available for inhalation therefore comprises a higher proportion of fine droplets <5 μm d_{ae} than would be the case for unrestricted atomization of a liquid stream [45]. The electronics associated with ultrasonic, SMI, and vibrating mesh/membrane nebulizers can be adjusted so that the aerosol is predominantly of fine droplets [40].

The measurement of APSD for nebulized suspension formulations is complicated by the fact that the relative size of the API particles to the droplets of nebulized liquid influences the measurement [46]. For example, a shift to finer droplet sizes caused by either a variation in critical components associated with the nebulizer itself, or more likely, an increase in driving gas pressure, will likely result in less efficient incorporation of the API particles, especially if their size is commensurate with the liquid droplets.

Solution formulations by definition are a homogeneous dispersion of the API, so that the resulting droplet APSDs determined by cascade impactor reflect only the operating conditions of the nebulizer (i.e., driving gas pressure and flow rate for pneumatic systems, vibration frequency for vibrating mesh/membrane systems, liquid pressure for SMIs), together with the local environment into which the aerosol is dispersed. In particular, control of relative humidity is important, particularly when the nebulized aerosol is not surrounded by saturated gas, as is the case with jet nebulizers without air entrainment that creates a locally saturated environment [47].

A further concern when evaluating nebulizing systems is the transfer of heat from the cascade impactor to the droplets, enhancing their evaporation, and thereby resulting in bias to finer sizes [48]. For this reason, the new normative chapter in the European Pharmacopoeia concerned with nebulizer testing recommends that the impactor be chilled [49, 50].

3.3.2 Aerosol Transport and APSD

Once the OIP aerosol has been created, there are several processes that affect its APSD during transport to either the CI or to the patient. These aerosols are in a continuous process of change as the relative motion of the individual particles combines them (agglomeration), inertial forces associated with their movement relative the flow streamlines of the surrounding support gas removes them by impaction onto adjacent surfaces (inertial/turbulent deposition), or the influence of gravity deposits them (gravitational sedimentation). The presence of electrostatic charge either intrinsic to the aerosol or present on adjacent nonconducting surfaces can also have a major effect on the APSD that gets measured.

3.3.2.1 Particle–Particle Agglomeration

Agglomeration is the process whereby individual particles collide to produce larger particles and is continuously occurring once an aerosol has been formed. Coagulation refers to the process of agglomeration in which the combined particles merge together by the process of coalescence immediately on contact and is confined to liquid droplets [51]. Thermal coagulation or agglomeration is the result of the random motion of particles caused by transfer of energy by collisions with adjacent gas molecules (Brownian diffusion) and is most evident with only the finest particles < ca. 1 μm d_{ae} produced from OIPs [15]. As well as thermal agglomeration/coagulation, particles may come together as the result of relative motion caused by kinematic processes, such as the result of differing settling velocities due to gravity. Whenever a flow velocity gradient exists, gradient or shear agglomeration/coagulation will occur, as in turbulent flow [51].

The classical theory for agglomeration was developed initially for monodisperse aerosols, based on Smoluchowski theory, and the reader is referred to the textbook by Hinds [51], for a full description of the processes involved, including exact solution of equations that enable time-dependent changes to particle number concentration and associated size distribution to be calculated.

In this chapter, it is necessary to focus on understanding time-dependent changes in OIP aerosol APSD that might be detected by the CI method; however, such aerosols are as a general rule polydisperse. An explicit mathematical solution, similar to that for monodisperse systems, describing the agglomeration process does not at present exist. It is therefore necessary to make some assumptions about the properties of the aerosol, the most common being that it is unimodal and lognormally distributed. Under such circumstances, Lee and Chen have described the process for agglomeration of an aerosol having a count median diameter (CMD) and geometric standard deviation (σ_g) by the relationship [52]:

$$K = \frac{2kT_{abs}}{3\eta}\left[1 + e^{\ln^2\sigma_g} + \left(\frac{2.49\lambda}{CMD}\right)\left(e^{0.5\ln^2\sigma_g} + e^{2.5\ln^2\sigma_g}\right)\right] \tag{3.1}$$

in which k is the Boltzmann constant, λ is the mean free path length of gas (air) molecules supporting an aerosol, T is the absolute temperature (degrees Kelvin), η is the gas (air) density, and K is the average agglomeration coefficient for the system. It should be noted that a simpler relationship exists for monodisperse aerosols, in which the coagulation coefficient becomes K_{mono}, and:

$$N(t) = \frac{N_0}{1 + N_0 K_{mono} T} \tag{3.2}$$

where N_0 is the initial number concentration (density) of the aerosol and $N(t)$ describes the time-dependent decrease due to agglomeration. Using this equation, if the aerosol number concentration is close to 10^{13} particles/m^3 (high in relation to the likely scenario in association with an OIP [53]) with CMD of 0.2 μm (a conservative estimate of the size effect) with GSD close to 2.0 (typical of the dispersity of OIP aerosols), after a 2 s delay, N_{2s} is only reduced to 9.98×10^{12} particles/m^3, and CMD has increased by less than 10%. Agglomeration is therefore unlikely to result in significant changes to the overall APSD of typical OIP aerosols, during the short transit time from creation to size fractionation, except perhaps in the immediate time interval after atomization takes place from MDI products, where aerosol number densities are at their highest.

3.3.2.2 Particle Inertia

The influence of inertia on particle motion is a significant factor in the development of the APSD both during inhalation by the patient [54] and during the transit of the aerosol to the measurement equipment in laboratory testing [55]. At its simplest, the effect of inertia can be related to how the motion of an individual gas-borne particle takes place relative to the suspending gas [usually air, oxygen, oxygen-enriched air, or occasionally a helium–oxygen (Heliox) mixture]. Finlay [53] has shown that the particle Stokes number (*St*) describes the influence of inertial force on particle motion in accordance with:

$$\frac{\tau \mathrm{U}_0}{\mathrm{D}} \frac{\mathrm{d}\mathbf{v}'}{\mathrm{d}t'} = \frac{\tau \mathrm{g}}{\mathrm{U}_0} \hat{\mathbf{g}} - \mathbf{v}'_{\mathrm{rel}} \tag{3.3}$$

in which U_0 is the mean velocity of the gas flow; D is a characteristic size related to that of the particle, typically the diameter of the airway through which it is moving as it passes through the HRT; and the differential term represents particle acceleration. $\hat{\mathrm{g}}$ is the non-dimensionalized gravitational term ($\mathbf{g}/g$) and $\mathbf{v}_{\mathrm{rel}}$ is the particle velocity relative to that of the surrounding gas. τ is the particle relaxation time, which is particle size-dependent, through the relationship:

$$\tau = \frac{\rho_{\mathrm{p}} d_{\mathrm{p}}^2 C_{\mathrm{c}}}{18\eta} \tag{3.4}$$

where ρ_{p} is the particle density, η is the gas (air) viscosity, and C_{c} is the Cunningham slip correction factor (close to unity for particles >2 μm d_{p}).

Both particle velocity ($\mathbf{v}'$) and time (t') are in nondimensional forms where:

$$\mathbf{v}' = \frac{\mathbf{v}}{U_0} \tag{3.5}$$

and

$$t' = \frac{t}{\left(\dfrac{D}{U_0}\right)} \tag{3.6}$$

St is the coefficient before the differential term in (3.3), and:

$$St = \frac{\tau U_0}{D} \quad (3.7)$$

or more familiarly by substitution into (3.4):

$$St = \frac{U_0 \rho_p d_p^2 C_c}{18\eta D} \quad (3.8)$$

Equation (3.8) [which is similar to (2.6) for the special case of a single-stage impactor] can be used to calculate particle Stokes numbers in order to assess the likelihood of inertial deposition to adjacent surfaces. From this equation, it can be seen that the square root of *St* is directly proportional to the square root of [$\rho_p d_v^2$] and therefore also to d_{ae} (assuming a dynamic shape factor of unity). As a practical guide, the probability of inertial deposition increases with larger particle size, since as *St* approaches zero, (3.3) indicates that the particle motion will converge to be identical with that of the supporting gas. Conversely, if *St* approaches unity or larger, the likelihood of the particle following the streamlines of the local gas flow decreases. In the round-nozzle CIs described in Chap. 2, it was mentioned that a value of $\sqrt{St}$ of 0.49 or greater is associated with inertial deposition [20], and similar estimates for this parameter have been made in modeling particle deposition in both models of the oropharynx and upper airways of the HRT [54, 56, 57] and with actual airways in vivo [58, 59].

3.3.2.3 Gravitational Sedimentation

Gravitational sedimentation is a potentially important external force affecting aerosol particle motion during transit from the inhaler to the patient or measurement equipment. Under Stokesian conditions, which apply to the sampling of OIP-generated aerosols, the drag force generated by passing through the support gas on the particle under consideration ($\mathbf{F}_d$) is counterbalanced by the relative motion of the particle with respect to the gas ($\mathbf{v}_{particle} - \mathbf{v}_{gas}$) in accordance with:

$$\mathbf{F}_d = -3\pi d_p \eta (\mathbf{v}_{particle} - \mathbf{v}_{gas}) \quad (3.9)$$

from which the particle sedimentation velocity (v_t) in a still gas is reduced to:

$$\mathbf{F}_d = -3\pi d_p \eta \mathbf{v}_t \quad (3.10)$$

The force on the particle, with volume V_p and density ρ_p, due to gravity is:

$$mg = \rho_p V_p g \quad (3.11)$$

Table 3.1 Sedimentation velocity (v_t) at selected sizes for spherical, unit density particles

	v_t (mm/s)	
d_p (μm)	Uncorrected for C_c	Corrected for C_c
0.1	0.00030	0.00086
0.5	0.0076	0.0100
1.0	0.0302	0.0348
2.0	0.121	0.134
5.0	0.749	0.782
10.0	3.014	3.060

For a spherical particle where $V_p = \pi d_p^3 / 6$, (3.11) becomes:

$$mg = \rho_p \pi d_p^3 g \quad (3.12)$$

When the force on the particle equals the drag force, the particle will travel at a constant settling velocity, $\mathbf{v}_t$, given by:

$$\mathbf{v}_t = \frac{\rho_p g d_p^2}{18\eta} \quad (3.13)$$

Values of $\mathbf{v}_t$ for unit density particles (cgs system) can be tabulated as a function of d_p in the size range of interest for OIP inhaled aerosols being size fractionated by cascade impactor (Table 3.1). The numerator of (3.13) is multiplied by the Cunningham slip correction factor (C_c) for particles <0.5 μm; the correction for air at ambient pressure and temperature increases the settling velocity by 34 % when d_p is 0.5 μm [53].

The velocity acquired through gravitational sedimentation is relatively unimportant in relation to typical sampling velocities involved with cascade impaction measurements, which approach 17 mm/s at the entry to the 19 mm internal diameter USP/Ph. Eur. induction port, when a typical flow rate of 28.3 L/min is used to sample the aerosol via an ACI. However, if the aerosol is collected by a VHC in connection with MDI testing with this type of add-on device, the test protocol is likely to be modified to include a delay interval of several seconds between MDI actuation and the onset of sampling in order to mimic how the device might be used by a patient [60]. The loss of aerosol due to gravitational sedimentation with elapsed time may become observable, but only after very long intervals greater than 10 s, as long as the confounding influence of electrostatic charge is eliminated [61]. In the testing of these devices recommended by a Canadian standard, the delay interval is kept to 2 s, as this is more typical of the elapsed time that might occur if a patient exhales instead of inhaling when actuating their inhaler [62]. Under these circumstances, the influence of gravity on both mass concentration and APSD can be treated as of minor consequence.

3.3.2.4 Molecular Diffusion (Brownian Motion)

Particle motion in relation to the motion of the supporting gas molecules (i.e., molecular diffusion or Brownian diffusion) becomes important at sizes <1–2 μm [53]. D_d, the diffusion coefficient, is given by the expression:

$$D_d = \frac{kTC_c}{3\pi\eta d_p} \tag{3.14}$$

Finlay has compared the random particle motion as the result of Brownian diffusion with the downward-directed motion due to gravitational sedimentation in accordance with the relationship:

$$\frac{p_{\text{diff}}}{p_{\text{sed}}} = \left[\frac{18\eta\sqrt{(2D_d t)}}{\rho_p g d_p^2 C_c t}\right] \tag{3.15}$$

in which p_{diff} is the root mean square displacement of a particle of diameter and d_p, due to molecular diffusion in time t. p_{sed} is the distance travelled by the same particle in the same time interval.

Molecular diffusion can be considered negligible if $p_{\text{diff}}/p_{\text{sed}}<0.1$ [53] and is therefore, in general, less important than either sedimentation or inertial behavior, except within the highest generations of the HRT in which the dimensions of the airways are finest or in cases where a breath-hold of several seconds takes place at the end of inhalation to extend the residence time of the particles in these airways and the alveolar sacs [53]. In this context, Landahl has reported that the influence of molecular diffusion becomes important in terms of deposition in the HRT only for particles ≤0.1 μm [63], significantly finer than the range of interest in connection with currently marketed OIPs. Its impact on CI-measured APSDs of OIP aerosols is also likely to be minimal, given the short residence times between the inhaler and size-fractionating apparatus, even when a VHC is present in the context of MDI testing.

3.3.2.5 Electrostatic Charge

Electrostatic charge-based particle motion is a significant and somewhat unpredictable contributor to APSD changes with almost all OIPs, except perhaps nebulizing systems. The likelihood that MDI-produced aerosols will carry significant intrinsic charge has already been discussed in connection with aerosol formation. This section is concerned with how electrostatic charging can affect the particle transport process from the inhaler. In contrast with the previous physical mechanisms, there are no analytical expressions that can reliably be used to describe the motion of ensembles of aerosol particles, most likely because of the complexity of evaluating particle behavior while considering both intrinsic charge and size distributions, as well as the influence of local electric fields, likely of both polarities associated with charges on nearby surfaces.

There are three mechanisms by which aerosol particles can acquire electrostatic charge [64]:

(a) Triboelectrification, whereby charge transfer takes place as each particle is separated from the bulk material or removed from a surface with different triboelectric properties
(b) Diffusion charging, where random collisions between particles and unipolar ions cause charge accumulation on the particles
(c) Field charging, where particles acquire charge from collisions with unipolar ions in an applied electric field

Diffusion and field charging are seldom encountered in inhalation therapy, except with specific liquid electro-hydrodynamic atomization systems, in which an applied electric field is used to charge the liquid stream containing medication emerging from an orifice or series of orifices.

Triboelectrification is a widespread phenomenon. It may be further subdivided into (1) contact charging, where there is an initial attachment between particles or particle–surface touching, followed by separation without rubbing together, and (2) frictional charging where relative movement of the two surfaces takes place while still in contact [65]. In practice, however, it is difficult to distinguish the two processes, and the term "triboelectrification" is therefore often applied to include both forms of static electrification.

The influence of electrostatic charge on the measurement of OIP aerosols is largest with systems not producing electrically conductive aerosol particles. Hence, the measurement of aerosols containing physiologically normal saline that are produced from wet dispersion systems (nebulizers and SMIs) is less likely to be affected, but charging effects have nevertheless been observed with nongrounded systems [66]. However, electrostatic phenomena are a much more significant issue to contend with when measuring DPI-generated aerosols, because they acquire bipolar charge during the dispersion process [65, 67]. Given that the materials of most DPIs are nonconducting and may therefore themselves acquire charge by contact electrification, the possibility of subtle changes in the aerosol dispersion and transport from the inhaler to the measurement apparatus may result in measurable shifts in APSD [68]. Sensible precautions to minimize the influence of such behavior are (a) controlled ambient conditions, especially relative humidity, and (b) electrically grounded workstations and operators undertaking the testing [38]. Triboelectrification is also a major consideration when measuring the APSDs of MDI-generated aerosols [69, 70], especially if nonconducting add-on devices are present [37, 71]. Similar precautions as those identified for DPI testing therefore apply equally with MDI-related evaluations.

3.3.2.6 Evaporation and Condensation Processes

Whenever there is either a volatile species (e.g., water, ethanol) present, or the presence of hygroscopic solid particles, consideration needs to be given to the effects of either evaporation or condensation (primarily from environmental moisture) upon the APSD of the OIP aerosol being measured. Generally, the flash evaporation of HFA or CFC propellants takes place rapidly on MDI actuation [72]. This process is largely complete by the time that the resulting aerosol has stabilized by eliminating the initial ballistic motion through drag on the individual particles by the surrounding gas molecules. From this time onwards, the particles retain the local gas velocity associated with inhalation or being sampled by the CI. Orifice design characteristics, specifically expansion orifice diameter, orifice jet length and expansion chamber sump depth, have been shown to be important in determining aerosol plume development associated with the propellant evaporation process [73]. However, these dimensions are fixed in early-stage product development, so that APSD changes originating from a manufactured inhaler design are likely to be small. However, if the atomized droplets contain a low-volatile cosolvent for the API, most usually ethanol, the process of evaporation to form the residual particles may take several seconds [74], so that the measurement process can be challenging to obtain reproducible results [34]. Unfortunately, the compendia currently do not provide guidance on how to compensate for incomplete cosolvent evaporation, and it is therefore likely that the process will not be complete until the aerosol has passed through the initial stages of the CI [74].

The evaporation of aqueous droplets formed from nebulizing systems has already been discussed in connection with the aerosol formation process (Sect. 3.3.1.3). The process can be extremely rapid, especially under dry conditions [53, 75], taking place within a few tens of milliseconds. The time scale for evaporation can be appreciated by referring to the data on time-dependent droplet diameter reduction at 15 L/min and 30 L/min, in which the ambient RH was 40%, presented in Fig. 3.4. Put in perspective, the time taken for an aerosol to pass through the 85 mL Ph. Eur./USP induction port [76] at 15 L/min and 30 L/min, assuming plug flow, would be of the order of 0.34 s and 0.17 s, respectively.

Droplet evaporation kinetics are strongly related to the initial droplet formation process, as well as the local relative humidity of the surrounding gas [53]. The example of droplet evaporation kinetics associated with sampling from a vibrating mesh nebulizer [initial droplet diameter (d_{ini}) = 4.3 μm, liquid feed rate (Q_l) = 0.296 ml/min], by an NGI at two different flow rates at room ambient conditions (T_{amb} and RH_{amb} of 21°C and 40%, respectively) as shown in Fig. 3.4, is helpful to understand the time scale of the evaporation process. However, it should not be seen as being more than an illustration of the scenario that can be anticipated under subsaturated conditions. Droplet evaporation is suppressed when the surrounding air is saturated (RH_{amb} = 100%), and this may be closer to conditions in the human oropharynx (RH_{op} ca. 75%) for oral breathing [77], than room ambient relative humidity values that are typically <50%. However, there are practical limitations to working under

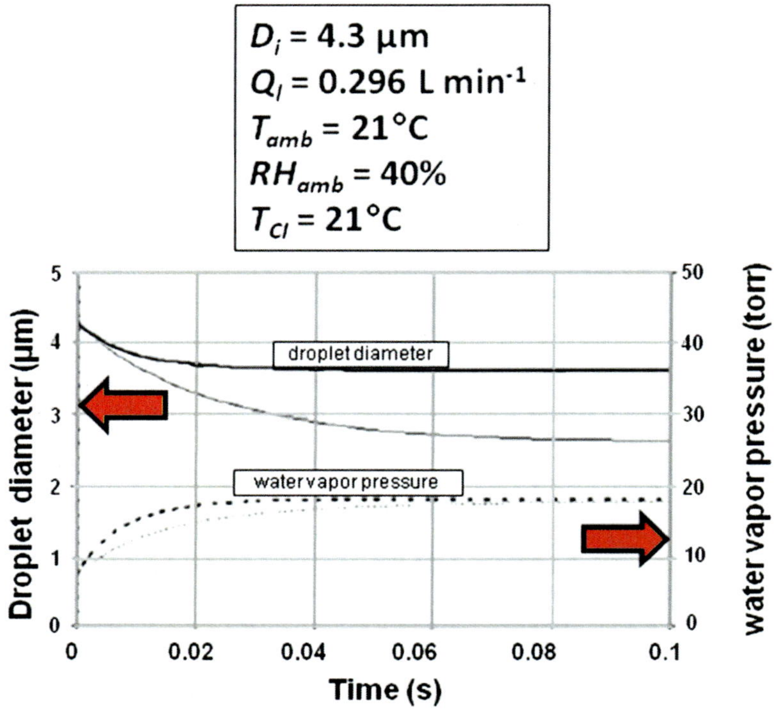

Q_a (L/min)	MMAD (µm)
15	3.6
30	2.6

Fig. 3.4 Predicted droplet size-time profiles for two different sampling flow rates using an evaporation model developed by Rao et al. (*From* [75]—*used with permission*)

saturated atmospheric conditions in the laboratory, which is why RH_{amb} is normally subsaturated. However, it follows that good controls of both ambient temperature and in particular relative humidity are important to obtain reproducible CI-measured APSDs for these systems.

Water vapor absorption onto hygroscopic powders is a well-known source of aerosol APSD shifts, predominantly in the direction of larger sizes, as the locally bound water increases particle–particle cohesiveness [78]. The processes involved include local dissolution and recrystallization, leading to irreversible aggregation through solid bridge formation [79]. While such processes are most likely to take place in storage, they may need to be considered with highly hygroscopic particles, such as cromolyn sodium, even when sampling the resulting aerosol [80]. The best

Table 3.2 Physical processes associated with aerosol transport and influence on CI-measured APSD; the *upward* and *downward arrows* indicate increases and decreases, respectively; the *sideway arrows* indicate variable/unpredictable influences on the magnitudes of *MMAD* and *GSD*

		Influence[a] on *MMAD* and *GSD*	
Process	Relevance to OIP aerosols	*MMAD*	*GSD*
Agglomeration/coagulation	Low, unless extra-fines present	↑	↑[a]
Inertial impaction	High	↓	↓
Gravitational sedimentation	High	↓	↓
Molecular diffusion	Low	≈	≈
Electrostatic charge	High and variable—mitigate where possible	↔	↔
Evaporation	High where volatile species present	↓	↓
Condensation/hygroscopic absorption of water	High if hygroscopic species are present	↑	↑

[a]Thermal agglomeration/coagulation by itself will increase GSD to about 1.35 and have no further effect thereafter. This change is therefore unimportant for OIP aerosols

precaution when making measurements of APSD with hygroscopic materials is to control local environmental conditions, especially keeping the relative humidity constant and preferably at a value where hygroscopic growth does not take place.

3.3.2.7 Summary

Table 3.2 is a qualitative summary of the effects on a hypothetical OIP APSD, assumed to be unimodal and lognormal, caused by the various processes that have been considered.

In this table, an upward facing arrow indicates an increase in the APSD property concerned, with a downward arrow showing the opposite behavior. The "≈" symbol indicates a small or insignificant effect, whereas the "↔" symbol represents a highly variable effect that can affect the indicated variable in either direction.

Agglomeration/coagulation is continuously present but is only likely to be important in the aerosol sampling and APSD measurement process by CI when extra-fines are present as a large portion of the total mass sampled. Inertial impaction takes place only when particles are accelerated or decelerated close to a nearby object in the flow path; this process dominates size classification in the CI, but is also important during the aerosol transport via the induction port and pre-separator (if used). Gravitational sedimentation, like agglomeration, is continuously present but is most important with the largest particles that enter the measurement system. Molecular (Brownian) diffusion is unimportant, except maybe in cases where extra-fine particles predominate in the APSD. Electrostatic charge has both a large and variable effect on aerosol APSD, depending upon the acquired charge distribution combined with the corresponding charge distribution on adjacent surfaces. It is therefore highly recommended that electrostatic charge is mitigated or avoided altogether, where the latter is feasible. The avoidance of low ambient relative humidity, the grounding of operators handling the apparatus, the use of electrically conductive and grounded surfaces, and, as a last resort, ionization of the air-stream containing the aerosol have each been found useful in this respect [64]. Evaporative changes to the APSD have to be considered where volatile species (i.e., water, ethanol) are present. Mitigation measures include cooling of the CI, a procedure that is recommended for the evaluation of aerosols derived from preparations for nebulization in the USP [81] and Ph. Eur. monograph chapters [49]. Finally, condensation of moisture on hygroscopic particles can be significant with certain OIPs, such as those containing cromones. This effect can be controlled by operating the CI system in an ambient environment set to a stable RH that is either kept as low as possible, if the object is to prevent hygroscopic growth as is likely to be the case in a QC environment or at conditions near to saturation, in order to simulate the environmental conditions in the upper airway upon inhalation.

It is important to stress at this stage in the book that none of these underlying physical processes is likely to give rise to the development of fine structure in the original APSD. Thus, the entire APSD is affected by each process. If discretized into a finite number of size bins (essentially the function of the CI as a size fractionator), the magnitude of changes to the API mass weighting assigned to each bin, moving from the smallest to largest size bins or vice versa, will always vary in a continuous but gradual way. Chapter 9 contains more information about possible changes to OIP APSD in the context of determining situations in which the related concept, effective data analysis (EDA), might fail.

3.3.3 Detectability of APSD Changes by the Cascade Impaction Method

In vivo aerodynamic performance is considered a Quality Target Product Profile (QTPP) property for OIPs [82], since the aerosol particle size characteristics affect delivery of drug to the respiratory tract (see Chaps. 2 and 12).

Currently, aerodynamic particle size testing by the cascade impaction method is the primary in vitro method in the regulatory environment used for assessing performance of OIP-generated aerosols (Chap. 2). The purpose of such in vitro testing is to act as a surrogate for in vivo testing with patients in order ultimately to provide assurance to stakeholders that such patients receive product within demonstrated acceptability limits derived from clinical trial data. This goal is ideally best achieved when the in vitro measurements can be related directly to in vivo aerodynamic performance that is in turn linked with intended outcomes for the disease modality being treated. Given the limited success thus far in the search for suitable IVIVRs/IVIVCs to link laboratory with clinical measures of OIP aerosol performance [16], this goal must be achieved indirectly. Under these circumstances, the impactor-based measurement approach to APSD characterization becomes the active control strategy tool for assessing the capability of an OIP to meet appropriately defined and established in vitro aerodynamic performance. The question of the capability of the CI method to meet this requirement is therefore of crucial importance, regardless of the foregoing discussion concerning the capability of EDA with or without AIM to detect changes in aerosol APSD that can be related in a meaningful way to in vitro performance change of the OIP.

The sensitivity of the CI method to detect real shifts in aerosol APSD to finer or coarser sizes is linked to the size-resolving capability of the system as a whole. Hence, at the simplest level, taking a hypothetical CI whose d_{50} values for the first and last stage are fixed at 10.0 and 0.5 μm aerodynamic diameter representing upper and lower bounds for a typical OIP APSD, resolving capability can be increased as more intermediate stages are inserted between the extremes, because the resulting APSD is defined in terms of the mathematical relationship between API mass and d_{50} at more locations. This reasoning partly drove the decision to design the NGI to have a minimum of five stages within the critical size range from 0.5 to 5.0 μm aerodynamic diameter throughout its operating flow rate range from 30 to 100 L/min for MDI and DPI APSD measurements [83]. On this basis, rather than simplify the CI to an AIM-based system, in which two or at most three size-measuring stages are present within this size range, the number of stages should be further increased as a means to improve resolution, much as is done with multichannel time-of-flight-based aerodynamic particle size analyzers [84]. However, there are at least two fundamental limitations:

1. The size-selectivity of the individual stages is finite. In Chap. 2, size-selectivity was defined in terms of GSD_{stage}, by analogy with the geometric standard deviation for a unimodal and lognormal APSD. At one extreme, a stage with ideal size-selectivity would have a GSD_{stage} of 1.0 and would be capable of resolving infinitesimally small shifts in APSD. The most well-designed CI stages from the aspect of their fluid-handling properties have GSD_{stage} values no smaller than 1.2. In practice, this limit restricts the number of stages in the critical range to 5, having optimally spaced individual stage d_{50} values at equidistant loci on a logarithmically scaled aerodynamic diameter axis. Attempts to increase this number of stages will result in loss of size-resolving power brought about by overlapping collection efficiency curves for adjacent stages.
2. The measurement precision based on the true stage d_{50} associated with each individual stage is also finite, being controlled largely through the degree of control over the nozzle diameter of a stage comprising an individual jet or the effective diameter (D_{eff}) for multi-jet stages [85]. D_{eff} is related in explicit form to the area mean (D^*) and area median (D_{median}) diameters of a multi-nozzle stage in accordance with the expression [84]:

$$D_{eff} = (D^*)^{2/3}(D_{median})^{1/3} \tag{3.16}$$

Roberts [85] reported stage acceptance values with the NGI for D_{eff} varying between 2.185 ± 0.02 mm for stage 3, 1.207 ± 0.01 mm for stage 4, 0.608 ± 0.01 mm for stage 5, 0.323 ± 0.01 mm for stage 6, and 0.206 ± 0.01 mm for stage 7. There were an insufficient number of nozzles for a statistically meaningful calculation of limits for stages 1 and 2, but it is reasonable to expect their tolerances to be at least as good as for lower stages based on the fact that the nozzles are larger and therefore easier to measure precisely. Roberts [85] also showed that D_{eff} and stage d_{50} are related through the expression:

$$D_{\mathrm{eff}} = \left(\frac{Q}{n}\right)^{1/3} \left(\frac{4C_{\mathrm{c}}\rho_{\mathrm{p}}}{9\pi\eta St_{50}}\right)^{1/3} (d_{50})^{2/3} \tag{3.17}$$

where Q is the volumetric flow rate, n is the number of nozzles for the stage in question, St_{50} is the particle Stokes number at which the stage collection efficiency is 50%, and the other terms have been previously defined. Importantly, in the present context, Roberts reported that the limiting precision associated with stage d_{50} can be kept close to 1.5% of the nominal value by keeping the uncertainty in flow rate to 3% of nominal operating value and at the same time as keeping the uncertainty in the nozzle diameters to 1% [85]. The ability to control flow rate within the limit indicated by Roberts is well within the capability of current flow measuring equipment. In an experimental estimation of the precision of commercially available optical image analysis systems used to stage mensurate CIs, Chambers et al. [86] confirmed that their overall capability was within the current pharmacopoeial stage specifications for two Andersen 8-stage "nonviable" cascade impactor "reference" stages that were representative of jet sizes for this instrument type (stages 2; $d_{\mathrm{eff}} = 0.914 \pm 0.0127$ mm and 7; $d_{\mathrm{eff}} = 0.254 \pm 0.0127$ mm). These findings confirm that the 1% uncertainty in this d_{eff} advocated by Roberts [85] is also a feasible proposition, in association with a regular program of stage mensuration for a given CI.

On the basis of these assessments, the capability of the CI method to resolve APSD shifts is dominated by limitation (1) above.

Changes to OIP aerosol APSDs may also occur in terms of an increase or decrease in the absolute magnitude of the mass of API that is collected within the size-fractionating portion of the CI system. Here, the capability of the method is controlled by the sensitivity of the recovery and assay method for the API [38]. It follows that for a given analytical sensitivity, the method will become less capable of resolving small differences as more stages are incorporated into the system. This trade-off is especially true when adding stages whose d_{50} values are located furthest from the MMAD of the aerosol being detected, as by definition they will capture the lowest mass of API per determination. Increasing the number of actuations per determination is one way to offset such loss of sensitivity, but this approach has been discouraged by regulators on the basis that the clinical dose of the OIP may be as small as one or two actuations [87]. If this option is not available, then an AIM-based approach, in which the number of intermediate stages is minimized, becomes an attractive proposition for detecting changes in APSD amplitude. The underlying rationale for introducing the AIM approach to the assessment of OIP APSDs is presented in Chap. 5, and case studies in which abbreviated impactors have been used successfully with all types of OIP are contained in Chap. 10.

3.4 Concluding Remarks

This chapter has reviewed the most likely physical causes underlying potential shifts in APSDs of aerosols emitted from the various forms of OIP and has provided some case studies that are indicative of how a multistage CI responds to such changes. In conclusion, it is important to appreciate that none of the causes of the shifts that have been described affect only a select portion of the APSD, in other words they are not especially size-selective. Hence, there are no processes envisaged that would selectively make the mass that is collected on one particular stage of a multistage CI grow, while the mass on a neighboring stage drops, without a concomitant change in the rest of the APSD profile.

References

1. Hickey AJ (2007) Inhalation aerosols: physical and biological basis for therapy, 2nd edn. Informa Healthcare, NY
2. Finlay WH (2001) The mechanics of inhaled pharmaceutical aerosols. Academic, NY
3. Bechtold-Peters K, Lüssen H (2007) Pulmonary drug delivery: basics, applications and opportunities for small molecules and biopharmaceutics. Editio Cantor Verlag, Aulendorf, Germany
4. Smyth HDC, Hickey AJ (2011) Controlled pulmonary delivery. Springer, NY
5. Morén F, Dolovich MB, Newhouse MT, Newman SP (1993) Aerosols in medicine: principles, diagnosis and therapy. Elsevier, Amsterdam
6. Byron PR (2004) Drug delivery devices: issues in drug development. Proc Am Thorac Soc 1(4):321–328
7. Dolovich MB, Ahrens RC, Hess DR, Anderson P, Dhand R, Rau JL, Smaldone GC, Guyatt G (2005) Device selection and outcomes of aerosol therapy. Chest 127(1):335–371
8. Leung K, Louca E, Coates AL (2004) Comparison of breath-enhanced to breath-actuated nebulizers for rate, consistency, and efficiency. Chest 126(5):1619–1627
9. Dolovich M, MacIntyre NR, Anderson PJ, Camargo CA, Chew N, Cole CH, Dhand R, Fink JB, Gross NJ, Hess DR, Hickey AJ, Kim CS, Martonen TB, Pierson DJ, Rubin BK, Smaldone GC (2000) Consensus statement: aerosols and delivery devices. Respir Care 45(6):589–596
10. Dolovich MB (2004) In my opinion – interview with the expert. Pediatr Asthma Allergy Immunol 17(4):292–300
11. International Commission on Radiological Protection (ICRP) (1994) Human respiratory tract model for radiological protection. Ann ICRP 24(1–3):36–52
12. Stocks J, Hislop AA (2002) Structure and function of the respiratory system. In: Bisgaard H, O'Callaghan C, Smaldone GC (eds) Drug delivery to the lung. Marcel Dekker, NY, pp 47–104
13. Heyder J, Gebhart J, Rudolf G, Schiller CF, Stahlhofen W (1986) Deposition of particles in the human respiratory tract in the size range 0.005–15 μm. J Aerosol Sci 17(5):811–825
14. Heyder J (2004) Deposition of inhaled particles in the human respiratory tract and consequences for regional targeting in respiratory drug delivery. Proc Am Thorac Soc 1(4): 315–320

15. Oberdörster G, Oberdörster E, Oberdörster J (2005) Nanotoxicology: an emerging discipline evolving from studies of ultrafine particles. Environ Health Perspect 113(7):823–839
16. Newman SP, Chan H-K (2008) *In vitro/in vivo* comparisons in pulmonary drug delivery. J Aerosol Med Pulm Drug Deliv 21(1):1–8
17. Patton JS, Fishburn S, Weers JG (2004) The lungs as a portal of entry for systemic drug delivery. Proc Am Thorac Soc 1(4):338–344
18. Hinds WC (1999) Aerosol technology: properties, behavior, and measurement of airborne particles, 2nd edn. John Wiley & Sons, NY
19. Islam N, Gladki E (2008) Dry powder inhalers (DPIs)—a review of device reliability and innovation. Int J Pharm 360(1–2):1–11
20. Hickey AJ, Concessio NM, VanOort MM, Platz RM (1994) Factors influencing the dispersion of dry powders as aerosols. J Pharm Technol 18:58–64
21. Hickey AJ, Crowder (2007) Next generation dry powder delivery systems. In: Hickey AJ (ed) Inhalation aerosols, 2nd edn. Informa HealthCare USA, NY, pp 445–460
22. Ashurst I, Malton A (2002) Passive dry powder inhalation technology. In: Rathbone MJ, Hadgraft J, Roberts MS (eds) Modified-release drug delivery technology. Informa HealthCare, NY, pp 867–877
23. Staniforth JN (1996) Pre-formulation aspects of dry powder aerosol. In: Dalby RN, Byron PR, Farr SJ (eds) Respiratory drug delivery. Interpharm, Buffalo Grove, IL, pp 65–73
24. Borgström L, Borgström L (1994) Deposition patterns with Turbuhaler®. J Aerosol Med 7S1:S49–S53
25. Duddu SP, Sisk SA, Walter YH, Tarara TE, Trimble KR, Clark AR, Eldon MA, Elton RC, Pickford M, Hirst PH, Newman SP, Weers JG (2002) Improved lung delivery from a passive dry powder inhaler using an engineered Pulmosphere® powder. Pharm Res 19(5):689–695
26. Olsson B (1995) Aerosol particle generation from dry powder inhalers: can they equal pressurized metered dose inhalers? J Aerosol Med 8S3:S13–S19
27. United States Pharmacopeial Convention (2012) USP 35-NF 30 Chapter 601: aerosols, nasal sprays, metered-dose inhalers and dry powder inhalers. United States Pharmacopeial Convention, Rockville, MD
28. European Directorate for the Quality of Medicines and Healthcare (EDQM) (2012) Preparations for inhalation: aerodynamic assessment of fine particles. Section 2.9.18 – European Pharmacopoeia [Apparatus B in versions up to 4th Edn. 2002], Council of Europe, Strasbourg, France
29. European Agency for the Evaluation of Medicinal Products (EMA) (1998) Note for Guidance on dry powder inhalers, CPMP/QWP/158/96, London, UK
30. Hoe S, Traini D, Chan H-K, Young PM (2009) Measuring charge and mass distributions in dry powder inhalers using the electrical next generation impactor (eNGI). Eur J Pharm Sci 38(2):88–94
31. Purewal TS (1998) Formulations of metered dose inhalers. In: Purewal TS, Grant DG (eds) Metered dose inhaler technology. CRC, Boca Raton, FL, pp 9–68
32. Smyth HDC, Evans RM, Hickey AJ (2007) Aerosol generation from propellant-driven metered dose inhalers. In: Hickey AJ (ed) Inhalation aerosols, 2nd edn. Informa HealthCare USA, NY, pp 399–416
33. Taylor G, Tran CH, Warren S, Thomas I, Marchetti G (2008) The Kemp HFA MDI valve for the delivery of novel budesonide/formoterol fumarate combination formulations. In: Dalby RN, Byron PR, Peart J, Suman JD, Farr SJ, Young PM (eds) Respiratory drug delivery–2008. Davis HealthCare Int. Publishing, River Grove, IL, pp 983–986
34. Stein SW (2008) Aiming for a moving target: challenges with impactor measurements of MDI aerosols. Int J Pharm 355(1–2):53–61
35. Lange CF, Finlay WH (2000) Overcoming the adverse effect of humidity in aerosol delivery via pressurized metered dose inhalers during mechanical ventilation. Am J Respir Crit Care Med 161(5):1614–1618

36. Peart J, Kulphaisal P, Orban JC (2003) Relevance of electrostatics in respiratory drug delivery. Business Briefing: Pharmagenerics, 84–87
37. Mitchell JP, Coppolo DP, Nagel MW (2007) Electrostatics and inhaled medications: influence on delivery via pressurized metered-dose inhalers and add-on devices. Respir Care 52(3):283–300
38. Bonam M, Christopher D, Cipolla D, Donovan B, Goodwin D, Holmes S, Lyapustina S, Mitchell J, Nichols S, Petterson G, Quale C, Rao N, Singh D, Tougas T, Van Oort M, Walther B, Wyka B (2008) Minimizing variability of cascade impaction measurements in inhalers and nebulizers. AAPS PharmSciTech 9(2):404–413
39. Rau JL, Ari A, Restrepo RD (2004) Performance comparison of nebulizer designs: constant-output, breath-enhanced, and dosimetric. Respir Care 49(2):174–197
40. Knoch M, Keller M (2005) Ultrasonics, mechanical pressure through ultrafine orifices (SMIs) or by applying mechanical vibration. Expert Opin Drug Deliv 2(2):377–390
41. Denyer J, Nikander K, Smith NJ (2004) Adaptive aerosol delivery (AAD) technology. Expert Opin Drug Deliv 1(1):165–176
42. Kesser KC, Geller DE (2009) New aerosol delivery devices for cystic fibrosis. Respir Care 54(6):754–768
43. Lefebvre AH (1989) Atomization and sprays. Hemisphere, New York
44. Nerbrink O, Dahlbäck M, Hansson HC (1994) Why do medical nebulizers differ in their output and particle characteristics? J Aerosol Med 7:259–276
45. Niven RW, Hickey AJ, Niven RW, Hickey AJ (2007) Atomization and nebulizers. In: Hickey AJ (ed) Inhalation aerosols, 2nd edn. Informa HealthCare USA, NY, pp 253–283
46. Berg E, Picard RJ (2009) *In vitro* delivery of budesonide from 30 jet nebulizer/compressor combinations using infant and child breathing patterns. Respir Care 54(12):1671–1678
47. Dennis JH (2007) Nebulizer efficiency: modeling versus *in vitro* testing. Respir Care 52(8):984–988
48. Finlay WH, Stapleton KW (1999) Undersizing of droplets from a vented nebulizer caused by aerosol heating during transit through an Andersen impactor. J Aerosol Sci 30(1):105–109
49. European Directorate for the Quality of Medicines and Healthcare (EDQM) (2012) Preparations for nebulisation. Section 2.9.44 – European Pharmacopoeia
50. Dennis J, Berg E, Sandell D, Ali A, Lamb P, Tservistas M, Karlsson M, Mitchell J (2008) Cooling the NGI – an approach to size a nebulised aerosol more accurately. PharmEur Sci Notes 1:27–30
51. Hinds WC (1993) Physical and chemical changes in the particulate phase. In: Willeke K, Baran PA (eds) Aerosol measurement: principles, techniques and applications. Van Nostrand Reinhold, NY, pp 41–53
52. Lee KW, Chen H (1984) Coagulation rate of polydisperse particles. Aerosol Sci Technol 3(3):327–334
53. Finlay WH (2001) The mechanics of inhaled pharmaceutical aerosols: an introduction. Academic, London, UK
54. Di Benedetto G, Clarke SW (1992) Inhalation therapy in asthma. J R Soc Med 85:3–5
55. Zhou Y, Sun J, Cheng Y-S (2011) Comparison of deposition in the USP and physical mouth-throat models with solid and liquid particles. J Aerosol Med Pulm Drug Deliv 24(6):277–284
56. Rader DJ, Marple VA (1985) Effect of ultra-Stokesian drag and particle interception on impactor characteristics. Aerosol Sci Technol 4(2):141–156
57. Cheng Y-S, Yazzie D, Zhou Y (2001) Respiratory deposition patterns of salbutamol MDI with CFC and HFA-134a formulations in a human airway replica. J Aerosol Med 14(2):255–266
58. Zhang Y, Finlay WH (2005) Experimental measurements of particle deposition in three proximal lung bifurcation models with an idealized mouth-throat. J Aerosol Med 18(4):460–473
59. Martonen TB, Lowe J (1983) Assessment of aerosol deposition patterns in human respiratory tract casts. In: Marple VA, Liu BYH (eds) Aerosols in the mining and industrial work environments, vol 1, Fundamentals and Status. Ann Arbor Science, Ann Arbor, MI, pp 151–164
60. Mitchell JP, Poochikian G, Hickey AJ, Suggett J, Curry P, Tougas T (2011) *In vitro* assessment of spacers and valved holding chambers used with pressurized metered-dose inhalers: the need

for a USP chapter with clinically relevant test methods. Pharm Forum 37(4). http://www.usppf.com/pf/pub/index.html
61. Bisgaard H, Anhøj J, Wildhaber JH (2000) Spacer devices. In: Bisgaard H, O'Callaghan C, Smaldone GC (eds) Drug delivery to the lung. Marcel Dekker, NY, pp 389–420
62. Canadian Standards Association (CSA) (2008) Spacers and holding chambers for use with metered-dose inhalers, CAN/CSA/Z264.1-02, Mississauga, ON, Canada. http://shop.csa.ca/en/canada/drug-labeling-and-delivery/cancsa-z2641-02-r2008/invt/27017422002/. Accessed 1 Sept 2011
63. Landahl HD (1963) Note on the removal of airborne particles by the human respiratory tract with particular reference to the role of diffusion. Bull Math Biophys 25:29–39
64. Kwok PCL, Chan H-K (2007) Electrostatic charge in pharmaceutical systems. In: Swarbrink J (ed) Encyclopedia of pharmaceutical technology, 3rd edn. Informa Healthcare, NY, pp 1535–1548
65. Hendricks CD (1973) Charging macroscopic particles. In: Moore AD (ed) Electrostatics and its applications. John Wiley & Sons, NY, pp 57–85
66. O'Leary M, Balachandran W, Chambers F (2008) Nebulised aerosol electrostatic charge explored using bipolar electrical mobility profiles. In: Industry Appl. IEEE-IAS Annual Meeting, Edmonton, AB, Canada, pp. 1–5. http://ieeexplore.ieee.org/xpl/freeabs_all.jsp?arnumber=4658914. Accessed 6 Jan 2012
67. Beleca R, Abbod M, Balachandran W, Miller PR (2010) Investigation of electrostatic properties of pharmaceutical powders using phase Doppler anemometry. IEEE Trans Ind Appl 46(3):1181–1187
68. Kulon J, Balachandran W (2001) The measurement of bipolar charge on aerosols. J Electrostat 51–52:552–557
69. Glover W, Chan H-K (2004) Electrostatic charge characterization of pharmaceutical aerosols using electrical low-pressure impaction (ELPI). J Aerosol Sci 35(6):755–764
70. Peart J, Magyar C, Byron PR (1998) Aerosol electrostatics: metered-dose inhalers (MDIs). In: Dalby RN, Byron PR, Farr SJ (eds) Respiratory drug delivery–VI. Interpharm, Buffalo Grove, IL, pp 227–233
71. Piérart F, Wildhaber JH, Vrancken I, Devadason SG, Le Souëf PN (1999) Washing plastic spacers in household detergent reduces electrostatic charge and greatly improves delivery. Eur Respir J 13(3):673–678
72. Martin AR, Finlay WH (2012) Aerosol drug delivery to the lungs. In: Gad SC (ed) Development of therapeutic agents handbook. John Wiley & Sons, NY, pp 565–588
73. Smyth H, Brace G, Barbour T, Gallion J, Grove J, Hickey AJ (2006) Spray pattern analysis for metered dose inhalers: effect of actuator design. Pharm Res 23(7):1951–1956
74. Myrdal P, Stein S, Mogalian E, Hoye W, Gupta A (2004) Comparison of the TSI model 3306 impactor inlet with the Andersen cascade impactor: solution metered dose inhalers. Drug Dev Ind Pharm 30(8):859–868
75. Rao N, Kadrichu N, Ament B (2010) Application of a droplet evaporation model to aerodynamic size measurement of drug aerosols generated by a vibrating mesh nebulizer. J Aerosol Med Pulm Drug Deliv 23(5):295–302
76. Copley M, Smurthwaite M, Roberts DL, Mitchell JP (2005) Revised internal volumes of cascade impactors for those provided by Mitchell and Nagel. J Aerosol Med 18(3):364–366
77. Primiano FP, Saidel GM, Montague FW, Kruse KL, Green CG, Horowitz JG (1988) Water vapour and temperature dynamics in the upper airways of normal and CF subjects. Eur Respir J 1(5):407–414
78. Telko MJ, Hickey AJ (2005) Dry powder inhaler formulation. Respir Care 50(9):1209–1227
79. Dunbar CA, Hickey AJ, Holzner P (1998) Dispersion and characterization of pharmaceutical dry powder aerosols. Kona Powder Part J 16:7–44
80. Hindle M, Makinen GM (1996) Effects of humidity on the *in-vitro* aerosol performance and aerodynamic size distribution of cromolyn sodium for inhalation. Eur J Pharm Sci 4S1:142S
81. United States Pharmacopeial Convention (2012) USP 35-NF 30 Chapter 1601: products for nebulization. United States Pharmacopeial Convention, Rockville, MD

82. US Federal Drug Administration (2009) Guidance for industry: Q8(R2) pharmaceutical development. Silver Spring, MD. http://www.fda.gov/downloads/Drugs/GuidanceCompliance RegulatoryInformation/Guidances/ucm073507.pdf. Accessed 8 May 2012
83. Marple VA, Roberts DL, Romay FJ, Miller NC, Truman KG, Van Oort M, Olsson B, Holroyd MJ, Mitchell JP, Hochrainer D (2003) Next generation pharmaceutical impactor. Part 1: Design. J Aerosol Med 16(3):283–299
84. Mitchell JP, Nagel MW (1999) Time-of-flight aerodynamic particle size analyzers: their use and limitations for the evaluation of medical aerosols. J Aerosol Med 12(4):217–240
85. Roberts DL (2009) Theory of multi-nozzle impactor stages and the interpretation of stage mensuration data. Aerosol Sci Technol 43(11):1119–1129
86. Chambers F, Aziz A, Mitchell J, Shelton C, Nichols C (2010) Cascade impactor (CI) mensuration: an assessment of the accuracy and precision of commercially available optical measurement systems. AAPS PharmSciTech 11(1):472–484
87. US Food and Drug Administration (FDA) (1998) CDER. Draft guidance for industry metered dose inhaler (MDI) and dry powder inhaler (DPI) drug products chemistry, manufacturing, and controls documentation, Rockville, MD. http://www.fda.gov/cder/guidance/2180dft.pdf. Accessed 6 Jan 2012

Chapter 4
Good Cascade Impactor Practices

Jolyon P. Mitchell

Abstract The CI-based methods for measuring the APSD properties of OIP-produced aerosols are complex, exacting, and laborious to undertake. Yet they are the only accepted methods by regulatory agencies worldwide for determining particle aerodynamic size-related properties. In 2003, a group within the Product Quality Research Institute (PQRI), a body set up by pharmaceutical industry, the FDA, and academia to explore complex scientific and regulatory problems, developed a guide to good cascade impactor practices (GCIP). This chapter contains a review of the essence of their work, augmented by developments that have taken place since the original article was published.

4.1 Intrinsic Variability Associated with CI Methodologies

The multistage CI is well known to be both a complex and labor-intensive apparatus, whatever compendial variant is used in OIP aerosol aerodynamic particle size assessment. In discussions within the European Pharmaceutical Aerosol Group (EPAG) following the development of the NGI, the comment was made on several occasions that a technician, well trained in the "art" of cascade impaction is highly sought after and difficult to replace. More seriously, the difficulty in replicating CI measurements undertaken in the same laboratory or when transferring a method to another laboratory is a significant hindrance to both OIP development and quality control processes. In 2004, Nichols reported a survey of members of the EPAG in connection with what they viewed as contributors to variability in CI-based measurements [1]. These estimates of causes of measurement variability that were based on individual experience rather than a formal work study and statistical analysis

J.P. Mitchell (✉)
Trudell Medical International, 725 Third Street, London, ON N5V 5G4, Canada
e-mail: jmitchell@trudellmed.com

T.P. Tougas et al. (eds.), *Good Cascade Impactor Practices, AIM and EDA for Orally Inhaled Products*, DOI 10.1007/978-1-4614-6296-5_4,

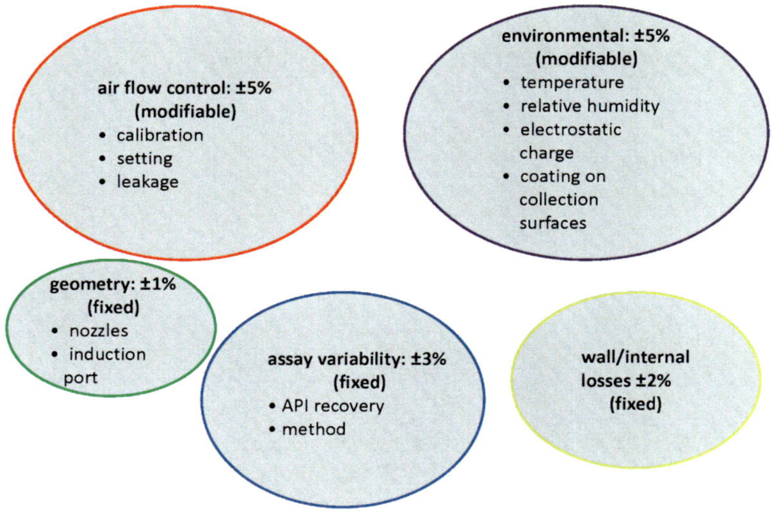

Fig. 4.1 Five principal causes of nonhuman factor-related CI method variability, showing their contribution to the overall variability of the method (*Adapted from* [1]—*used by permission*)

(Fig. 4.1) are probably unsurprising. Not considering human-related factors, airflow and laboratory environmental controls were estimated to contribute ±10% of total variability between them, followed by impactor geometry causes (mainly nozzle size control) at ±2%, and finally internal losses in the CI providing ±2% to the measures of OIP performance. CI-related contributors were only part of the overall measurement variability process, with a further ±20% estimated to come from the inhaler itself in the form of actuation-to-actuation reproducibility through life, as well as a contribution estimated to be ±3% from the recovery and assay of the API.

4.2 Assessment of Factors Contributing to CI Measurement Variability

Bonam et al. [2], in a comprehensive review of sources of measurement variability in the CI method, were able to break down the causes of CI method variability in more detail (Fig. 4.2). They established that there are four major contributors to overall variability in CI measurements:

1. Man—the CI operator/analyst
2. Machine—the CI System
3. Measurement—the API recovery and analysis procedure
4. Material—the inhaler including its drug product, including both device and formulation

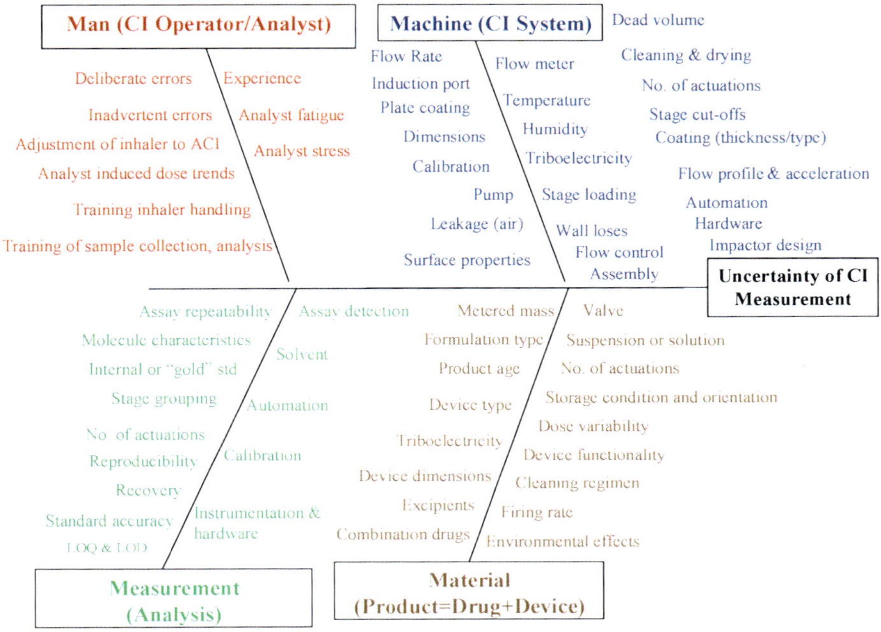

Fig. 4.2 Ishikawa cause and effect diagram for CI method variability (*From* [2]—*used by permission*)

4.2.1 *Contribution to Variability from "Man"*

Performing a CI measurement requires several manual operations, such as impactor assembly, connections to vacuum valves and pump, adjustment of airflow rate, inhaler manipulation and actuation, quantitative recovery of the deposited API, dilution for subsequent HPLC–spectrophotometric assay of API, or other types of analysis. Human involvement in these steps, both due to imperfect technique and as a result of unintentional mistakes, increases the likelihood of bias and increased variability even when adequate precautions are taken [3]. Semi- and fully automated systems are part of the answer to minimize operator-based variability, but they are relatively expensive, especially full automation, and so unlikely to be available to all except the largest organizations.

Bonam et al. went on to conclude that despite these alternatives, the conventional ACI, NGI, and other manual systems based on highly laborious methods are likely to remain in use by most organizations for the foreseeable future, especially for regulatory purposes [4]. They also emphasized that man-related variability can be expected to continue to be a significant contributing factor to the overall CI test outcomes. Key issues relating to this contributor to overall method inconsistency were identified as follows:

1. CI Assembly: Bonam et al. noted that operator training in impactor assembly is crucial, in particular with stack designs such as the ACI, due to the need to assure correct stage order as well as ensuring that there is a proper seal between impactor stages [2]. They proposed that under such circumstances, a gantry-type system could be used to ensure that the stack is properly aligned. Proper sealing could be checked via leak tests and/or pressure-drop tests or by including a differential mass flowmeter between the induction port and the impactor exit. They proposed that a checklist may be a helpful aid, particularly for an inexperienced operator. However, even with proper training, it is evident from discussions with laboratory management that occasional errors in the assembly of ACI-type impactors are likely, given the complexity of the process, leading to "failed" CI tests. For example, incorrect ordering (switching) of CI stages or misalignment of collection plates in an ACI are inadvertent errors that may happen even to a well-trained operator. In the case of the NGI, a complete set of collection cups can be fixed permanently into a carrier that makes the process of loading and unloading the impactor more efficient. Furthermore, for this CI, the stage order is fixed due to the integral nature of the impactor body itself [5].
2. Impactor handling and sample introduction: In 2001, Purewal, reporting on an EPAG-based assessment of test methods to check OIP performance under both normal and unintentional use conditions, reported that proper training in sample collection is critical because of a wide variety of different test methods, including variations in inhaler handling and introduction of the sample to the CI [6]. Even with proper training, individual differences between operators (e.g., inhaler shaking frequency and intensity, delay between shaking and actuation, alignment of inhaler to ACI, actuation) may go unnoticed but may result in different systematic biases (e.g., dose-through-use trends) and also may contribute to the seemingly random variability when results are compared from several operators or even from the same operator on different occasions. For example, Purewal commented that a different rate of actuating an MDI may lead to different cooling of the canister and, therefore, different evaporation behavior of the propellant, leading to variations in measured APSD, even when the inhaler units are identical. Using a bench timer may help standardize this aspect of the method and minimize the associated variability. Stewart et al. [7] and Miller et al. [8] have observed that automation of some of the steps in the CI measurement process may further reduce this type of variability.
3. API analytical method (commonly HPLC with either UV-visible spectrophotometric or fluorescence detection): When HPLC is used for API assay, Bonam et al. noted that an analyst might introduce an "individual" bias into APSD results by preparing the HPLC standard at the upper or lower limit of the predefined range. Under such circumstances, all recoveries from the CI stages would either be slightly over- or underestimated. It is self-evident that similar considerations would apply with newer assay methods, such as ultrahigh-pressure liquid chromatography (UPLC) that is frequently combined with mass spectrometric API identification.

4. CI sample collection and subsequent analysis for API: From an analyst's perspective, the procedure for API recovery from the CI and subsequent quantification of mass deposited in each component part of the system offers multiple opportunities for errors. Furthermore, the potential for loss of API on disassembly of the impactor prior to API recovery should not be ignored [2]. Therefore, careful handling of the assembled CI, together with collection surfaces containing deposited API upon disassembly, is highly important. In addition, a robust technique for recovering API material is critical, as is proper organization of samples to prevent wrong dilutions and incorrect sample vial filling [2]. It is well known that the operation of pipetting aliquots of recovery solvent will also contribute to random method variability, since more careful pipetting will result in a complete discharge of liquid, whereas less careful operation is likely to result in some liquid left over in the pipette. Bonam et al. therefore suggested that the use of automated pipettes should be considered as standard practice [2]. Another potential source of variability they identified is due to the use of different volumes for stage dissolution or for dilution of recovered samples, e.g., in order to obtain the same concentration for all API stage-by-stage samples to be quantified using HPLC/UPLC analysis. They therefore recommended that potential errors and increased variability should be weighed against potential benefits of this procedure. Evaporation of solvent during sample recovery will give an overestimation of the API amount and its effects should be minimized, e.g., by inclusion of an internal chemical standard in the solvent used for recovery. If the HPLC/UPLC method has good linearity, it is also possible to dispense the same volume of internal standard solution to each impactor stage/cup/throat and to the standard used for API quantitation. In this case, the sample solution is transferred directly to the LC vial and the exact volume transferred is of no importance. Bonam et al. noted that the exact concentration of the internal standard solution is not critical as long as the internal standard dispenser is repeatable and the same dispenser is used for both samples and standard solutions [2].
5. Operator stress and fatigue: It is self-evident that during complex operations, such as those associated with the CI measurement process, stress and fatigue always deteriorate human performance. The consequence is an inevitable increased incidence of errors. Stress and fatigue can be minimized by limiting the number of samples to be analyzed per operator per working day. In addition, mechanical aids and procedural steps might be introduced to counteract potential operator-related errors. For example, to prevent delivery of an incorrect number of doses to the impactor, Bonam et al. suggested that the use of "counters" linked to actuation of the airflow valve may be considered for DPIs, or the weighing of MDIs could be used before and after the CI test to verify the correct number of doses actuated [2]. As another example, to enable detection of a mix-up of vials from different stages during API analysis, for unimodal and lognormal APSD distributions, plotting log-probability versus log-particle-size and calculating the regression coefficient could be used as a system suitability tool. Furthermore, mixed-up dilutions could be avoided by using the same amount of solvent for each impactor stage or by eliminating the dilution step altogether if the detection sensitivity and API detection method linearity allow.

6. Bias introduced by the experienced operator: It is perhaps less well documented than the previous cause of unintentional variability, but even well-trained and experienced operators can develop habits that lead to repetitive mistakes in technique. Such errors can often be small, but taken together they can accumulate so that the rate of measurement failures increases above typical. Bonam et al. proposed that biases from individual operators could be detected through use of control charts (e.g., mass of API reported on key stages tracked as a function of time), and upon further investigation, they might be traced to one of the errors listed above [2]. Control charts are particularly useful with repeated testing of established products. However, broader metrics, such as API mass balance, may be more useful at tracking operator-introduced bias when testing OIPs in early development.

4.2.2 Contribution to Variability from "Machine"

An APSD measurement is never "absolute" but depends on the technique [9–12], instrument calibration [13], its intrinsic size resolution [14], as well as the dynamic nature of the aerosol cloud as it interacts with the environment when it enters and passes through the CI (e.g., evaporation of volatile species or hygroscopic growth [15]). Therefore, the outcome of an APSD measurement should always be reported along with the specific instrument and technique used.

Compared to other types of analytical measurement, a CI does not have an inhaler-aerosol-like reference material that is ultimately traceable to the international length standard, such as a standardized polydisperse aerosol, with which to validate the method [16]. In that sense, every CI measurement is unique. Monodisperse certified reference particles do exist as an alternative to a polydisperse standard [16], but their valid application is a slow and difficult process [17], which is unsuited to being used on a regular basis to verify CI performance [18]. This lack of a simple-to-do calibration procedure utilizing standard particles of known aerodynamic size has made CI data prone to bias and increased variability prior to the introduction of stage mensuration as an alternative and fully traceable approach to CI performance verification [18, 19].

The application of stage mensuration is therefore recommended in the compendial procedures for OIP aerosol aerodynamic particle size analysis in both European and US pharmacopeias [20, 21]. This process is the automated measurement of individual nozzles of each stage by computer-aided image analysis and has now become the current technique of choice [18]. Recently published data confirm that bias introduced by several widely used and commercially available mensuration systems is small [18]. Furthermore, a collective measure of average diameter for a multi-nozzle stage, termed effective diameter, D_{eff}, calculated in terms of the area-weighted median (D_{median}) and area-weighted mean (D^*) diameters of the group of nozzles in accordance with the expression [19]

$$D_{\mathrm{eff}} = (D^*)^{2/3}(D_{\mathrm{median}})^{1/3} \tag{4.1}$$

can be used to define the in vitro performance of that stage in terms of its D_{50} size [22]. However, errors in mensuration using stop-go pins for the larger mm-sized nozzles have been reported [23], although these pins, if used carefully, could be useful for cleaning impactor jets. Recently, the determination of the flow resistance (equivalent to pressure drop) across each stage has been proposed by Millhomme et al., as an "in-use" test and a surrogate for stage mensuration [24]. This is because the pressure drop ($\Delta P_{\mathrm{stage}(i)}$) across an impactor stage "i" with the CI operated at flow rate, Q, can be described well by a Bernoulli-style equation:

$$\Delta P_{\mathrm{stage}(i)} = \left[\frac{\rho_g}{2}\right]\left[\frac{Q}{kA_t}\right]^2 \tag{4.2}$$

where ρ_g is the support gas (i.e., air) density and A_t is the total area of the nozzle array. This group also showed that a change in the area-mean jet diameter of a given CI stage, D^* from D^*_0 to D^*_1, can be related to a variation in $\Delta P_{\mathrm{stage}(i)}$, in accordance with the following relationship:

$$\frac{D^*_1}{D^*_0} = \left[\frac{Q_1}{Q_0}\right]^{1/2}\left[\frac{\rho_{gas(1)}}{\rho_{gas(0)}}\right]^{1/4}\left[\frac{\Delta P_0}{\Delta P_1}\right]^{1/4} \tag{4.3}$$

in which the subscripts "0" and "1" represent the initial and final states, respectively, for $\Delta P_{\mathrm{stage}(i)}$. For well-maintained impactors, characteristic of good laboratory pharmaceutical practices, there is little difference between the area-mean and effective jet diameters [19] so that (4.3) may be rewritten on a stage-by-stage basis in terms of the effective diameter for that stage, D_{eff} [(4.1)], given by

$$D_{eff(1)} = \left[\frac{Q_1}{Q_0}\right]^{1/2}\left[\frac{\rho_1}{\rho_0}\right]^{1/4}\left[\frac{\Delta P_0}{\Delta P_1}\right]^{1/4} D_{eff(0)} \tag{4.4}$$

In principle, if the discharge coefficient for the nozzle set remains at a constant value stage pressure drop may be conveniently performed immediately *before* every APSD measurement, as a system suitability test, thereby guaranteeing that the CI is fit for purpose. The potential of this approach requires further investigation. This situation directly contrasts with stage mensuration, which is a comparatively complex task requiring that the stages are individually inspected and is therefore only likely to be undertaken on an infrequent (i.e., annual) basis. In the event that D_{eff} is found to be outside acceptable limits (most likely the manufacturer tolerances associated with nominal nozzle diameter), the validity of an indefinite number of APSD determinations that have taken place since the previous inspection could be called into question. Furthermore, the measurement of the pressure drop across the entire CI system (ΔP_{CI}) may also be used as an in-use test of leak tightness before each measurement.

Even with the CI-based method including validation of the CI system itself optimized, unnoticed biases and variability may remain, making it difficult to comply with an a priori specification which disregards data from that particular impactor and method. Careful method development work should therefore be done in an attempt to try to identify and counteract all major sources of imprecision and bias.

Bonam et al. [2] went on to describe the various sources of impactor-related variability in more detail:

(a) Jet dimensions, stage cutoffs, calibration, and mensuration: The process of converting API mass-weighted deposition data obtained from a multistage CI into an APSD depends on the established magnitudes of the stage d_{50} sizes, as has been described in Chap. 2. Although ideally every CI of a given design should have an identical d_{50} for a given stage, several studies have shown that stages of so-called identical CIs have often slightly different nozzle sizes, due either to manufacturing variations or by wear, corrosion, or accumulation of debris [25–27]. It is self-evident that time- and use-dependent processes that result in partial plugging, wear, and corrosion will change the actual size of the nozzles of a given stage and, therefore, its d_{50} value. These effects can ultimately result in shifts of API mass between stages and, therefore, introduce bias accompanied by increased variability of APSD measurements, especially when data from several impactors are used together [25].

(b) Flow rate, flow profile, acceleration, and control: It is well known that CI stage d_{50} sizes are affected by the flow rate at which the measurement system is operated, decreasing as flow rate increases and vice versa [28]. Relatively speaking, nozzle-diameter-caused variability in d_{50} is likely to be small in terms of its overall impact on performance and is relatively easily monitored [29]. Flow-rate-induced variability, however, is likely to be more significant and less tractable, given the fact that in the pharmacopeial method, the specified flow control is typically no better than ±5% of the nominal flow rate [20, 21]. An alternative to the pharmacopeial method, using a flowmeter calibrated for the entering, rather than exiting, flow rate has been shown to yield similar performance [30].

    Flow-rate variability therefore is an important source of APSD measurement uncertainty in addition to changes in stage d_{50} values. In contrast to the latter which tends to change only gradually through repeated use, the flow-rate setting may vary from one instance of using a given CI to the next instance of using the same impactor. For MDIs, stage d_{50} sizes are determined by the magnitude of the (constant) flow rate achieved during testing; for DPIs, in addition, the flow-time profile (rise time, acceleration) affects stage d_{50} values and consequently the measured APSD. The magnitude and direction of these effects depend on the CI design and internal geometry, in particular the magnitude of the internal dead volume [31, 32]. It is therefore important to define the vacuum tube length for repeated use of the same impactor type in order to keep the dead volume as constant as possible. The airflow rise time could also be measured as part of the installation checks of a new instrument setup. In summary, flow-rate bias can be minimized if flowmeters are well maintained, properly calibrated, and regularly qualified.

(c) Internal dead volume in the CI system: In the context of DPI testing, the magnitude of the internal dead volume (i.e., the open space within the assembled CI, including accessories such as the induction port or pre-separator and vacuum tubing) will influence both the time taken for particles to traverse the system and be collected and the rise time for the flow rate to increase to the final value when the flow-control solenoid valve is opened to initiate sampling [33]. Since 2005, accurate measures of the dead volumes of the ACI, NGI, and multistage liquid impinger (MSLI) have been available, with and without the various accessories [34]. This basic information about the impactor system configuration provides a useful starting point when selecting a CI for a new product. Bias introduced to APSD measurements by dead volume will be fixed for a given impactor type and configuration, and ideally the internal volume should be as small as possible. However, the tradeoff between potential bias arising from dead volume and other more important constraints, such as ease of (semi) automation or improved aerodynamic size-separating characteristics, may result in the choice of a system with a higher internal volume.

(d) Air leakage: Air leakage into the CI can arise from incorrectly placed or defective seals or misaligned stages, since a partial vacuum always exists within the flow channel through the CI once flow is enabled through the system. The problem is particularly prevalent with the standard O-rings used with ACIs, which are prone to cracking with repeated use and exposure to solvents. Defective seals are most significant when they occur at stages closest to the impactor exit, where the internal vacuum is at its greatest with respect to the surrounding atmosphere. It therefore follows that although the APSD may be shifted (in some cases significantly), the magnitude of the displacement being dependent on leak location and size, the mass balance (MB) for the recovered API should likely be unaffected. However, if the leakage caused an increase on inter-stage deposition, MB might be reduced unless internal losses are included in the analysis [3]. However, broader metrics, such as API mass balance, may be useful at tracking operator-introduced bias, when testing OIPs in early development. Air leaks are detected and prevented through periodic visual inspection and replacement of defective seals, and careful attention to system assembly. A final system suitability check can also be made by comparing the volumetric flow rate at the impactor inlet (at ambient atmospheric pressure) with that measured downstream of the system at the vacuum pump (corrected for the local reduced pressure) [30] or by the measurement of flow resistance (pressure drop) across the entire CI system (Sect. 4.3.2). However, even if these checks are made, it is possible for a small leak between stages to go unnoticed. As part of the CI method development, Bonam et al. observed that it would be prudent to study the effect of leaks on the APSD of the product through the use of designed experiments, e.g., by introducing controlled leaks through small cuts in O-rings and observing the changes in APSD resulted from such simulated failures [2]. This information could later be used for root-cause analysis when atypical APSD profiles are observed, in combination with reassessment of the tested unit to confirm that the failure is not related to the product.

(e) Environmental conditions, temperature, and humidity: It is widely known that the local environmental conditions surrounding the aerosol as it is generated by the inhaler may influence variability of the aerosol and thereby contribute to measured product variability. However, local temperature and relative humidity also may influence the aerosol APSD during the process of measurement. The first type of influence is not considered in this chapter because it is product-specific. The second type of effect could, however, be categorized as CI (method)-related variability. In particular, the relative humidity will influence electrostatic (i.e., triboelectric) effects that are discussed later. Together with temperature, local humidity will also affect droplet growth or evaporation throughout the CI system [35–39], especially for aqueous droplets such as those produced by SMIs and the various types of nebulizer. Unless corrected for, ambient temperature variations may additionally bias measured volumetric flow rate through the system and thus further affect the APSD measurement. Bonam et al. observed that usually, temperature and nowadays relative humidity are both normally monitored and controlled to minimize these effects [2]. The extent of environmental influences and the needed controls depend on the product type (API, carriers, propellants, cosolvents, etc.), as well as on the CI system and its other operating conditions (e.g., whereas the flow rate may affect the rate of particle growth and evaporation, the CI temperature may affect heat transfer and droplet evaporation).
(f) Electrostatic charge and triboelectrification: Several groups have demonstrated that aerosolized particles acquire and carry intrinsic electrostatic charges (of either sign). This acquisition of electric charge is chiefly by triboelectric effects (contact electrification). It takes place always during aerosol formation from either DPIs [40–45] or MDIs [45–52]. However, triboelectrification is unlikely with aerosols from SMIs and nebulizers, as the preparations used with these classes of OIP nearly always contain physiologically normal saline which is fully ionized. Horton et al. have shown that electrostatic charge may also be acquired by CI components that are nonconducting or electrically insulated and can significantly influence particle collection behavior [53], thereby adding to the variability of the measured APSD and MB. Unless the OIP is a nebulizer, exploratory studies are likely to be warranted to optimize the CI method.
(g) Collection surface coating material and coating thickness: Several groups, including Nasr et al. [54], have shown that grease-coating CI plates reduces or eliminates particle bounce and reentrainment altogether. At the micron-size scale, the addition of a tacky coating modifies the impaction surface such that the incoming kinetic energy associated with the particles whose trajectories intersect with the surface is absorbed more efficiently than would be the case with a smooth and hard surface that has a relatively high coefficient of restitution. In 2003, Mitchell surveyed EPAG member organizations, reporting that almost all practice some form of collection surface coating to reduce bias and variability of CI measurements [55]. This effect is particularly important for DPI formulations [56, 57] and may also occur with some MDI products [58]. However, it may be unimportant for the collection of liquid droplets arising from nebulizing systems

[59], including metered liquid inhalers. The choice of the most appropriate coating material and thickness appears to be inhaler type/drug product-dependant. At the present time, it is not known whether coating prolongs or shortens collection surface-life by either preventing or causing corrosion, though prevention is more likely if the coating presents an impermeable barrier to aqueous ionic species. Bonam et al. observed that (semi)-automation or use of a hand-operated tool that ensures a uniform depth of coating across the collection surface could be considered as part of CI method development in order to standardize both coating depth and coverage [2].

(h) Stage loading and number of actuations of the inhaler into the CI: Kamiya et al. have shown that it is important to consider potential accumulation of material on CI stages to the point at which further incoming particles bounce when working with formulations requiring multiple actuations for CI testing and containing a high mass loading of particulates [60]. The amount of API mass depositing on stages needs to be low enough to prevent collected particulate becoming reentrained in the airflow and transferred to stage(s) further in the CI system. Furthermore, the mass per stage of excipient (and also carrier particles for some DPIs) should be taken into account [2]. In the extreme, excessive actuations may result in stage overloading that could affect the jet-to-plate distance. If the number of actuations used per a single CI test is too large, an apparent shift to finer particle sizes will be observed as the result of particle resuspension from the stage at which they should have been collected and reentrainment in the flow to more distal stages of the CI [58, 61, 62]. Merrin et al. have shown that the type of formulation may also play a role with high unit dose MDI products emitting >1 mg/actuation [63]. Such sources of bias should be studied and eliminated in method development, prior to any method validation, by undertaking measurements with progressively increased numbers of actuations.

(i) Collection surface properties: Bonam et al. also considered the fact that different roughness of uncoated stage surfaces at the micro-size scale commensurate with incoming particle dimensions may influence ballistic behavior of aerosol particles and therefore measured APSD variability [2]. This effect may depend on the surface properties of the formulation (i.e., material hardness), wear in use experienced by a particular CI, as well as the cleaning and drying procedures in use. From a survey of EPAG users, Mitchell, in 2006, concluded that the use of collection cups or plates that have surface roughness that is 100% inspected will assist with uniform coating and that any that are scratched, bent, or dented should not be used [64]. Currently, there are no published data in which the effect of surface roughness on APSD has been systematically investigated, but both ACI and NGI vendors [65], as well as the pharmacopeias [20, 21], now describe tolerances for surface roughness. In the case of the compendia, these are ~0.4 μm for the inner surfaces of the induction port, and entrance cone to the ACI, and between 0.5 and 2 μm for the collection cups of the NGI. However, these values are currently informative, rather than normative in nature.

(j) Cleaning/drying of CI components: It is self-evident that thorough cleaning and drying of impactor surfaces as part of the maintenance process between measurements will keep them as close to their specifications as possible [2]. Appropriate cleaning/drying procedures should therefore be determined during method development. In this context, users should be aware that variations in the cleaning/drying procedures, such as the use of a new cleaning solvent or a move from hand washing to an automated system may potentially affect variability of APSD measurements. Such an outcome is obvious from the standpoint of minimizing corrosion in long-term use, as well as maintaining the nozzles fully unobstructed before each measurement in day-to-day use. At the present time, however, there are no published data systematically studying the effect of different cleaning/drying regimens on APSD results. For the new user or for cases where the cleaning regimen is suspected as being the cause of more than desirable variability in measurements, the EPAG has published the results of a survey on CI cleaning methods [66].

(k) Internal (wall) losses of API within the CI: Mitchell et al. have shown, through a calibration of an ACI with monodisperse particles of different aerodynamic diameters in which internal losses were systematically investigated as a function of particle size, that there is a strong association between these two variables [67]. This is especially true when d_{ae} exceeds 5 μm. It follows therefore that internal losses of API to the non-collection surfaces within a CI (wall losses) may be an important source of error to be considered in method development [2]. Mitchell et al. further showed that losses in the standard pre-separator supplied with this impactor for use at 28.3 L/min were especially significant [67]. The pharmacopeial compendia have for some time required that internal losses not accountable for in API mass recovery be <5% of the delivered mass ex inhaler [20, 21]. If this limit is breached, the material from the entire stage including jets and walls (and not only the impaction plate) should be collected and added to the total dose recovered from the CI as a compromise position [20, 21]. Since such losses result from grazing interactions between airborne particles and the internal surfaces not associated with normal collection, their magnitude almost certainly depends on the physicochemical properties of the emitted aerosol (linked to the formulation of the product) as well as the CI preparation method (e.g., coating type, number of actuations delivered per measurement).

(l) Choice of induction port: The induction port (IP) is particularly important for MDI testing, due to the need to capture the ballistic component (i.e., the fraction of the metered dose that is expelled rapidly in projectile-like motion from the inhaler mouthpiece, due to propellant flash evaporation), separately from the fraction that is intended to penetrate beyond the oropharynx into the lower respiratory tract [20, 21, 68]. The IP presents the first impaction surface for the moving aerosol once it has left the inhaler (and any add-on device), and therefore, it strongly affects the proportion of the emitted mass entering the CI. For this reason, Bonam et al. pointed out that the use of adapters to align the MDI actuator with the IP is important for ensuring correct angle of entry of the

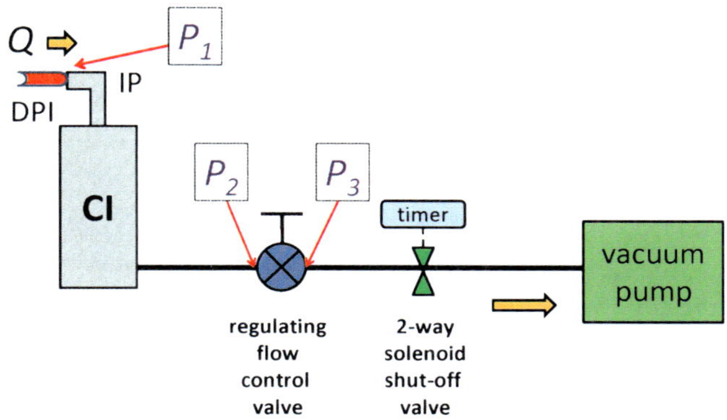

Fig. 4.3 Schematic diagram showing the principle of compendial methodology for DPI testing with CI

aerosol plume emitted from the inhaler mouthpiece [2]. This plume initially moves faster than the sampling flow into the IP, resulting in limited time for evaporation and for the reduction in emitted droplet sizes. It therefore follows that deviations from nominal dimensions and associated tolerances will influence APSD measurements for any given design of IP [69–71].

The compendia advise that consideration should also be given to the internal surface roughness as its magnitude may influence measurement variability (see collection surface properties). It should be noted that although identified as part of the CI system in one compendial monograph [68], an IP is not strictly necessary for the testing of droplets emitted by nebulizing systems, including the soft-mist plume from SMIs, because of the absence of the ballistic fraction.

(m) Vacuum pump: Different pumps may produce different time-dependent flow-rate profiles, which are important for DPI testing [73]. In the compendial methodologies for evaluating this class of OIPs, the pump-capacity-related variability is minimized by operating the impactor with the flow set by means of a critical orifice (Fig. 4.3) and by measuring the critical flow ($P_3/P_2<0.5$, where P_2 and P_3 are pressures upstream and downstream of the flow-control valve) [20, 21]. P_1 is the upstream near-to-ambient pressure at the inhaler-induction port, and Q is the volumetric flow rate through the CI system.

Because some pumps heat up and lose efficiency over time if left running, flow-rate setting/verification and APSD measurement should be conducted as quickly as possible, if needed.

(n) Impactor design: Bonam et al. also identified the design of a CI as another fixed contributor to the overall method variability [2]. Detailed descriptions of the design characteristics of each of the CI systems identified in the pharmacopeial monographs covering aerodynamic particle size analysis have therefore been provided as a reference source in Chap. 2. Bonam et al. noted that although this aspect of the test method is often fixed at the outset by the availability of equipment and preference within the organization, the choice of CI may be driven by the type of OIP being characterized, as well as the preferences of regulatory agency involved with the product submission [2]. For example, the NGI operated at 15 L/min has been identified for the assessment of preparations for nebulization in both Ph. Eur. monograph 2.9.44 [68] and its equivalent in 2012 in development for incorporation as Chapter 1601 in the US Pharmacopeia [72]. As well as affecting its size-discriminating capability, the CI design influences the internal airflow characteristics and, therefore, the magnitude of nonideal deposition on internal surfaces. These processes in turn influence both the variability and potential bias of CI results.

4.2.3 Contribution to Variability from the Measurement and API Analysis

HPLC or UPLC combined with some form of spectroscopy (UV-visible light detection, fluorescence, or mass spectrometry) is widely applied to assay API deposited on CI stages. Fluorescence detection is limited to APIs that fluoresce at detectable wavelengths, and MS, though highly sensitive and discriminating, is a relatively sophisticated and expensive technique, requiring careful sample preparation, especially if volatile species are present.

Although UPLC, being a newer technique, is in the process of becoming more widely adopted, HPLC-UV-visible spectrophotometry is a well-characterized and easily standardized technique that is available in almost all laboratories carrying out OIP testing. Despite its familiarity, this is another area contributing to both random and systematic uncertainty of CI measurements. For example, one of the difficulties of using this type of assay for API recovery and quantitation may be poor chromophore properties of the API. Often, low amounts of the API are recovered for assay because the collection stage under consideration has its particular d_{50} size located far from the center of the APSD. Under such circumstances, the mass of API recovered could be near to its limit of detection (LOD) or limit of quantitation (LOQ) established for a given method. Another difficulty can be the high number of dilutions and wide ranges of concentrations for a given API delivered from multiple strengths of an inhaled product. In a given dataset encompassing maximum and minimum strengths, the mass of API recovered from each stage of the set of stages contained in the CI may range from the LOQ for the lowest strength tested for stages with d_{50} values distant from the aerosol MMAD to 150% of the highest deposition for the highest strength tested collecting on stages having d_{50} values closest to the MMAD. The number of actuations used per determination might need to be adjusted to enable an accurate and precise detection method. Additionally, there may be product-specific complications contributing to the overall method variability, for example, use of highly volatile solvents (which may be necessary for certain APIs), leading to erratic results due to solvent evaporation during pipetting. Careful method development, involving iterative evaluation and optimization, may help minimize variability arising from the API recovery and analysis procedures.

4.2.4 Contribution to Variability from the Drug Product (Material)

Bonam et al. observed that the APSD of the aerosol delivered by an OIP will have an intrinsic variability that is truly attributable to the product [2]. They identified a number of factors potentially influencing this true APSD variability, including the following attributes:

(a) Product orientation during storage
(b) Formulation interaction with excipients
(c) Interaction with components of container closer system
(d) Sensitivity to moisture
(e) Product age
(f) Temperature and humidity during storage
(g) Tolerances of device components
(h) Electrostatic effects during de-aggregation of some DPI formulations
(i) Static electrification due to interaction of MDI aerosol particles with nonconducting elastomers used in MDI valve manufacture [40, 73, 74]

All such factors should be studied during product development and are not considered here in detail because they are specific to the product rather than the CI method, which is the focus of this chapter.

In addition, Bonam et al. noted that the physicochemical characteristics of certain OIPs may make the CI method more variable [2]. For example, suspensions by nature will tend to produce more variable aerosols than originate from homogeneous solutions. This outcome happens because of the intrinsic tendency for particle sedimentation from the suspending fluid (i.e., sinking) or particle–particle coalescence above the suspending fluid (i.e., creaming) in suspensions, thereby adding to product variability. Furthermore, in the case of suspension-formulated MDIs, the magnitude of the elapsed time between shaking and actuation (which is part of the CI assessment "method") may influence the measured APSD variability. The accompanying drop in the MDI canister temperature immediately following actuation may affect the measured APSD from a subsequent actuation if insufficient time is allowed to elapse. 30 s is a widely practiced minimum elapsed time between replicate actuations to avoid variability in aerosol delivery characteristics arising from this source.

Bonam et al. also noted that OIPs containing more than one API can present special challenges for the development of a precise and accurate CI method [2]. For example, one of the API components could be a strong chromophore and the other a weak one, or the particular API might produce a high-density (mass concentration) aerosol, whereas the other(s) are present in lower mass concentrations in the formulation. In consequence, the mass of the relatively highly concentrated API may rapidly reach the limit for stage loading. At the same time, the mass of the relatively lower concentration API(s) present on the same stage might only be close to or at the LOD level for the same number of inhaler actuations. Alternatively, one

API might be a hydrophilic compound and the other a hydrophobic one, resulting in the potential for different collection characteristics if the surface coating of the collection substrate on the stages is also markedly hydrophobic or hydrophilic.

4.2.5 Contribution to Variability from Interactions Between Contributing Causes

Bonam et al., in concluding their assessments on causes of CI method variability, also considered the possibility of interactions between the four factors previously considered separately (material, man, machine, and analysis) [2]. From Fig. 4.2, it is evident that some factors appear more than once, that is, they arise in more than one category of influences governing the overall variability of the CI method. For example, the number of inhaler actuations may conceivably affect the APSD uncertainty through man (e.g., due to repetitive stress or differences in delay times when firing multiple actuations), machine (e.g., due to stage overload and particle bouncing), measurement (e.g., due to the limit of detection when the number of actuations is small), and material (e.g., due to suspension resettling time). In another example, incorrect device cleaning may induce particle growth, giving rise to a man–material interaction component of variability. Specifically designed experiments might be conducted to separate variability components. However, the total mass collected from the CI system (mass balance) may be used more routinely as a diagnostic tool to distinguish between a method-related and product-related abnormality when deviating APSD results are observed [3, 75].

Bonam et al. advised method developers to be aware of this degree of potential complexity either when optimizing a given CI-based measurement procedure or conducting an investigation for deviating results [2].

4.3 Summary of GCIP

Recognizing the complexity of the CI methodology and its importance in the assurance of OIP quality, the Product Quality Research Institute (PQRI), a forum for technical interaction between the USFDA, the pharmaceutical industry, and academia, formed the APSD-Mass Balance Working Group in 2002 to address issues associated with API recovery from multistage CI measurements [76]. Although the main focus of this activity was on resolving the appropriate specification for mass balance from multistage CI measurements, during the course of discussions, it became evident that the complexity of the compendial methods was one of the most important contributors to the variability associated with this measure of system suitability.

The GCIP guidance document developed from this working group [3] was its main achievement and provided the basis for subsequent investigations into how the methodology could be simplified.

Table 4.1 Factors to be considered in CI method development (*Adapted from* [3]—*used by permission*)

Method development			
Factor potentially affects	Cause(s)	MB	APSD
API recovery solvent	Measurement	Yes	Yes
Quantitation lower limit	Measurement	Yes	Yes
Use of collection surface coating	Machine	Yes	Yes
Recovery techniques	Man/measurement	Yes	Yes
Use of a pre-separator	Machine	No, except for carrier based DPIs	Yes
Cleaning procedure	Man/material	Yes	Yes
Electrostatic charge	Machine/material	Yes	Yes
Environmental factors (barometric pressure, temperature, humidity)	Material	Yes	Yes
Use of a backup filter	Machine	Yes	Yes

GCIP can be considered the sum of two basic parts:

1. Factors affecting the performance of the CI system that should be considered in method development
2. Factors that should be considered in day-to-day use

As might be expected, most of the issues raised in this guidance are the same as those already discussed largely from the subsequent assessment of Bonam et al. [2].

The factors identified in the GCIP document as influencing APSD as well as MB are summarized in Tables 4.1 and 4.2, respectively. These lists provide a systematic guide to those involved in CI method development and routine use as to signposts for potential causes of deviations in both MB and aerosol APSD.

These tables serve to highlight the desirability for a simpler-to-perform yet as precise as possible process for determining OIP quality based on emitted aerosol particle size-related data. These factors are each linked to one or more of the causes of variability discussed in the previous section.

4.3.1 CI Method Failure Investigation Tree

The compilers of the GCIP document were primarily focused on addressing the issue of how to manage MB values that are outside the current ±15% label claim dose limit set in the 1998 draft CMC Guidance for OIPs by the FDA [4]. An issue of significance is the lack of a fallback method in the event that an out-of-specification (OOS) or out-of-trend (OOT) MB is determined. The MB failure investigation tree, which might also be termed as CI method failure investigation tree (Fig. 4.4), was a significant achievement, in that for the first time, a process for addressing OOS/OOT events associated with CI/MSLI-based measurements was established that provides a workable pathway for the analyst faced with such circumstances.

Table 4.2 Factors that should be considered in CI day-to-day use (*Adapted from* [3]—*used by permission*)

Routine CI setup and operation			
Factor potentially affects	Cause(s)	MB	APSD
Locating collection surfaces	Man/machine	Yes	Yes
Accounting for collection surfaces and final filter	Man	Yes	Yes
Assertion of stage order	Man	No	Yes
Air leakage into CI/MSLI	Man/machine	Yes, unless wall losses are accounted for	Yes
Poor seal and orientation between induction port/ pre-separator/CI	Man/machine	No	No, unless the leak is massive
Improper alignment between inhaler mouthpiece and induction port	Man	No	Yes
Inadequate liquid volume or liquid missing from liquid-based collection surfaces	Man	Yes, due to analytical procedure differences	Yes
CI/MSLI flow rate	Man	Not normally, but consider for highly flow-rate-dependent DPIs if tested at flow rates different to dose content uniformity testing	Yes
Timer operation of two-way solenoid valve for DPI testing and MDIs with integrated spacers	Man/machine	Yes	Yes
Cleaning of stage nozzles	Man	Yes	Yes
Worn/corroded stage nozzles	Machine	No	Yes
Insufficient or excessive number of inhaler actuations	Man	Yes	Yes
Improper sample recovery	Man	Yes	Yes
Electrostatic effects	Machine	Yes	Yes

The need for a failure investigation tree for the cascade impaction method reinforces the complexity of the procedure and brings into question why an alternative approach has not become the accepted norm by stakeholders involved in the testing of OIPs. The lack of an alternative measurement technique having similar capability in terms of traceability to the aerodynamic diameter size scale and, at the same time, specificity for mass of API are the main reasons why this relatively historic method for aerosol analysis remains the standard approach. The search for more rapid and easy-to-use measurement apparatuses together with a more effective data analysis technique has been one of the driving forces underlying the development of the methodologies that are discussed later in this book. Although alternatives to CI-based measurement of OIP APSD exist and are widely used, particularly in the research and development environment, with few exceptions, none of the techniques

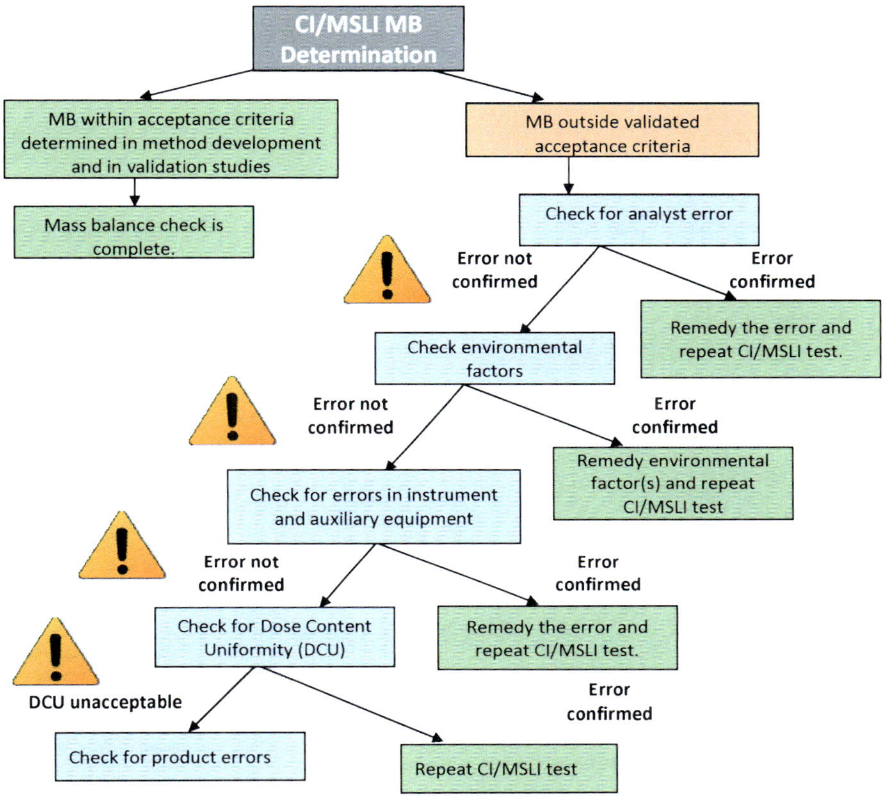

Fig. 4.4 A CI method failure investigation tree (*Adapted from* [3]—*used by permission*)

developed thus far meet the needs of regulatory agencies to provide data that are unambiguous in terms of both recovered API mass and aerodynamic diameter [77]. This situation, however, may change, as API-specific measurement techniques, such as Raman spectroscopy, are combined with microscopy-image analysis, as a means of quantifying particle size distributions for systems comprising compact, near-to-spherical particles. In such circumstances, the dynamic shape factors (χ) are close to unity, making it possible to approximate to the aerodynamic diameter scale from microscopy-measured circle/volume-equivalent diameter (d_v), provided that the particle density is close to unity (see Chap. 2).

4.3.2 The Potential in CI Methodology Simplification

The advantages of CI method simplification become obvious when the number of operations associated with a typical eight-stage full-resolution CI measurement (Fig. 4.5 shows the situation for the ACI as an example system) is compared with

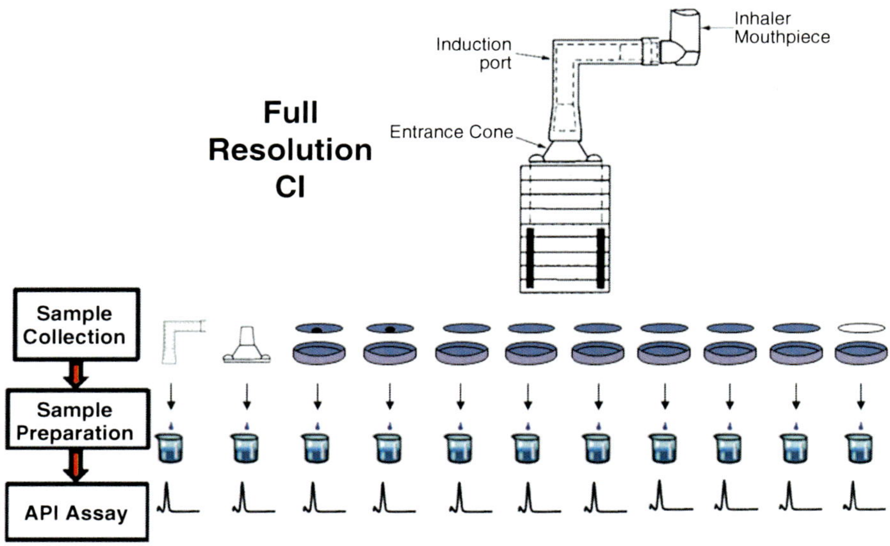

Fig. 4.5 Potential steps involved in API recovery and assay with eight-stage full-resolution CI including backup filter

those needed if the number of stages is reduced to one plus the backup filter in an abbreviated impactor (Fig. 4.6). However, even with method simplification, the principles of GCIP, which have been outlined earlier in this chapter, are equally applicable to ensure the most accurate and reproducible measurements.

Much of the remainder of this handbook is concerned with the development and validation of concepts based on the AIM concept, augmented by the adoption of new ways of working more effectively with the data that are produced by both abbreviated and full-resolution CI systems (EDA).

4.4 Good CI Data Analysis Practices

Although not covered by the GCIP article, the development of good CI data analysis principles for APSD assessments was later summarized by Mitchell and Dunbar [78] and later augmented by Christopher et al. [79]. These concepts are described below in order to complete the process of defining practices that are likely to optimize the overall CI methodology for OIP assessments.

4.4.1 Representation of CI Data

The mass of API recovered from each component of the CI system, from the induction port to the backup filter constitutes the raw data that are obtained by the cascade

Fig. 4.6 Potential steps involved in API recovery and assay with abbreviated CI based on ACI design

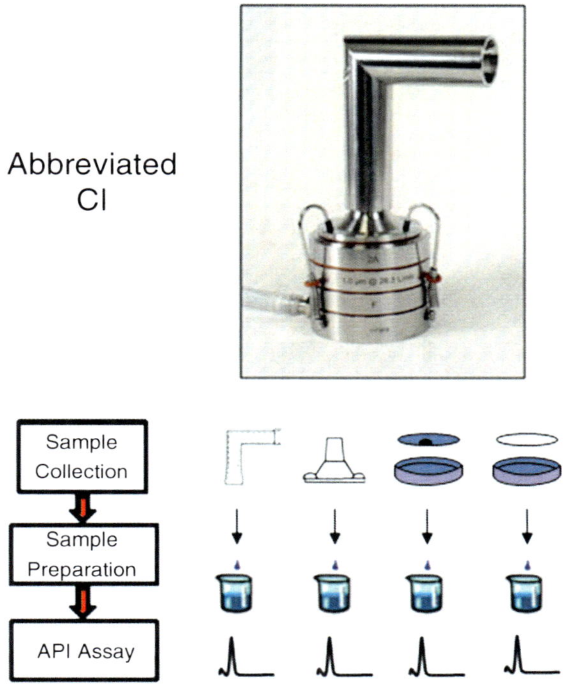

impaction procedure. These data may be handled in several different ways in order to arrive at meaningful measures of OIP aerosol performance. There are at least three distinctly different ways in which to represent the raw data (Table 4.3).

Mitchell and Dunbar noted that the simplest approach is to treat the API mass distribution as a nominal function of CI stages and the non-sizing auxiliary components, such as the induction port and pre-separator (if used) [78]. These values are related to the locations of the discrete stages and auxiliary components in the entire CI system by name only and without order (left-hand illustration in Fig. 4.7).

There is no size scaling whichever representation method is chosen, and it is not possible to compare findings from different CI systems meaningfully in this way. The data are commonly rank-ordered (right-hand illustration in Fig. 4.7), for comparison with equivalent results obtained by the same system.

It therefore becomes necessary to explore other perhaps more informative ways of looking at the descriptive statistics that are available from the raw data provided by CI measurements and derive inferential statistical relationships from metrics such as SPM, LPM, EPM, FPM, CPM, and their related mass fractions, as well as ISM and the total mass entering the system, including the non-sizing components.

As the first step in this process, the mass of API can be presented as an ordinal function of CI size-separation stage range (Fig. 4.8), the mass of API collected from a particular stage, i, is linked to the d_{50} value of the preceding stage $(i-1)$.

Table 4.3 Representation of data from full-resolution CI measurements

Approach	Advantage	Application
API mass on each component including non-sizing parts, such as the induction port	Simplicity	OIP quality control
API mass on each size-fractionation stage of the CI	Basic aerodynamic size-related information	OIP development and in bioequivalence
API mass mathematically linked to aerodynamic diameter range of sizing stages	Aerodynamic size-related information comparable from one CI system to another	OIP development and in bioequivalence; this approach is essential for input to models predicting respiratory tract deposition

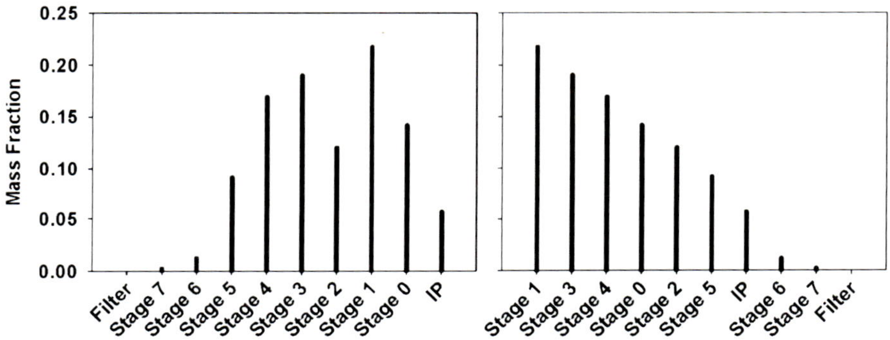

Fig. 4.7 Representation of CI raw data by mass deposition in both sizing and non-sizing components (*From* [78]—*used by permission*)

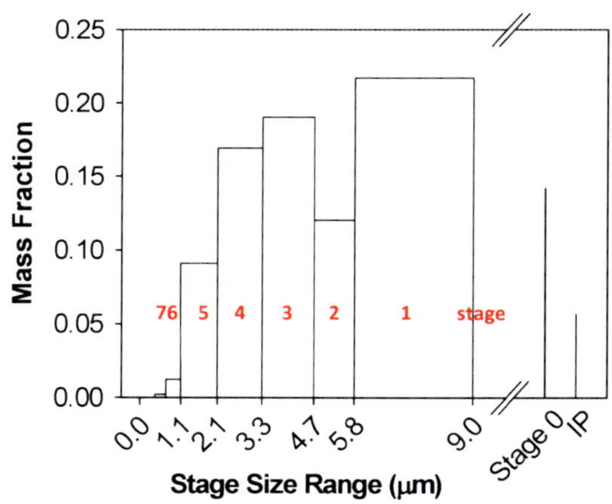

Fig. 4.8 API mass on CI size-separating stages as an ordinal function of stage d_{50} size: example from ACI operated at 28.3 L/min (*From* [78]—*used by permission*)

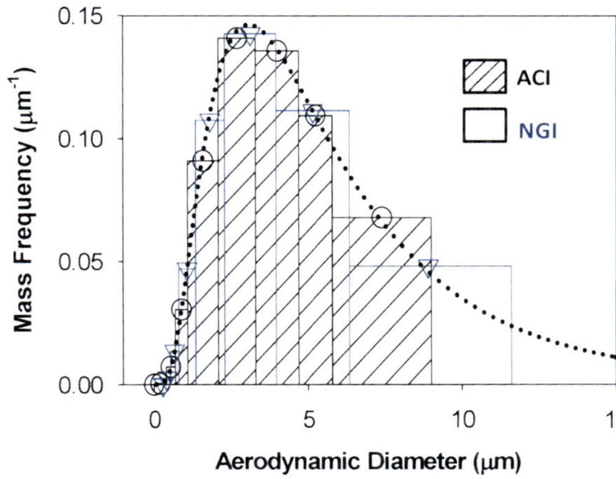

Fig. 4.9 Differential mass-weighted APSD showing how data for the same aerosol (MMAD = 5.0 µm; GSD = 2.0) can be meaningfully represented from two different CI systems (ACI and NGI): the *dotted line* is the true APSD (*From* [78]—*used by permission*)

In this instance, API mass is related to the aerodynamic diameter scale through the d_{50} values of the size-separating stages, but the non-sizing components (in this case stage 0, which has no upper bound defined, the induction port and pre-separator) are represented by thin lines rather than a size range, with the size scale broken between the data from the two categories to designate the difference in properties with respect to aerodynamic diameter. This representation is popular if the measurements are all made using the same CI system and/or the OIP aerosol is captured predominantly in the size-separating portion of the system.

However, Mitchell and Dunbar showed that its weakness is the lack of a more formal link between API mass and size, making it impossible to compare data from two different CI systems in a readily accessible way [78].

To achieve this goal, it is necessary to invoke the formalism of the differential mass-weighted APSD, in which the mass frequency of API on each stage of the size-fractionating part of the CI system is obtained by dividing the absolute mass by the width of the size range associated with that particular stage (Fig. 4.9), in accordance with the relationship:

$$F_{\mathrm{m}}(d_{\mathrm{ae},i}) = \frac{m_i}{M_{\mathrm{EM}}} \frac{1}{\Delta d_{50,i}} \tag{4.5}$$

in which $d_{\mathrm{ae},i}$ represents the aerodynamic diameter of the ith stage, m_i is the mass of API recovered from that stage, M_{EM} is the emitted mass from the inhaler (or spacer/valved holding chamber patient interface in the case of pMDI testing with these add-on devices), and $\Delta d_{50,i}$ is the size width associated with the stage of interest.

Under these circumstances, it is possible to represent APSDs from different CI systems on the same graph for direct comparison, and the midpoint of each bar of each histogram corresponds to the place at which the curve of the true APSD (i.e., size-resolved to an infinitesimally small amount) intersects.

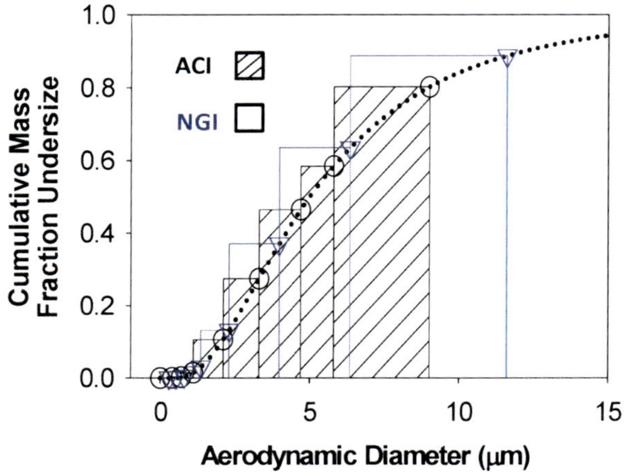

Fig. 4.10 Same data from Fig. 4.9 presented in cumulative mass-weighted format (*From* [78]—*used by permission*)

Mitchell and Dunbar also noted that the same data can be represented, perhaps more conveniently, in cumulative mass-weighted format (Fig. 4.10). It is this approach that is most widely used to represent CI-based data because subcomponents of the aerosol, such as extra-fine (EPF), fine (FPF), and coarse (CPF) particle mass fractions, are readily obtainable from the ordinate scale (cumulative mass % undersize) at the appropriate aerodynamic sizes, using the assumed form of the APSD with infinitesimally fine resolution (dotted line). As this form of the APSD considers only the impactor-sized mass (ISM), it is also possible to derive the mass fractions that correspond to small (SPF) and large particles (LPF) in the context of EDA-based data analysis. However, it is easier to obtain the ratio metric LPM/SPM and ISM from the raw mass/stage data (Fig. 4.7), and these measures are the most useful for EDA analysis.

The cumulative mass-weighted APSD shown in Fig. 4.10 will not be helpful, however, in cases in which an add-on device (spacer or VHC) is being used in conjunction with an OIP, because the influence of the add-on cannot be assessed unless the mass fractions are normalized to the total mass emitted from the inhaler, including that captured by the non-sizing components, principally the induction port.

In the case of spacers and valved holding chambers used with pMDIs, the add-on eliminates the ballistic fraction of the aerosol emitted from the inhaler mouthpiece that would normally be captured in the induction port [80]. The format illustrated by Fig. 4.11, which includes all the mass of API emitted from the actuator mouthpiece of the pMDI entering an ACI as the example CI, is more appropriate for comparisons with and without add-on device. In this particular configuration, stage 0 has no upper size limit, so that the APSD comparison is based on the total mass passing beyond this stage compared with the total mass that deposits in the entire system including the induction port.

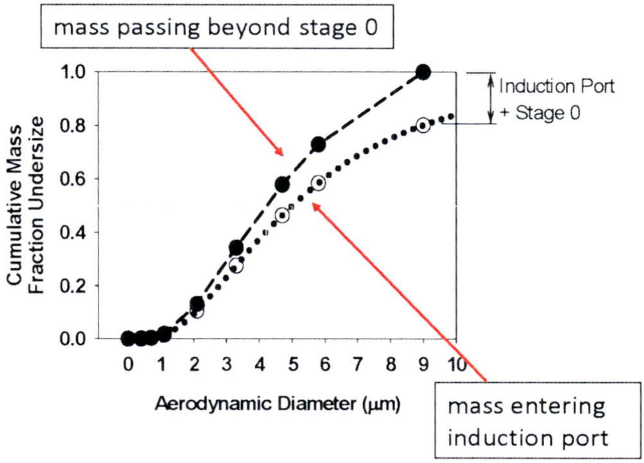

Fig. 4.11 Cumulative mass-weighted APSDs considering emitted mass of API entering the size-fractionating stages of an ACI having an upper size limit in relation to the mass entering the induction port of this measurement system (*From* [78]—*used by permission*)

4.4.2 Data Reduction

4.4.2.1 Initial Inspection of the APSD

The APSD by itself, though useful in understanding the behavior of the whole population of the aerosol particles that are size-fractionated within the operating limits of the CI, is awkward as a means of representing OIP performance parameters that are meaningful both in the product QC and clinical environments.

This difficulty arises because the true APSD is a continuous variable function. However, data coming from a given CI-based measurement discretizes the APSD into a multivariable form with increasing complexity as the number of size-fractionating stages in the CI system as a whole becomes greater. Thus, treating the APSD as a collection of seven or more independent measurements increases the overall error in decision-making due to the effect of multiplicity. Furthermore, the time and resources required to arrive at a decision concerning OIP quality from such data become more complex than it need be. It therefore becomes necessary to adopt a data reduction strategy that is appropriate for the application for which the CI data are required. This approach is at the heart of EDA and will be discussed in depth in later chapters. However, at this stage when considering good data analysis practices, it is a good idea to check for the following when inspecting data obtained from the size-fractionating stages of the CI:

1. More than one mode may be present in the APSD, as processes such as atomization of liquids often produce bimodal distributions [81] as the result of Rayleigh breakup of the liquid stream via ligament formation into primary and satellite

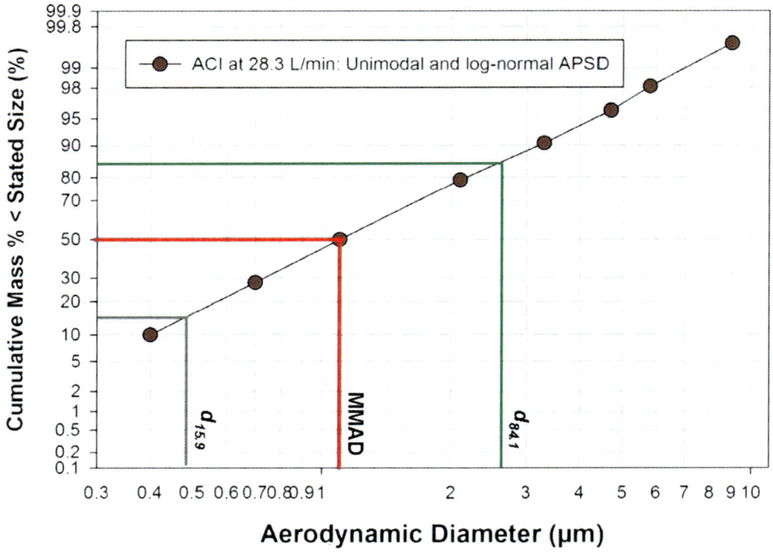

Fig. 4.12 Near-normal to lognormal APSD from a pMDI product presented in log-probability scaled format

droplets, each with its own APSD. The elimination of one of these modes (e.g., the coarser primary droplets) by impaction on baffles within jet nebulizers is one way to present a unimodal APSD to the CI system.

2. If auxiliary moments, such as GSD, are being used to define the spread of the APSD, check for deviations from lognormal by plotting the cumulative mass on log-probability scaling and ensuring that the relationship between mass of API and aerodynamic diameter is both symmetric about the MMAD and linear or near to linear (Fig. 4.12).
3. The GSD is determined from the relationship

$$GSD = \sqrt{\frac{d_{84.1}}{d_{15.9}}} \tag{4.6}$$

in which $d_{15.9}$ and $d_{84.1}$ are the sizes corresponding to the 15.9th and 84.1st percentiles of the cumulative mass-weighted APSD. If the plot similar to that illustrated by Fig. 4.12 is not linear or near to linear in shape, then it is recommended that for regulatory reporting, at least, that the fact that the APSD is not lognormal be acknowledged and simply the spread of the distribution (S_{APSD}) reported as a ratio, where

$$S_{\text{APSD}} = \frac{d_{90}}{d_{10}} \tag{4.7}$$

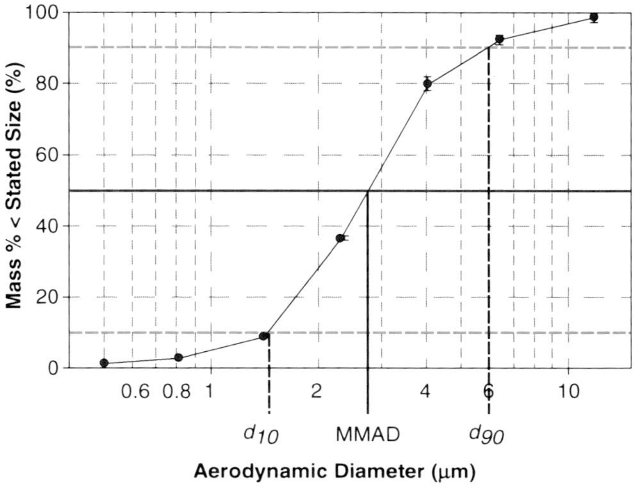

Fig. 4.13 Cumulative mass-weighted APSD, showing relationships between span-related metrics $d_{10,}$ d_{90} and *MMAD*

where d_{10} and d_{90} are the 10th and 90th percentiles of the cumulative mass-weighted APSD, respectively.

Assuming the APSD is unimodal, the moment of central tendency, MMAD, is identical to the size that corresponds to the 50th mass percentile, *whether or not the distribution is lognormal*, so that a relative span factor (RSF), based on the relationship

$$RSF = \frac{[d_{90} - d_{10}]}{MMAD} \tag{4.8}$$

can be a useful way to report the spread of the APSD (Fig. 4.13).

4.4.2.2 Estimation of the MMAD

Seldom will the APSD provide an exact value of the MMAD that can be obtained directly from the plotted data because of alignment of one of the stage d_{50} sizes with the 50th mass percentile of the distribution. Under most circumstances, it is therefore necessary to estimate this value. The method described in the pharmacopeias relies on the underlying assumption of lognormality, which is frequently not the case with OIP-generated aerosols. In 2010, the CI Working Group of IPAC-RS published a Stimulus to Revision article, in which they provided evidence from an extensive database of OIP APSDs that two multipoint curve-fitting methods [Chapman–Richards (CR) and Mercer–Morgan–Flodin (MMF) Models], as well as the linear two-point interpolation of data points straddling the 50th mass percentile, all yielded MMADs that were in good agreement with each other [79]. More importantly, these estimates of MMAD more accurately fitted the actual data than did values derived using the USP method. Usefully, the alternative approaches neither require log-transformation of the raw data nor rely on underlying assumptions about the form of the APSD, other than requiring that it be unimodal. True lognormal distributions are also adequately addressed by any of the proposed generalized approaches. The linear two-point interpolation method, based on the closest data points straddling the MMAD value, is the simplest of these options to implement and therefore the preferred choice.

4.4.2.3 Other Derived Metrics

There are many other metrics beyond the moments that can be derived from CI-measured APSDs. These additional measures are conveniently classified into two subsets: metrics related to EDA and those associated with the more conventional approach to data reduction that focuses on providing measures that may be predictive of size-related deposition in the HRT. Table 4.4 summarizes the options for size-related metrics that are commonly chosen, together with those that can be determined by AIM-based methods in which an APSD is not obtainable as part of the process.

Each mass subfraction (*LPF*, *SPF*, *EPF*, *FPF*, and *CPF*) can be determined directly from a cumulative mass-weighted APSD as the mass fraction corresponding to the chosen size limits for the subfraction of interest. Likewise the corresponding metrics based on absolute mass (*LPM*, *SPM*, *EPM*, *FPM*, and *CPM*) are obtained from the original raw data expressed as mass per stage. Stage groupings may be used to establish these absolute mass values, since more than one stage is usually involved in the collection of each mass subfraction. However, EDA is more efficient, in that only LPM and SPM are needed to undertake the assessment of the CI size-fractionated portion of the OIP aerosol.

It is self-evident from Table 4.4 that the AIM-based approach is capable of providing the same degree of flexibility with regard to all of these derived metrics, regardless of the eventual application.

4.4.3 CITDAS and Other Software for Assessing APSD Data from CI Measurements

A laboratory may develop its own computerized techniques for assessing CI APSDs. However, commercially available software has been available for some time in order to assist those in the quite complex process of deriving the most appropriate measures to report. CITDAS® (Copley Scientific Ltd, Nottingham, UK) is the most tailored product for OIP-related applications. Lewis has recently reported that this software is both versatile and easy to use with an intuitive user interface [82].

CITDAS allows standardized data processing for the operating conditions of four alternative impactors: ACI, MSLI, MMI, and NGI. In the current release (CITDAS

Table 4.4 Metrics in common use with APSDs from OIPs

	Related to OIP QC		Related to particle deposition in HRT	
Metric	Full resolution	AIM	Full resolution[a]	AIM
EPM or EPF			Yes	Yes
FPM or FPF			Yes	Yes
CPM or CPF			Yes	Yes
LPM or LPF	Yes	Yes		
SPM or LPF	Yes	Yes		
ISM	Yes	Yes		
IM[b]			Yes	Yes
TEM[c]	Yes	Yes	Yes	Yes
MB[d]	Yes	Yes	Yes	Yes

[a]EPM or FPM or CPM can be established from groupings of stages. In the case of FPM, the grouping may be defined with both an upper and lower bound (i.e., from 1.1 to 4.7 μm aerodynamic diameter from stages 3 to 5 of the ACI operated at 28.3 L/min), or just with the upper bound defined (i.e., <4.7 μm aerodynamic diameter from stages 3 to the backup filter in the example quoted beforehand)
[b]IM includes the mass collected in the first stage of the CI where the upper bound size is undefined (e.g., stage 0 in the ACI configuration for use at 28.3 L/min)
[c]TEM includes all non-sizing components of the CI system (i.e., the induction port, pre-separator (if used), and the first stage of the CI)
[d]MB includes IM + the mass retained by the OIP and is only included to identify the need for a system suitability test in CI-based determinations of OIP performance

version 3.10), data from the NGI operated at 15 L/min can be interpreted, thereby making this software useful for those testing nebulizing systems in accordance with compendial guidance to operate the CI at this relatively low flow rate, as well as with MDIs and DPIs. In addition to plotting the raw mass/component data, CITDAS reports the cumulative mass-weighted APSD and associated descriptive statistics (i.e., MMAD, GSD). An interesting new feature is the ability to specify up to five different interpretations of subfraction mass, defined either by impactor stage locations or in terms of aerodynamic particle diameter range. Therefore, in addition to reporting FPM, the reported delivered mass ex OIP can be subdivided into as many as this number of discrete subpopulations. CITDAS determines the profile of each subpopulation by interpolation, which means that it is possible for users to process multimodal particle size distributions as a series of lognormal subpopulations.

CITDAS has certain features that make it attractive to those who need to use CI data but do not want to get deeply involved in understanding all the complexities associated with good data analysis practice. For instance, it carries out the following data integrity checks:

1. Limit of detection (LOD) is reported for measures of FPD and FPF if the following criteria are met:
   (a) Values are associated with a cumulative drug mass <2% of the total mass.
   (b) The cumulative mass on fewer than three stages is greater than 1% of the total mass.

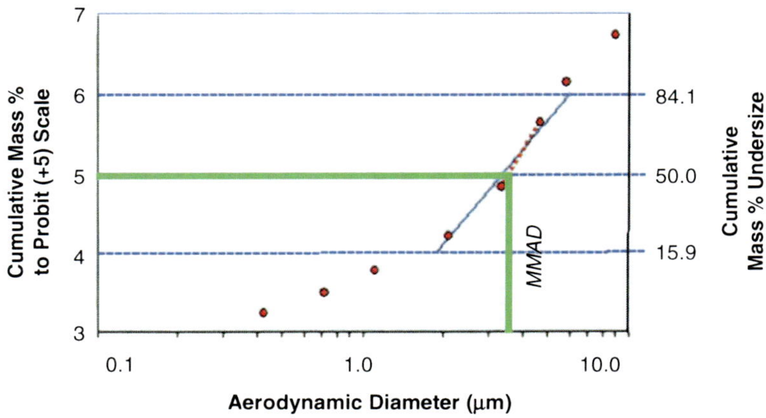

Fig. 4.14 APSD presented using CITDAS software with log-probit (+5) scaling for *left-hand ordinate axis*: The linear relationship with this scaling used to estimated GSD by (4.6) is shown on *right-hand ordinate axis*

2. Not applicable (NA) is reported for MMAD and GSD if the following situations occur:

   (a) Greater than 50% of the cumulative drug mass is deposited on the lowest impaction stage (backup filter or MOC (in the case of the NGI—if used)).

   (b) Fewer than three stages have a cumulative drug mass >1% of the total mass collected.

In addition, GSD is calculated after applying a test such that the APSD distribution is determined to be lognormal within ±1 standard deviations of the MMAD. CITDAS calculates whether or not to report GSD by performing a linear regression of the cumulative mass % undersize plot (applying the log-probit scale) between the probit (+5) values 4 and 6 (Fig. 4.14). If the R^2 coefficient for the regression exceeds the user defined value (default is 0.95), GSD is reported; otherwise, the value for GSD appears as "NA."

The "import CSV file" function in CITDAS provides a powerful feature with the potential to enable this software to link up with other applications, such as EXCEL (Microsoft Corporation, Redmond, WA, USA). This capability allows files exported from CITDAS in CSV format to be modified/edited and reimported. Use of this function with appropriate verification of data integrity streamlines data input into CITDAS, potentially reducing manual data entry requirements in repetitive situations.

DISTFIT 2008® (Chimera Technologies Inc., Forest Lake, MN, USA) is an alternative software package, and though not specifically tailored for OIP aerosol measurement applications, has considerable capabilities in being able to handle a much wider range of particle size, as well as count (number), surface area and volume (mass) weightings. Its main advantage is the ability to join particle size distributions from different measurement techniques into a contiguous whole for subsequent data manipulation. Much of its content is based on the chapter on particle size statistics in the textbook by Hinds on the topic Aerosol Technology [83].

References

1. Nichols S (2004) Particle size distribution parameters using the next generation pharmaceutical impactor. In: Dalby RN, Byron PR, Peart J, Suman JD, Farr SJ (eds) Respiratory drug delivery-IX. Davis HealthCare International Publishing, River Grove, IL, pp 485–487
2. Bonam M, Christopher D, Cipolla D, Donovan B, Goodwin D, Holmes S, Lyapustina S, Mitchell J, Nichols S, Petterson G, Quale C, Rao N, Singh D, Tougas T, Van Oort M, Walther B, Wyka B (2008) Minimizing variability of cascade impaction measurements in inhalers and nebulizers. AAPS PharmSciTech 9(2):404–413
3. Christopher D, Curry P, Doub B, Furnkranz K, Lavery M, Lin K, Lyapustina S, Mitchell J, Rogers B, Strickland H, Tougas T, Tsong Y, Wyka B (2003) Considerations for the development and practice of cascade impaction testing including a mass balance failure investigation tree. J Aerosol Med 16:235–247
4. US Food and Drug Administration (FDA) (1998) CDER. Draft guidance for industry metered dose inhaler (MDI) and dry powder inhaler (DPI) drug products chemistry, manufacturing, and controls documentation, Rockville, MD. Accessed 6 Jan 2012 at http://www.fda.gov/cder/guidance/2180dft.pdf
5. Marple VA, Roberts DL, Romay FJ, Miller NC, Truman KG, Van Oort M, Olsson B, Holroyd MJ, Mitchell JP, Hochrainer D (2003) Next generation pharmaceutical impactor. Part 1: design. J Aerosol Med 16(3):283–299
6. Purewal TS (2001) Test methods for inhalers to check performance under normal use and unintentional use conditions. In: Drug delivery to the lungs-12. The Aerosol Society, London, pp 92–98
7. Stewart E, Holt J, Fitzgerald C, Bell P, Popow J (2006) Impact of using an automated shake-fire system on the shot weight and dose content uniformity of an HFA metered dose inhaler. In: Dalby RN, Byron PR, Peart J, Suman JD, Farr SJ (eds) Respiratory drug delivery-2006. Davis Healthcare International Publishing, River Grove, IL, pp 581–583
8. Miller NC, Roberts DL, Marple VA (2002) The 'Service Head' approach to automating the next generation pharmaceutical impactor: proof of concept. In: Dalby RN, Byron PR, Peart J, Farr SJ (eds) Respiratory drug delivery VIII. Davis Horwood International, Raleigh, NC, pp 521–523
9. Holzner PM, Muller BW (1995) Particle size determination of metered dose inhalers with inertial separation methods: apparatus A and B (BP), four stage impinger and Andersen Mark II cascade impactor. Int J Pharm 116(1):11–18
10. Mitchell JP, Nagel MW, Wiersema KJ, Doyle CC (2003) Aerodynamic particle size analysis of aerosols from pressurized metered dose inhalers: comparison of Andersen 8-stage cascade impactor, next generation pharmaceutical impactor, and model 3321 Aerodynamic Particle Sizer aerosol spectrometer. AAPS PharmSciTech 4(4):article 54. Accessed 10 Jan 2012 at: http://www.aapspharmscitech.org/view.asp?art=pt040454&pdf=yes
11. Taki M, Zeng XM, Marriott C, Martin G (2006) Comparison of deposition profiles of drugs from a combination dry powder inhaler using the Andersen cascade impactor (ACI), multi-stage liquid impinger (MSLI) and next generation impactor (NGI). In: Dalby RN, Byron PR, Peart J, Suman JD, Farr SJ (eds) Respiratory drug delivery-2006. Davis Healthcare International Publishing, River Grove, IL, pp 659–662
12. Jozwiakowski J, Lor X, Paulson S, Schultz D (2006) Comparison of Andersen cascade impactor and next generation impactor performance of beclomethasone pMDIs with oligolactic acid. In: Dalby RN, Byron PR, Peart J, Suman JD, Farr SJ (eds) Respiratory drug delivery-2006. Davis Healthcare International Publishing, River Grove, IL, pp 357–359
13. Mitchell JP, Nagel MW (2004) Particle size analysis of aerosols from medicinal inhalers. KONA Powder Part 22:32–65
14. Roberts DL, Mitchell JP (2011) Influence of stage efficiency curves on full-resolution impactor data interpretation. Drug delivery to the lungs-22. The Aerosol Society, Edinburgh, pp 181–184. Available at: http://ddl-conference.org.uk/index.php?q=previous_conferences. Visited 4 Aug 2012

15. Finlay WH, Stapleton KW (1999) Undersizing of droplets from a vented nebulizer caused by aerosol heating during transit through an Andersen impactor. J Aerosol Sci 30(1):105–109
16. Mitchell JP (2000) Particle standards: their development and application. KONA Powder Part 18:41–59
17. Marple VA, Olson BA, Santhanakrishnan K, Mitchell JP, Murray SC, Hudson-Curtis BL (2003) Next generation pharmaceutical impactor (a new impactor for pharmaceutical inhaler testing)—Part 2: archival calibration. J Aerosol Med 16(3):301–324
18. Chambers F, Ali A, Mitchell J, Shelton C, Nichols S (2010) Cascade impactor (CI) mensuration—an assessment of the accuracy and precision of commercially available optical measurement systems. AAPS PharmSciTech 11(1):472–484
19. Roberts DL, Romay FJ (2005) Relationship of stage mensuration data to the performance of new and used cascade impactors. J Aerosol Med 18(4):396–413
20. European Directorate for the Quality of Medicines and Healthcare (EDQM). Preparations for inhalation: aerodynamic assessment of fine particles. (2012) Section 2.9.18—European Pharmacopeia [—Apparatus B in versions up to 4th edn. 2002] Council of Europe, 67075, Strasbourg, France
21. United States Pharmacopeial Convention (USP) (2012) Chapter 601: Aerosols, metered-dose inhalers, and dry powder inhalers. USP35-NF30, Rockville, MD
22. Roberts DL (2009) Theory of multi-nozzle impactor stages and the interpretation of stage mensuration data. Aerosol Sci Technol 43(11):1119–1129
23. Svensson M, Pettersson G, Asking L (2005) Mensuration and cleaning of the jets in Andersen cascade impactors. Pharm Res 22(1):161–165
24. Milhomme K, Dunbar C, Lavarreda D, Roberts D, Romay F (2006) Measuring changes in the effective jet diameter of cascade impactor stages with the flow resistance monitor. In: Dalby RN, Byron PR, Peart J, Suman JD, Farr SJ (eds) Respiratory drug delivery-2006. Davis Healthcare International Publishing, River Grove, IL, pp 405–407
25. Stein SW, Olson BA (1997) Variability in size distribution measurements obtained using multiple Andersen mark II cascade impactors. Pharm Res 14(12):1718–1725
26. Stein SW (1999) Size distribution measurements of metered dose inhalers using Andersen mark II cascade impactors. Int J Pharm 186(1):43–52
27. Kadrichu N, Rao N, Sluggett G, Fong B, Jones G, Perrone T, Seshadri S, Shao P, Williams G, Zhang J, Bennett D (2004) Sensitivity of Andersen cascade impactor response to stage nozzle dimensions. In: Dalby RN, Byron PR, Peart J, Suman JD, Farr SJ (eds) Respiratory drug delivery-IX. Davis Healthcare International Publishing, River Grove, IL, pp 561–564
28. Marple VA, Rubow KL, Olson BA (2001) Inertial, gravitational, centrifugal, and thermal collection techniques. In: Baron PA, Willeke K (eds) Aerosol measurement: principles, techniques and applications, 2nd edn. Wiley, New York, NY, pp 229–260
29. Asking L, Mitchell J, Nichols S (2008) Air flow meters used at testing of inhalation products—an inter-laboratory comparison, Drug delivery to the lungs-19. The Aerosol Society, Edinburgh, UK, pp 42–44, Available at: http://ddl-conference.org.uk/index.php?q=previous_conferences. Visited 4 Aug 2012
30. Olsson B, Asking L (2002) Methods of setting and measuring flowrates in pharmaceutical impactor experiments, 13th edn, Drug delivery to the lungs-13. The Aerosol Society, London, pp 205–208
31. Wiktorsson B, Asking L (2002) Comparison between flowmeters used to set flows in pharmaceutical inhaler testing, 13th edn, Drug delivery to the lungs-13. The Aerosol Society, London, pp 168–171
32. Van Oort M, Downey B, Roberts W (1996) Verification of operating the Andersen cascade impactor at different flowrates. Pharm Forum 22(2):2211–2215
33. Mitchell JP, Nagel MW (2003) Cascade impactors for the size characterization of aerosols from medical inhalers: their uses and limitations. J Aerosol Med 16(4):341–377
34. Copley M, Smurthwaite M, Roberts DL, Mitchell JP (2005) Revised internal volumes of cascade impactors for those provided by Mitchell and Nagel. J Aerosol Med 18(3):364–366

35. Stein S, Myrdal PB (2006) The relative influence of atomization and evaporation on metered dose inhaler drug delivery efficiency. Aerosol Sci Technol 40(5):335–347
36. Peng C, Chow A, Chan CK (2000) Study of the hygroscopic properties of selected pharmaceutical aerosols using single particle levitation. Pharm Res 17(9):1104–1109
37. Finlay WH (1998) Estimating the type of hygroscopic behavior exhibited by aqueous droplets. J Aerosol Med 11(4):221–229
38. Byron PR, Davis SS, Bubb MD, Cooper P (1977) Pharmaceutical implications of particle growth at high relative humidities. Pesticide Sci 8(5):521–526
39. Martin AR, Finlay WH (2004) Effect of humidity on size distributions of MDI particles exiting a mechanical ventilation holding chamber. In: Proceedings of international conference on MEMS, NANO and Smart Systems, 2004. ICMENS, Banff, Alberta, Canada, pp 280–283. Available at: http://www.computer.org/portal/web/csdl/doi/10.1109/ICMENS.2004.57. Visited 10 Jan 2012
40. Byron PR, Peart J, Staniforth JN (1997) Aerosol electrostatics I: properties of fine powders before and after aerosolization by dry powder inhalers. Pharm Res 14(6):698–705
41. Murtomaa M, Mellin V, Harjunen P, Lankinen T, Laine E, Lehto VP (2004) Effect of particle morphology on the triboelectrification in dry powder inhalers. Int J Pharm 282(1–2):107–114
42. Carter PA, Cassidy OE, Rowley G, Merrifield DR (1997) Triboelectrification of fractionated crystalline and spray dried lactose. Pharm Pharmacol Commun 4:111–115
43. Carter PA, Rowley G, McEntee NJ (1997) An investigation of experimental variables during triboelectrification studies on powders. J Pharm Pharmacol 49(S4):23
44. Murtomaa M, Strengella S, Lainea E, Bailey A (2003) Measurement of electrostatic charge of an aerosol using a grid-probe. J Electrostat 58(3–4):197–207
45. Ramirez-Dorronsoro J-C, Jacko RB, Kildsig DO (2006) Chargeability measurements of selected pharmaceutical dry powders to assess their electrostatic charge control capabilities. AAPS PharmSciTechnol 7(4):article 103 (2006), Available at: http://www.aapspharmscitech.org/view.asp?art=pt0704103. Accessed 10 Jan 2012
46. Kwok PCL, Glover W, Chan HK (2005) Electrostatic charge characteristics of aerosols produced from metered dose inhalers. J Pharm Sci 94(12):2789–2799
47. Glover W, Kwok P, Chan HK (2004) Electrostatic charges in metered dose inhalers. In: Dalby RN, Byron PR, Peart J, Suman JD, Farr SJ (eds) Respiratory drug delivery-IX. Davis Healthcare International Publishing, River Grove, IL, pp 829–832
48. Glover W, Chan HK (2004) Electrostatic charge characterization of pharmaceutical aerosols using electrical low-pressure impaction (ELPI). J Aerosol Sci 35(6):755–764
49. Crampton M, Kinnersley R, Ayres J (2004) Sub-micrometer particle production by pressurized metered dose inhalers. J Aerosol Med 17(1):33–42
50. Glover W, Chan HK (2004) Electrostatic charge characterization of pharmaceutical aerosols. In: Dalby RN, Byron PR, Peart J, Suman JD, Farr SJ (eds) Respiratory drug delivery-IX. Davis Healthcare International Publishing, River Grove, IL, pp 825–826
51. Kwok P, Chan HK (2004) Measurement of electrostatic charge of nebulised aqueous droplets with the electrical low pressure impactor. In: Dalby RN, Byron PR, Peart J, Suman JD, Farr SJ (eds) Respiratory drug delivery-IX. Davis Healthcare International Publishing, River Grove, IL, pp 833–836
52. Peart P, Magyar C, Byron PR (1998) Aerosol electrostatics—metered dose inhalers (MDIs): reformulation and device design issues. In: Dalby RN, Byron PR, Farr SJ (eds) Respiratory drug delivery-VI. Interpharm Press, Buffalo Grove, IL, pp 227–233
53. Horton KD, Ball MHE, Mitchell JP (1992) The calibration of a California measurements PC-2 quartz crystal cascade impactor. J Aerosol Sci 23(5):505–524
54. Nasr MM, Ross DL, Miller N (1997) Effect of drug loading and plate coating on the particle size distribution of a commercial albuterol metered dose inhaler (MDI) determined using the Andersen and Marple–Miller cascade impactor. Pharm Res 14(10):1437–1443
55. Mitchell J (2003) Practices of coating collection surfaces of cascade impactors: a survey of members of EPAG, 14th edn, Drug delivery to the lungs-14. The Aerosol Society, London, pp 75–78

56. Byron PR (1994) Compendial dry powder testing: USP perspectives. In: Byron PR, Dalby RN, Farr SJ (eds) Respiratory drug delivery-IV. Interpharm Press, Buffalo Grove, IL, pp 153–162
57. Dunbar CA, Hickey AJ, Holzner P (1998) Dispersion and characterization of pharmaceutical dry powder aerosols. KONA Powder Part 16:7–45
58. Kamiya A, Sakagami M, Hindle M, Byron PR (2004) Locating particle bounce in the next generation impactor (NGI). In: Dalby RN, Byron PR, Peart J, Suman JD, Farr SJ (eds) Respiratory drug delivery-IX. Davis Healthcare International Publishing, River Grove, IL, pp 869–871
59. Berg E, Svensson JO, Asking L (2007) Determination of nebulizer droplet size distribution: a method based on impactor refrigeration. J Aerosol Med 20(2):97–104
60. Kamiya A, Sakagami M, Hindle M, Byron PR (2004) Aerodynamic sizing of metered dose inhalers: an evaluation of the Andersen and next generation pharmaceutical impactors and their USP methods. J Pharm Sci 93(7):1828–1837
61. Nasr MM, Allgire JF (1995) Loading effect on particle size measurements by inertial sampling of albuterol metered dose inhalers. Pharm Res 12(11):1677–1681
62. Feddah MR, Davies NM (2003) Influence of single versus multiple actuations on the particle size distribution of beclomethasone dipropionate metered-dose inhalers. J Pharm Pharmacol 55(8):1055–1061
63. Merrin C, Lee S, Needham M, Chambers F (2003) Evaluation of NGI performance with high dose pMDIs, 14th edn, Drug delivery to the lungs-14. The Aerosol Society, London, pp 184–187
64. Copley M (2007) Understanding cascade impaction and its importance for inhaler testing. Copley Scientific Ltd Technical Briefing, Available at: http://www.copleyscientific.co.uk/documents/ww/Understanding%20Cascade%20Impaction%20White%20Paper.pdf. Visited 10 Jan 2012
65. MSP Corporation (2007) NGI User's Guide. NGI-0170-6001, Revision C. Available at: http://www.epag.co.uk/Download2.asp?DID=902. Visited 10 Jan 2012
66. Mitchell JP (2006) Cleaning: a survey of members of the European Pharmaceutical Aerosol Group (EPAG), 17th edn, Drug delivery to the lungs-XVII. The Aerosol Society, Edinburgh, pp 197–199
67. Mitchell JP, Costa PA, Waters S (1987) An assessment of an Andersen Mark-II cascade impactor. J Aerosol Sci 19(2):213–221
68. European Directorate for Quality of Medicines and Healthcare (EDQM) (2012) Preparations for nebulisation. Section 2.9.44—European Pharmacopeia, Council of Europe, 67075 Strasbourg, France
69. Dolovich M, Rhem R (1998) Impact of oropharyngeal deposition on inhaled dose. J Aerosol Med S1:S112–S121
70. Harris D, Chaudhry S, Chaudry I, Li S, Sequeira J, Wyka B (1996) Influence of entry-port design on drug deposition in cascade-impactor from metered-dose inhalers. AAPS Annual Meeting. AAPS Poster Session, 1996
71. Van Oort M, Downey B (1996) Cascade impaction of MDIs and DPIs: proposal of induction port, inlet cone, and pre-separator lid designs for inclusion in general chapter <601> Pharm Forum 22(2):2204–2210
72. US Pharmacopeial Convention (2010) Chapter 1601: Products for nebulization—characterization tests. In Process Revision, Pharm Forum 36(2):534–538
73. Keil JC, Reshima K, Peart J (2006) Using and interpreting aerosol electrostatic data from the electrical low pressure impactor. In: Dalby RN, Byron PR, Peart J, Suman JD, Farr SJ (eds) Respiratory drug delivery-2006. Davis Healthcare International Publishing, River Grove, IL, pp 267–277
74. Peart J, Orban JC, McGlynn P, Redmon M, Sargeant CM, Byron PR (2002) MDI electrostatics: valve and formulation interactions which really make a difference. In: Dalby RN, Byron PR, Peart J, Farr SJ (eds) Respiratory drug delivery-VIII. Davis Healthcare International Publishing, River Grove, IL, pp 223–230
75. Bagger-Jörgensen H, Sandell D, Lundbäck H, Sundahl M (2005) Effect of inherent variability of inhalation products on impactor mass balance limits. J Aerosol Med 18(4):367–378

76. Product Quality Research Institute (2012) Information about PQRI as well as reports of WG meetings can be found at: http://www.pqri.org/. Visited 10 Jan 2012
77. Mitchell JP, Bauer R, Lyapustina S, Tougas T, Glaab V (2011) Non-impactor-based methods for sizing of aerosols emitted from orally inhaled and nasal drug products (OINDPs). AAPS PharmSciTech 12(3):965–988
78. Mitchell JP, Dunbar C (2005) Analysis of cascade impactor mass distributions. J Aerosol Med 18(4):439–451
79. Christopher JD, Dey M, Lyapustina S, Mitchell JP, Tougas TP, Van Oort M, Strickland H, Wyka B (2010) Generalized simplified approaches for mass median aerodynamic determination. Pharm Forum 36(3):812–823
80. Mitchell JP, Nagel MW (2000) Spacer and holding chamber testing *in vitro*: a critical analysis with examples. In: Dalby RN, Byron PR, Farr SJ, Peart J (eds) Respiratory drug delivery-VII. Serentec Press, Raleigh, NC, pp 265–274
81. Lefebvre AH (1989) Atomization and sprays. Taylor and Francis, Hemisphere Publishing Corporation, New York, NY
82. Lewis DA (2008) New cascade impactor software. Inhalation 2(4):7–10
83. Hinds WC (1998) Properties, behavior, and measurement of airborne particles, 2nd edn. Wiley-Interscience, New York, NY, pp 75–110

Chapter 5
The AIM and EDA Concepts: Why They Are Needed and How They Fit Together

Jolyon P. Mitchell and Terrence P. Tougas

Abstract AIM and EDA concepts were developed to address the high variability and susceptibility to error of the conventional full-resolution CI methods in OIP quality control. Abbreviated measurements allow increased throughput during product development and routine quality control testing. Efficient data analysis simplifies data analysis (by using only two metrics for making decisions about presence or absence of APSD changes while in most cases, reducing the rate of false-positive and false-negative decisions). Each of these aspects is explored in more detail in later chapters. This chapter presents an overall rationale for the development of these alternative approaches, with emphasis placed on how they could fit into the OIP life cycle.

5.1 Current Experience with CI Measurements: The Need for Strict Controls

The previous chapter highlighted the fact that full-resolution multistage CI measurements, when carried out to best current practices, are labor-intensive, involving numerous steps with the possibility of error likely even when stringent precautions are taken [1]. In 2003, an across-industry survey of CI users was undertaken by the Particle Size Distribution Mass Balance Working Group of the Product Quality Research Institute (PQRI) to assess the frequency of CI-based

J.P. Mitchell (✉)
Trudell Medical International, 725 Third Street, London, ON N5V 5G4, Canada
e-mail: jmitchell@trudellmed.com

T.P. Tougas
Boehringer Ingelheim Pharmaceuticals Inc., 900 Ridgebury Road, Ridgefield, CT, USA 06877-0368
e-mail: terrence.tougas@boehringer-ingelheim.com

T.P. Tougas et al. (eds.), *Good Cascade Impactor Practices, AIM and EDA for Orally Inhaled Products*, DOI 10.1007/978-1-4614-6296-5_5,

measurement failures [2]. Replies were received from 14 different organizations representing both small and large pharmaceutical companies. There were 21 instances of simultaneous mass balance (MB) with APSD failure, 10 of which were assignable (9 to product, 1 to analyst); 261 instances of APSD failure with acceptable MB, of which 71 were assignable (65 to impactor, 6 to analyst); and 33 instances of MB failure with acceptable APSD, of which 19 were assignable (5 to product, 14 to analyst). Although not a large number of failures, considering that 4,300 individual CI measurements were represented, each instance would have resulted in an out-of-specification investigation with associated inconvenience, delay, and possible rejection of *good* product. In essence, this survey confirmed that the multistage CI method is over-complex for the purpose of inhaler quality control (QC). Alternative approaches are therefore necessary to mitigate or preferably avoid altogether errors of the sort identified by this survey. The ultimate goal for the CI method is to be able to develop a methodology that is as sensitive as possible to detect APSD shifts that are truly due to the product and not to some other confounding cause.

5.2 AIM and EDA: A *Road Map*

It is worthwhile reviewing the context in which CI-based measurements are acquired and assessed before discussing the details of how AIM and EDA concepts can assist in the simplification of the process of determining OIP aerosol metrics that are meaningful as descriptors for either quality of the product or for likely deposition behavior in the HRT. The relationships between EDA metrics and their determination methods, as well as the primary purpose and method of determination of each pertinent CI-derived metric, were introduced in Chap. 1. This material is built upon by Fig. 5.1, which contains the *road map* for the process of data collection and analysis as a whole. It begins with presenting the underlying rationale for making OIP aerosol particle-size measurements (A). It then moves to the mechanics of the measurement process (B), followed by the establishment of the nature of the APSD as a continuous multivariate function between mass of API and aerodynamic diameter that is constructed from the raw data of mass of API per size-fractionating stage in the CI (C). If the APSD is unimodal, as most OIP aerosols are, its *MMAD*, as the prime measure of central tendency, is readily derived from the size at which the cumulative mass-weighted APSD reaches 50% (D). Furthermore, *GSD* can be used as a descriptor of spread between 15.9 and 84.1 mass percentile values, if the APSD is lognormal in shape (or close to this mathematical description) (D). The map continues by introducing the various metrics that are derived mathematically from the continuous form of the cumulative mass-weighted APSD (E), distinguishing them from those metrics that can be obtained directly from the raw data (F). These metrics as a whole provide a framework for assessing OIP aerosol performance changes throughout the product life cycle, as will be seen in Chap. 6.

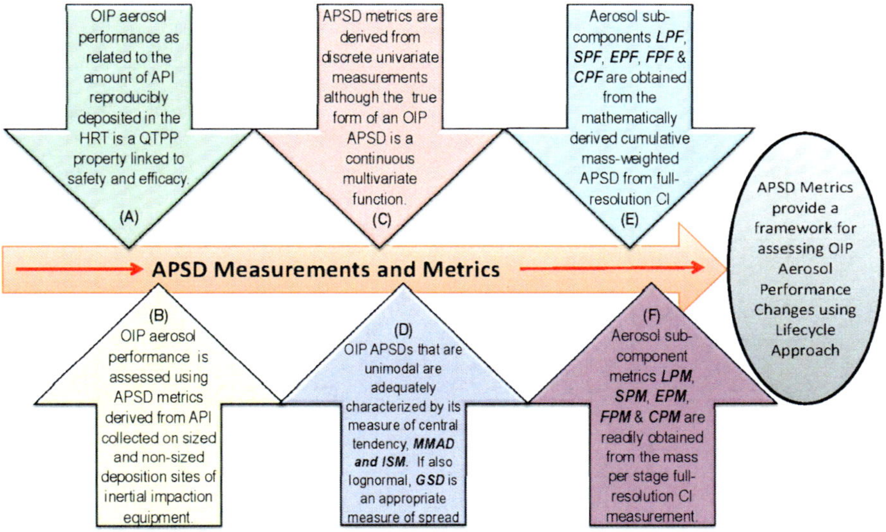

Fig. 5.1 OIP aerosol performance measurements and metrics

5.3 How an AIM-Based Method May Help Simplify the Process of Determining Metrics Related to OIP APSD

Ideally, a measurement with the relative simplicity of the dose content uniformity apparatus would be the perfect solution to the problem of reducing the complexity of the cascade impactor-based measurement process.

In practice, however, at the minimum, it is likely to be necessary to distinguish between the fine mass fraction of particles that have the potential to carry API(s) to target receptors in the respiratory tract where a therapeutic benefit may be obtained [3–7] and the coarse mass fraction that will likely not penetrate much beyond the oropharyngeal region [8, 9]. There may even be a requirement to quantify the extra-fine fraction less than ca. 1 μm aerodynamic diameter; as such particles are prone to being exhaled without depositing in the lungs, due to their lack of susceptibility to the forces governing their movement (predominantly Brownian diffusion) from the air stream to sites of deposition [10].

Early work with a variety of two-stage impaction systems in the 1980s paved the way towards simplicity in distinguishing solely the coarse from fine particle mass fractions of OIP-generated aerosols [11–13]. However, in 1992, Miller et al. pointed out that the Twin Impinger (and by implication, other simple two-portion classifiers) would be incapable of distinguishing unimodal and lognormal APSDs with particular *MMAD* and *GSD* combinations based on the size selectivity and location of the collection efficiency curve defining the divide between the two fractions [14]. This study was important because it provided a timely warning to those intending to size-characterize OIP aerosols using such equipment and likely influenced the

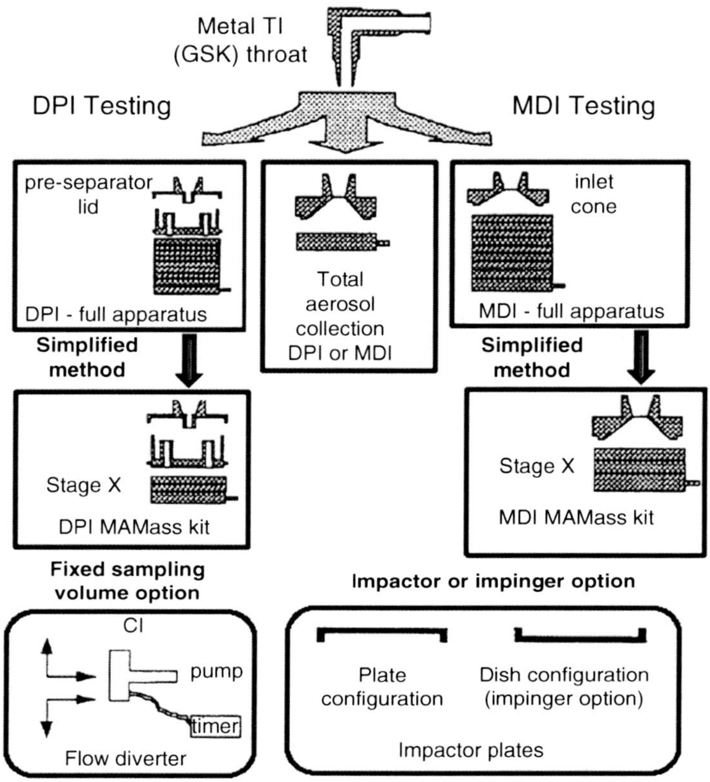

Fig. 5.2 Abbreviated impactor measurement approach to DPI and MDI testing pioneered by Van Oort and Roberts *(adapted from ref [15]—used with permission)*

thinking of the regulatory agencies, particularly in the USA. However, they did not go further by recommending that a two-stage size fractionator might be used in conjunction with a full-resolution CI system that could be used to verify that the simplified measurement system was sensitive to the changes in APSD of relevance to a particular OIP.

Pioneering work by Van Oort and Roberts [15] in the mid-1990s pointed the way forward in terms of setting out a hierarchy of reduced stack Andersen cascade impactor (ACI) measurements, supported by the full-resolution eight-stage system (Fig. 5.2). However, the regulatory approach at that time favored the inclusion of at least five stages with cut-point sizes between 0.5 and 5.0 μm aerodynamic diameter, and this requirement was a significant contributor to the design of the NGI [16].

In the mid-2000s, the abbreviated impactor measurement (AIM) concept and related efficient data analysis (EDA) approach were simultaneously developed out of the need to reduce method complexity. In addition, there was an increasing recognition by stakeholders involved with the regulatory process for OIPs that

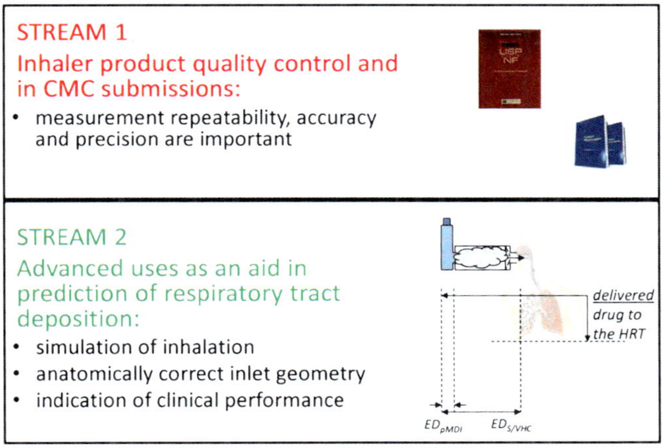

Fig. 5.3 Two streams for laboratory measurements of OIP aerosol performance

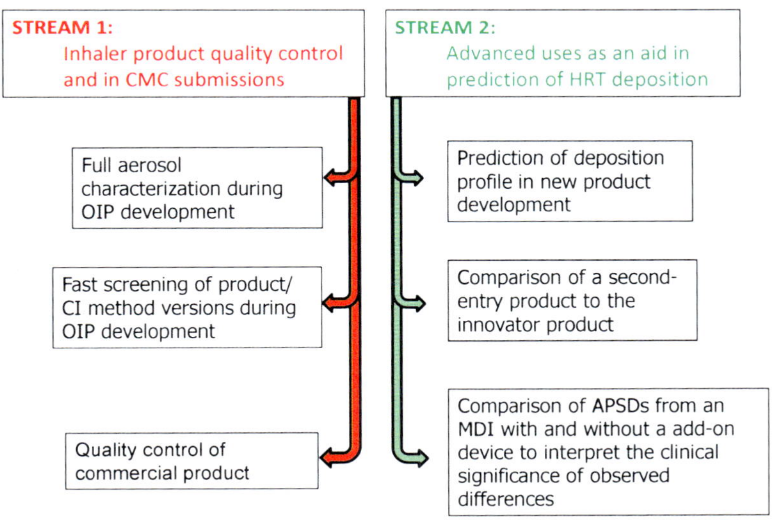

Fig. 5.4 Purposes for making CI-based measurements in either OIP quality control or to predict HRT deposition

the multistage CI, although adequate for size-fractionating aerosol particles, is not an analogue of the particle deposition processes that take place in the HRT [17] (see Chap. 2).

Early on in the process of developing the two concepts, an understanding developed that there are two fundamentally different reasons why OIPs are evaluated in the laboratory (Fig. 5.3). Each stream can be further subdivided into a series of different purposes (Fig. 5.4).

The description of the approaches taken to obtain the most appropriate in vitro OIP performance metrics for these different purposes forms the main topic for the remainder of the handbook. At this introductory stage, though, it is important to appreciate that the provision to stakeholder confidence that the inhaler will deliver medication safely and efficaciously is common to all these types of laboratory evaluation.

5.4 AIM Applied in OIP Quality Assessment

Tougas et al. have claimed that an approach involving an abbreviated CI measurement has the potential to offer at least equivalent and possibly greater sensitivity than that achievable by multistage impactor in the OIP quality control environment [18]. In the so-called AIM–QC approach, they recommended that the process of measurement be simplified to its maximum extent (Fig. 5.5), namely, the determination of the size fractions related to large (*LPM*) and small particle mass (*SPM*) that are sensitive to shifts in the APSD.

In the example shown, the size boundary between small and large mass fractions does not necessarily have to be fixed at a physiologically relevant particle size, such as 5 μm, that is defined as the bound between fine and coarse fractions in the European Pharmacopoeia [19]. In fundamental terms, changes in a mass-weighted APSD obtained from full-resolution CI measurements can be reduced to those associated with position of the mass distribution profile on the abscissa (size) scale and with its area under the curve or amplitude position of the mode(s) on the ordinate (mass) scale [18]. Tougas et al. showed that two metrics, namely, the ratio, *LPM*/*SPM* and the sum, *LPM* + *SPM*, are foundational to the EDA concept [18], explained in more detail in Chap. 6. It is important to note that the sum, *LPM*+*SPM*, is identical with the impactor-sized mass (*ISM*) and is the total mass of API collected by size-fractionating stages of the CI that have a defined upper bound size limit.

An essential aspect of the EDA approach is that it is as applicable to full-resolution CI data as to the assessment of results from an abbreviated system.

In addition to reducing measurement complexity, it was realized that there is the potential to undertake more determinations in a given time period of a batch being QC tested, thus achieving greater coverage, associated with increased statistical power. This benefit, of course, only applies if the time saved is allocated to making more replicate determinations.

Importantly, due to elimination of confounding variables, EDA-based measures of OIP aerosol APSD should be less prone to method and analyst error (which may lead to wrong decision-making) [20]. This benefit of EDA is expanded upon in Chap. 7.

As well as its more obvious use in OIP QC, the AIM–EDA approach should also be an excellent tool in exploratory stability studies which are performed at the early product and formulation development stages. This aspect is discussed further in Chap. 6, and some case studies are presented in Chap. 9. Since the identification of

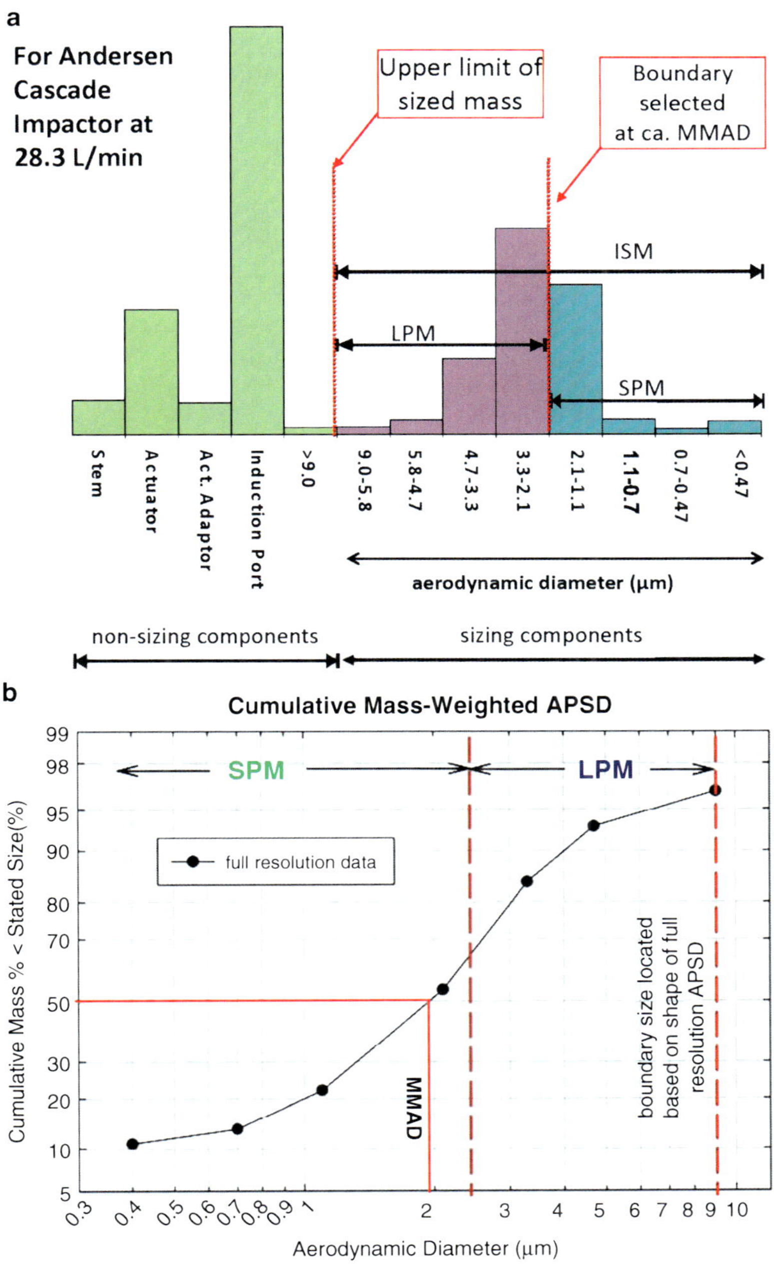

Fig. 5.5 Nomenclature for a hypothetical abbreviated impactor based on the Andersen nonviable CI operated at 28.3 L/min for potential OIP QC application (AIM–QC system). (**a**) Mass of API by component. (**b**) Cumulative API mass-weighted APSD

changes relating to OIP APSD as a function of elapsed time relative to the properties measured at stability start is the main objective, a well-designed stability study involving rapid AIM–EDA-based assessments offers the potential as a fit-for-purpose methodology in this application. Finally, the ability to make measurements in a given time that are more useful in the batch disposition decision-making process than the determination of equivalent metrics from fewer full-resolution CI measurements could be beneficial in the context of defining design and control spaces in the quality-by-design paradigm [21, 22] that is currently being encouraged by regulatory agencies [23].

As a significant by-product, AIM-based methods, by virtue of their relative simplicity, have the potential to improve product batch disposition decision-making by allowing more samples from the lot (coverage) to be evaluated in a given time period and, importantly, by improving the signal-to-noise ratio, thanks to EDA metrics [24]. AIM-based methods also overcome the problem that for products with narrow particle-size distributions (small *GSD*), several stages of full-resolution CIs capture little or no active pharmaceutical ingredient (API), reducing the overall precision of the method [25]. Furthermore, Tougas et al. observed that there is the potential that accurate and precise measurements may be made possible with the clinical dose (typically 1–2 actuations of the inhaler), rather than after several actuations, as is typical with full-resolution systems to acquire sufficient drug deposits on stages collecting particles at the periphery of the APSD to permit acceptably accurate and precise assay of the API collected at these locations [18]. However, the variability will be altered when the dose number changes, so this must be handled with care.

In summary, moving away from full-resolution CI measurements and associated stage groupings towards product-specific EDA metrics for routine quality control, Tougas et al. observed that the latter approach has the following advantages [24]:

1. Easier operation of an AIM system.
2. Similar sensitivity to APSD changes compared to current methods.
3. Fewer false-positive results.
4. More sensitive measures are available for detecting shifts in position and amplitude (area under the curve) of an APSD, leading to better diagnostic capability and predictability.
5. Fewer inhaler actuations per CI measurement are possible due to the acquisition of sufficient mass in fewer subfractions, which has the potential to reduce errors, experimental uncertainty, and makes it potentially possible to test APSD with the prescribed dose, at least for moderate- and low-potency formulations.
6. Less time is required per CI measurement, making it possible to design sufficiently powerful experiments for assessing product and CI method variability on a sound statistical basis.

These advantages, together with others that have become apparent since the original assessment, are summarized in Table 5.1. However, since the article by Tougas et al. was published, it has become evident that many pharmaceutical companies are also examining the potential for AIM-based measurements for rapid screening of

Table 5.1 Advantages of AIM and EDA approaches

AIM	EDA with AIM	EDA with full-resolution CI
Simpler apparatus	Similar sensitivity to APSD changes compared to current methods	Complete APSD—useful in diagnosis in event of OOS/OOT with AIM system
Fewer samples for assay	More sensitive measures of APSD shifts than with stage groupings from full-resolution CI	
Greater mass of API per sample than with single stages in full-resolution CI, therefore improved sensitivity per inhaler actuation	Potential for fewer false-positive results in batch release	
Less recovery solvent volume (*Green Chemistry* compatible)		
Less time to make measurement	Less time per measurement in data processing with potential for more powerful experiment designs improving *coverage* of the batch	

potential candidate formulations. Chapter 10 contains examples of many such studies. The general observation is that AIM-based measurements can be successfully made, provided that certain precautions, described in detail in Chap. 10, are taken. The ability to make more rapid measurements is a distinct advantage, even though such preliminary work needs later to be supported by full-resolution CI data for candidate products that pass the screening process.

5.5 AIM Applied in Research and Development as a Tool for Rapid Assessment of Likely Particle Deposition Behavior in the HRT

In the research and development environment, where the focus of attention is shifted towards predicting the likely clinical response related to particle deposition behavior in the lungs, the process of measurement is still simplified with respect to that associated with the full-resolution CI, but an additional stage is added to enable the proportion of extra-fine particles <ca. 1 μm aerodynamic diameter to be determined (Fig. 5.6).

In Fig. 5.6, the quantity *IM* refers to the impactor-collected mass, which differs slightly from *ISM*, in that the mass on the first stage of the CI is included, even if its upper size limit is undefined, since the purpose here is to quantify the total mass of API that penetrated beyond the induction port that serves to model the oropharynx. Terms *Ex-ACT* and *Ex-MVM* refer to the total mass of API ex actuator mouthpiece

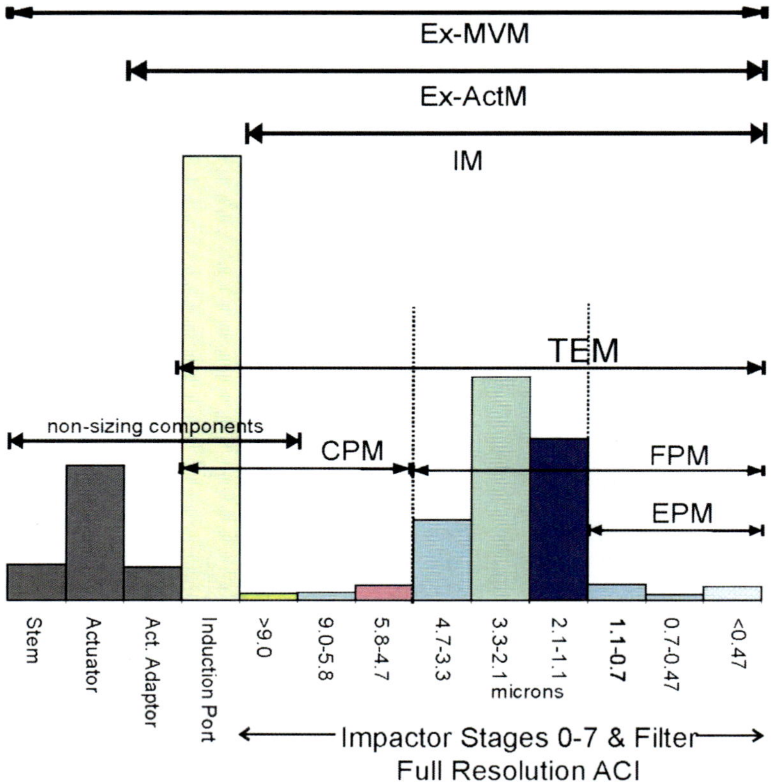

Fig. 5.6 Nomenclature for a hypothetical abbreviated impactor system based on the Andersen CI operated at 28.3 L/min for potential HRT application (AIM–pHRT system)

and total mass of API ex metering valve, in this example that is based on the evaluation of a pMDI, where it is necessary to distinguish between the two quantities, since an add-on device may be interposed between the inhaler mouthpiece and the entry to the CI system. It is only necessary to define the total mass ex inhaler mouthpiece for other types of OIP.

In the so-called AIM–pHRT apparatus, the boundary size delineating fine from coarse mass fractions is fixed at 5 μm aerodynamic diameter, in accordance with monographs 2.9.18 and 2.9.44 of the Ph. Eur. [19, 26] (Fig. 5.6). It is important to note that the standard Ph. Eur./USP induction port may be replaced by an idealized upper airway, such as that developed by Finlay and colleagues [27], or an anatomically accurate model airway cast to add further realism in terms of the fluid dynamics associated with aerosol entry from the inhaler into the measurement system [28]. The development of AIM–pHRT systems is described in Chap. 12, where it will be seen that prototype systems have been successfully evaluated side by side with a full-resolution CI as the reference technique.

5.6 Selection of an AIM System

In practical terms, when selecting an appropriate technique, it is important to understand that the AIM concept is not confined to one particular configuration of impactor (Fig. 5.7). Instead, many options exist, from potentially the Twin Impinger to reduced versions of the Andersen cascade impactor (ACI), semiautomated ACI, fast-screening impactor (FSI), and modified versions of the NGI. The choice of AIM platform should ultimately depend on the familiarity of the testing laboratory with the equipment and its limitations, the preference for (semi)automation, as well as the nature of the product being evaluated (MDI with or without add-on device, DPI, or nebulizing system), as the latter will dictate important operating variables, in particular flow rate through the system.

The adoption of AIM together with EDA concepts by regulatory agencies and eventually the pharmacopeial compendia will depend upon the assembly of both experimental and theoretical evidence that can be developed in support of a paradigm shift from multistage CI-based measurements.

Work continues to understand and characterize AIM and EDA approaches [29, 30]. AIM and EDA topics were presented and extensively discussed in a workshop held by the USP in December 2011.

In Europe, the Inhalanda Committee that has oversight of the aerosol-based monographs of the European Pharmacopoeia took on the AIM concept as a new work item in 2011 and is seeking high-quality validated data sets with currently marketed OIPs to support moving forward with this concept. In the long term, it is to be hoped that both

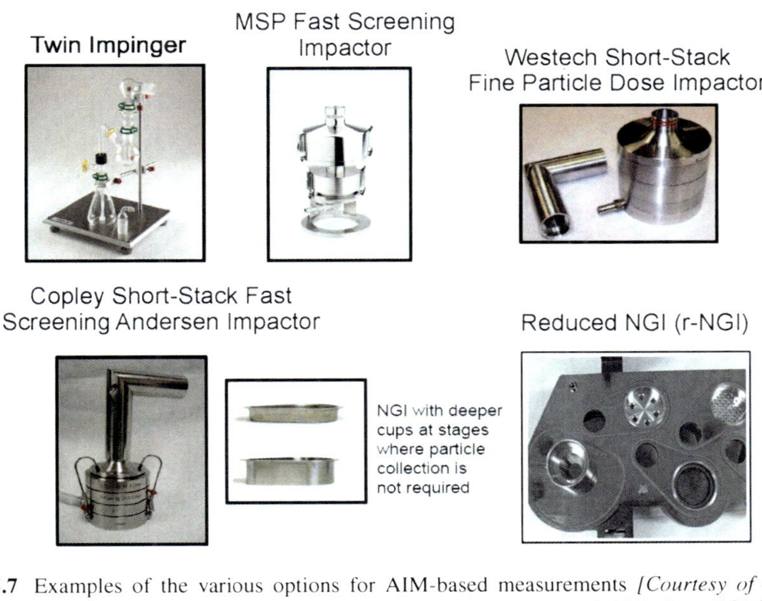

Fig. 5.7 Examples of the various options for AIM-based measurements *[Courtesy of Copley Scientific Ltd., Westech Instrument Services Ltd., MSP Corporation, and GSK plc (r-NGI)]*

European and United States pharmacopeias can harmonize their approaches, certainly with respect to the adoption of AIM approaches, even if the acceptance of EDA as an alternative to current European practice takes longer to achieve. Regulatory and compendial aspects associated with AIM and EDA are discussed in more detail in Chap. 11.

5.7 The Right Approach for the Task in Hand

This chapter has presented what could be seen as a bewildering set of options for those contemplating AIM and/or EDA implementation, so Table 5.2 contains guidance developed by Mitchell et al. [31] on questions that should be addressed at the outset when considering these concepts within the OIP life cycle.

Table 5.3 then addresses the practicalities that follow when responses to these questions have been made. A detailed assessment of the approaches that are potentially available during the OIP life cycle is provided in Chap. 6.

Table 5.2 Applications for full-resolution and AIM-based CI measurements in the OIP product life cycle: questions

	APSD measured for		
	Product characterization and development	Product quality control	In vitro assessment for equivalence comparisons
What stage in OIP life cycle?	Product under development (either brand name or generic)	Approved and understood product (either brand name or generic)	Generic OIP or brand-name OIP if needed for bridging studies during development or post-approval
What are the questions to be answered?	What is APSD (typical distribution, typical variability, etc.)? What factors affect the distribution? How does it change?	Is the APSD of a product from a given manufacturing run essentially the same as in the pivotal clinical trials?	Are there any clinically important differences between distributions from the two products being compared? *Note that the definition of a "clinically important difference in APSD" should be set, ideally, from clinical considerations*
How soon is the decision needed?	Within days of a test	Within hours of a test	Within days of a test
How frequently is the test conducted?	As needed while the product is under development	After each manufacturing run and on stability	As needed while the product is under development. Typically many tests over several months are needed
For how long are the tests conducted?	Typically many tests over several months are needed	During entire commercial phase	Typically many tests over several months are needed
What is the goal for these tests?	Difference between different versions of a candidate product	Disposition of a given batch	Reject null hypothesis that two products (test and reference) have different APSDs

Table 5.3 Applications for full-resolution and AIM-based CI measurements in the OIP product life cycle: implementation

	APSD measured for		
	Product characterization and development	Product quality control	In vitro assessment for equivalence comparisons
Impactor system to use	Full-resolution CI for initial screening candidate formulations, then AIM–QC on short-listed candidates	AIM–QC, supported by full-resolution CI (e.g., in the context of OOS evaluation)	Full resolution or AIM–pHRT With an anatomically correct or idealized inlet possible
Proposed metrics to use	Initially, *EPM*, *FPM*, and *CPM* with size ranges related to likely deposition in the HRT; later, EDA metrics to develop QC specification	EDA metrics, namely, *SPM* and *LPM* with boundary near to *MMAD*: *LPM*/*SPM* and *ISM* (*SPM* + *LPM*)	Full-profile comparisons or *EPM*, *FPM*, and *CPM* with size ranges related to likely deposition in the HRT
Statistical approaches	A number of approaches related to the APSD characterization and the development of the product specification	Tests to detect significant changes in APSD	Generic statistical equivalence testing

The following considerations also apply:

1. The full-resolution CI, ideally an apparatus of the same type as the abbreviated system, is always present as the reference apparatus, in the event that discrepancies between full- and AIM-based systems arise.
2. EDA is as applicable to full resolution as to abbreviated CI data.
3. The implementation of AIM or EDA approaches will be more likely to be successful when discussed in advance with the appropriate regulatory agencies for the OIP, and where a good correlation between AIM and full-resolution CI results can be demonstrated.

Finally, the choice of AIM platform should ultimately depend on the familiarity of the testing laboratory with the equipment and its limitations, the preference for (semi)automation, as well as the nature of the product being evaluated (MDI with or without add-on device, DPI, or nebulizing system), as the latter will dictate important operating variables, in particular flow rate through the system.

References

1. Christopher D, Curry P, Doub W, Furnkranz K, Lavery M, Lin K, Lypapustina S, Mitchell J, Rogers B, Strickland H, Tougas T, Tsong Y, Wyka B (2003) Considerations for the development and practice of cascade impaction testing including a mass balance failure investigation tree. J Aerosol Med 16(3):235–247

2. Mitchell JP (2003) Regarding the development and practice of cascade impaction testing, including a mass balance failure investigation tree. J Aerosol Med 16(4):443
3. Zanen P, Go LT, Lammers JWJ (1994) The optimal particle size for beta-adrenergic aerosols in mild asthmatics. Int J Pharm 107(3):211–217
4. Zanen P, Go LT, Lammers JWJ (1995) The optimal particle size for parasympathicolytic aerosols in mild asthmatics. Int J Pharm 114(1):111–115
5. Zanen P, Go LT, Lammers JWT (1996) Optimal particle size for beta-agonist and anticholinergic aerosols in patients with severe airflow limitation. Thorax 51(10):977–980
6. Usmani OS, Biddiscombe MF, Nightingale JA, Underwood SR, Barnes PJ (2003) Effects of bronchodilator particle size in asthmatic patients using monodisperse aerosols. J Appl Physiol 95(5):2106–2112
7. Usmani OS, Biddiscombe MF, Barnes PJ (2005) Regional lung deposition and bronchodilator response as a function of beta-2 agonist particle size. Am J Respir Crit Care Med 172(12):1497–1504
8. Roland NJ, Bhalla RK, Earis J (2004) The local side effects of inhaled corticosteroids: current understanding and review of the literature. Chest 126(1):213–219
9. Dubus JC, Marguet C, Deschildre A, Mely L, Le Roux P, Brouard J, Huiart L (2001) Local side effects of inhaled corticosteroids in asthmatic children: influence of drug, dose, age and device. Allergy 56(10):944–948
10. Laube B, Janssens HM, de Jongh FHC, Devadason SG, Dhand R, Diot P, Everard ML, Horvath I, Navalesi P, Voshaar T, Chrystyn H (2011) What the pulmonary specialist should know about the new inhalation therapies. Eur Respir J 37(6):1308–1331
11. Hallworth GW, Westmoreland DG (1987) The twin impinger: a simple device for assessing the delivery of drugs from metered dose pressurized aerosol inhalers. J Pharm Pharmacol 39(12):966–972
12. Vidgren M, Silvasti M, Vidgren P, Sormunen H, Laurikaininen K, Korhonen P (1995) Easyhaler® multiple dose powder inhaler: practical and effective alternative to the pressurized MDI. Aerosol Sci Technol 22(4):335–345
13. Geuns ERM, Toren JS, Barends DM, Bult A (1997) Decrease of the stage-2 deposition in the twin impinger during storage of beclomethasone dipropionate dry powder inhalers in controlled and uncontrolled humidities. Eur J Pharm Biopharm 44(2):187–194
14. Miller NC, Marple VA, Schultz RK, Poon WS (1992) Assessment of the twin impinger for size measurement of metered-dose inhaler sprays. Pharm Res 9(9):1123–1127
15. Van Oort M, Roberts W (1996) Variable flow-variable stage-variable volume strategy for cascade impaction testing of inhalation aerosols. In: Dalby RN, Byron PR, Farr SJ (eds) Respiratory drug delivery V. Interpharm, Buffalo Grove, IL, pp 418–420
16. Marple VA, Roberts DL, Romay FJ, Miller NC, Truman KG, Van Oort M, Olsson B, Holroyd MJ, Mitchell JP, Hochrainer D (2003) Next generation pharmaceutical impactor (a new impactor for pharmaceutical inhaler testing) – Part 1: Design. J Aerosol Med 16(3):283–299
17. Mitchell JP, Dunbar C (2005) The interpretation of data from cascade impactors. J Aerosol Med 18(4):439–451
18. Tougas TP, Christopher D, Mitchell JP, Strickland H, Wyka B, Van Oort M, Lyapustina S (2009) Improved quality control metrics for cascade impaction measurements of orally inhaled drug products (OIPs). AAPS PharmSciTech 10(4):1276–1285
19. European Directorate for the Quality of Medicines and Healthcare (EDQM) (2012) Preparations for inhalation: aerodynamic assessment of fine particles. Section 2.9.18 – European Pharmacopeia [Apparatus B in versions up to 4th Edn. 2002], Council of Europe, Strasbourg, France
20. Christopher D, Dey M (2011) Detecting differences in APSD: efficient data analysis (EDA) vs. stage grouping. In: Dalby RN, Byron PR, Peart J, Suman JD, Young PM (eds) Respiratory drug delivery Europe 2011. Davis Healthcare International, River Grove, IL, pp 215–224
21. Yu L (2006) Implementation of quality-by-design: OGD initiatives, FDA Advisory Committee for Pharmaceutical Science Presentation (October 2006). Available at http://www.fda.gov/ohrms/dockets/ac/06/slides/2006-4241s1_8.ppt. Accessed 10 Jan 2012

22. Nasr MM (2006) Implementation of quality by design (QbD): status, challenges and next steps, FDA Advisory Committee for Pharmaceutical Science Presentation (October 2006). Available at http://www.fda.gov/ohrms/dockets/ac/06/slides/2006-4241s1_6.ppt. Accessed 10 Jan 2012
23. Mitchell JP, Copley M (2010) Accelerating inhaled product testing. Pharma Mag Jan/Feb:22–25. Available at http://www.copleyscientific.com/documents/ww/COP%20JOB%20098_Accelerating%20inhaled%20product%20testing.pdf. Accessed 5 Aug 2012
24. Tougas TP, Christopher D, Mitchell J, Lyapustina S, Van Oort M, Bauer R, Glaab V (2011) Product lifecycle approach to cascade impaction measurements. AAPS PharmSciTech 12(1):312–322
25. Bonam M, Christopher D, Cipolla D, Donovan B, Goodwin D, Holmes S, Lyapustina S, Mitchell J, Nichols S, Petterson G, Quale C, Rao N, Singh D, Tougas T, Van Oort M, Walther B, Wyka B (2008) Minimizing variability of cascade impaction measurements in inhalers and nebulizers. AAPS PharmSciTech 9(2):404–413
26. European Directorate for the Quality of Medicines and Healthcare (EDQM) (2012) Preparations for nebulisation: characterisation. Section 2.9.44 – European Pharmacopeia, Council of Europe, Strasbourg, France
27. Zhang Y, Finlay WH (2005) Experimental measurements of particle deposition in three proximal lung bifurcation models with an idealized mouth-throat. J Aerosol Med 18(4):460–473
28. Zhou Y, Sun J, Cheng Y-S (2011) Comparison of deposition in the USP and physical mouth-throat models with solid and liquid particles. J Aerosol Med Pulm Drug Deliv 24(6):277–284
29. Tougas T (2011) Efficient data analysis in quality assessment. In: Dalby RN, Byron PR, Peart J, Suman JD, Young PM (eds) RDD Europe-2011. Davis Healthcare International, River Grove, IL, pp 209–213
30. Mitchell J, Copley M (2010) EPAG-sponsored workshop on abbreviated impactor measurement (AIM) concept in inhaler testing: overview of AIM-EDA. Drug Delivery to the Lungs-21, The Aerosol Society, Edinburgh, UK, 21:370–373. Available at http://ddl-conference.org.uk/index.php?q=previous_conferences. Accessed 4 Aug 2012
31. Mitchell JP, Tougas T, Christopher JD, Lyapustina S, Glaab V (2012) The abbreviated impactor measurement and efficient data analysis concepts: why use them and when. In: Dalby RN, Byron PR, Peart J, Suman JD, Young PM (eds) Respiratory drug delivery 2012. Davis Healthcare International, River Grove, IL, pp 731–736

Chapter 6
Product Life Cycle Approach to Cascade Impaction Measurements

Richard Bauer, J. David Christopher, Volker Glaab, Svetlana A. Lyapustina, Jolyon P. Mitchell, and Terrence P. Tougas

Abstract Over the OIP life cycle (from development to commercial production, to the development of generic/follow-on products), APSD measurements are used extensively but for different purposes. The analytical and statistical approaches to the measurements and data analyses at these stages must therefore be different, depending on the specific questions pursued in a given situation. For some of those questions, full-resolution CIs are the instrument of choice. For others, an AIM system can provide all the needed information and support decision-making. This chapter describes how the utilization of specific measurement approaches changes

R. Bauer
MannKind Corporation, Aerosol Analysis, One Casper Street, Danbury, CT 06810, USA
e-mail: rbauer@mannkindcorp.com

J.D. Christopher
Merck Research Laboratories, Nonclinical and Pharmaceutical Sciences Statistics, 770 Sumneytown Pike, WP37C-305, West Point, PA 19486-0004, USA
e-mail: j.david.christopher@merck.com

V. Glaab
Boehringer Ingelheim, Respiratory Drug Delivery, Binger Strasse 173, Ingelheim am Rhein 55216, Germany
e-mail: volker.glaab@boehringer-ingelheim.com

S.A. Lyapustina, Ph.D.
Drinker Biddle & Reath LLP, 1500 K Street NW, Washington, DC 20005-1209, USA
e-mail: svetlana.lyapustina@dbr.com

J.P. Mitchell
Trudell Medical International, 725 Third Street, London, ON N5V 5G4, Canada
e-mail: jmitchell@trudellmed.com

T.P. Tougas (✉)
Boehringer Ingelheim Pharmaceuticals Inc., 900 Ridgebury Road, Ridgefield, CT 06877-0368, USA
e-mail: terrence.tougas@boehringer-ingelheim.com

T.P. Tougas et al. (eds.), *Good Cascade Impactor Practices, AIM and EDA for Orally Inhaled Products*, DOI 10.1007/978-1-4614-6296-5_6,

over the product life cycle, and how the entire body of APSD data is interlinked to enable those transitions (e.g., by establishing correlations or by establishing *typical* profile parameters), thereby enhancing product knowledge, appropriate control, and comparisons between innovator and generic OIPs.

6.1 Choosing an AIM System Suitable for Purpose

The reasons for making APSD measurements vary throughout the product life cycle, as has been described in detail in Chap. 5, and a summary of the goals and recommended CI system for different life cycle stages was presented in Table 5.2. These considerations include discovery and screening of early candidate formulations, followed by development and characterization of lead candidates, introduction of the technique for QC in commercial production, and finally, demonstration of in vitro equivalence for modified versions of the original OIP or follow-on (generic) copies of the product. At each of the life cycle stages, a suitable APSD measurement system, together with data analysis procedure, must be employed, while appropriate continuity should be built in, to allow the justification of, and to enable switches between, different systems.

The present chapter addresses the considerations for choosing an AIM-based approach, recognizing at the outset that such a system would be used to augment and not replace the need for some APSD measurements by full-resolution CI. There are many different AIM-based configurations that are currently available, some of which are available "off the shelf," such as the FSA (Copley Instruments Ltd, Nottingham, UK), FPD (Westech Instruments Services, Upper Stondon, Beds., UK), and the FSI (MSP Corp., St Paul, MN, USA). Other systems must be constructed from existing components of the full-resolution system, as can be done with the ACI, or by relocating stages within the existing flow path, as is possible with the NGI. Even when an off-the-shelf apparatus is selected, there are still further considerations to be made before use, such as the selection of the most appropriate cut-point size(s) for the stage(s) in the reduced system. These, and other related decisions, are the topic of the remainder of the chapter.

6.2 Selecting Size Ranges for an AIM-Based Apparatus

When selecting an AIM apparatus, OIP-appropriate size ranges should be chosen, based on the full-resolution APSD data that are already available for the product, and to the extent possible, be directed by the clinical importance of specific size ranges of that product [1]. If an AIM-based system is being considered for candidate formulation screening in early stage OIP development, it may be necessary to experiment with more than one configuration, depending upon the expected shifts in APSD that may be encountered in such preliminary work.

For an AIM–QC system, where the goal is to maximize sensitivity to any APSD change (increased area under the curve, shift in *MMAD*, and/or change in shape of the APSD profile), the boundary between *LPM* and *SPM* should be chosen as near as possible to the *MMAD*. However, as will be seen in the following chapters on the EDA concept, precise alignment with the *MMAD* is not essential if the choice of stage cut point available at the chosen flow rate for these measurements is restricted by the hardware availability. If there is a particular (e.g., clinical) need to control some other type of change encountered in a given product, the boundary could be placed elsewhere, as determined and justified during product development. One alternative boundary placement that could be considered is around 5 μm aerodynamic diameter, which would be in accordance with the current European pharmacopeial method [2], and therefore aligned with the appropriate EMA guidance [3].

Although the work by Tougas et al. has shown that the boundary between LPM and SPM can be anywhere between 0.3 and 3.0 times the *MMAD* value without sacrificing significant sensitivity to detect changes in *MMAD* [4], setting a single boundary may not work for all OIPs. In such circumstances (which would be for the product developer to decide, based on the evidence from full-resolution CI measurements), it may be necessary to introduce a second boundary to distinguish between fine and extra-fine particles. Under such circumstances, EDA could still be used to assess movements in *MMAD* based on the ratio of coarse to fine particle fractions (analogous to *LPM* and *SPM*), but the measure of extra-fine particle mass would be separate. Since extra-fine particle mass, by being a component of the fine particle fraction, and fine particle mass are linked measures, an EDA-type approach based on these two metrics would be inappropriate. However, in principle, it may be possible, by combining fine and coarse particle masses as a single metric termed, for convenience, super-micron mass, to compare extra-fine particle mass and super-micron mass by the EDA approach to follow changes in *MMAD*. However, to the best knowledge of the authors, this approach has not been done thus far, and caution is therefore advocated.

For an AIM–pHRT impactor, published research has focused on extra-fine and fine particle fractions defined as <1.1 μm and <4.7 or <5.0 μm aerodynamic diameter, respectively [5–8], which seem to be relevant in light of available clinical evidence that is discussed in Chap. 12. The specific sizes appropriate for a given product need to be considered based on the APSD profile of the product and the clinical importance of various size ranges for the active ingredient in question. The question of clinical relevance remains an area of active research with no single answer yet, so the sponsor is advised to review all pertinent current literature and sponsor's own studies to guide the selection of size ranges. Other considerations, such as the use of alternative induction port/throat geometries and breath simulation that are appropriate with this type of abbreviated system, are described in Chap. 12.

6.3 Qualifying AIM-Based Systems Against an Appropriate Full-Resolution CI

Regardless of the abbreviated or full-resolution systems that are chosen, there are common concerns, such as the magnitude of internal wall losses, particle bounce, and re-entrainment, all of which require attention as potential sources of bias, as

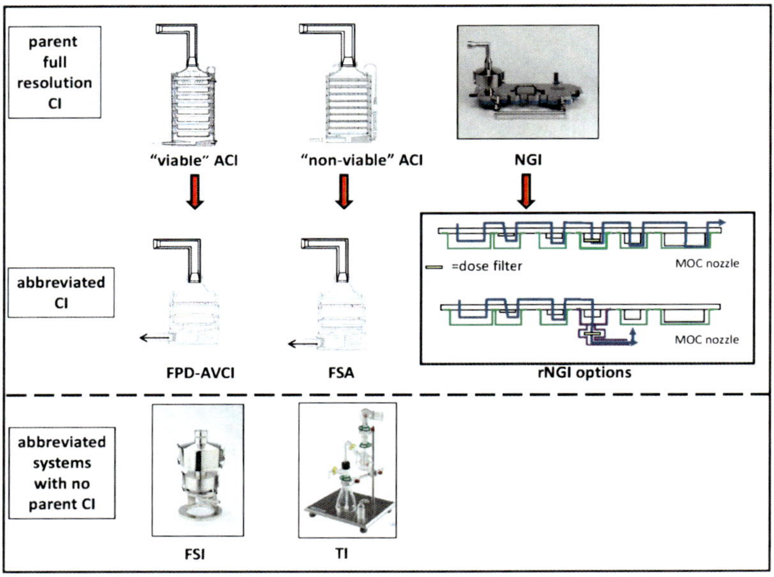

Fig. 6.1 AIM CIs in relation to parent full-resolution systems (*Courtesy Copley Scientific Ltd, MSP Corp., Astra Zeneca and Westech Instrument Services Ltd*)

have been reviewed in Chap. 4. Most of these considerations have already been explored with all types of abbreviated CIs, and outcomes of such experimental studies are reviewed in Chap. 10. This chapter should therefore be consulted before choosing an AIM-based approach.

Whether a sponsor selects their own size ranges or uses one or more of the ranges recommended in Sect. 6.3, the chosen size ranges need to be not only justified but also qualified against the only widely accepted *gold standard*, i.e., full-resolution CI data for the given product. This exercise will form a sound basis for future use of the chosen system by the sponsor and any potential troubleshooting or investigations of deviations. It may also alert the sponsor to underappreciated or overlooked factors influencing accuracy and precision of CI measurements.

The issue of which "parent" full-resolution CI to choose as the reference apparatus against which to validate measurements made by the chosen AIM-based system will in many cases be obvious (Fig. 6.1). For example, the designs of the fast-screening Andersen impactor (FSA) and the fine particle dose-abbreviated system (FPD-AVCI) are based on those of the nonviable and viable forms of the ACI, respectively (see Fig. 10.28 and associated explanatory text in Chap. 10 for a comparison between the two types of Andersen impactor). Likewise, the full-resolution NGI would be the obvious choice as reference CI, if modifying this system into an abbreviated apparatus. However, some abbreviated CIs, in particular the fast-screening impactor (FSI), have no obvious parent full-resolution impactor, as they were developed from the outset as AIM-based systems. Likewise, the twin impinger (TI) was an abbreviated system from its inception.

It is not strictly necessary to choose an abbreviated impactor that has a parent full-resolution system, since several groups have reported good results matching data obtained by the FSI with the NGI, as can be seen from the data reviewed in Chap. 10. However, in some instances (e.g., when using the rNGI as the abbreviated impactor configuration), it may be more convenient in terms of the ability to use the same components that have already been qualified in terms of stage mensuration. However, in this context, it is desirable, particularly in the case of DPI testing, to match apparatus-specific variables, such as the magnitude of the internal volume (also called *dead volume*), between abbreviated and full-resolution CI when there is no obvious parent system.

6.4 The Role of APSD in OIP Life Cycle Management

6.4.1 Overview

APSD testing is used for a variety of purposes during the life cycle of an OIP. In the various stages of development, the sponsor studies safety and efficacy of the product and establishes the target APSD with associated metrics and specifications.

By contrast, in commercial production, QC testing is meant to confirm whether the APSD is the same as that of the clinical batches. Because the intent of QC testing is not to repeat the extensive studies required for safety and efficacy studies, the most reasonable and practical goal of QC testing is to ascertain that the APSD is within the specifications established for the product for release of the clinical batches and during stability studies.

Developers of add-on devices for OIPs rely on the already established safety and efficacy profiles of the approved drug product. Therefore, their purpose of determining APSD-related data is related to the need to minimize the undesirable coarse particle mass that deposits in the oropharyngeal region while maintaining the amount of emitted fine particles ideally equivalent to that from the OIP device without add-on [6, 9]. Consequently, once the behavior of the add-on in this respect has been established by full-resolution CI, there is no need to reestablish or retest the entire detailed APSD profile in future measurements of in vitro performance.

6.4.2 Management Strategy

It is highly desirable to have a defined strategy for optimizing the use of the different variants of CI measurements throughout a product's life cycle. Figure 6.2 and the following subsections of 6.4.2 contain the outline developed by Tougas et al. [4], showing how such a management strategy might be developed where the need for

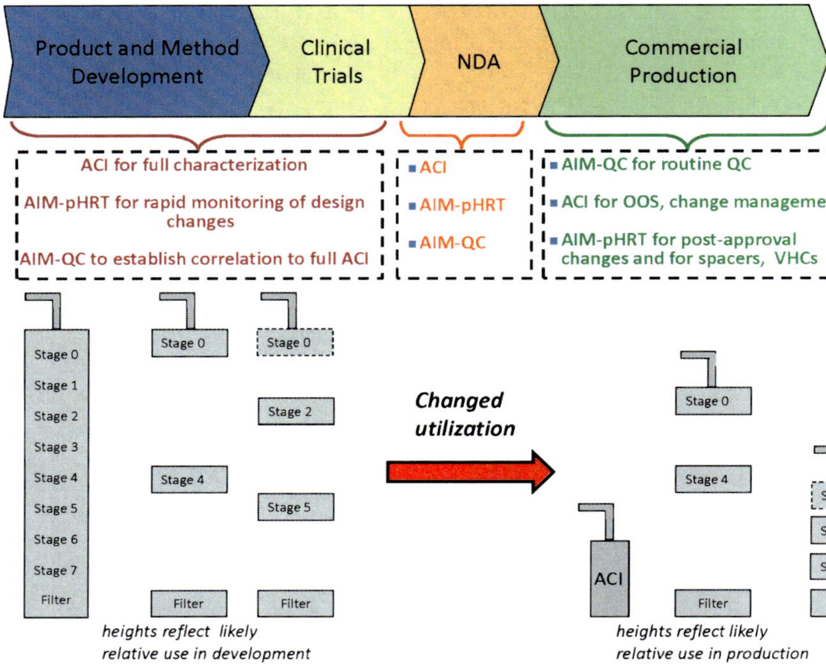

Fig. 6.2 Scheme for full-resolution CI and abbreviated impactor configurations together with likely relative use during development and production phases of the OIP life cycle (*From* [4]—*used by permission*)

APSD-pertinent data is identified in relation to several distinct but complementary processes. While this example uses the ACI for full characterization, the approach would apply equally to the NGI and the related rNGI or FSI abbreviated systems.

6.4.2.1 During Product Development

1. Select an appropriate AIM apparatus (guidance on the performance of the different options is given as part of Chap. 10).
2. Use either the AIM–pHRT or the AIM–QC CIs as screening tools in early formulation development, noting that the AIM–QC system may provide greater sensitivity for detecting important changes in the APSD profile while taking advantage of higher throughput. The AIM–pHRT configuration could be used to obtain additional resolution if an in-vivo, in-vitro (IVIV) relationship has already been established. Note, however, that once the formulation and delivery system (MDI, DPI, etc.) have been developed, the full-resolution CI would still be used to define product's APSD characteristics for the clinical batches.
3. Establish the full-resolution APSD profile of the OIP with full-resolution CI based measurements. This process would require multiple determinations representative

of the product, for example from different units, different batches, different life stages through individual inhaler content testing (as a minimum from beginning and end of unit) and at various times during stability testing, in numbers sufficient to obtain adequate statistical power.

4. Choose *LPM*, *SPM* values and correlate AIM–QC CI-based measurements of both metrics to their equivalents determined by full-resolution CI measurements after: (1) selecting an optimum particle size boundary between *LPM* and *SPM*; and (2) demonstrating preferably that a linear relationship exists between *LPM/SPM* and *MMAD*.

   Note the following considerations:

   (a) The appropriate boundary between *LPM* and *SPM* must be determined by full-resolution CI.
   (b) The traditional coefficient of determination (R^2) may not be appropriate for all cases, when establishing the correlation between AIM- and full resolution-based metrics. For instance, when the range of *MMAD* values for a given product is narrow, the value of R^2 may appear low relative to other products possessing higher variability in *MMAD*, even though their correlation is just as good. This coefficient is therefore more appropriate for comparisons of distributions with similar ranges of *MMAD*, and not as an absolute indicator of goodness of fit. The root mean square error (*RMSE*) divided by the slope of the linear regression (b) is an alternative goodness-of-fit statistic that may be more robust in terms of predictive power [4].
   (c) Release batches against specifications based on *MMAD*, together with EDA metrics, *LPM/SPM* and *ISM*, after correlation between EDA metrics obtained from the full resolution and AIM systems has been established, and the target profile has been created. Establishing the correlation could occur either in development or after approval (depending on when a sufficient number of batches is available to justify the proposed approach—a sponsor company may make that decision based on its own risk assessment).
   (d) Based on their regulatory strategy, the sponsor company will also have to determine if they are going to include the AIM method as the primary APSD method in the NDA submission (or a similar appropriate application for product registration in Europe, Canada or other countries), to support the registration stability program, or whether, the switch to the AIM method will be a post-approval submission. It is recommended to obtain prior approval from regulatory agencies if a matrix approach is to be utilized in the NDA registration stability studies.
   (e) For the near term, determination of appropriate acceptance limits for *LPM/SPM* ratio and *ISM*, could be accomplished by developing limits that produce operating characteristic (OC) curves that match existing approaches (i.e., groupings) with respect to type II error (false acceptance) consistent with limits for approved products, to achieve the same minimum acceptable quality standard. Longer term, QbD is likely to drive the desire for limits driven by some relationship to product performance, i.e., OIP efficacy and/or safety.

(f) Establish suitable precision for both *LPM/SPM* and *ISM* determinations (which is also necessary for full-resolution CI). Many replicate measurements by both full-resolution CI and AIM–QC system, ideally from several different batches or at a minimum from widely spaced intervals during manufacture of a single batch, will be needed in order to assess how small of a change in *MMAD* and total mass entering the sizing part of the impactor can be detected by the EDA metrics for that particular product.

(g) Quantify the minimum number of batches required to achieve a statistically robust correlation between EDA metrics obtained from the full-resolution and the chosen AIM systems. In line with the QbD philosophy, the OIP sponsor will likely need data from many different batches (sufficient to represent adequately sources of variability in the manufacturing process, input materials, analysis and stability effects), with multiple CI runs per batch (sufficient to assess within-batch variability) in order to get a good estimate of the target distribution (mean and variability), which would be representative of future production batches at release and for stability testing. The following outline provides an idea of what may be required:

   (i) Establish the APSD of the product, which will require a large body of full-resolution CI data.

   (ii) In both the product release and stability programs run abbreviated impactor based measurements in parallel to the full-resolution systems to establish correlation between EDA metrics from both abbreviated and full-resolution systems to enable use of AIM for routine control later on.

   In totality, the data required in part (ii) will likely amount to hundred(s) of CI measurements spread over numerous batches, reflecting different lots of API, device components, etc. However, the cost of this upfront work should be more than offset later by the option to use EDA in conjunction with abbreviated CI measurements in the QC environment. This sequence of events mirrors the principle underlying the Quality-by-Design approach, in which the full resolution CI essentially maps out the "design space" for the product APSD, with the abbreviated CI working within the "control space." Such a regimen will also improve decision making by virtue of enabling more samples from the batch under consideration to be assessed for a given expenditure in terms of effort and equipment.

However, despite the advantage of the approach just outlined, it is recognized that some companies may chose to collect these data only after the product has been approved. In such instances, the switch from full-resolution CI measurements with traditional data assessment to AIM-based CI determinations coupled with EDA could still take place. Nevertheless, it should be recognized that delaying this decision could be associated with some business risk, because complete understanding of the APSD-properties of the product in both measurement regimens (with sufficiently different batches) is also postponed.

5. Establish acceptable limits and associated acceptance criteria for *ISM*, *LPM* and *SPM* for the product with the same AIM–QC CI procedure that will be used later in product quality control.
6. In designed experiments undertaken during product and method development, use full resolution CI data to identify possible in-vitro failure modes of the sort that have been identified by looking at underlying physical causes and through case study assessments, described in Chaps. 3 and 9 respectively. In other words, undertake the following:
   (a) Establish ways that an APSD could potentially change and determine associated root causes. Such sources might include manufacturing trends, dimensions of the device components, analytical instrumentation, and methods, etc.
   (b) Develop control strategies to mitigate identified risks and potential failure modes, and evaluate the ability of the chosen QC (EDA) metrics to detect significant changes. These insights should be helpful for setting product-appropriate specifications for the EDA metrics, and later on, during commercial production, for OOS investigations.
7. Use full-resolution CI based measurements as part of an in-depth investigation of any OOS results as well as when any changes are introduced.

6.4.2.2 During the OIP Commercial Phase

1. Release the commercial product against the already established QC specifications based on *LPM/SPM* and *ISM*.
2. Continue stability testing of the product using the QC metrics and specifications for *LPM/SPM* and *ISM*.
3. Bring in full-resolution CI measurements for an OOS investigation, i.e., to explore the nature of a change that was detected by EDA, or any time that unexpected or unusual trends are observed (e.g., increase in variability). Note that since the EDA metrics have the ability to detect changes quickly (due to the high sensitivity of EDA to changes), they can serve as an efficient trigger for such action.

6.4.2.3 Post-approval and Device Changes

1. Use full-resolution APSD measurements and possibly AIM–QC and/or AIM–pHRT systems as part of the change management process, for instance, when substantial changes to the device, formulation, or manufacturing process are considered or introduced. The option and choice of the AIM system would be determined by the sponsor's risk assessment of the change.

2. A developer of OIP add-on devices, or a pharmaceutical manufacturer interested in including such add-on device information in their product label, would also determine characteristic values of coarse (*CPM*), fine (*FPM*), and extra-fine (*EPM*) particle mass of the product ideally by means of an AIM–pHRT approach, likely using either the ACI or NGI as the full-resolution CI benchmark apparatus (see Sect. 6.4). Ultimately, it is anticipated that these metrics would be correlated to clinical response if an adequate IVIVC or IVIVR for product efficacy can be demonstrated, although it is recognized that such correlations are notoriously difficult to attain for OIP for a variety of reasons [10]. In the absence of an established IVIVR, the add-on device developer would have to resort to correlating AIM–pHRT-based measurements with their equivalents obtained by full-resolution CI to provide baseline data for comparisons if any changes are introduced post-approval or if add-on devices are to be recommended with the OIP.
3. Use the AIM–pHRT system to manage uses with either bespoke or commercially available add-on devices (e.g., spacers, VHCs), which are well known to attenuate and therefore modify the oropharyngeal deposition of aerosol particles emitted from an OIP.

6.4.2.4 Summary of Approaches in OIP Life Cycle Management

Table 6.1 (a–d) together maps out a series of considerations as an aid in the understanding of the AIM/EDA life cycle approach. Its underlying purpose is to ensure a cascading flow of information as the product and associated aerosol particle size measurement equipment are moved through the process of development, characterization, approval, and manufacturing.

6.5 Additional Considerations Concerning AIM and EDA Approaches in the Product Life Cycle

A number of additional considerations may be of concern to individual organizations considering the implementation of AIM with or without EDA. These matters tend to be misunderstandings concerning either or both concepts, or product specific, making it more appropriate to deal with as a series of topics here, rather than in the previous section, where issues likely to be of more general concern were addressed.

6.5.1 Use of MMAD as One of APSD Quality Metrics

In discussions with stakeholders, there has been some hesitancy regarding the use of *MMAD* as an indicator of the OIP quality based on APSD. Such hesitation may have

Table 6.1 Uses of full resolution and AIM-based CI systems in OIP lifecycle management: the right impactor for the right purpose. (a) In product development (*From* [11]—*used with permission*); (b) in commercial production (*Developed based on* [11]); (c) in post-approval changes (*Developed based on* [11]); (d) in add-on device development (*Developed based on* [11])

Impactor type/lifecycle stage (goal of APSD testing)	Full resolution impactor	AIM–QC system with EDA metrics (*LPM/SPM* and *ISM*)	AIM–pHRT system (*CPM*, *FPM* and *EPM*)
(a) Product development (*Goal: establish a target APSD profile for a safe and effective product*)	Define APSD of product for the clinical batches Identify failure modes for the APSD Use full-resolution for OOS investigations and justification of changes In order to prepare for routine use of AIM–QC, also conduct studies to select optimal boundary between *LPM* and *SPM* (e.g., near *MMAD* or to maximize sensitivity to significant failure modes)	Use as a screening tool in early development In order to prepare for routine use of AIM–QC, also conduct studies to: 1. Establish correlation of *LPM* and *SPM* to full-resolution APSD 2. Determine appropriate specifications for *LPM/SPM* and *ISM* (e.g., as 90 % confidence intervals on the *LPM/SPM* and *ISM* values observed with a representative sample of product)	Use as a screening tool in early development Conduct studies to establish correlation of *CPM*, *FPM*, and *EPM* to full-resolution APSD. This information will provide support for the subsequent use of the selected AIM–pHRT configuration by establishing an IVIVR or IVIVC
(b) Commercial production (*Goal: confirm that the APSD of commercial batches is the same as that of clinical batches*)	Use full-resolution APSD for OOS investigations	Use AIM–QC for routine quality control (release and stability)	
(c) Post-approval changes (e.g. supplier change) (*Goal: demonstrate sameness of APSD*)	If, for product registration, it has already been proven that data from the AIM–QC were equivalent to those from full resolution CI, simply demonstrate that the results are still equivalent Use full-resolution APSD for complete justification of change only if the regulatory agency requires such a demonstration	Use AIM–QC as a rapid indicator of the same quality	Once an IVIVR or IVIVC has been established, use AIM–pHRT as a rapid indicator that clinically relevant fractions have not changed
(d) Add-on device development (*Goal: minimize oropharyngeal deposition*)	Provides baseline data to compare with OIP alone		Use AIM–pHRT to optimize add-on devices, ensuring clinically equivalent performance to that of OIP without add-on device

been due in part to the mistaken view that *MMAD* would be the only quality metric. Systematic movements in *MMAD* (indicative of shifts in APSD in terms of the aerodynamic diameter scale) must be controlled *together* with the total mass that is collected by the impactor (*IM*).

Systematic movements in *IM* are indicative of shifts in mass output from the OIP and may or may not be related to movements in APSD. If detecting shifts in APSD is important (while also controlling *IM*), then detecting changes in *MMAD* is important since they are directly related. If this premise is accepted, then it follows that EDA is a superior approach because it provides a better way to detect *MMAD* shifts than either stage groupings from full-resolution CI measurements or mass deposited in a particular size range (be it a particular grouping or related to a particular mass fraction of potential clinical relevance, such as *FPM*).

In some situations, it is believed that *MMAD* could be used directly as a QC metric without the need for EDA—namely, when full-resolution CI data are available. If, however, the additional goal is to reduce labor and time resources for QC testing, then AIM–QC would be more efficient than full-resolution ACI testing, and consequently, the EDA ratio *LPM/SPM* (which is directly linked to *MMAD*) would be the metric of choice.

6.5.2 CI Data for Clinical Significance Versus Product Quality Confirmation

In Chap. 5, it was shown that for the effective application of EDA, the location of the *LPM*-to-*SPM* boundary size should be selected based on the underlying purpose of the test. If the goal is to maximize the ability of metrics *ISM* and *LPM/SPM* to detect changes in APSD, then setting that boundary equal to MMAD should provide the optimal outcome in terms of method sensitivity.

The question can be asked: Are these EDA-based metrics and the associated changes in APSD that are detected clinically relevant? The answer is a qualified *yes*, to the extent that the entire APSD profile is clinically relevant, since all impactor-sized particles are likely to deposit somewhere within the respiratory tract (ignoring losses upon exhalation). However, it is important to appreciate that the underlying intent for these metrics is to serve as best possible tools to provide assurance that the APSD of the clinical batches matches the target specifications and in addition, to confirm quickly and reliably in the QC environment whether a given OIP has an APSD within agreed specifications. By themselves, therefore, these metrics and associated particle subfractions *do not claim to be* and *do not need to be* reflective of the API deposition profile in precisely specified regions of the HRT or of the ultimate clinical response due to drug-receptor interaction. In this context, it is worth noting the precedent that current APSD metrics based on grouped stages from full-resolution CI data have not been directly linked to clinical performance either. Given the large inter-patient variability in clinical trials whose intent has been to elicit dose–response relationships, in addition to the seldom considered added variability introduced with disease modifying patency of airways in the respiratory tract, small shifts in mass within the size ranges related to *LPM* and *SPM* subfractions are unlikely to have measurable clinical consequences [12]. This situation may be true even when a convincing IVIVR is established, as could be argued is potentially possible for some bronchodilator-based formulations [10, 13]. Put in another way, the precision of existing CI-based methods for determining these QC metrics greatly exceeds the precision available to the clinician for the corresponding clinical metrics such as forced expiratory volume in 1 s (FEV_1), forced expiratory flow from 25 to 75% of vital capacity ($FEF_{25-75\,\%}$), and similar indicators of airway patency obtainable from well-established spirometric measurements to assess obstructive disease [14]. The higher precision of in vitro methods is likely to become even more apparent for other therapeutic modalities such as anti-inflammatory products, where IVIVRs are not yet fully established [15]. In view of these considerations, caution is advised when utilizing CI-generated APSDs in the development of IVIVRs or IVIVCs. Further consideration of this topic is covered in Chap. 12, in which the type of additional measures that should be considered is discussed in the context of making the CI-based measurement process approximate more closely to actual OIP use.

Another clinically related question that has been asked is as follows: Could one use historical therapeutic-class information to set population-based specifications for APSD? In response, EDA-based specifications would allow for such a universal approach with OIPs across a particular therapeutic class. Importantly, however, data from grouped stages using full-resolution CI measurements will not be useful. This potentially counterintuitive outcome arises because the decisions about which stages to group and how to set those specifications will depend on the specific APSD profile for each OIP, irrespective of API.

6.6 Concluding Thoughts

Both full-resolution and abbreviated CI measurements should be employed to fulfill different goals in support of the various phases through the life cycle of an OIP. This chapter has provided a *road map* to assist the developer wishing to implement AIM as part of an ongoing process to improve productivity in the measurement laboratory and at the same time, to retain sensitivity to important changes in APSD.

The establishment of strong correlations between particle size-related metrics that are obtained from the selected abbreviated CI to the corresponding particle size-related data from a full-resolution system is an important goal. This target should ideally be achieved as early as possible in the OIP development process, ideally developing specifications common to both techniques.

In the commercial phase, it should be possible to release product using abbreviated impactor measurements combined with EDA to interpret the data. This type of data interpretation represents a simpler, yet statistically more powerful approach to analyze the APSD data in a quality control setting (see Chaps. 7 and 8). It also has intrinsic resource-saving potential that is maximized when combined with an AIM–QC system. However, it is important to appreciate that EDA is also appropriate to interpret data obtained from a full-resolution impactor. There is no *one-size-fits-all* definition for the boundary size selection process. Each sponsor must therefore do the due diligence to demonstrate the sensitivity of EDA to detect important changes in a given product, using their in-house particle size data obtained during the product development phase. Such an approach, if undertaken effectively, has the inbuilt advantage that the full-resolution CI is always available for the management of investigations, change control, and troubleshooting. A combination of AIM and EDA can ultimately optimize resource allocation in the product QC environment.

An approach making use of abbreviated impactors based on the AIM–pHRT design would be more appropriate, if the development of a robust IVIVR in cases where the clinically relevant sizes are known. Such an outcome could be the ultimate goal of CI-based measurements for the OIP, where the clinical performance of the product is the primary focus of interest. Such measurements could serve as a quick indicator that clinically relevant size fractions associated with the product have not altered during the transfer from development to production phases and also in the management of intentional changes post-approval.

References

1. European Medicines Agency (EMA) (2009) Requirements for clinical documentation for orally inhaled products (OIP) including the requirements for demonstration of therapeutic equivalence between two inhaled products for use in the treatment of asthma and chronic obstructive pulmonary disease (COPD) in adults and for use in the treatment of asthma in children and adolescents. London, UK. CPMP/EWP/4151/00 Rev 1 Available at URL: http://www.ema.europa.eu/docs/en_GB/document_library/Scientific_guideline/2009/09/WC500003504.pdf. Visited 27 Sep 2012
2. European Directorate for the Quality of Medicines and Healthcare (EDQM). Preparations for inhalation: aerodynamic assessment of fine particles. (2012) Section 2.9.18—European Pharmacopeia [– Apparatus B in versions up to 4th edn 2002] Council of Europe, 67075 Strasbourg, France
3. European Medicines Agency (EMA) (2006) Guideline on the Pharmaceutical Quality of Inhalation and Nasal Products. London, UK, EMEA/CHMP/QWP/49313/2005 Final. Accessed 20 Jan 2012 at: http://www.ema.europa.eu/pdfs/human/qwp/4931305en.pdf
4. Tougas TP, Christopher D, Mitchell JP, Strickland H, Wyka B, Van Oort M, Lyapustina S (2009) Improved quality control metrics for cascade impaction measurements of orally inhaled drug products (OIPs). AAPS PharmSciTech 10(4):1276–1285
5. Mitchell JP, Newman SP, Chan H-K (2007) *In vitro* and *in vivo* aspects of cascade impactor tests and inhaler performance:a review. AAPS PharmSciTech. 8(4):article110. Available at URL: http://www.aapspharmscitech.org/articles/pt0804/pt080411O/pt080411O.pdf. Visited 30 June 2012
6. Mitchell JP, Dolovich MB (2012) Clinically relevant test methods to establish in vitro equivalence for spacers and valved holding chambers used with pressurized metered dose inhalers (pMDIs). J Aerosol Med Pulm Drug Deliv 25(4):217–242
7. Usmani OS, Biddiscombe MF, Nightingale JA, Underwood SR, Barnes PJ (2003) Effects of bronchodilator particle size in asthmatic patients using monodisperse aerosols. J Appl Physiol 95(5):2106–2112

8. Usmani OS, Biddiscombe MF, Barnes PJ (2005) Regional lung deposition and bronchodilator response as a function of beta-2 agonist particle size. Am J Respir Crit Care Med 172(12): 1497–1504
9. Mitchell JP, Nagel MW, MacKay H, Avvakoumova VA, Malpass J (2009) Developing a "universal" valved holding chamber (VHC) platform with added patient benefits whilst maintaining consistent *in vitro* performance. In: Dalby RN, Byron PR, Peart J, Suman JD, Young PM (eds) Respiratory Drug Delivery-Europe 2009. Davis Healthcare International Publishing, River Grove, IL, pp 383–386
10. Newman SP, Chan H-K (2008) *In vitro/in vivo* comparisons in pulmonary drug delivery. J Aerosol Med Pulm Drug Deliv 21(1):77–84
11. Tougas T, Christopher D, Mitchell J, Lyapustina S, Van Oort M, Bauer R, Glaab V (2011) Product lifecycle approach to cascade impaction measurements. AAPS PharmSciTech 12(1):312–322
12. Mitchell JP, Newman SP, Chan H-K (2007) In vitro and in vivo aspects of cascade impactor tests and inhaler performance: a review. AAPS PharmSciTech 8(4): article 24. Accessed on 3 July 2012 at: http://www.aapspharmscitech.org/view.asp?art=pt0804110
13. Evans C, Cipolla D, Chesworth T, Agurell E, Ahrens R, Conner D, Dissanayake S, Dolovich M, Doub W, Fuglsang A, Garcia-Arieta A, Golden M, Hermann R, Hochhaus G, Holmes S, Lafferty P, Lyapustina S, Nair P, O'Connor D, Parkins D, Peterson I, Reisner C, Sandell D, Singh GJP, Weda M, Watson P (2012) Equivalence considerations for orally inhaled products for local action—ISAM/IPAC-RS European Workshop Report. J Aerosol Med Pulm Drug Deliv 25(3):117–139
14. Hegewald MJ, Crapo RO (2010) Pulmonary function testing. In: Mason RJ, Broaddus VC, Martin TR, King T Jr, Schraufnagel DMD, Murray JF, Nadel JA (eds) Murray and Nadel's textbook of respiratory medicine, 5th edn. Saunders Elsevier, Philadelphia, PA, Chapter 24
15. Barnes PJ, Pedersen S, Busse WW (1998) Efficacy and safety of inhaled corticosteroids: new developments. Am J Respir Crit Care Med 157(3Pt2):S1–S53

Chapter 7
Theoretical Basis for the EDA Concept

Terrence P. Tougas and Jolyon P. Mitchell

Abstract Efficient Data Analysis (EDA) was designed specifically to address quality control (QC) decisions with respect to the CI-measured APSD from an OIP. The general goal of QC testing is to confirm that the batch in question is of suitable quality. In the case of EDA, this testing is intended to confirm that the OIP in question generates an aerosol with expected particle size characteristics to deliver drug to the human respiratory tract. Note that this process necessarily takes the form of sampling a relatively small number of units, measuring properties of the aerosols generated by these samples, and making a decision concerning the quality of the sampled batch. This practice leads to three primary considerations:

1. The properties measured should be relevant to detecting significant abnormalities from the expected APSD.
2. The measurements should possess sufficient precision and accuracy over the range of interest.
3. The decision process based on the measurements should reliably make correct inference about the quality of the batch by appropriately minimizing and balancing the risk of decision errors, i.e., judging a batch suitable when it is not suitable and conversely judging a batch unsuitable when it is suitable.

This chapter will briefly introduce the latter two considerations, but will primarily focus on the first. A detailed discussion of the evaluation of measurements and the decision-making process is the topic of Chap. 8.

T.P. Tougas (✉)
Boehringer Ingelheim Pharmaceuticals Inc., 900 Ridgebury Road, Ridgefield, CT 06877-0368, USA
e-mail: terrence.tougas@boehringer-ingelheim.com

J.P. Mitchell
Trudell Medical International, 725 Third Street, London, ON N5V 5G4, Canada
e-mail: jmitchell@trudellmed.com

T.P. Tougas et al. (eds.), *Good Cascade Impactor Practices, AIM and EDA for Orally Inhaled Products*, DOI 10.1007/978-1-4614-6296-5_7,

7.1 Introduction

It is often stated that the aerodynamic particle size characteristics of orally inhaled products (OIPs) are critical to their performance [1]. What remains ill defined is exactly what is meant by particle size characteristics. In a simplistic sense, aerosol particles need to be small enough to reach the intended deposition site in the human respiratory tract, but not so small that they are exhaled and not deposited [2]. There is a further consideration that aerosol particles are not too large and consequently end up in the GI tract posing a potential safety concern [3]. While there has been some discussion concerning monodisperse aerosols and specific targeting of sites within the human respiratory tract [4–8], the current commercial technologies employed in OIPs are limited to producing polydisperse aerosols. Further complicating the situation are the technical challenges posed in obtaining in vivo/in vitro data specifically elucidating the relationship between aerosol properties and receptor locations that are presumed to be related to the in vivo response (safety and efficacy) associated with OIPs [9, 10].

The basic premise of EDA is that the fundamental critical quality attribute (CQA) of an OIP with respect to particle size is the multivariate APSD that is characteristic of a particular product [11]. In other words, OIPs are designed to deliver an aerosol with a nominal APSD containing the appropriate chemical composition of active pharmaceutical ingredient (API) and associated excipients (if the latter are present as part of the formulation).

Given that basic premise, EDA was designed specifically to address quality control (QC) decisions with respect to the APSD [11]. In other words, it was intended to be sensitive enough to be able to detect small changes to the APSD. In general terms, the goal of end-product QC testing is to confirm that a particular batch of OIP product in question is of suitable quality. In the case of EDA, the testing is intended to confirm that the particular OIPs under assessment, when actuated under normal conditions of use, generate aerosols possessing expected APSD characteristics defined from the outcome of product development efforts and subsequently approved by regulatory authorities. Note that this process of assurance necessarily takes the form of sampling a relatively small number of units and measuring properties of the aerosols seen to be critical to the performance of the OIP in question. Thus, relatively few aerosols generated by these samples are actually assessed, and the decision concerning the quality of the entire sampled batch is ultimately made from the outcome of these assessments. This process leads to three primary considerations:

1. The properties measured should be relevant to detecting significant abnormalities from the expected APSD.
2. The measurements should possess sufficient precision and accuracy over the range of interest.
3. The decision process based on the measurements should reliably make correct inference about the quality of the batch by appropriately minimizing and balancing the risk of decision errors, i.e., judging a batch suitable when it is not suitable and conversely judging a batch unsuitable when it is suitable.

The first bullet relies on risk assessment, i.e., what can go wrong that will impact APSD and if it did would the QC test detect the resulting abnormality (see Chap. 9 for a more complete discussion of risk assessment). This present chapter will briefly introduce the latter two considerations. A detailed discussion of the evaluation of measurements and the decision-making process will follow in Chap. 8.

7.2 Measurement Theory and Evaluation

In the design and implementation of any measurement, it is important to consider the purpose of the measurement as this informs both the design and evaluation of the effectiveness of the measurement. For example, there are different considerations for a measurement intended to characterize or describe an attribute of a particular object versus one intended to make a decision about a batch of objects with respect to a particular characteristic based on representative samples. Wheeler has described this concept in more detail [12].

Measurements can be classified into four categories based on the general purpose of measurement:

1. Description
2. Characterization
3. Representation
4. Prediction

Description refers to measurements that inform about the attributes of the item being measured. *Characterization* is similar to description except that it also involves comparison of the measurements to some expectation for the particular object studied, i.e., a requirement or limit. *Representation* involves using measurements on a representative sample to make inference about the population the sample is intended to represent. This is in essence the QC application where a batch is released or rejected on the basis of testing performed on a sample(s) taken from the batch in question and comparing the measurement results to some requirement. Finally, *prediction* is based on using measurements of samples from current batches to predict the attributes of future batches. Since EDA is proposed for QC purposes where batch disposition is decided based on a representative sample, it is classified as a *representation* measurement.

The adequacy of a particular measurement with respect to precision should inherently consider the variability of the measurement versus the variability of product being measured or the tolerances imposed on the product. This is the fundamental essence of measurement system analysis (MSA) [13], which typically employs ANOVA designs to estimate measurement and product variances. These types of designs are also collectively known as gage repeatability and reproducibility (Gage R&R) studies [14]. Application of MSA concepts to compare the relative performance of the EDA metrics with CI stage groupings in the assessment of OIP APSDs is covered in Chap. 8.

7.3 QC Testing: Purpose and Limitations

As an element of the quality control strategy, end-product testing schemes should reflect an in-depth understanding of the required performance characteristics of the product. There should be thoughtful selection of those critical tests that are best performed at the end-product stage. It is important to avoid redundant testing of the same quality attribute through multiple tests either at the end-product stage or in-process. There should be an evaluation of the overall control scheme to avoid multiplicity issues. By "multiplicity issues," we mean testing the same attribute multiple times and not considering the statistical consequences in the setting of specification limits [15]. Finally, statistical design and evaluation of the tests and acceptance criteria should be an integral part of the development process [i.e., the end product in process analytical technology (PAT)].

It is also important to understand the limitations of end-product QC testing. This topic has been discussed in some detail by Tougas [15]. First and foremost of these limitations is that the outcome of end-product testing is primarily limited to an "accept" or "reject" decision. If a product exhibits substandard characteristics at the end of the manufacturing process, there is typically little recourse, but to reject the batch. This is particularly true of pharmaceutical manufacture where the impact is related to human health. A further limitation arises if the end-product testing is destructive in nature (common in pharmaceutical manufacture). Destructive testing relies on testing a relatively small number of units that are representative of the batch or lot. The results from such testing can be used to make inference about the batch characteristics (mean or variability of measured characteristic), but are ineffective at detecting low-frequency aberrant units that arise from an intermittent failure mode. This is illustrated in Fig. 7.1 which depicts batch distributional characteristics that are nominal, abnormal, and the case of intermittent failure modes.

This figure depicts four ways in which the frequency (number)-weighted distribution of a quality attribute related to the batch may lie with respect to quality limits, in this example fixed at 98% and 102% of the nominal value. In the first instance, A, the mean of the symmetrical distribution is centered within the quality limits, and the variance is such that essentially no units are outside the quality limits. This represents a batch with acceptable quality. Testing of this batch via representative samples should result in a decision to accept this batch as being of suitable quality. Case B illustrates a batch distribution where the variance is suitable, but the batch mean is abnormal such that a significant portion of the units are outside the quality limits, and therefore, QC testing should result in a decision that the batch is not of acceptable quality. Case C illustrates a batch distribution with a nominal mean, but an abnormal variance resulting in a significant fraction of units (both high and low) outside the quality limits. As in case B, QC testing should result in a decision to not accept this batch as being of suitable quality. In the final example, D, the overall distribution is similar to case A, but the overall distribution contains additional modes such as might arise from some intermittent failure mode. QC testing based on representative samples is unlikely to be effective at detecting these types of failures.

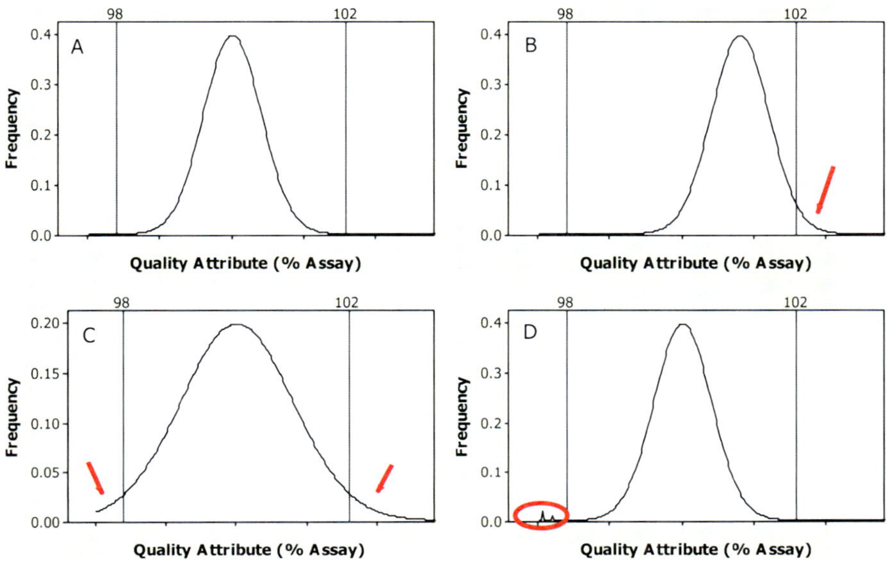

Fig. 7.1 Potential distributional properties of a quality attribute in relationship to quality limits: case A reflects a nominal distribution of a batch with acceptable quality, case B illustrates a quality failure due to an abnormal batch mean, case C reflects a quality failure due to abnormal variability, and case D illustrates a situation where conventional (representation) end-product testing is unlikely to detect the quality failure

More sophisticated schemas can include tiered testing or simultaneous evaluation of mean and variance. In all cases, the decision-making capability of any particular schema can be evaluated through an operating characteristic curve (OCC [16]). In essence, an OCC is a transfer function that relates the probability of a particular decision (accept or reject) to true values of the quality attribute being evaluated. "True" in this context refers to the population parameter that is estimated by measurement of samples. OCCs are also discussed in more detail in Chap. 8.

7.4 Fundamental Properties of the APSD

Back in Chap. 2, the idea was presented that data from a CI are not to be linked directly to specific regions within the HRT. However, CIs by virtue of having a number of stages in series, each of which acts as a size fractionator to the incoming aerosol particles, are capable of providing API-linked, mass-weighted APSDs when combined with appropriate analytical assay method(s) for the API(s) emitted by the OIP whose in vitro performance is being investigated [17].

It is important to note several limitations and trade-offs with using the CI approach to characterizing particle size. First, the resolution of the histogram

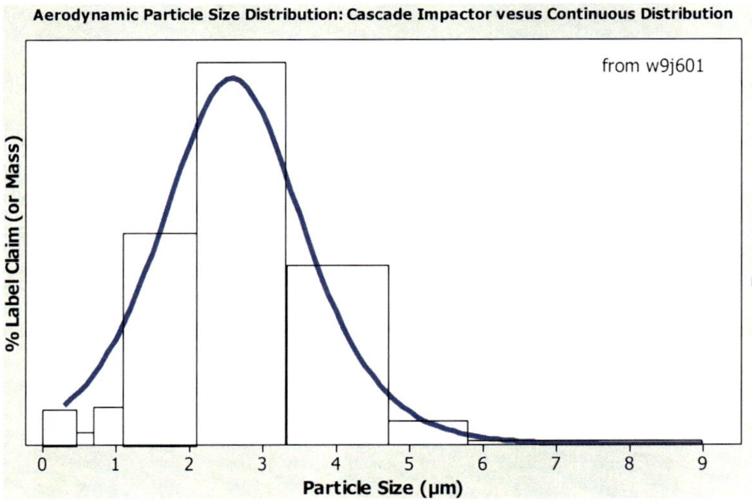

Fig. 7.2 Comparison of a continuous APSD (frequency distribution) with a corresponding mass-weighted histogram obtained from an 8-stage ACI operated at 28.3 L/min

estimating the APSD is directly related to the number of impactor stages. For practical reasons (i.e., to minimize errors due to overlapping stage collection efficiency curves with respect to the aerodynamic diameter axis), the resolution between the important size range for OIP aerosols from 0.5 to 5.0 µm aerodynamic diameter is limited to five measurements of API mass [18].

This outcome arises because the CI-generated APSD is not a continuous frequency distribution, as shown by the blue line in Fig. 7.2, but is a series of discrete values of API mass that are connected by the appropriate stage cut sizes (d_{50}) values for the CI at the flow rate (Q) at which it is being operated (black histogram in Fig. 7.2). The width (size range, $\Delta d_{50,i}$) for each value of API mass is determined by the relationship between adjacent stage d_{50} sizes, whose values for the CIs listed in the pharmacopeial compendia are given in Chap. 2. This relationship for stage "i" is described by the equation

$$\Delta d_{50,i} = (d_{50,i-1} - d_{50,i}) \tag{7.1}$$

in which "$i-1$" represents the immediately preceding stage in the CI.

The assumption is normally made that the collection efficiency $E_{\text{stage},i}$ of a given stage is a step function at its d_{50} size. In reality, $E_{\text{stage},i}$ is a smooth function of aerodynamic diameter, transitioning from its minimum to maximum values about the aerodynamic size at which the stage is 50% efficient (solid line in Fig. 7.3). This relationship is often assumed to be symmetrical, in that the mass of particles larger than $d_{50,i}$, but which penetrate the stage in question, is exactly compensated by the mass associated with particles which are finer than this size, but which are collected thereon [19]. Thus, $d_{50,i}$ can still be defined as a single-valued constant for operation

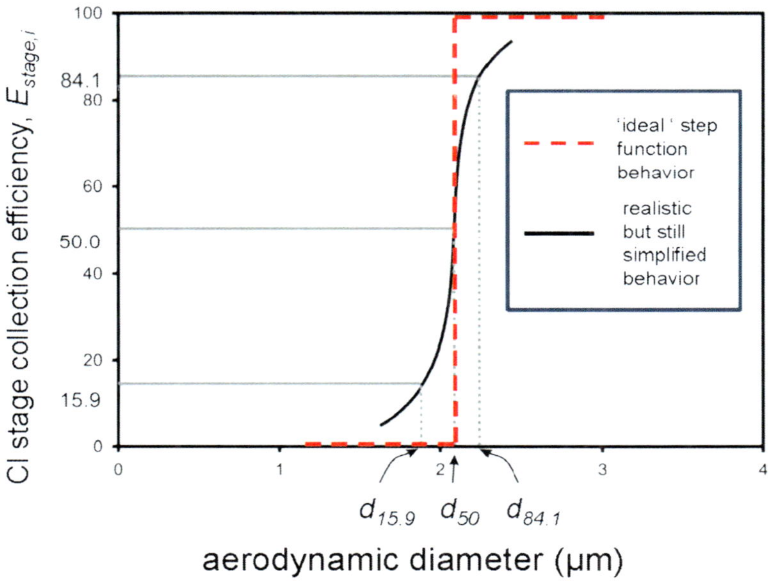

Fig. 7.3 CI stage collection efficiency curve showing "ideal" step function case (*red dashed line*) and realistic but simplified case for establishing cut point size (d_{50}); the square root of the ratio of the sizes corresponding to the 84.1st and 15.9th percentiles for $E_{stage,i}$ is the stage geometric standard deviation by analogy with the properties of a unimodal, log-normal distribution for this variable

of the CI at a fixed flow rate, *Q*. This simplification avoids the need to invoke mass-per-stage data inversion measures that would require the shape of the response function for each stage of the CI to be defined mathematically [20].

In Chap. 2, the effect of making this assumption on the accuracy of measurements using the ACI and NGI has been investigated further. The main conclusion from this analysis is that any inaccuracy introduced is sufficiently small to be of no consequence for full resolution CIs, but the simplification concerning stage collection efficiency may require consideration for AIM-based apparatuses, particularly those derived from the ACI.

Figure 7.2, which is taken from product ***w9j601*** (CFC-suspension MDI) of the blinded IPAC-RS database of APSDs from marketed OIPs, alludes to the fact that the mass-weighted APSD from the CI experiment cannot be directly related to a continuous APSD since there are two different *y*-axes in the plot. The latter is inherently a frequency distribution that can only be estimated from a mass-weighted cumulative APSD.

An estimate of the mass-weighted cumulative APSD is directly derived from the mass-weighted CI results. This is illustrated in Fig. 7.4. The continuous APSD is then the derivative of the continuous cumulative APSD (Fig. 7.5).

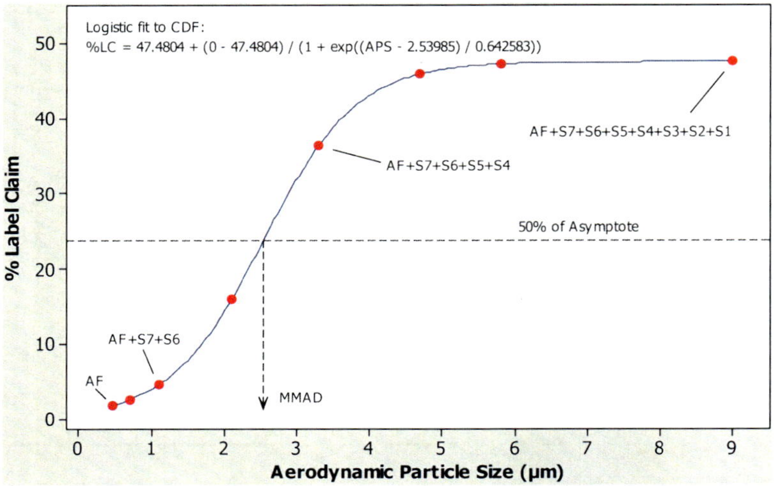

Fig. 7.4 Derivation of the cumulative mass-weighted APSD from the mass API-per-stage results following a CI measurement

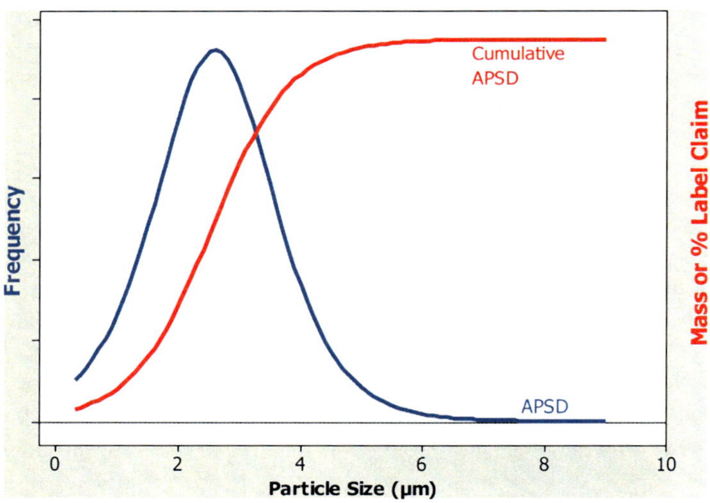

Fig. 7.5 Relationship between the frequency APSD and the cumulative mass-weighted APSD

Internationally the current requirements are varied with respect to the metrics specified. In the USA, the FDA expects that results from multistage cascade impactors are evaluated based on groupings of the individual stage results [21]. Acceptance criteria are established for each of the 3–4 expected groupings, and samples from a particular batch must meet the acceptance criteria for each grouping independently in order for the product to be considered "fit for purpose." In contrast, the expectation in the European Union and Canada is that acceptance criteria are established for "fine particle dose/mass" where *FPM* is defined as the mass of active ingredient in the collected size fraction below about 5 μm aerodynamic diameter [22, 23].

EDA was developed primarily as a quality control tool with the intended purpose of detecting aberrant aerodynamic particle size distributions (APSD) [11]. In other words, EDA was designed primarily to address the QC decision: accept or reject a batch (with respect to APSD). In order to characterize the performance of EDA, the Cascade Impactor Working Group (CI-WG) has utilized CI data gathered by IPAC-RS and used three fundamental statistical approaches: measurement system analysis (MSA), operating characteristic curves (OCCs), and principal components analysis (PCA). Basic background information on measurement processes and these two statistical approaches is given in the following section to aid the reader in assessing the subsequent material that compares the performance of EDA to current approaches.

Making good quality decisions relies on the following principles:

1. Measuring the right quality attributes
2. Understanding how well one can measure those attributes and apply the measurements to quality decisions

The work of the CI-WG has paralleled these concepts. To understand what to measure, the group has explored from a risk assessment perspective what impacts the aerosol properties of OIPs. To characterize the measurement capabilities, the group has evaluated the accuracy and precision of proposed measurements and applied the findings from these investigations and other related information to undertake formalized statistical analyses evaluating the ability to make the correct quality decision concerning the release, rejection, or recall of a batch of pharmaceutical product. This approach quantifies the fundamental uncertainties in the CI measurement system that give rise to type I and II errors, cast in terms of QC decisions as the probability of rejecting acceptable product and releasing unacceptable product, respectively. Thus, the capability of a QC test of metrics derived from CI-generated APSD data can be judged based on a characterization of these error rates.

7.5 Defining the EDA Metrics and Their Background

The EDA metrics were developed out of a need to have a practical set of metrics for OIP QC purposes. Mathematically, APSD is a multivariate measurement (i.e., it requires an array of numbers to describe it). For QC purposes, it is desirable to have univariate metrics (i.e., single-numbered) that sufficiently describe the distribution and that are sensitive to variation in the original APSD (Fig. 7.6).

These ideas have driven the selection of the two EDA metrics, with the underlying intent that both metrics can be easily obtained from CI-generated data.

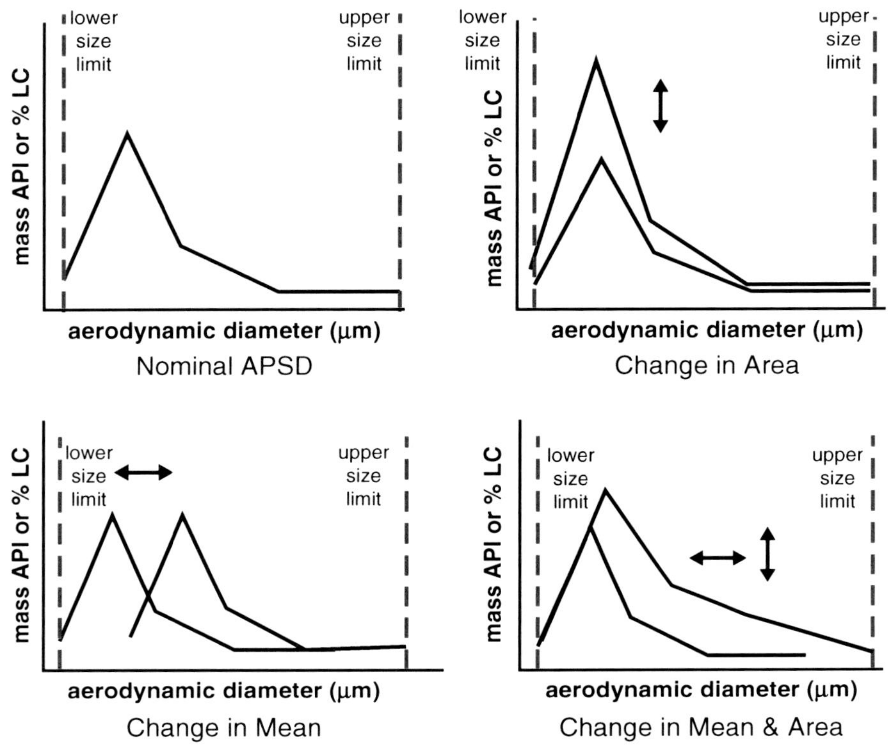

Fig. 7.6 Possible variation types in a unimodal OIP APSD

The ratio metric, large to small particle mass (*LPM*/*SPM*) is highly correlated with the mass-weighted mean of the APSD (represented customarily by its *MMAD*), but independent of the area of the APSD, defined by the *AUC* (Chap. 1) [11]. The second metric, impactor-sized mass (*ISM*) is related to the area defined by the APSD when expressed in differential mass-weighted format, but independent of the mean of the distribution [11]. Both metrics can be readily obtained directly either from full resolution or abbreviated CI measurements (the latter is discussed in Chap. 5).

In order to characterize the performance of EDA, the CI-WG has utilized CI data from marketed OIPs previously gathered by the parent IPAC-RS organization into a blinded database, employing two fundamental statistical approaches in its assessment: measurement system analysis (MSA) and operating characteristic curves (OCCs). Background information on measurement processes and these two statistical approaches is provided here in order to aid the reader in understanding the theoretical principles that underlie EDA, before assessing the subsequent material that is presented in Chap. 8.

7.6 What Constraints Do Various QC Metrics Impose on CI-Determined APSD?

The cumulative mass-weighted form of the APSD is a unique and direct transformation of the frequency form of the distribution [24]. The cumulative form is the best way to understand how the various metrics attempt to control the APSD. Figure 7.7 illustrates three different approaches to APSD assessment [EDA, "grouped stages," and "fine particle dose" (*FPD*)]. In the case of EDA, the ratio metric constrains the (sigmoidal) cumulative distribution near to its point of inflection, i.e., it controls the measure of central tendency (*MMAD*). It should be noted that the value of *ISM* constrains the sigmoid at its upper asymptote. By comparison, the grouped stages approach and *FPD* constrain the amplitude of the APSD at three different chosen size locations and at one size location defined as 5 μm aerodynamic diameter, respectively. The "grouped stages" approach indirectly constrains the *ISM* through the value of the sum of the mass of API allocated within the groups, while *FPD* does not.

The logic that drove the selection of the two EDA metrics was as follows [11]:

1. Both can be easily obtained.
2. The ratio metric, *LPM/SPM*, is highly correlated with the mass-weighted mean of the APSD (represented customarily by the *MMAD*), but independent of the *AUC* of the APSD.
3. The other metric, *ISM*, is related to the *AUC* of the APSD, but independent of the mean of the distribution.

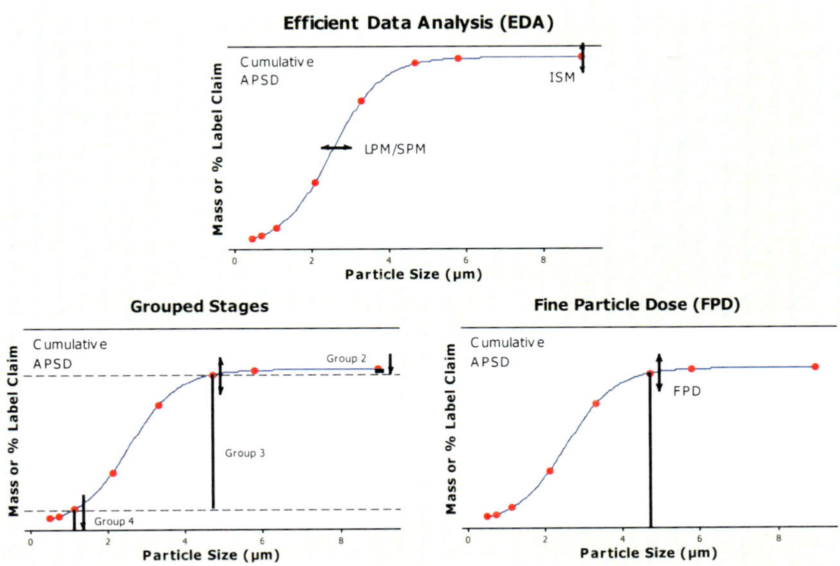

Fig. 7.7 Comparison of constraints to the APSD imposed by various QC approaches

A significant change from the typical mean aerodynamic particle size should therefore be detectable as a change in the *LPM/SPM* ratio, and a significant change in the sized portion of the inhalable dose should be reflected in a change in the *ISM*. In addition, any significant change in the APSD impacting both the mean and the area under the APSD should be detectable as changes in both metrics.

Tougas et al. [11] showed that the aerodynamic particle size boundary differentiating *LPM* from *SPM* can be selected based on the characteristics of the target (i.e., nominal) APSD and depends on the product being tested. Therefore, the boundary does not have to be a universal value for all OIP types and products. Furthermore, it should not be considered as a necessity for it to have clinical significance, although it may be chosen to be related to a clinically meaningful particle size established in prior clinical trials on a particular product. Ideally, this boundary should be selected so as to maximize the sensitivity of these metrics to meaningful changes in APSD from the perspective of measuring product quality.

Even though the boundary size demarcating *LPM* and *SPM* is potentially unique to every OIP, the impaction equipment used for the proposed testing need not be unique to every product. Importantly, the proposed method is relatively robust to the choice of the boundary (Fig. 7.8). In addition, the CI could be operated at a different flow rate to adjust the boundary between *LPM* and *SPM* using the simple and well-defined relationship between flow rate through the system and stage cutoff size that is described in Chap. 2. Furthermore, the range of possible *MMAD* values (and therefore selection of boundaries) is not large for inhalation products since they are all intended to target the lung. For example, among the eight products from the IPAC-RS database, which were purposely selected to be as diverse as possible (Table 7.1), Tougas et al. showed that only three different boundaries (2.1, 3.3, and 4.7 μm aerodynamic diameter) were required. It is likely that these locations will prove to be suitable for all OIPs, and consequently only two or three versions of an AIM-type instrument/method would be needed.

Figure 7.8 provides experimental evidence from analysis of CI results taken from the IPAC-RS database concerning selection of the boundary. The precision of the relationship between the ratio metric *LPM/SPM* and *MMAD* is insensitive to the placement of that boundary over a wide range. Each point in this figure characterizes the quality of the correlation between *LPM/SPM* and *MMAD* for a given boundary location associated with a specific OIP. The boundary location in this analysis was moved such that the corresponding ratio varied over four orders of magnitude around unity. The value of unity for *LPM/SPM* by definition represents the boundary location that is coincident with the *MMAD* for the APSD in question. Both panels of this illustration depict the goodness of fit of the regression between ratio *LPM/SPM* and *MMAD* as a function of the average ratio for a given product. Using average ratio as the abscissa allows comparison across all OIPs studied in a single plot based on how close the boundary is selected relative to the *MMAD* for each product.

The upper plot shows the relationship with respect to a conventional goodness-of-fit statistic [coefficient of determination (R^2)] for each regression as a function of the average ratio. Due to some shortcomings of this statistic when applied to survey

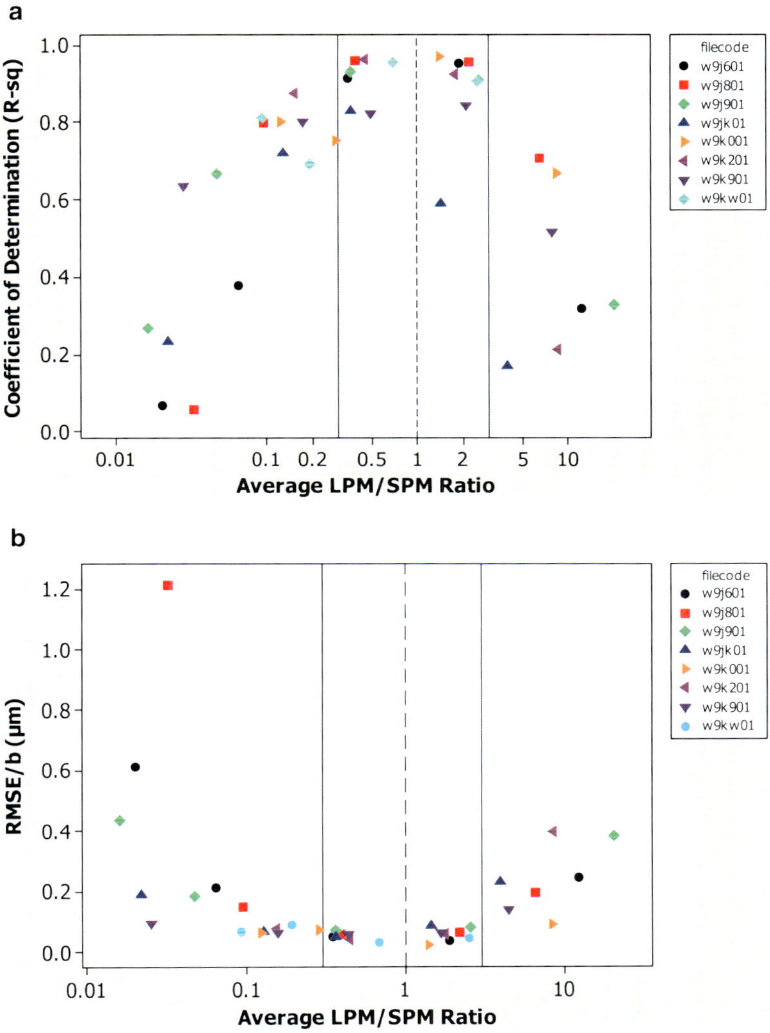

Fig. 7.8 Analysis of the goodness-of-fit statistics as a function of boundary size; the *dashed line* depicts the hypothetical optimum for the ratio selected at the target *MMAD*, and the *solid lines* illustrate arbitrarily defined bounds for the region of acceptable performance. (**a**) Relationship with respect to the coefficient of determination (R^2) for each regression as a function of the average ratio *LPM/SPM*. (**b**) Root-mean-square error (*RMSE*) of the individual regressions divided by the slope of the regressions (*b*) versus the average *LPM/SPM* ratio (*From* [11]—*used with permission*)

data where there is no control of the span of the data, an alternative goodness-of-fit approach was taken, resulting in the lower plot. This plot depicts the root-mean-square error (*RMSE*) of the individual regressions divided by the slope of the regressions (*b*) versus the average *LPM/SPM* ratio. The *RMSE* is the standard deviation

Table 7.1 Regression analysis and goodness-of-fit statistics for *LPM/SPM* ratio versus *MMAD*

OIP filecode[a]	Product type	CI runs (*n*)	Optimum boundary[b] (µm)	Average *MMAD* (µm)	Slope (*b*)	*RMSE*[c]	Coefficient of determination, R^2 (%)	*RMSE/b* (µm)
w9k201	HFA suspension MDI	80	4.7	3.91	0.4071	0.0162	96.4	0.040
w9j901	HFA suspension MDI	39	3.3	2.57	0.4959	0.0350	93.4	0.071
w9j801	HFA solution MDI	201	2.1	1.50	0.7155	0.0421	96.2	0.059
w9jk01	Dry powder inhaler	279	3.3	2.66	0.4319	0.0201	83.0	0.047
w9k901	Dry powder inhaler[d]	279	2.0	2.59	2.3831	0.1278	84.3	0.054
w9j601	CFC suspension MDI	43	2.1	2.54	2.4548	0.0872	95.5	0.036
w9k001	CFC suspension MDI	272	3.3	3.54	1.6127	0.0330	97.3	0.020
w9kw01	CFC suspension MDI	272	3.3	2.86	0.7046	0.0198	95.8	0.028

[a] The filecodes are unique, randomly-generated alpha-numerical labels assigned to specific products in the IPAC-RS database
[b] Here, optimum boundary is based on results from the eight-stage Andersen Impactor
[c] *RMSE*—root-mean-square error
[d] Modified Andersen nonviable CI used for this product

of the residuals about the individual regressions (*LPM/SPM* vs. *MMAD*), and by definition reflects the precision of the fitted regression. *RMSE* is transformed by dividing by the slope of the regression to express this statistic in terms of precision of the estimated *MMAD*.

Both of these plots do the following:

1. They verify that setting the *LPM/SPM* boundary at the target *MMAD* provides optimum precision.
2. They confirm that the precise selection of the boundary location with respect to the *MMAD* value is not critical and can vary by an order of magnitude in the ratio (from 0.3 to 3.0, as indicated by the placement of the solid vertical lines for ratio in both plots), without serious degradation of precision.

7.7 Experimental Evidence for Ratio *LPM/SPM* as Measure of Mean of the APSD

The most important aspects of applying abbreviated data acquisition and analysis strategies for OIPs are the initial determination of the full-resolution APSD profile for each product in a robust manner and subsequent confirmation that an AIM system with a particular chosen boundary between *LPM* and *SPM* provides acceptable predictive capability for *MMAD*.

Overall, Tougas et al. [11] showed in their investigations of OIP APSDs that the *LPM/SPM* ratio appears to be capable of detecting small changes in *MMAD* of the order of tenth(s) of microns. This finding is reflected in the magnitude of the goodness-of-fit statistic (R^2) and [*RMSE*/b], obtained for regressions of the *LPM/SPM* ratio versus *MMAD* reported in Table 7.1.

Tougas et al. observed that the relationship between *MMAD* and the *LPM/SPM* ratio was approximately linear for every OIP type studied, illustrated by the magnitudes of the coefficient of determination and *RMSE/b* goodness-of-fit statistics [11]. They also noted that a small degree of systematic deviation from linearity was observed in some cases and observed that such behavior is consistent with the expectations for the ratio metric *LPM/SPM*. Thus, as values of *MMAD* approach the lower bound (finest particles) of the size range of the CI, the *LPM/SPM* ratio trends towards zero. Similarly, as values of *MMAD* approach the upper bound (coarsest particles) of the size range, this ratio trends towards infinity.

The results in Table 7.1 reflect outcomes for the *LPM/SPM* boundary placement that provided the best correlation between *LPM/SPM* ratio and *MMAD* (denoted as the optimum boundary in this table). Figure 7.9 illustrates the nature and quality of these regressions for two cases selected as examples (OIPs ***w9k001*** and ***w9k901***). The 95% prediction bounds at the mean *LPM/SPM* ratio were projected onto the *x*-axis (aerodynamic diameter in μm). Tougas et al. noted that the difference between these projections of the upper and lower prediction intervals reflects the ability of the *LPM/SPM* ratio to detect differences in *MMAD* and indicates that changes of a

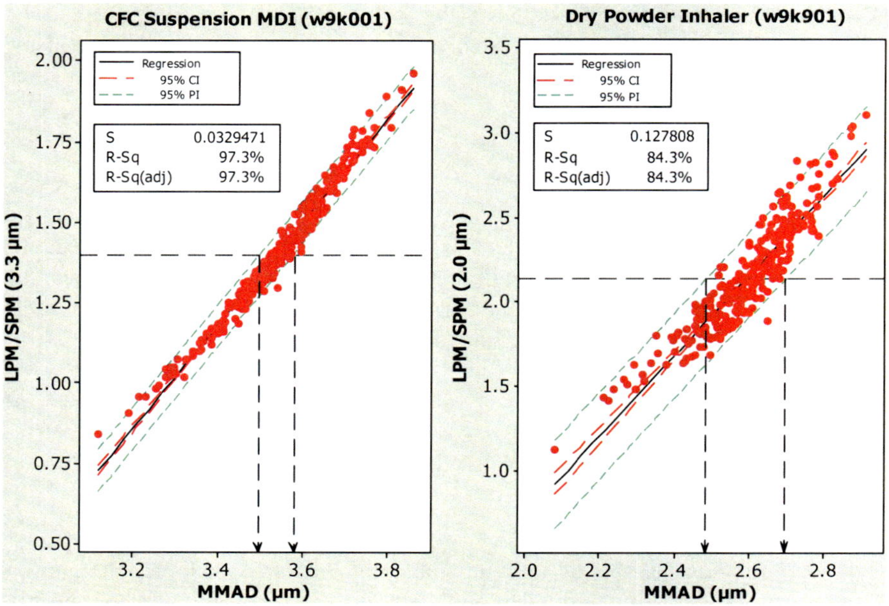

Fig. 7.9 Example regression plots for *LPM/SPM* ratio versus *MMAD* (*From* [11]*—used with permission*)

few tenths of a micron are easily detected. Note that the value of the goodness-of-fit statistic *RMSE/b* is directly proportional to the difference between the projected prediction bounds at the mean *LPM/SPM* ratio by a factor related to the selected confidence level.

While both goodness-of-fit statistics (R^2 and *RMSE/b*) are in general agreement about the quality of the correlation between *LPM/SPM* ratio and *MMAD*, they do not rank order the products in exactly the same way. Note, for example, that the two DPIs (***w9jk01*** and ***w9k901***) have the lowest values of R^2 yet the corresponding *RMSE/b* values are in the middle of the range of results at 0.047 and 0.054 µm, versus an overall range from 0.020 to 0.071 µm. This apparent discrepancy arises primarily from the survey nature of this study and the inherent characteristics of the particular products included. Individual *MMAD* values from these two products exhibited two of the three smallest variations in *MMAD* (interquartile range from 0.133 to 0.159 µm) among all eight products (total range of the interquartile ranges is from 0.133 to 0.444 µm). A smaller range of *MMAD* values results in more uncertainty in the estimation of the regression parameters and hence poorer R^2 values for a given *RMSE*. Conversely a better R^2 would result for a wider range of *MMAD* values with a similar *RMSE*. Thus, at a given level of *RMSE*, the value of R^2 is a function of the range of values in the dataset. In contrast, the *RMSE/b* statistic is a measure of the uncertainty in estimated *MMAD* values at the mean *LPM/SPM* of the

Table 7.2 Regression analysis and goodness-of-fit statistics for stage groupings versus *MMAD*

OIP filecode	Product type	Stage grouping	Slope (*b*)	*RMSE*	Coefficient of determination, R^2 (%)	*RMSE/b* (μm)
w9k201	HFA suspension MDI	>9.0	22.1	4.69	48.2	0.212
		9.0–4.7	12.3	1.28	79.4	0.104
		4.7–2.1	−2.93	2.56	5.2	−0.874
		<2.1	−2.81	0.986	25.5	−0.351
w9j901	HFA suspension MDI	>9.0	3.25	5.72	0.5	1.760
		9.0–3.3	8.57	0.624	74.5	0.073
		3.3–1.1	11.3	3.55	13.6	0.314
		<1.1	0.096	0.767	0.0	7.990
w9j801	HFA solution MDI	>9.0	21.8	9.04	33.9	0.415
		9.0–3.3	5.21	1.08	67.0	0.207
		3.3–1.1	2.53	6.10	1.0	2.411
		<1.1	−24.1	3.28	82.5	−0.136
w9jk01	DPI	>8.6	−9.98	6.09	2.8	−0.610
		8.6–4.4	8.58	0.733	59.2	0.085
		4.4–1.1	13.1	4.42	8.5	0.337
		<1.1	0.391	0.960	0.2	2.455
w9k901	DPI	>8.6	0.990	5.85	0.2	5.909
		8.6–4.4	18.0	1.43	91.8	0.079
		4.4–1.1	−10.2	3.47	37.7	−0.340
		<1.1	−1.34	0.43	13.4	−0.317
w9j601	CFC suspension MDI	>9.0	2.02	8.93	0.1	4.421
		9.0–4.7	2.30	0.376	50.1	0.163
		4.7–1.1	0.09	4.03	0.0	44.778
		<1.1	−1.71	0.413	31.5	−0.242
w9k001	CFC suspension MDI	>10	11.3	3.78	12.0	0.335
		10–4.7	14.9	1.52	59.4	0.102
		4.7–2.1	−2.68	3.49	0.9	−1.302
		<2.1	−9.54	1.02	57.1	−0.107
w9kw01	CFC suspension MDI	>10	7.62	2.97	10.6	0.390
		10–4.7	9.65	1.01	62.3	0.105
		4.7–2.1	5.68	3.19	5.4	0.562
		<2.1	−18.0	1.29	77.9	−0.072

particular data set. The *RMSE/b* values for both DPIs indicate that the *LPM/SPM* ratio is about average in performance with respect to detecting changes in *MMAD*. Based on these considerations, Tougas et al. came to the conclusion that the *RMSE/b* statistic is believed to be the better predictor of the relative performance of the *LPM/SPM* ratio among the product types surveyed.

In contrast, the performance of regressions of stage groupings versus *MMAD* reported in Table 7.2 was significantly inferior with respect to this statistic. The results in Table 7.2 are believed to reflect the performance of the current practice of constructing stage groupings based on empirical inspection. Besides exhibiting inferior correlation with *MMAD*, there is no apparent approach to selection of stage groupings that optimize this correlation or is even predictive of positive or negative correlation.

The slope of plots of *LPM/SPM* ratio versus *MMAD*, with *LPM/SPM* ratio as the directly measured dependent variable, reflects the sensitivity of this metric towards detecting changes in *MMAD* (the steeper the slope, the higher the sensitivity). Thus, slight changes in *MMAD* resulted in magnified variations in *LPM/SPM* ratio when the slope was steep.

Tougas et al. [11] concluded that the ratio metric *LPM/SPM* is superior to using either the separate variables *LPM*, *SPM*, or grouped stages as individual metrics, since the ratio removes the confounding influence of *AUC* of the APSD in trying to detect changes in *MMAD*. In this context, it should be noted that *ISM*, which, as has already been mentioned, is directly related to the *AUC*, is determined simultaneously as the sum of *LPM* and *SPM*. *LPM*, *SPM*, or metrics derived from grouped stages are each influenced by both changes in *MMAD* and *AUC*. In contrast, the *LPM/SPM* ratio *used in conjunction with* the sum of *LPM* and *SPM* will detect separately changes in either *MMAD* or *AUC*.

Tougas et al. [11] also verified the lack of influence of *AUC* on the *LPM/SPM* ratio by performing regression analysis of this ratio versus *ISM*. Table 7.3 summarizes the results of these regression analyses and compares goodness-of-fit statistics for the *LPM/SPM* ratios versus *ISM* to the ratios versus *MMAD*. The results, correlating *LPM/SPM* versus *ISM*, exhibited poorer coefficients of determination (R^2 and *RMSE/b* values). For instance, the R^2 values for *LPM/SPM* versus *ISM* were 2.5–240 times smaller than those from the corresponding *LPM/SPM* versus *MMAD* correlations. Likewise, *RMSE/b* values for *LPM/SPM* versus *ISM* were 2–3 orders of magnitude larger than the corresponding *RMSE/b* results from the *LPM/SPM* versus *MMAD* correlations.

The good correlation between *LPM/SPM* and *MMAD*, taken together with the absence of a correlation between this ratio and *ISM*, is further illustrated graphically in Fig. 7.10, by comparing representative plots for a selected OIP product, ***w9kw01*** (CFC suspension MDI).

7.8 Conclusions

The EDA metrics have a solid theoretical basis that has been confirmed experimentally with consistent performance over a wide range of different types of OIP-generated aerosols. They are simple to apply to CI raw data, yet have the potential to be very sensitive to changes in the APSD in terms of both central tendency (*MMAD*) and *AUC*.

Table 7.3 Regression analysis and goodness-of-fit statistics for the *LPM/SPM* ratio versus *ISM*

		Regression analysis: *LPM/SPM* ratio versus *ISM*				Goodness-of-fit: *LPM/SPM* ratio versus *MMAD* (at optimum boundary)	
OIP filecode	Product type	Slope (*b*)	*RMSE*	Coefficient of determination, R^2 (%)	*RMSE/b* (μm)	Coefficient of determination, R^2 (%)	*RMSE/b* (μm)
w9k201	HFA suspension MDI	0.005	0.082	7.4	16.4	96.4	0.040
w9j901	HFA suspension MDI	0.003	0.136	1.0	45.3	93.4	0.071
w9j801	HFA solution MDI	−0.012	0.185	27.2	−15.4	96.2	0.059
w9jk01	DPI	0.003	0.045	15.0	15.0	83.0	0.047
w9k901	DPI	0.039	0.263	33.9	6.8	84.3	0.054
w9j601	CFC suspension MDI	0.017	0.406	2.9	23.9	95.5	0.036
w9k001	CFC suspension MDI	0.003	0.202	0.4	67.3	97.3	0.020
w9kw01	CFC suspension MDI	−0.003	0.096	1.7	−32.0	95.8	0.028

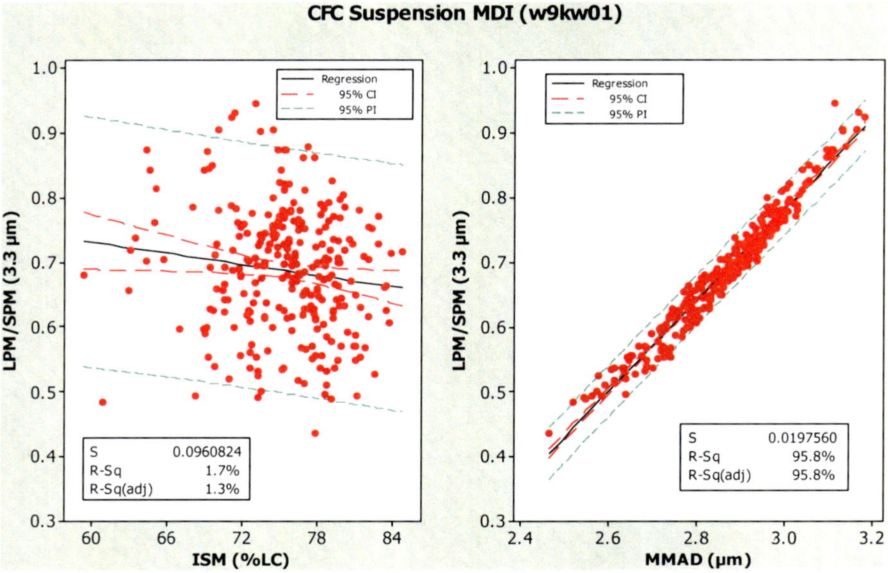

Fig. 7.10 Example regression plots for *LPM/SPM* ratio versus *ISM* and *MMAD* (OIP ***w9kw01***, CFC Suspension MDI) (*From* [11]—*used with permission*)

References

1. Newman SP, Chan H-K (2008) In vitro/in vivo comparisons in pulmonary drug delivery. J Aerosol Med 21(1):1–8
2. Heyder J, Svartengren MU (2002) Basic principles of particle behavior in the human respiratory tract. In: Bisgaard H, O'Callaghan C, Smaldone GC (eds) Drug delivery to the lung. Marcel Dekker, New York
3. Dolovich M (2002) Airway delivery devices and airways/lung deposition. In: Schleimer R, O'Byrne PM, Szefler S, Brattsand R (eds) Inhaled steroids in asthma. Marcel Dekker, New York, NY, pp 169–212
4. Usmani OS, Biddiscombe MF, Nightingale JA, Underwood SR, Barnes PJ (2003) Effects of bronchodilator particle size in asthmatic patients using monodisperse aerosols. J Appl Physiol 95:2106–2112
5. Usmani OS, Biddiscombe MF, Barnes PJ (2005) Regional lung deposition and bronchodilator response as a function of β2-agonist particle size. Am J Respir Crit Care Med 172(12): 1497–1504
6. Zanen P, Go LT, Lammers JWJ (1994) The optimal particle size for beta-adrenergic aerosols in mild asthmatics. Int J Pharm 107:211–217
7. Zanen P, Go LT, Lammers JWJ (1995) The optimal particle size for parasympathicolytic aerosols in mild asthmatics. Int J Pharm 114:111–115
8. Zanen P, Go LT, Lammers JWJ (1996) Optimal particle size for beta-agonist and anticholinergic aerosols in patients with severe airflow limitation. Thorax 51:977–980
9. Newman SP (1998) How well do in vitro particle size measurements predict drug delivery in vivo? J Aerosol Med 11(S1):S97–S104

10. Newman SP, Wilding IR, Hirst PH (2000) Human lung deposition data: the bridge between in vitro and clinical evaluations for inhaled drug products? Int J Pharm 208:49–60
11. Tougas TP, Christopher D, Mitchell JP, Strickland H, Wyka B, Van Oort M, Lyapustina S (2009) Improved quality control metrics for cascade impaction measurements of orally inhaled drug products (OIPs). AAPS PharmSciTech 10(4):1276–1285
12. Wheeler DJ (2006) EMP (evaluating the measurement process) III: using imperfect data. SPC, Knoxville, TN
13. AIAG (Automotive Industry Action Group) (2010) Measurement system analysis, reference manual, 4th edn. AIAG, Southfield, MI, USA, ISBN#: 978-1-60-534211-5
14. Bower KM, Touchton ME (2001) Evaluating the usefulness of data by gage repeatability and reproducibility. Asia Pacific Process Engineer. http://www.minitab.com/en-US/training/articles/default.aspx. Accessed 22 July 2012
15. Tougas TP (2006) Considerations of the role of end product testing in assuring the quality of pharmaceutical products. Process Anal Technol 3:13–17
16. Summers DCS (1997) Quality. Prentice Hall, Upper Saddle River, NJ
17. Dunbar C, Mitchell JP (2005) Analysis of cascade impactor mass distributions. J Aerosol Med 18(4):439–451
18. Marple VA, Roberts DL, Romay FJ, Miller NC, Truman KG, Van Oort M, Olsson B, Holroyd MJ, Mitchell JP, Hochrainer D (2003) Next generation pharmaceutical impactor. Part 1: Design. J Aerosol Med 16:283–299
19. Mitchell JP, Nagel MW (2003) Cascade impactors for the size characterization of aerosols from medical inhalers; their uses and limitations. J Aerosol Med 16:341–376
20. O'Shaughnessy PT, Raabe OG (2003) A comparison of cascade impactor data reduction methods. Aerosol Sci Technol 37(2):187–200
21. US Food and Drug Administration (FDA) (1998) CDER. Draft guidance for industry metered dose inhaler (MDI) and dry powder inhaler (DPI) drug products chemistry, manufacturing, and controls documentation, Rockville, MD, USA. http://www.fda.gov/cder/guidance/2180dft.pdf. Accessed 15 July 2012
22. European Medicines Agency (EMA) (2006) Guideline on the pharmaceutical quality of inhalation and nasal products. London, UK, EMEA/CHMP/QWP/49313/2005 Final. http://www.ema.europa.eu/pdfs/human/qwp/4931305en.pdf. Accessed 20 Jan 2012
23. Health Canada (2006) Guidance for industry: pharmaceutical quality of inhalation and nasal products. File Number file number: 06-106624-547. http://www.hc-sc.gc.ca/dhp-mps/alt_formats/hpfb-dgpsa/pdf/prodpharma/inhalationnas-eng.pdf. Accessed 20 Jan 2012
24. Hinds WC (1999) Aerosol technology: properties, behavior, and measurement of airborne particles, 2nd edn. John Wiley & Sons, New York

Chapter 8
Performance Characterization of EDA and Its Potential to Improve Decision Making in Product Batch Release

J. David Christopher, Helen Strickland, Beth Morgan, Monisha Dey, Alan Silcock, Terrence P. Tougas, Jolyon P. Mitchell, and Svetlana A. Lyapustina

Abstract In this chapter, APSD data are examined from real products using several different strategies to compare the ability of EDA to detect APSD changes with grouped stages from full-resolution CI measurements. These comparisons were made

J.D. Christopher (✉)
Nonclinical and Pharmaceutical Sciences Statistics, Merck Research Laboratories, WP37C-305, 770 Sumneytown Pike, West Point, PA 19486-0004, USA
e-mail: j.david.christopher@merck.com

H. Strickland
GlaxoSmithKline, N226, 1011 North Arendell, Zebulon, NC 27597, USA
e-mail: helen.n.strickland@gsk.com

B. Morgan
GlaxoSmithKline, Zebulon Manufacturing and Supply, 1011 N. Arendell Avenue, Zebulon, NC 27597, USA
e-mail: beth.e.morgan@gsk.com

M. Dey
Merck & Co. Inc., One Merck Dr, Whitehouse Station, NJ 08889-0100, USA
e-mail: monisha.dey@merck.com

A. Silcock
Nicoventures Ltd, 25 Wootton Street, London SE1 8TG, UK
e-mail: Alan.silcock@nicoventures.co.uk

T.P. Tougas
Boehringer Ingelheim Pharmaceuticals Inc., 900 Ridgebury Road, Ridgefield, CT 06877-0368, USA
e-mail: terrence.tougas@boehringer-ingelheim.com

J.P. Mitchell
Trudell Medical International, 725 Third Street, London, ON N5V 5G4, Canada
e-mail: jmitchell@trudellmed.com

S.A. Lyapustina
Drinker Biddle & Reath LLP, 1500 K Street NW, Washington, DC 20005-1209, USA
e-mail: svetlana.lyapustina@dbr.com

T.P. Tougas et al. (eds.), *Good Cascade Impactor Practices, AIM and EDA for Orally Inhaled Products*, DOI 10.1007/978-1-4614-6296-5_8,

relative to decision making associated with OIP disposition in the QC environment. The strategies involve (1) measurement system analysis (MSA), (2) operating characteristic curves (OCC), and (3) principal component analysis (PCA). A general description of these techniques and their basic concepts is provided in the first part of the chapter, while the computational details and results for each of the employed strategies are given following the same order in the later part of the chapter. All results point to the conclusion that compared to grouped stages, the *LPM/SPM* ratio from EDA is more accurate and more sensitive to APSD changes. Each of the examination strategies used different statistical methodologies, based on different assumptions, and one of the approaches used a different set of data independent of the other two. Yet the same qualitative conclusion validating the superior decision-making ability of the EDA concept was reached, regardless of approach.

8.1 Introduction

The current practice generally used by the FDA for assessing changes in APSD is based upon selected stage groupings from full-resolution, multistage CI testing. Each grouping represents a different portion of the APSD profile, in which the total mass in each stage grouping is compared to specified acceptance limits [1]. The grouped-stage approach typically includes the deposition sites of the impactor system that are non-sized (e.g., actuator, induction port (throat), pre-separator), plus three groupings that represent the upper, middle, and lower size fractions of the sized material captured in the CI. In practice, the "coarse tail" grouping may or may not include the first impactor stage (sometimes referred to as stage 0), for which, in the case of the ACI, there is no upper cutoff limit and, therefore, for which there can be no definitive size range. In the latter case, the first stage (S0) is usually included in the non-sized grouping, and this has been done in this chapter. The remaining stages were allocated to three sized groupings for the present analysis (groups 2, 3, and 4, as defined in Table 8.1), which is consistent with the current FDA practice [2].

The reader will recall from previous chapters that the two metrics used in EDA assessment are (1) the large-to-small particle (*LPM/SPM*) *ratio* of API mass and (2) the *sum* of *LPM* and *SPM* collected in the impactor-sized portion of the emitted dose from the inhaler (Fig. 8.1).

Table 8.1 Typical stage groupings adopted by FDA [2] and used in the comparisons within this chapter; identities for each grouping are based on the 8-stage nonviable ACI and are described in Fig. 8.1

Grouping	ACI stages
1	Group 1 (actuation adapter-S0)
2	Group 2 (S1–S3)
3	Group 3 (S4–S6)
4	Group 4 (S7-filter)

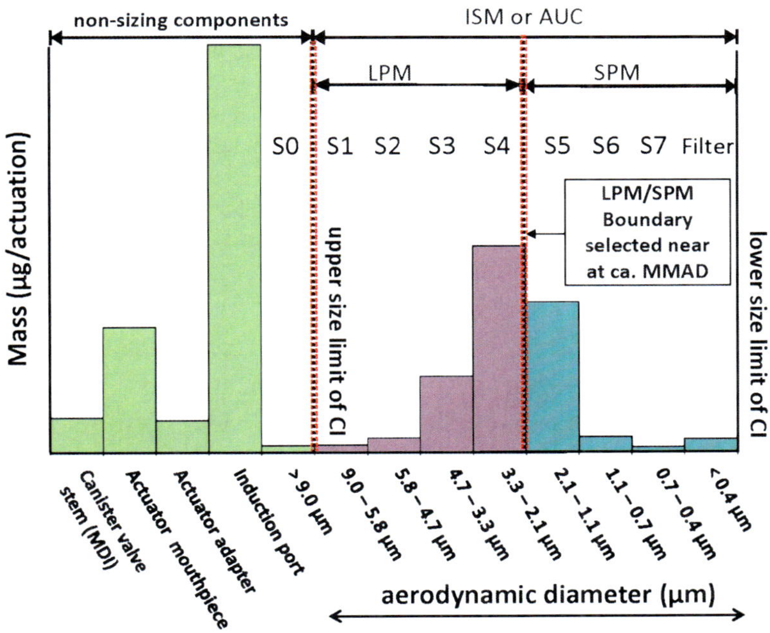

Fig. 8.1 EDA concept illustrated for a hypothetical CI using the ACI at 28.3 L/min to represent the full-resolution measurement system; the pre-separator, if used, would provide a data bar between the induction port and S0

The sum (*LPM* + *SPM*) is numerically equal to the total mass collected on all stages of the impactor, which have an upper-bound size, and is termed the impactor-sized mass (*ISM*). The values of these two metrics, *ISM* and *ratio*, defined as *LPM*/*SPM*, can be *mathematically* independent: thus, the total mass of API collected in the CI can change without affecting the relative masses contained in the large- and small-sized fractions. Even though there could be physical causes that lead the two metrics to change "in synchrony" and thus to appear correlated, mathematically, the *ratio* can vary without *ISM* being changed, and vice versa. In other words, a change in one metric does not require a concomitant variation in the other. This means that the effects detected by these metrics can be measured independently (i.e., a shift of the entire profile or a shift of mass between stages—by the *ratio* and total area under the curve—by the value of *ISM*).

Even when there is physical correlation between the two effects, a change in one does not affect the ability to measure or interpret the other. This is not the case for the grouped-stage approach, in which there is a mathematical relationship that, for any change in total mass, requires a change in mass assigned to one or more of the stage groupings. However, it is possible for there to be changes in two or more groupings due to a shift that may or may not affect the total mass collected. Furthermore, since decisions for grouped stages are made for each grouping separately, it may be difficult to correctly interpret the meaning of a change in a stage grouping (e.g., an increase in mass assigned to one stage grouping could either be due to mass transfer from one group to another group, because of a shift in APSD, or it could be due to an increase in both groupings, due to an increase in total mass collected). The magnitude of *ratio*, which measures the relative proportions of API assigned to coarse and fine fractions, is not affected by changes in the total amount collected. *Ratio* has been shown to be more capable of detecting smaller shifts in APSD than can be achieved by the current practice of grouped stages (2), and this conceptual reasoning is exemplified by the quantitative analysis presented in this chapter.

8.2 A "Road Map" for the Comparative Assessment of EDA Versus Grouped Stages for the Interpretation of CI-Measured Aerosol APSDs in the Context of OIP Quality Control

The remainder of this chapter is concerned with testing the power of the EDA metrics compared with the grouped-stage approach to discriminate changes in CI-measured OIP APSDs that might be used to come to a decision regarding the quality of a lot in the context of QC. Figure 8.2 is a "road map" that describes the different ways in which such data have been assessed. Each of the avenues "A"

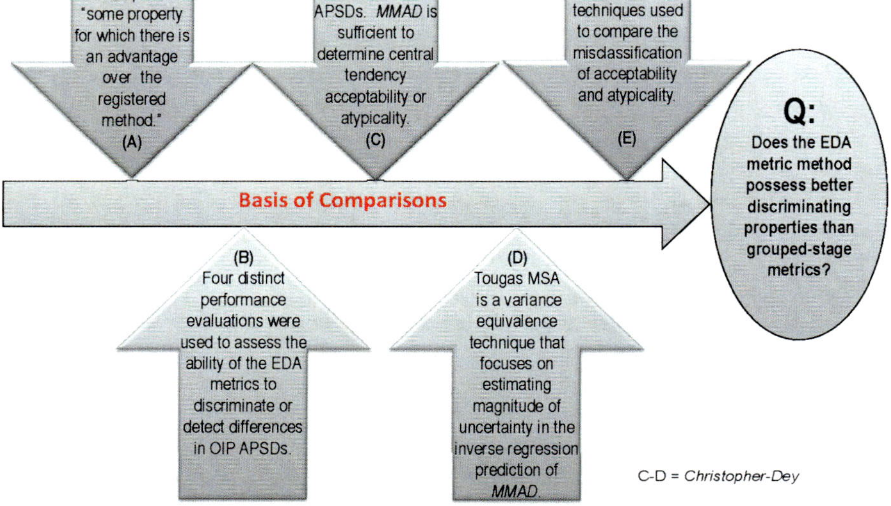

Fig. 8.2 "Road map" for APSD assessments: posing the question

through "E" will be explored in turn, with the ultimate objective of finding a definitive answer to the question posed concerning the relative discriminating ability of the two approaches. Importantly, these assessments have been made with real OIP data, not computer-generated APSDs, so that the variability seen (Fig. 8.3) is likely to be typical of the real world in terms of the quality of the data obtainable for these products by the current compendial methods that involve full-resolution CIs.

8.3 Underlying Data for Metric Comparisons

The data sets for eight products analyzed using operating characteristic curves (OCC) and measurement system analysis (MSA) came from the IPAC-RS APSD database from OIP performance measurements that were collected in 2000. For that database collection, pharmaceutical companies were asked to submit aerodynamic particle size data for individual CI determinations, on a stage-by-stage basis, for as many OIPs as possible, obtained from batch release testing and/or from real-time stability studies. Data were presented as a percent of the amount of API stated in the label claim (%LC). No other manipulations or normalizations were applied. To avoid bias, it was recommended that companies submit either:

1. All available data for the product
2. Data for a random selection of batches
3. Data for all batches manufactured during a defined time span

The database does not have information about regulatory specifications applicable to particular products; therefore, data could be included from batches that "passed" as well as "failed" the regulatory APSD requirements. The data are therefore considered representative of real manufacturing capability and variability. Companies were asked, however, not to include any batches that had clear quality problems or manufacturing faults, as would have been detected by other (non-APSD) tests and investigations. Each product was tested following its analytical method.

The original database encompassed 34 OIPs [3]. For this analysis, eight products were selected that had the largest amount of data per product. The selected OIPs represent different product types (CFC suspension MDIs, HFA suspension MDIs, an HFA solution MDI, and two DPIs) as well as having different APSD profiles, as can be seen from the "spaghetti" plots of all available data points (Fig. 8.3a, b).

The six MDI products were tested with ACI, and the two DPI products were tested with a modified ACI. The number of CI runs available for each product is listed in Table 8.2.

The assessments based on principal component analysis (PCA) described later in this chapter used a data set that was unrelated to the IPAC-RS database. The PCA data set consisted of 1,990 individual NGI measurements for variations of a real (blinded) product throughout the development process, of which 252 measurements were derived from the dose-ranging clinical batches. The

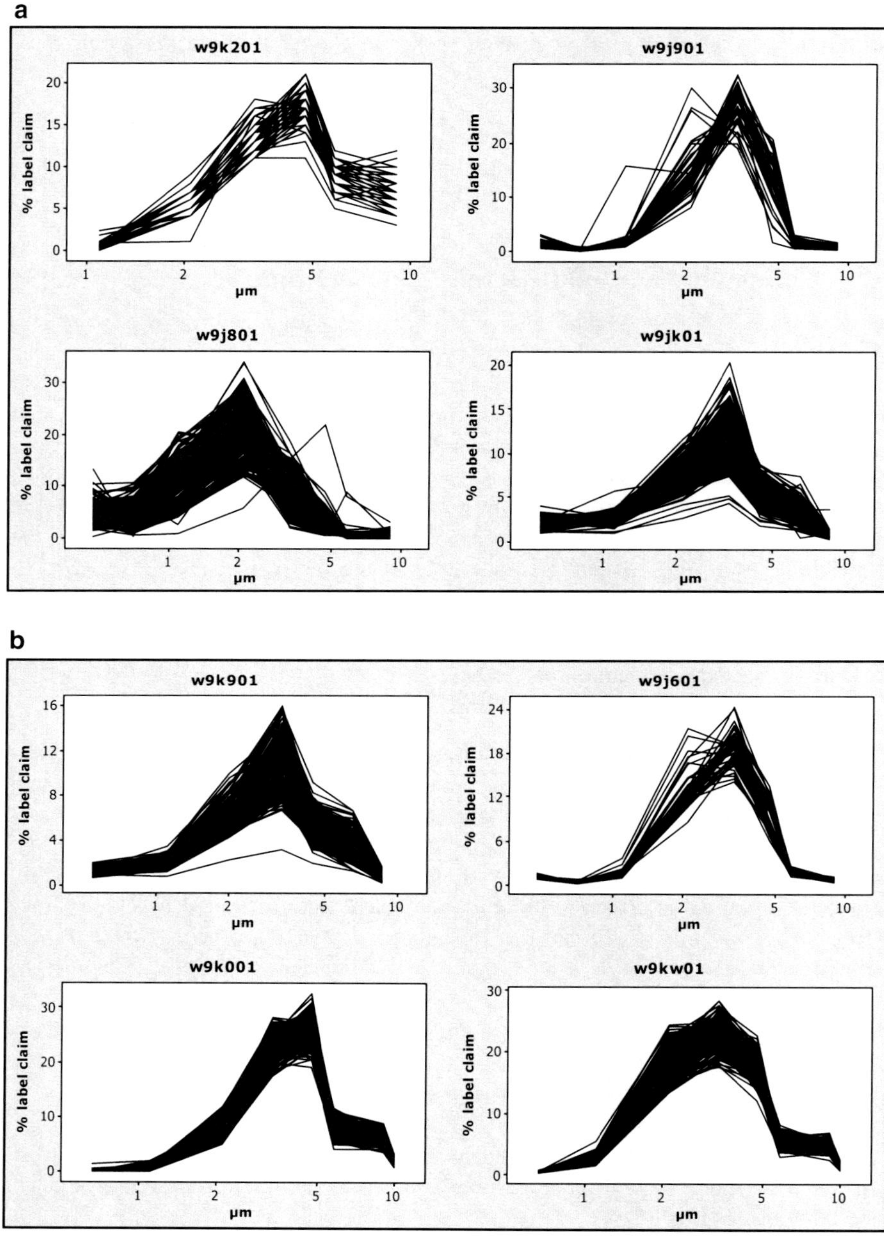

Fig. 8.3 APSD profiles of eight OIPs used for comparisons analysis; each *line* corresponds to a single CI determination. (**a**) OIPs ***w9k201***, ***w9j901***, ***w9j801***, and ***w9jk01***. (**b**) OIPs ***w9k901***, ***w9j601***, ***w9k001***, and ***w9kw01*** (*From* [3]—*used by permission*)

Table 8.2 List of products used for MSA and OCC comparisons in this chapter

File code	Product type	Number of CI determinations (*n*)
w9k201	HFA suspension MDI	80
w9j901	HFA suspension MDI	39
w9j801	HFA solution MDI	201
w9jk01	Dry powder inhaler	279
w9k901	Dry powder inhaler	279
w9j601	CFC suspension MDI	43
w9k001	CFC suspension MDI	272
w9kw01	CFC suspension MDI	272

product in question was a dry powder formulation of the API in association with lactose carrier particles. The product has been used in dose-ranging clinical studies at four nominal strengths, which were all designed to be dose proportional in terms of respirable dose (≡*FPM*). The test method applied to control the product was NGI (with pre-separator) at a flow rate of 60 L/min with data collected on a single actuation of the device. Final quantitative analysis of each of the NGI stages was conducted by reverse phase HPLC using an external reference standard. For the purposes of the PCA work, the data for each of the stages of the NGI was normalized as a percentage of the nominal dose to allow all four product strengths to be evaluated together. The data from the mouthpiece, induction port (MT), and pre-separator (PS) were excluded from the PCA model for the following reasons:

1. Only the profile of the respirable dose was of interest since this would be indicative of lung distribution.
2. Inclusion of MT and PS data did not change the model but did increase the noise level.

SIMCA (Umetrics AB, Umeå, Sweden) was used to conduct the principal component analysis with the data being mean-centered throughout.

8.4 Approaches to the Evaluation of CI-Derived Metric Performance in Assessment of OIP Quality

8.4.1 Introduction to MSA, OCC, and PCA Approaches

There are many facets of a test or measurement that are usually evaluated prior to adoption for a specific measurement situation. These include ease of use, cost of instrumentation, acceptance (by regulators and wider industry), and finally the quality of results obtained. It is this last item, the quality of results, which is the focus of this chapter. Specifically, the intent is to characterize the performance of the EDA metrics from the perspective of their use within a quality control strategy.

In addition to simplify the characterization of this approach, EDA performance will be compared to the current regulatory data reduction procedure used for quality control (QC) purposes related to OIP aerosol APSD, which makes use of grouped impactor stages.

As was mentioned in Chap. 6, in the design and implementation of any measurement system, it is important to consider the purpose of the measurement as this informs both the design and evaluation of the effectiveness of the measurement and associated system. For example, there are different considerations for a measurement intended to characterize or describe an attribute of a particular object versus one intended to make a decision about a batch of objects with respect to a particular characteristic and based on representative samples. Wheeler [4] has described this concept in more detail.

Measurements can be classified into four categories based on the general purpose of measurement:

1. *Description*
2. *Characterization*
3. *Representation*
4. *Prediction*

Description refers to measurements that inform one about the attributes of the item being measured. *Characterization* is similar to description except that it, in addition, involves comparison of the measurements to some expectation for the particular object studied (i.e., a requirement or limit). *Representation* involves using measurements on a representative sample to make inference about the population the sample is intended to represent. The category of representation is in essence the QC application where a batch is released or rejected on the basis of testing performed on a sample(s) taken from the batch in question and comparing the measurement results to some requirement. Finally, *prediction* is based on using measurements of samples from current batches to predict the attributes of future batches (i.e., in skip-lot testing). Since EDA is proposed for QC purposes, where batch disposition is decided based on a representative sample, it is classified as a representation measurement.

The adequacy of a particular measurement with respect to precision should consider the variability of the measurement versus the variability of product being measured or the tolerances imposed on the product. This is the fundamental essence of measurement system analysis (MSA) [5], which typically employs analysis of variance (***ANOVA***) designs to estimate measurement and product variances. These types of designs are also collectively known as gage repeatability and reproducibility (Gage R&R) studies.

Besides an evaluation of the adequacy of measurements employed for QC, it is important to consider the adequacy of the schema used to make the QC decision. In practice, measurements are made on some number of samples. The results are then used in accordance with a previously established specification to make an inference about the batch quality and decide on the disposition of the batch (either release or reject). There are many possible "protocols" (schemas and factors) for using sample measurements to reach a disposition decision. Some simple examples include:

- Averaging a specified number of results and comparing averages to limits
- Comparing all individual results to limits
- Comparing both sample mean and sample standard deviation to limits

In all of the above cases, the sample size has an impact on the inference that can be drawn on the related population (batch). More sophisticated schemas can include tiered testing or simultaneous evaluation of mean and variance. In all cases, the decision-making capability of any particular schema can be evaluated through an OCC.

Finally but importantly, given the context of this chapter, PCA is one possible technique that can be used to compare data in the form of distributions, multivariately. The purpose of including PCA is to compare the ability of the multivariate classification achieved through this technique with the allocations obtained through the use of univariate metrics in either MSA- or OCC-based techniques.

8.4.2 MSA Approach: Definitions and Basic Concepts

MSA can be used to characterize measurements, to analyze total variance, and to separate product- and measurement-related sources of variance. Fundamentally, this approach views obtaining a measurement as a process, the "product" of which is the measurement value obtained. The information obtained through MSA therefore speaks to the precision of the measurement. In turn, this allows evaluations of the adequacy of measurement and measurement system relative to their intended purpose.

It is useful to start a discussion of MSA and the approach used in the evaluation of EDA with some definition of terms. In their manual describing MSA, the Automotive Industry Action Group (AIAG) has published the following definitions pertinent for the current discussion [5]:

- *Measurement* is defined as "the assignment of numbers [or values] to material things to represent the relations among them with respect to particular properties." This definition was first given by Eisenhart [6]. The process of assigning the numbers is defined as the measurement process, and the value assigned is defined as the measurement value.
- *Gage* is any device used to obtain measurements, frequently used to refer specifically to the devices used on the shop floor, and includes go/no-go devices (also, see [7]).
- *Measurement system* is the collection of instruments or gages, standards, operations, methods, fixtures, software, personnel, environment, and assumptions used to quantify a unit of measure or fix assessment to the feature characteristic being measured; the complete process used to obtain measurements.

If we consider a typical pharmaceutical QC situation, where multiple units from a large batch are sampled and tested, then the variance of these multiple measurements can be modeled as the sum of variances associated with the units (i.e., product related) and the variance associated with the measurement system. The latter can, in principle, be partitioned among components of the measurement system (e.g., according to components outlined in the AIAG definition). An MSA for this situation might consist of an experiment designed to allow estimation of these variances with the goal of understanding if the measurement system is appropriate. For the QC situation, "appropriate" means that the measurements could be reliably used as the basis for QC decisions. This definition implies that a product deemed unacceptable could be detected through the measurements. It also follows that some type of comparison of the estimated variance of the measurement system is made to either the variance of the product or to some limits that distinguish between acceptable and unacceptable product.

The ideal situation for MSA is where measurements can be made repeatedly on the same object. This provides a simple direct means of distinguishing variability due to the measurement system from the variability of the actual objects being measured. This situation is possible when testing is nondestructive but is problematic when samples are destroyed by the measurement process. The latter is typical in chemical analysis and is the case in the specific situation associated with APSD determinations obtained by cascade impaction.

Destructive measurements are very common in pharmaceutical analysis, where often a sample must be transformed (e.g., dissolved, diluted, reacted, etc.) to make a chemical or physical measurement. The normal recourse for an MSA is to seek an experimental design that minimizes the influence of sample-to-sample variability. For example, a larger composite sample might be generated to enable drawing multiple measurement samples from a homogeneous blend, or product is sampled from a discrete portion of the batch in which production variation is minimized.

The data, analyzed in the MSA section, represent an example of destructive testing. OIPs generate an aerosol that is intended to be inhaled by the patient. Each actuation of a given inhaler results in a unique aerosol influenced by both the formulation and the delivery device (e.g., MDI, DPI, or other inhalation device). CI testing consists of actuating the OIP, size fractionating and collecting the resulting aerosol particles on different stages of the apparatus, and subsequently performing a chemical assay of the mass deposited on these stages and related accessories.

The cascade impaction data for the MSA were not generated from a planned experimental design, as might be typical for an MSA design. The data represent historical cascade impaction results from the IPAC-RS database, and thus, the analysis could be considered a historical MSA. The statistical approaches for this section are intended to characterize, using regression techniques, the variability in the EDA measurement and stage grouping measurement relative to the *MMAD* values from the full-resolution CI results. Hence, even though there was only one physical process of performing the testing that resulted in the original measurements from individual stages, four additional metrics—the values of *MMAD*, the *LPM*/*SPM* ratio, *ISM*, and the stage groupings—can be calculated from those original measurements. *MMAD* in this instance was determined according to the approach described by Christopher et al. via construction of the cumulative mass-weighted APSD from the individual stage data [8].

The statistical approach presented in the MSA section uses regression techniques to relate the *LPM/SPM* ratio and stage groupings to *MMAD*. This, in turn, allowed an estimation of measurement system variability of either *LPM/SPM* ratio or grouped stages and a comparison of this variability relative to changes in *MMAD*.

8.4.3 OCC Approach: Definitions and Basic Concepts

The concept of operating characteristic curves derives from the view of testing as a decision-making process. Typical pharmaceutical quality control (QC) testing involves making measurements on representative samples of a batch and using the results as the basis to judge the quality of the material being tested either against some predetermined quality standard (which is conceptually and statistically the preferred way) or against a predetermined set of rules (which has been regulatory practice to date) [9]. This assessment of the quality is used to decide on some action, i.e., release or reject the batch. In the statistical sense, such decision-making is a form of hypothesis testing. As with all hypothesis testing, one strives to design the statistical test to be as reliable as possible in terms of mitigating decision-making error, within the practical constraint of obtaining a sufficient number of sample measurements. In the presence of some level of measurement noise and variability, however, a finite probability of making an incorrect decision exists. The terminology for corresponding error types is illustrated in Fig. 8.4.

Type I and II errors (which could also be considered "misclassifications" rather than errors) arise out of the uncertainty in the estimation of the true value of a quality attribute associated with the batch being tested. In general, there are two ways to

	TRUE SITUATION	
DECISION	Material is of Acceptable Quality	Material is NOT of Acceptable Quality
Accept	Correct Decision	Type II Error (β) (Consumer Risk)
Reject	Type I Error (α) (Producer Risk)	Correct Decision

Fig. 8.4 The product QC decision process, possible outcomes, and potential error types

minimize these errors. One could either improve the fundamental measurement by reducing the variability of the measurement or increase the number of measurements in order to improve the precision of the estimate of the quality attribute.

As discussed earlier, in the context of Wheeler's classification of measurements, the QC testing discussed here corresponds to "characterization" and "representation." It is important, however, to make the distinction between these two categories. In the former case, the decision is strictly with respect to the measured entity. Thus, in a practical sense, characterization implies both nondestructive testing and 100% inspection. The much more common situation for pharmaceutical products is representation. The typical batch release process involves testing a relatively small number of samples drawn from the batch and then releasing or rejecting the batch on the basis of the measurement results based on some a priori sampling plan and decision-making process.

While MSA focuses on characterizing the precision of the measurement, an OCC considers and characterizes the entire schema used to make the QC decision, including the sampling plan and the rules used to arrive at the final decision. The OCC approach recognizes that the quality of decisions made based on testing is a function of not only the measurement characteristics (precision and accuracy) but also the sampling plan and decision rules. The construction of this decision-making process (i.e., "design of the test") allows one to adjust the relative anticipated rates of type I (false rejection) versus type II (false acceptance) error, although there is always some trade-off between the two types of error. This may be important depending on relative consequences of making either of these errors. In the pharmaceutical world, there is in general more emphasis on minimizing type II errors at the expense of type I errors.

Simply stated, an OCC is the determination of the probability of acceptance (or rejection) as a function of the true value of the quality attribute being estimated. In essence, an OCC is a transfer function that relates the probability of a particular decision (accept or reject) to true values of the quality attribute being evaluated. The "true" value of a population parameter can never be known in practice because even a 100% inspection would yield only an estimate, whose accuracy and precision is influenced by the analytical uncertainty and other limitations of the measurement technique. The OCC approach requires simulating, or modeling, of the underlying data distribution (so that "true" attributes and parameters are known exactly) and evaluating the ability of a test (i.e., a combination of limits and a specified schema) to make correct reject/accept decisions, as a function of the (assumed) quality of the material under consideration.

Consider the hypothetical case of a tablet assay where the desired absolute batch mean is between 90% and 110% LC. The ideal situation would be where the true value of the assay for any batch (all tablets) is known with certainty, with no measurement error and consequently no type I or II errors. The resulting OCC would consist of a step function where the probability of acceptance would be zero for all assay values above 110% and below 90% and unity for all values between 90% and 110%.

Now consider the impact of measurement error. For batches where the true assay was well below 90% or well above 110%, the probability of acceptance would

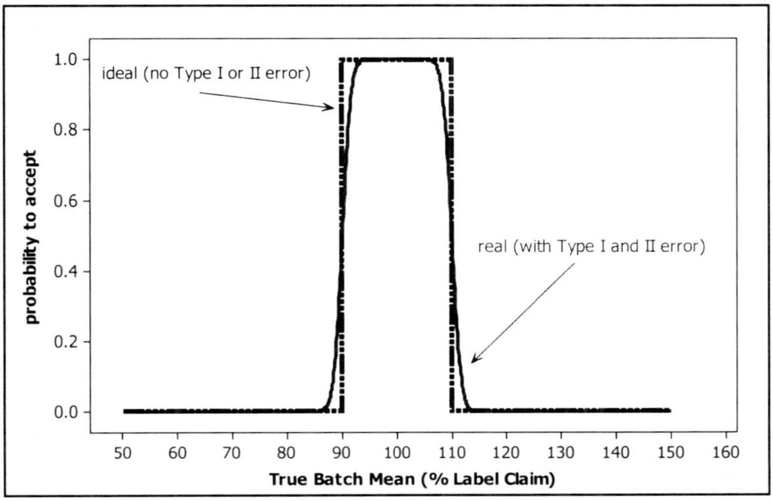

Fig. 8.5 Comparison of an ideal step function (*dashed black*) versus a real (*solid red*) OCC for a two-sided limit

remain essentially zero. Likewise, for batches where the true assay was well within the 90–110% critical values, the probability for acceptance would be essentially unity. However, as the true assay approaches the critical values of 90% or 110%, the impact of indeterminate measurement error would be to cause deviation from the ideal step function such that there would be some finite probability of accepting a batch with a true assay slightly below 90% or slightly above 110% and, conversely, a finite probability of rejecting a batch that had an assay slightly above 90% or slightly below 110%. This is illustrated in Fig. 8.5. The uncertainty of the mean estimate due to the finite sample size (i.e., less than 100% inspection) in this example will work in the same way as the analytical measurement uncertainty.

The example described above employed a two-sided limit, i.e., an upper and lower bound. There are many examples of one-sided limits, where the quality attribute (in some specified representation, e.g., as a sample average, or a batch mean estimate, or individual measurements in a sample) is required to be either below an upper bound or above the lower bound. An example of the former might be an upper limit on moisture content or the level of an impurity. An example of the latter might be a minimum allowable level for an antioxidant. The same principles apply to the one-sided case as two-sided cases, although the shapes of the OCCs will change. Examples of ideal and real one-sided OCC are illustrated in Fig. 8.6.

The general interpretation of OCCs is that the closer the OCC is to the ideal step function, the better the decision-making process, i.e., better discrimination among batches, particularly those close to the limits. OCCs provide information about type I (false rejection) and II (false acceptance) error rates as a function of the true value of the quality attribute under consideration. This is also illustrated in Fig. 8.6.

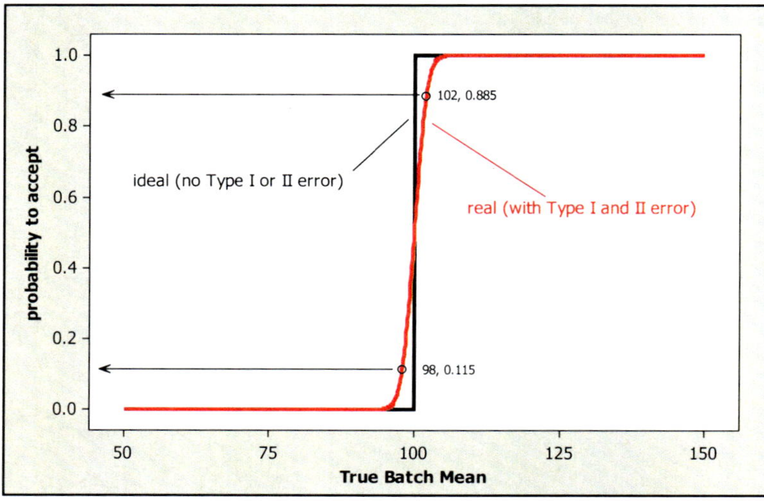

Fig. 8.6 Comparison of an ideal step function (*solid black*) versus a real (*solid red*) OCC for a one-sided limit

The degree to which a real OCC deviates from the ideal step function is influenced by the magnitude of the indeterminate measurement error, the number of sample measurements obtained, and the decision rules used to make the disposition decision. For example, Fig. 8.7 illustrates the impact of increasing the number of replicate measurements and comparing the average of the results to the same limits.

If a different decision rule were employed (e.g., instead of comparing the average of results as above, the decision rule required that all individual results must meet the limits), the consequence would be a different set of OCCs (Fig. 8.8).

The following sections in this chapter focus on applying the concepts of OCCs to evaluating the performance of the EDA ratio metric (*LPM/SPM*) and conventional stage groupings. Three approaches to generating OCCs were considered in this work, differing primarily in the level of sophistication and underlying assumptions. Fundamentally, to construct an OCC, one needs to specify the proposed decision-making process together with critical limits and to have or obtain information concerning the measurement precision. From this information, one can estimate the probability of a particular decision (accept or reject) as a function of assumed true values of the quality attribute.

As described above, the OCC work presented here relied on CI results from the IPAC-RS database. One difference from establishing an OCC in a real QC situation was that the actual regulatory requirements for APSD of these blinded products were unknown. For the purpose of comparing the performance of the various metrics studied, limits were therefore assumed. As a consequence, the results presented here reflect the ability of a specified metric to discriminate among product batches on the basis of those assumed limits (in an "apples-to-apples" comparison).

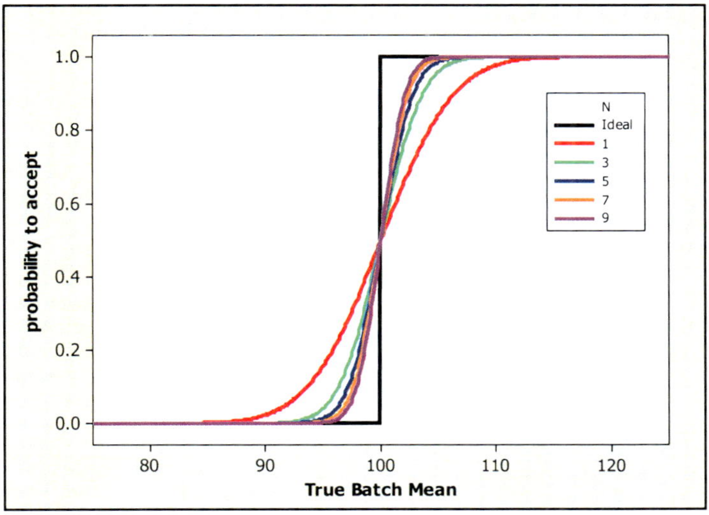

Fig. 8.7 Effect of increasing the number of replicate measurements on the steepness of the OCC for a one-sided limit; decision rule based on comparing the average of results

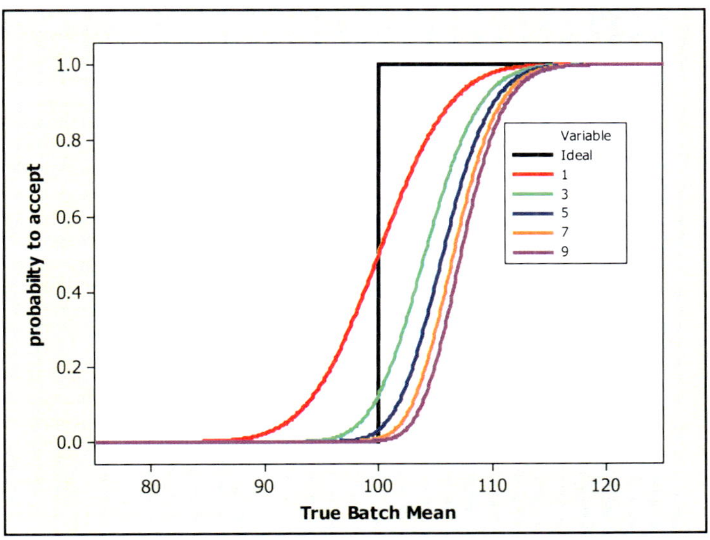

Fig. 8.8 Effect of increasing the number of replicate measurements on the steepness of the OCC for a one-sided limit; decision rule required that all individual results must meet the limits

However, it is important to appreciate that these results have no bearing on the actual quality of any product or particular batch.

Another important difference between the use of OCCs in this chapter and the customary use of OCCs in industrial practice is that here, we are conducting an

Table 8.3 Differences between the construction and meaning of OCCs in this chapter examples and in the customary QC use

OCC in this chapter examples	OCC in customary use
Demonstrates and compares the performance of different metrics (*LPM/SPM* and grouped-stage depositions) to detect changes in APSD by the ability to predict *MMAD*	Demonstrates and compares performance of a test (i.e., a combination of metrics, decision-making rules, sampling plan, and acceptance criteria) or different tests to make correct decisions
The goal is to characterize the metric's ability to make correct decisions regarding shifts in *MMAD*	The goal is to characterize the ability of the entire test(s) to make correct QC decisions about batch disposition
False rejections and false acceptances are "misclassifications" by the metric	False rejections and false acceptances are errors in decision-making
"Limits" for the metric's values are chosen to enable a consistent and fair comparison	"Limits" or "acceptance criteria" are given as part of the test that is being studied

evaluation of different metrics using the same specification criteria applied to the same data set. By contrast, customarily, OCCs are used to evaluate a different specification criteria or different test designs using a range of simulated data sets (Table 8.3). The metrics being compared here are the *LPM/SPM* ratio and the grouped-stage API mass depositions in the specified groupings, both of which reflect a breakdown of the mass of API into their specified sub-fractions.

8.4.4 PCA Approach: Definitions and Basic Concepts

Both QC testing and in vitro equivalence testing of APSD share the essential goal of comparing a range of numbers (i.e., the APSD profile) against another range of numbers and determining their "sameness." For QC, that "another" range of numbers represents APSD of clinical pivotal batches. For equivalence testing, it is the original product or initial version of a product. Comparing APSD profiles could be done simplistically by overlaying the APSD profiles and "eyeballing" them to determine similarity or differences; however, devising a more objective means of determining whether a difference in the shape of a profile is statistically relevant or not presents a more complex challenge. PCA is one way of accomplishing such a comparison.

Principal component analysis is mathematically defined as an orthogonal linear transformation that transforms the data to a new coordinate system such that the greatest variance by any projection of the data comes to lie on the first coordinate (called the first principal component), the second greatest variance on the second coordinate, and so on [10].

PCA operates on the principle of reducing the dimensionality of a given data set containing a large number of variables. PCA has been used to deal with distributional data successfully in a number of fields [11]. For example, PCA has been used as a means of assessing near-infrared (NIR) spectroscopic data sets [12], in the newspaper industry for process monitoring [13], as a means of assessing stationary phase selectivity in HPLC analysis [14] and to classify beer taste profiles [15]. More closely related to the subject of this book is the work of Sandler and Wilson, where PCA was used as a means of comparing particle size and shape data from laser diffraction analysis [16]. In all of these cases, PCA was utilized in order to deal with a large data set, where each sample consists of multiple variables, which would be impossible to cross-correlate in a univariate fashion.

PCA generally consists of two types of graphical presentation: a scores plot and a loadings plot:

- The scores plot is a Cartesian scatter plot where the abscissa axis contains a user-selected principal component (PC). The ordinate axis contains another user-selected PC. The plot contains points that represent the original samples projected onto the user-selected PCs. By default, the scores plot shows data on the first two PCs (PC1 versus PC2).
- The loadings plot is another Cartesian scatter plot that displays the individual elements of the PCs. Since each PC is a vector, it has constituent elements which are called the coefficients or loadings. By mathematical definition, the vector length of each PC is 1. The loadings of a given PC represent the relative extent to which the original "variables" influence the PC.

An APSD profile derived from either an ACI or NGI is very similar to the data of Sandler and Wilson [16]. For the study described further on in this chapter, a model was established that was based on 252 clinically relevant batch measurements (described in Sect. 8.2). All other measurements were then assessed in relation to this model. A comparison was subsequently made between the EDA and grouped-stage approaches using the same data set. The number of errors was evaluated using the PCA as the reference point.

8.5 Results and Assumptions from the Three Approaches to the Evaluation of CI-Derived Metric Performance in Assessment of OIP Quality

8.5.1 MSA Approach

A detailed measurement system analysis characterizing and comparing the performance of the metric *LPM/SPM* to stage groupings was conducted using cascade impactor results from IPAC-RS database. The MSA performed on the IPAC-RS data set benefits from the nature of the data employed. This data set includes stage-by-stage CI results for the eight products studied. Normally, for destructive testing, as is the case with CI testing, the design would attempt to minimize sample-to-sample variability and treat them as replicate measures. In this instance, the goal was to determine the ability of EDA to discriminate or detect differences in APSD on the basis of *MMAD* as the measure of central tendency of each APSD. *MMAD* by itself does not describe the APSD. However, once such a distribution is given (as in examples used in this study), the *MMAD* value is an important characteristic as well as a key indicator of any change to that APSD.

The advantage arising from using full-resolution CI results is that both the EDA metrics and stage groupings can be compared to *MMAD* values determined on the *exact same* CI run. This situation eliminates the need to obtain replicate measurements, and the measurement variability can be directly estimated by assuming all the measurement error resides with the EDA or stage grouping metrics when applied to the estimation of *MMAD*. This latter assumption is reasonable considering the expected precision of determining *MMAD* from the cumulative APSD obtained through the CI-based measurement. Further, since one goal was to compare the performance of EDA to stage groupings, any observed difference in the relative performance of the metrics should be unaffected by any slight noise contribution from the *MMAD* determination.

In practice, values of *MMAD* were determined for each individual CI run according to the method outlined by Christopher et al. [8]. A logistic model was fit to the cumulative CI stage data and the resulting model was used to determine the *MMAD* of that particular measurement. Next, the *LPM/SPM* ratio and stage groupings were also computed for each corresponding individual CI measurement. The *LPM/SPM* ratio was obtained at a previously determined optimum boundary between *LPM* and *SPM* (see Chap. 7).

The particular stage groupings were preselected consistent with FDA recommendations outlined in Table 8.1 (Sect. 8.1). Regression analysis was then performed on the metrics versus *MMAD*, on a product-by-product basis. The relationship between *LPM/SPM* and *MMAD* was already known to be nonlinear, and therefore, simple power functions were observed to provide an adequate fit to the data. An *a priori* model for the relationship between stage groupings and *MMAD* was unknown. Linear models appeared to provide an adequate fit and were used to fit stage grouping results to *MMAD*. In all cases, 95% prediction bands were computed for each regression. An example of the results is depicted in Figs. 8.9, 8.10, and 8.11 for a CFC suspension MDI (IPAC-RS blinded database code: ***w9kw01***).

Table 8.4 is a summary of the full results of the regression analyses for all eight products that were evaluated. The expressions for the power and linear models were of the form

$$\text{metric} = b_0[MMAD]^{b_1} \tag{8.1}$$

and

$$\text{metric} = b_0 + b_1[MMAD] \tag{8.2}$$

respectively.

An examination of Table 8.4 yields some general observations. In all cases, the model fit for the ratio metric was superior to any grouped stage (group 1–group 4) for all products, as evidenced by the relative magnitudes of the coefficients of determination (R^2).

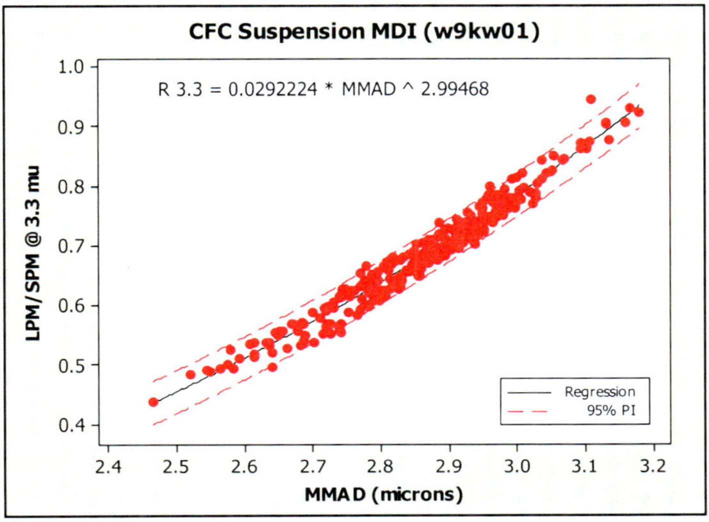

Fig. 8.9 Ratio metric *LPM/SPM* versus *MMAD* from individual CI APSD determinations for a CFC-formulated MDI

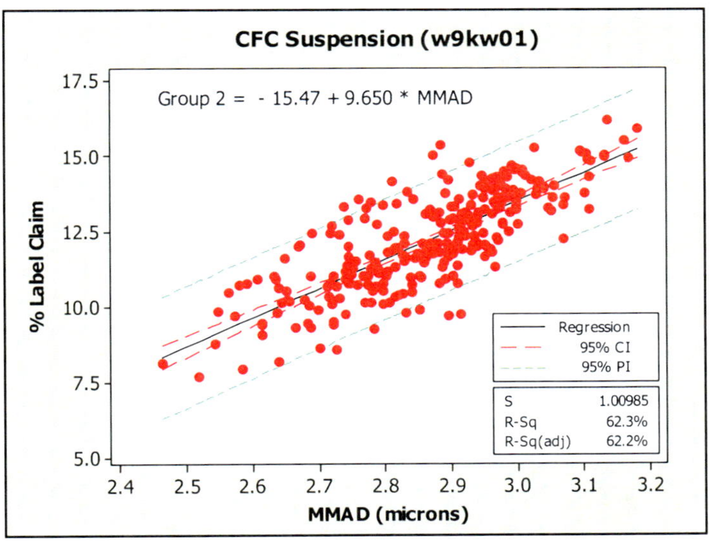

Fig. 8.10 CI stage group 2 versus *MMAD* from individual CI APSD determinations for a CFC-formulated MDI

Further, the data for all OIPs exhibited consistent monotonic trends between the ratio metric, *LPM/SPM*, and *MMAD*. Based on a general lack of correlation, neither group 1 (unsized particles) nor group 3 appears to contain any significant information concerning the mean of the APSD (taken as the *MMAD*). With one exception (product

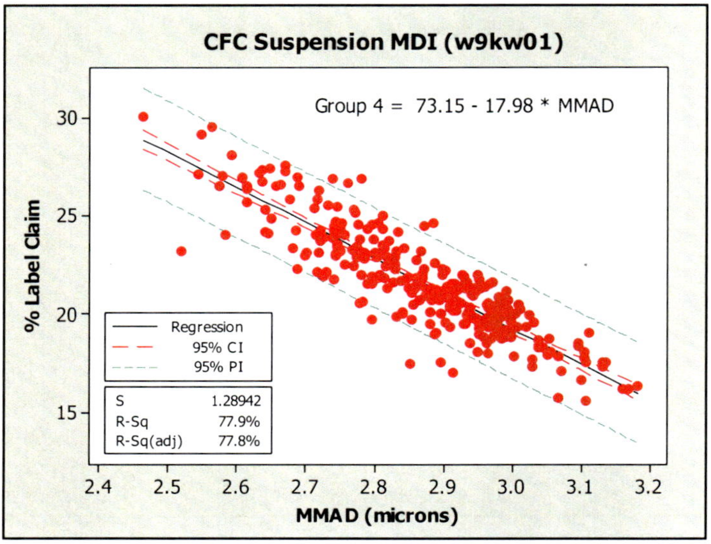

Fig. 8.11 CI stage group 4 versus *MMAD* from individual CI APSD determinations for a CFC-formulated MDI

w9jk01), there was a consistent pattern concerning groups 2 and 4. In general, group 2 increased, while group 4 decreased as a function of *MMAD*.

To perform an MSA, the prediction bands determined through the regression analysis were in turn used to construct figures of merit that reflect the measurement capabilities of the respective metrics. Prediction bands are an estimate of the indeterminate error associated with a regression model. In the present case, this is an approximation of the measurement error associated with estimating *MMAD* by the particular metric considered (i.e., *LPM/SPM* or stage grouping).

Calculations based on the prediction intervals, called figures of merit, were constructed from the regression models. These figures of merit reflect the measurement capabilities and the measurement variability of the respective metrics, EDA, and stage groupings. In the present case, based on inverse regression techniques, this is an estimate of the measurement error associated with estimating *MMAD* by the particular metric considered (i.e., *LPM/SPM* or stage grouping). The figures of merit constructed from the regression models are for comparative purposes to give an indication of the performance of EDA compared to grouped stages. These were not intended to make statistically exact absolute statements about the population; they were intended to be used in a relative sense to compare performance between two different metrics.

One such figure of merit was constructed based on inverse regression. As an example, from Fig. 8.9, at the *LPM/SPM* ratio corresponding to the average *MMAD*, the limits of the prediction interval were projected onto the *MMAD* axis.

Table 8.4 Regression analysis of eight OPIs from the blinded IPAC-RS APSD database

Product	Metric	b_0	b_1	Model	R^2 (%)
w9k201	**Ratio**	**0.00284634**	**3.69211**	**Power**	**99.9**
	Group 1	−40.84	22.076	Linear	48.2
	Group 2	−32.049	12.2654	Linear	79.4
	Group 3	41.951	−2.929	Linear	5.2
	Group 4	16.653	−2.8122	Linear	25.5
w9k901	**Ratio**	**0.118678**	**3.0304**	**Power**	**99.7**
	Group 1	58.888	3.252	Linear	0.5
	Group 2	−17.71	8.5686	Linear	74.5
	Group 3	−7.11	11.301	Linear	13.6
	Group 4	6.8964	−1.3431	Linear	13.4
w9j901	**Ratio**	**0.00618604**	**4.24202**	**Power**	**99.8**
	Group 1	42.328	0.995	Linear	0.2
	Group 2	−30.489	18.0471	Linear	91.8
	Group 3	68.16	−10.151	Linear	37.7
	Group 4	17.988	−5.67	Linear	28.9
w9j601	**Ratio**	**0.0684496**	**3.53278**	**Power**	**99.8**
	Group 1	65.34	2.018	Linear	0.1
	Group 2	−2.9706	2.2962	Linear	50.1
	Group 3	41.459	0.088	Linear	0.0
	Group 4	7.97	−1.7094	Linear	31.5
w9j801	**Ratio**	**0.10997**	**2.85358**	**Power**	**99.6**
	Group 1	30.549	20.56	Linear	31.9
	Group 2	−3.5972	6.4721	Linear	59.3
	Group 3	56.861	−10.406	Linear	15.1
	Group 4	25.0026	−11.1332	Linear	73.5
w9k001	**Ratio**	**0.00731676**	**4.1528**	**Power**	**99.9**
	Group 1	−20.109	11.304	Linear	12.0
	Group 2	−36.46	14.9147	Linear	59.4
	Group 3	58.554	−2.684	Linear	0.9
	Group 4	41.804	−9.5447	Linear	57.1
w9jk01	**Ratio**	**0.0150038**	**3.24843**	**Power**	**99.3**
	Group 1	90.81	−9.985	Linear	2.8
	Group 2	−18.585	8.5812	Linear	59.2
	Group 3	−9.537	13.078	Linear	8.5
	Group 4	6.522	0.3908	Linear	0.2
w9kw01	**Ratio**	**0.0292224**	**2.99468**	**Power**	**99.9**
	Group 1	−4.411	7.623	Linear	10.6
	Group 2	−15.472	9.6498	Linear	62.3
	Group 3	25.338	5.684	Linear	5.4
	Group 4	73.147	−17.9754	Linear	77.9

This represents an interval expected to contain 95% of the time or with 95% confidence, an individual *MMAD* value associated with a particular ratio value.

The projected prediction intervals from the different regression models are reported in Table 8.5. The prediction intervals for the ratio are narrower than for the grouped stages, highlighting the fact that the ratio is able to detect changes in *MMAD* with more precision compared to grouped stages.

Table 8.5 MSA comparisons of EDA ratio metric (*LPM/SPM*) to results from stage groupings for eight OPIs from the blinded IPAC-RS APSD database

Product	Metric	CI Stage d50(μm)	average MMAD (μm)	95% Prediction Interval Lower limit	95% Prediction Interval Upper limit	Discrimination μindex(m-1)	Precision/ Total Variability (%)
w9k201	**Ratio**	**4.7**	**3.91**	**3.836**	**3.980**	**6.96**	**18.0**
	Group 1			3.484	4.336	1.17	106.6
	Group 2			3.700	4.120	2.39	52.5
	Group 3			2.158	5.662	**0.29**	438.6
	Group 4			3.207	4.613	**0.71**	175.9
w9j901	**Ratio**	**3.3**	**2. 57**	**2.495**	**2.638**	**6.99**	**13.9**
	Group 1			−9.494	14.632	**0.04**	2343.8
	Group 2			2.407	2.733	3.07	31.7
	Group 3			1.868	3.272	**0.71**	136.4
	Group 4			1.714	3.426	**0.58**	166.3
w9j801	**Ratio**	**2.1**	**1.50**	**1.415**	**1.577**	**6.18**	**13.9**
	Group 1			0.643	2.357	**0.58**	147.8
	Group 2			1.014	1.986	1.03	83.8
	Group 3			0.109	2.891	**0.36**	240.0
	Group 4			1.148	1.852	1.42	60.7
w9jk01	**Ratio**	**3.3**	**2.66**	**2.567**	**2.747**	**5.55**	**44.7**
	Group 1			1.457	3.863	**0.42**	597.1
	Group 2			2.492	2.829	2.97	83.6
	Group 3			1.993	3.327	**0.75**	330.8
	Group 4			−2.184	7.505	**0.10**	2404.4
w9k901	**Ratio**	**2.0**	**2.59**	**2.490**	**2.682**	**5.21**	**39.4**
	Group 1			−0.879	6.058	**0.14**	1425.0
	Group 2			2.446	2.734	3.48	59.0
	Group 3			1.971	3.209	**0.81**	254.2
	Group 4			1.965	3.215	**0.80**	256.6
w9j601	**Ratio**	**2.1**	**2.54**	**2.474**	**2.602**	**7.83**	**20.1**
	Group 1			−6.497	11.576	**0.06**	2844.1
	Group 2			2.206	2.874	1.50	105.3
	Group 3			−91.011	96.114	**0.01**	29448.5
	Group 4			2.047	3.034	1.01	155.4

(continued)

Table 8.5 (continued)

Product	Metric	CI Stage d50(μm)	average MMAD (μm)	95% Prediction Interval Lower limit	Upper limit	Discrimination index(μm-1)	Precision/ Total Variability (%)
w9k001	**Ratio**	**3.3**	**3.54**	**3.504**	**3.575**	**13.96**	**14.8**
	Group 1			2.880	4.200	**0.76**	272.8
	Group 2			3.339	3.741	2.48	83.3
	Group 3			0.974	6.105	**0.19**	1060.7
	Group 4			3.329	3.751	2.36	87.4
w9kw01	**Ratio**	**3.3**	**2.86**	**2.807**	**2.910**	**7.43**	**19.6**
	Group 1			2.091	3.630	**0.65**	292.1
	Group 2			2.654	3.066	2.42	78.4
	Group 3			1.753	3.967	**0.45**	420.3
	Group 4			2.718	3.001	3.53	53.7

Discrimination Index	Color Code	Range
	(dark green)	>10
	(light green)	>5
	(yellow)	>1
	(red)	<1

An alternative approach commonly used in MSA is to consider the levels of discrimination possible with a particular measurement within a critical range, typically the specification requirements. In the present case, *MMAD* did not have specific requirements, but a reasonable range can be assumed and applied to all metrics (i.e., EDA and grouped stages), leading to a valid relative comparison of performance. A range of 1 μm aerodynamic diameter was assumed, and figures of merit were computed as 1.0 μm divided by the width of the transposed 95% prediction interval at the mean value of *MMAD* from the inverse regression approach above. The resulting values are tabulated in Table 8.5. The discrimination index was much higher for the ratio metric, *LPM/SPM*, compared to the grouped stages, again showing the ability of the ratio metric to detect smaller shifts or changes of the *MMAD*, compared to grouped stages.

The precision to total variability figure of merit was also examined through a graphical representation. In Fig. 8.12, 95% prediction intervals for these metrics were compared to intervals constructed to encompass the total variability for each

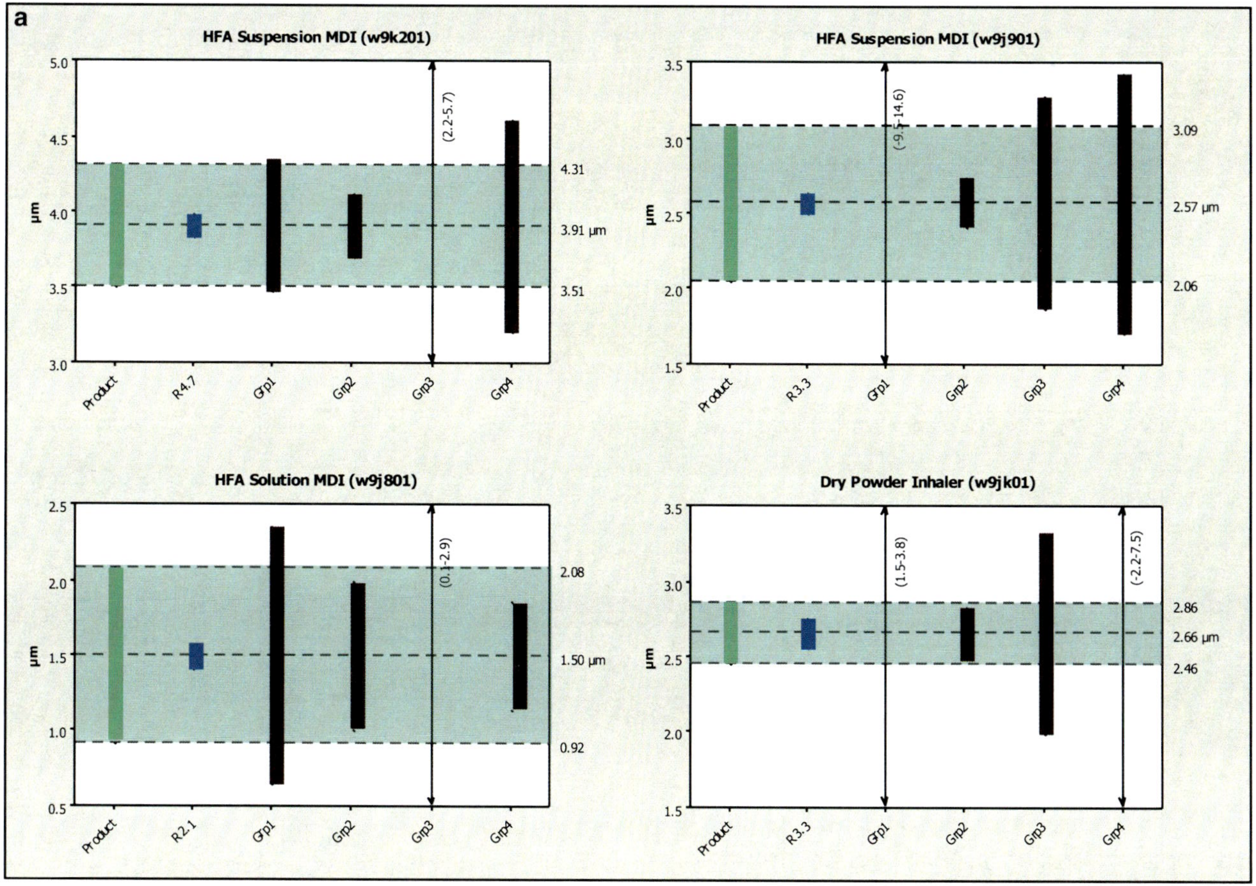

Fig. 8.12 Graphical comparison of measurement precision to product variability. (**a**) OIPs ***w9k201***, ***w9j901***, ***w9j801***, and ***w9jk01***. (**b**) OIPs ***w9k901***, ***w9j60***1, ***w9k001***, and ***w9kw01***

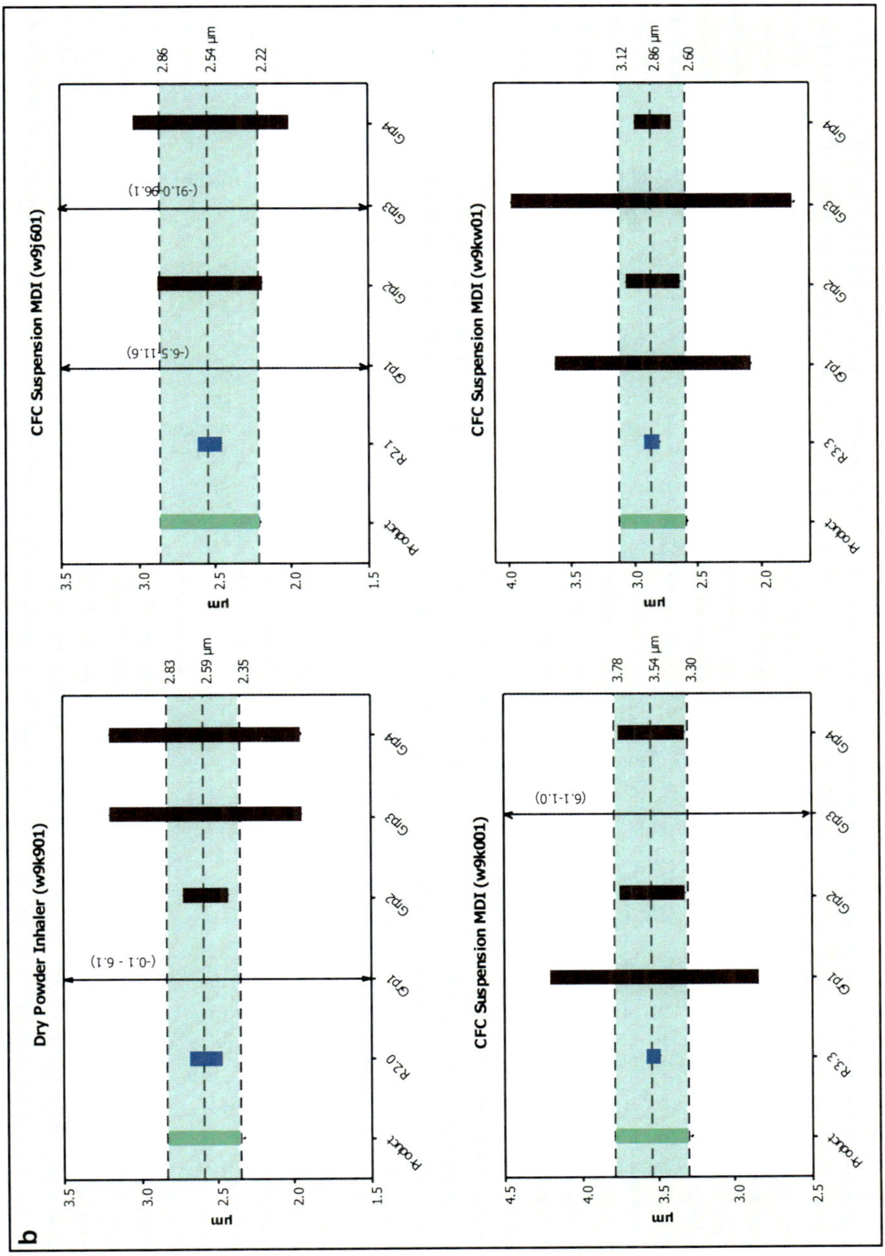

Fig. 8.12 (continued)

of the products. The intervals that represent the total variability were just under 4σ units in width, σ representing the estimated standard deviation (SD) and μ representing the estimated mean in *MMAD* for that product. This interval was specifically defined as

$$\mu + (1.96\sigma) - [\mu - (1.96\sigma)] = 3.92\sigma \tag{8.3}$$

in which 1.96σ is the width of an interval that would contain, approximately, 95% of the individual *MMAD* values. This figure of merit is a comparison of the variability associated with estimating the average *MMAD* to the total variability in the individual *MMAD* measurements. In all cases, the variation in estimating the average *MMAD* through the *LPM/SPM* ratio was consistently severalfold smaller than the variability expected from individual measurements, indicating the consistent ability to adequately detect difference in *MMAD* within the normal observed variability of the product.

In contrast, the grouped stages consistently exhibited poorer performance. Groups 1 and 3 were consistently unable to detect difference in *MMAD* within this range, and groups 2 and 4 exhibited inconsistent performance with respect to being able to detect differences. Even in those cases where one of these metrics might be judged adequate to detect changes in *MMAD*, its performance was inferior to the corresponding *LPM/SPM* metric.

The results in Table 8.5 and Fig. 8.12 confirm that for the eight products examined, the *LPM/SPM* ratio consistently out-performed all stage groupings with respect to a measurement related to the central tendency of the APSD (as reflected in the values of *MMAD*). Note that the CI stage d_{50} in this table refers to the cut-point size of the stage separating *LPM* from *SPM*. The two MSA figures of merit also confirmed the early observations concerning groups 1 and 3, i.e., that these stage groupings contain virtually no information related to the central position of the APSD.

The values obtained for the discrimination index indicate that the *LPM/SPM* ratio is consistently able to detect differences in *MMAD* of the order of tenths of microns. While groups 2 and 4 exhibited some ability to detect changes in *MMAD*, the performance was both inconsistent and always inferior to the *LPM/SPM* ratio for the same product.

Another graphical approach to comparing performance is to plot *MMAD* predicted by the metric through the regression model versus the actual *MMAD* as determined through the full-resolution CI results. The comparison was limited to comparing the ratio metric to only the group 2 and group 4 results since the other two groupings consistently appear unable to offer any information about the *MMAD*. The results of this analysis are depicted in Figs. 8.13, 8.14, 8.15, 8.16, 8.17, 8.18, 8.19, and 8.20 and again clearly demonstrate the consistent superiority of the ratio metric over stage groups with respect to measuring changes in the *MMAD*, i.e., any shift in the average particle size of the emitted aerosol from any OIP.

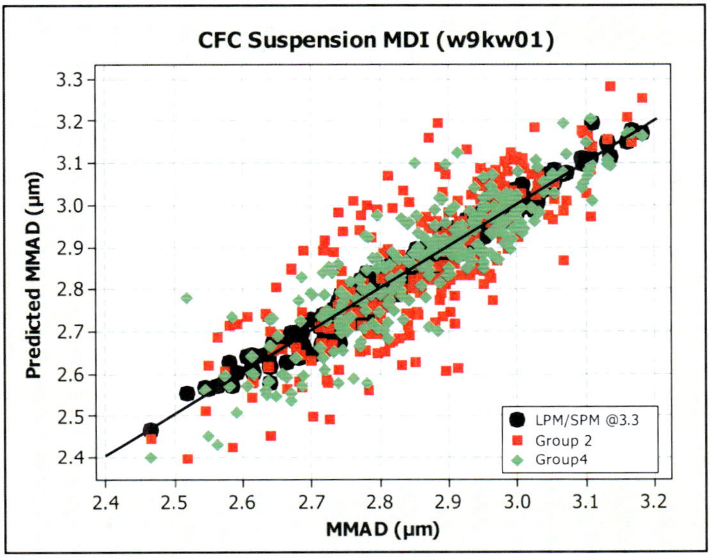

Fig. 8.13 Comparison between predicted (by metric) versus actual *MMAD* from CI data: CFC suspension product code, ***w9kw01***

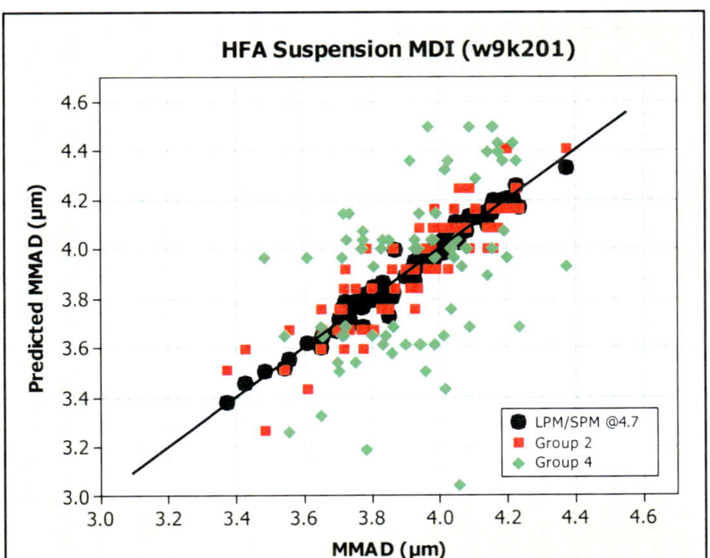

Fig. 8.14 Comparison between predicted (by metric) versus actual *MMAD* from CI data: HFA suspension MDI product code, ***w9k201***

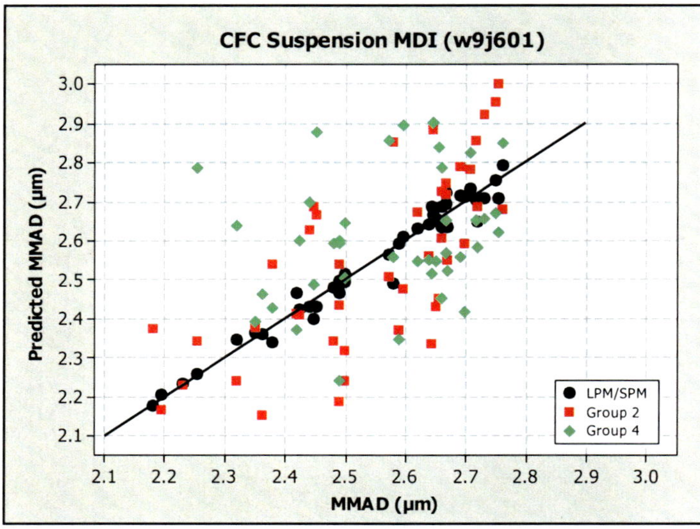

Fig. 8.15 Comparison between predicted (by metric) versus actual *MMAD* from CI data: CFC suspension MDI product code, ***w9j601***

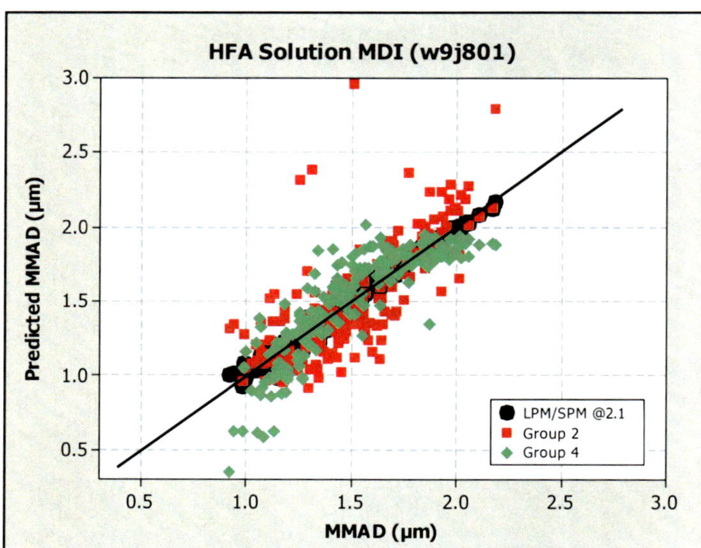

Fig. 8.16 Comparison between predicted (by metric) versus actual *MMAD* from CI data: HFA solution MDI product code, ***w9j801***

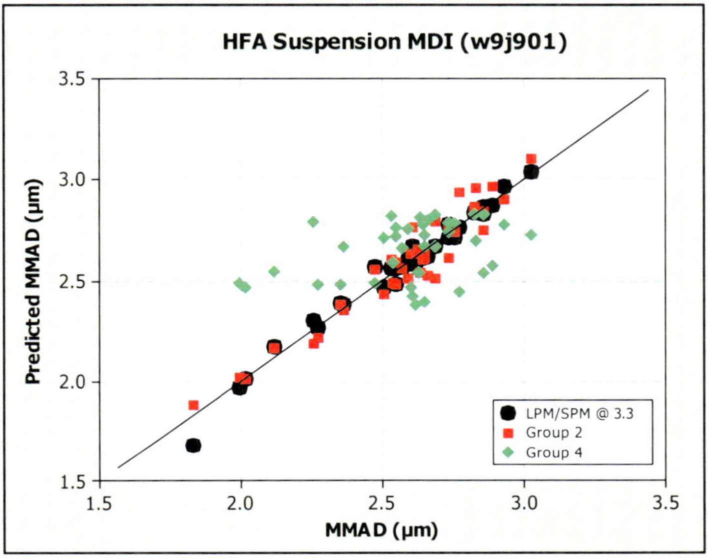

Fig. 8.17 Comparison between predicted (by metric) versus actual *MMAD* from CI data: HFA suspension MDI product code, ***w9j901***

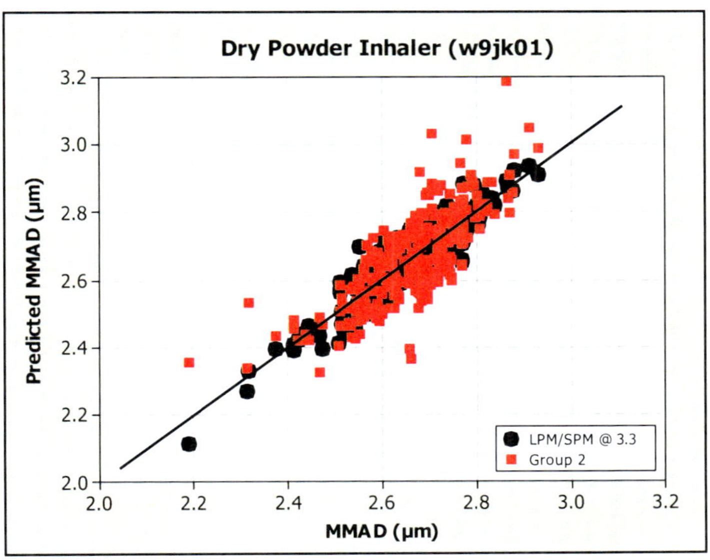

Fig. 8.18 Comparison between predicted (by metric) versus actual *MMAD* from CI data: DPI product code, ***w9jk01***; group 4 results excluded due to exceptionally poor correlation with actual *MMAD*

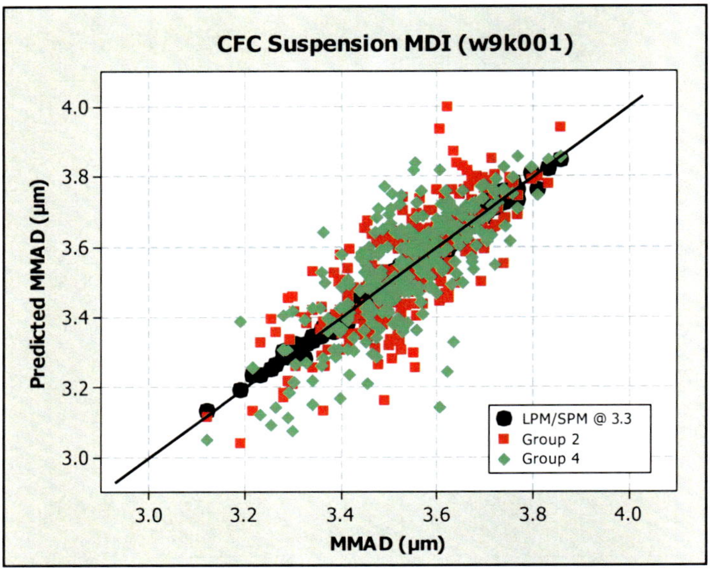

Fig. 8.19 Comparison between predicted (by metric) versus actual *MMAD* from CI data: CFC suspension MDI product code, ***w9k001***

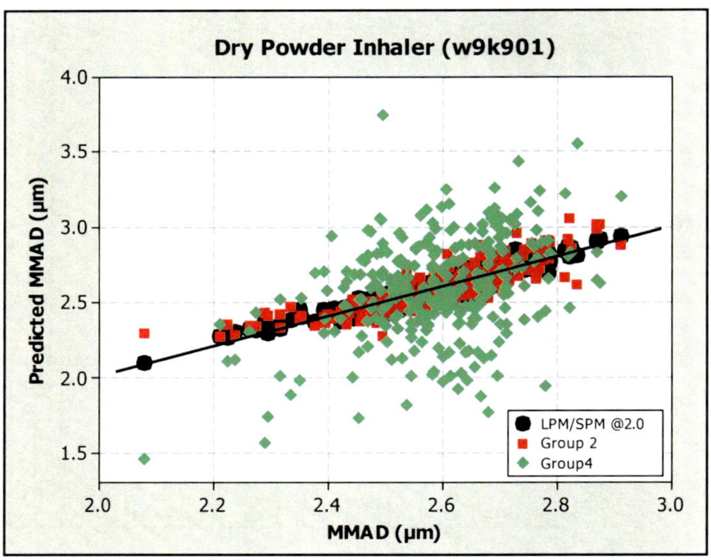

Fig. 8.20 Comparison between predicted (by metric) versus actual *MMAD* from CI data: DPI product code, ***w9k901***

8.5.2 *OCC Approaches*

Two different strategies were used to develop OCC curves to compare the relative capability of the grouped-stage metrics versus EDA to make the correct decision about shifts in *MMAD*. The first strategy described below, developed by Tougas and termed the "Tougas" approach, is based on a more simplistic model with basic assumptions. The second strategy, developed by Christopher and Dey and termed the "Christopher–Dey" approach, uses the same concepts but incorporates a more sophisticated modeling of the actual CI data with simulation relying on fewer assumptions and driven more by characteristics of the actual CI data. The two different assessment strategies, however, lead to the same conclusion that the EDA approach, even under conservative assumptions, has a higher probability of correct decisions.

The purpose of this exercise was not to compare or assess product quality. Consequently, the values of acceptance limits were selected for the purpose of comparing the two approaches. These limits were based on the body of data for eight different OIP cases available from the IPAC-RS blinded database, considered to represent process capability. However, actual specification limits for those products are unknown and may well be wider than the limits used here.

In this assessment, the stage groupings included only those impactor stages with upper cutoff limits (groups 2–4 from Table 8.1). This is to ensure that both the EDA and grouped-stage approaches are equivalent in terms of the total mass of sized material (i.e., possessing the same *ISM*).

8.5.2.1 Overview of "Tougas" Strategy

In the Tougas approach, individual stage means were calculated from actual CI data for each product considered, and a cumulative particle size distribution was fit to establish a reference, or "ideal" APSD.

As discussed above, the direct determination of an OCC requires both limits and an estimate of measurement precision. To construct OCCs that reflect the ability to discriminate based on changes in *MMAD*, a series of assumed limits for *MMAD* were selected, and overall measurement precision values were applied to all eight OIP cases. The basis of the selected measurement precision (*LPM*/*SPM* and grouped stages) was the observed precision within the studied products. The actual values applied to this analysis were 0.05% and 1% LC for *LPM*/*SPM* and grouped stages, respectively. Normally, the computation of an OCC would also require the test plan and decision schema. For simplicity and a fair comparison, it was assumed in the present case that the decision was based on the result from a single measurement.

The overall procedure can be summarized as follows:

1. Consider a particular OIP.
2. Calculate the means of all individual stages for all available runs.
3. Fit cumulative particle size distribution to stage means. This defines the reference/ideal APSD (a 4-parameter logistic model with the minimum forced to 0 was used in the described work, according to the method described by

Christopher, et al. [8]). Note that the fitted parameters in this model directly relate to the *MMAD* value, maximum size limit, minimum size limit, and "steepness" of the cumulative mass-weighted form of the APSD. Besides *MMAD* and steepness, the parameters include the two asymptotes (i.e., the minimum and maximum mass), noting that the former is forced to zero.

4. Define two cumulative particle size distributions that bracket the reference/ideal cumulative APSD fit in step 3. These two cumulative particle size distributions represent distributions that are considered equivalent/similar to the reference/ideal. The degree of sameness/similarity to the reference/ideal APSD was established by computing a goodness-of-fit statistic between the displaced cumulative APSD and the reference/ideal APSD. That procedure is described below. Region of similarity to the target APSD moving the distribution below and above—self-imposed similarity—reproducible methodology for characterizing minimum and maximum batches with similar APSD properties and *MMAD* value.
5. From the displaced cumulative APSD distributions, limits that represent equivalency to the reference/ideal APSD curve could be calculated for *MMAD*, *LPM*/*SPM* ratio, and grouped stages. This procedure[1] established "goalposts" across all three metrics of equivalency to the reference/ideal cumulative APSD metric. Label these equivalency "goalposts" θ_L and θ_H.
6. Based on the parameter estimates from step 3, create a series of cumulative APSDs where only the *MMAD* was varied.
7. Compute the corresponding APSD metrics (*LPM*/*SPM* ratio and grouped stages) from the series of cumulative APSDs generated in step 6.
8. Assuming standard deviations for *LPM*/*SPM* of ±0.05 and ±1% LC for each stage grouping and an acceptance criterion based on a single measurement, calculate the probability of meeting the limits established; i.e., calculate the following probability:

$$\text{prob}(\theta_L) \le X \le \left(\frac{\theta_H}{X}\right) \approx \text{normal}(\mu, \sigma = 0.05 \text{or} \sigma = 1\%) \tag{8.4}$$

where *X* represents either *LPM*/*SPM* ratio or grouped stages and μ represents the appropriate metric value computed in step 7 from the series of cumulative APSDs that generated varying *MMAD*. The observed data, either ratio or grouped stages, were unimodal and approximately symmetric; therefore, normality is not an unreasonable approximation from which to base the comparison.

9. The final OCCs are a plot of probability acceptance or probability of the appropriate metric producing values within the equivalency goalposts, plotted as a function of *MMAD*.

[1]It should be noted that the assumed limits were established by computing a goodness-of-fit statistic between each displaced cumulative APSD and the target cumulative APSD. Several levels of this statistic (0.9, 0.8, 0.7, and 0.6) were applied to each studied product, and the resulting limits are provided in Table 8.6. While this provided a systematic process for establishing assumed limits, the actual values are not critical. The important constraint for comparing the performance of the two metrics is that each particular limit was converted to equivalent requirements with respect to each metric.

Table 8.6 Assumed limits used to construct operating characteristic curves comparing the performance of the ratio metric *LPM/SPM* to "grouped stages"

Goodness-of-fit criterion (R^2)	Assumed limits for *MMAD* (μm)			
	w9j601	***w9j801***	***w9j901***	***w9jk01***
0.9	2.03–3.04	1.14–1.83	2.25–3.00	2.11–2.97
0.8	1.81–3.24	1.02–1.98	2.09–3.15	1.94–3.14
0.7	1.62–3.40	0.93–2.08	1.96–3.27	1.81–3.27
0.6	1.47–3.54	0.85–2.17	1.86–3.36	1.71–3.38
	w9k001	***w9k201***	***w9k901***	***w9kw01***
0.9	3.17–4.04	3.49–4.31	2.26–3.10	2.56–3.36
0.8	2.99–4.23	3.33–4.47	2.08–3.27	2.39–3.52
0.7	2.85–4.37	3.19–4.59	1.95–3.40	2.26–3.66
0.6	2.73–4.49	3.08–4.69	1.84–3.51	2.15–3.77

8.5.2.2 Overview of "Christopher–Dey" Strategy

Using the same stage groupings as those chosen by Tougas (Table 8.1), Christopher–Dey established, for the purpose of this assessment, a range considered to represent acceptable values of *MMAD* and determined the relationship between each of the grouped stages and values of *MMAD* over that range. From this, acceptance limits for each of the grouped stages were determined, which were confirmed to be consistent with typical FDA limits. Applying the same technique, the relationship between the metric *LPM/SPM* ratio and *MMAD* was established, along with corresponding acceptance limits for *LPM/SPM*, thus providing a basis for an "apples-to-apples" comparison of the grouped-stage and EDA approaches. It was necessary to expand the range of CI results while maintaining characteristics of the actual CI data in order to obtain robust estimates of performance characteristics at and beyond the range of acceptable values for *MMAD*. To accomplish this objective, the original data were shifted toward either higher or lower values of *MMAD* in a way that is consistent with physical mechanisms that govern and constrain APSD shifts in OIPs.

8.5.2.3 Results from "Tougas" Strategy

The generation of OCCs by this approach started with defining a target APSD for the specific product under consideration. Figure 8.21 illustrates the cascade impactor results from the IPAC-RS database for one of the products studied, a CFC suspension MDI identified as ***w9j601***. This figure is a superimposition of all individual CI determinations ($n=43$) and presents individual stage data as a function of aerodynamic particle size. Note that the process of defining a target APSD in a real OIP development situation would depend on gathering sufficient representative results that adequately characterize the product in question. In this case, all available data from the IPAC-RS database were utilized. The target APSD was selected from these data, by first averaging the individual mass API-per-stage data on a stage-by-stage basis, then converting to cumulative weighting, and finally fitting a

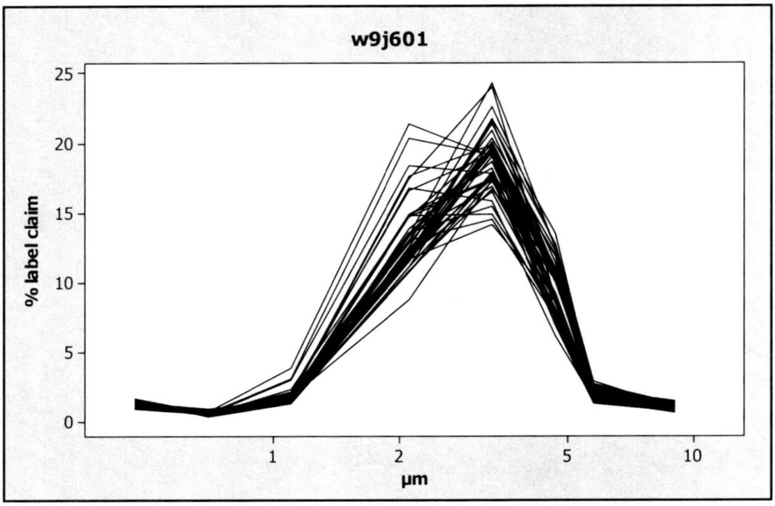

Fig. 8.21 Superimposed CI-measured APSDs for CFC suspension MDI (***w9j601***) from the IPAC-RS database

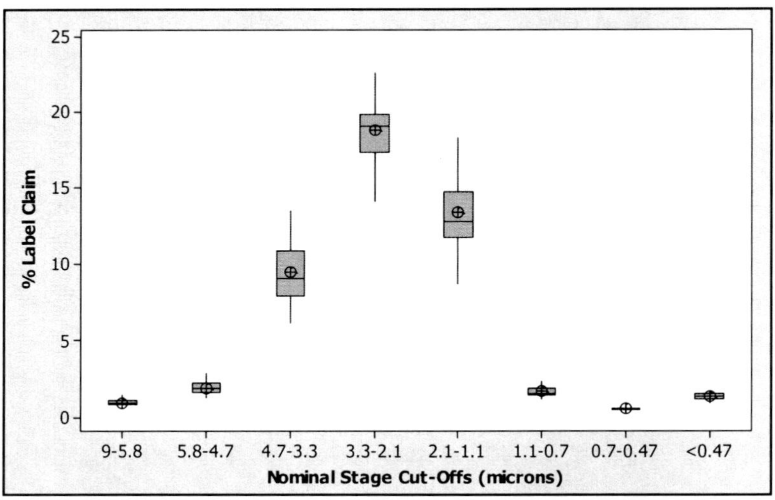

Fig. 8.22 Stage-by-stage box plots of CI results for CFC suspension MDI ***w9j601***; the *encircled* "plus" signs represent the stage mean values for all $n = 43$ results

logistic sigmoidal model to the cumulative APSD. Figures 8.22 and 8.23 illustrate the outcomes for product ***w9j601***.

Limits were established by first constructing a goodness-of-fit statistic based on the concept of a coefficient of determination (R^2). The target APSD was displaced by its

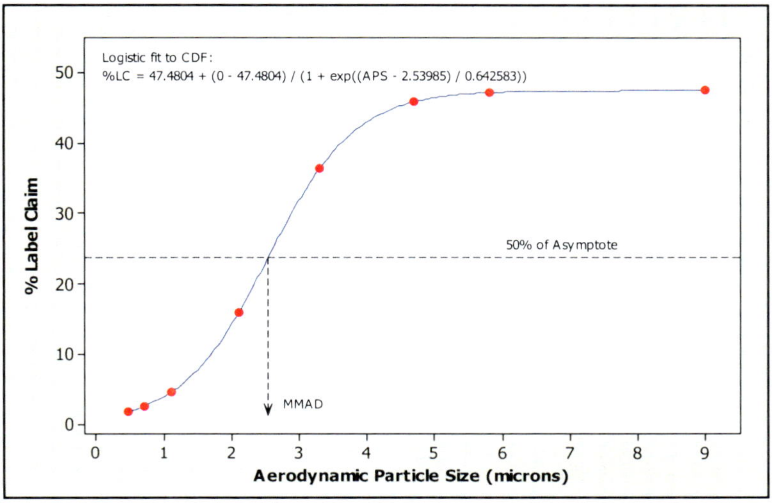

Fig. 8.23 Identification of target cumulative APSD for CFC suspension MDI ***w9j601***: fit of logistic model to cumulative mass-weighted APSD; *%LC* % label claim; "APS" in equation = aerodynamic particle size (μm)

MMAD value, and its pseudo R^2 values were computed as a function of the displaced *MMAD*. Limits in terms of *MMAD* were then established by selecting arbitrary R^2 values (0.9, 0.8, 0.7, and 0.6) and selecting the corresponding *MMAD* values as the assumed limits. The calculation for the pseudo R^2 is given by Eq. (8.5) for clarity.

$$R^2 = \frac{\sum_{i=1}^{n}(y_i - y_i)^2}{\sum_{i=1}^{n}(y_i - \overline{y}_i)^2} = \frac{\sum_{i=1}^{n}(\%\mathrm{LC}_{\mathrm{Displaced}} - \%\mathrm{LC}_{\mathrm{Reference}})^2}{\sum_{i=1}^{n}(\%\mathrm{LC}_{\mathrm{Displaced}} - \%\overline{\mathrm{LC}}_{\mathrm{Displaced}})^2} \tag{8.5}$$

It is important to note that the differences were calculated for particular fixed particle sizes as measures of distance between the curves vertically. This is one approach for determining a bound around the target/reference APSD curve that defines a range of acceptable *MMAD* values.

Again the reader is reminded that these selections are not connected in any manner to the actual quality limits for the studied products. They were selected exclusively to study the characteristic of the APSD-related metrics. Figure 8.24 illustrates this process for OIP ***w9j601*** when R^2 is set at 0.9. The entire set of limits (Table 8.6) employed for constructing OCCs were obtained via this process.

Using these assumed limits for the *MMAD* values, both cumulative APSDs (Fig. 8.25) and APSDs plotted in the form of individual values of API per stage (Fig. 8.26) could be constructed for hypothetical APSDs at the assumed limits. Note that this exercise considers the case of a shift in *MMAD* of the distribution *without* a change in its shape.

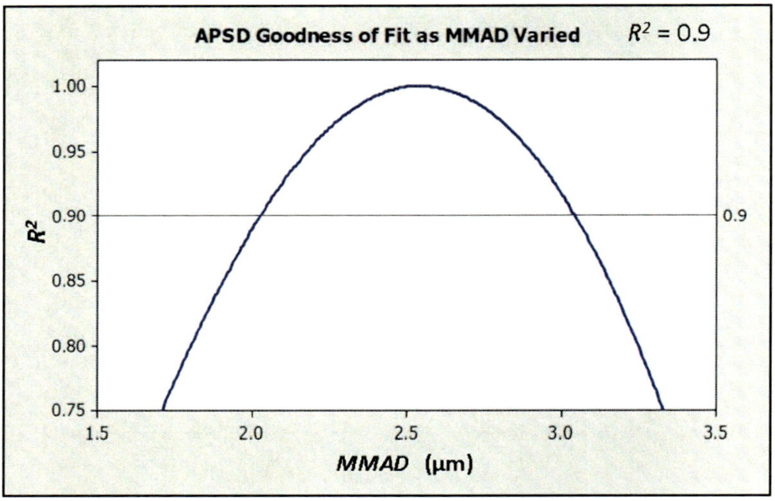

Fig. 8.24 Process for establishing assumed limits for construction of OCCs: the limit is defined by the value of *MMAD* corresponding to the intersection of the selected R^2 value (in this case 0.9) and the goodness-of-fit curve

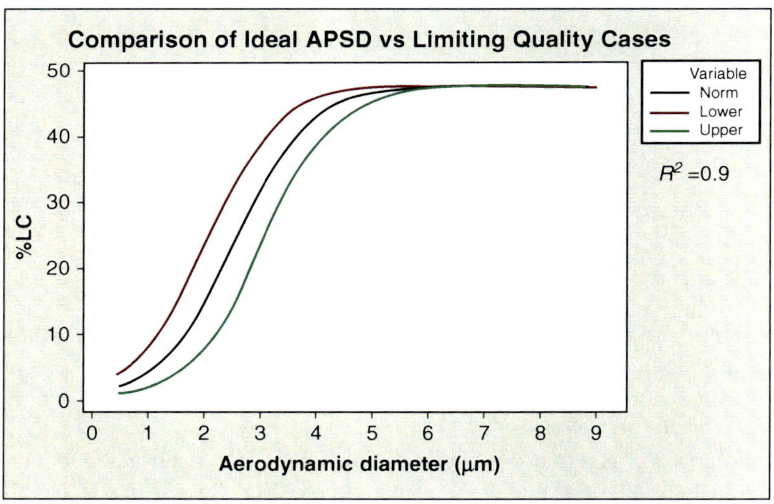

Fig. 8.25 Cumulative mass-weighted APSDs for OIP product ***w9j601*** illustrating target and limiting APSDs derived with R^2 set at 0.9

Corresponding and equivalent limits for the metrics of interest in the context of EDA and "stage groupings" could be established primarily from these limiting cumulative APSDs, once the limits were established in terms of *MMAD* and the corresponding extreme cumulative APSDs were computed. Figure 8.27 tracks the ideal

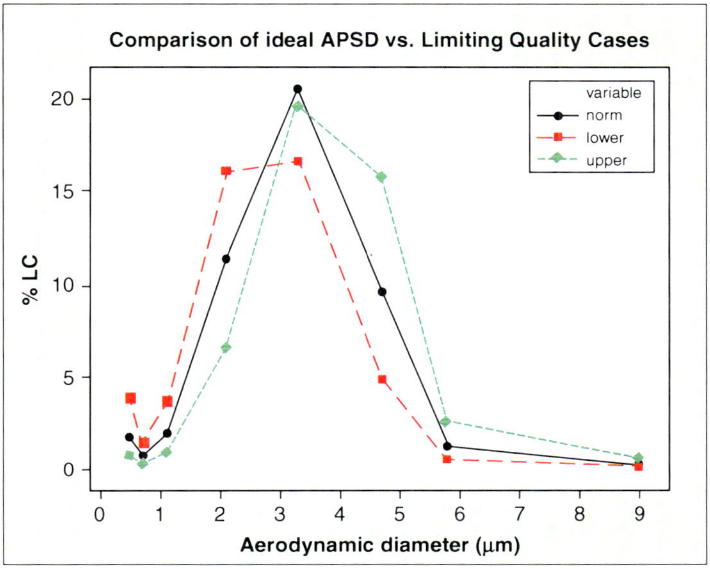

Fig. 8.26 Individual API mass API-per-stage values derived from target and limiting cumulative APSDs for OIP product ***w9j601*** whose cumulative mass-weighted APSD is shown in Fig. 8.25

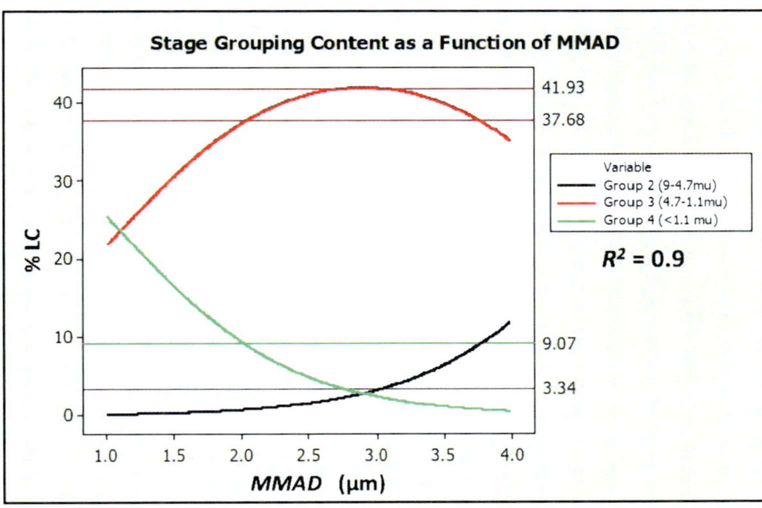

Fig. 8.27 Expected ideal behavior of CI stage groups as a function of shifting the *MMAD* value of the target APSD for OIP product ***w9j601***. The *horizontal lines* represent one-sided limits that were established for the groupings related to the upper and lower tail of the APSD (groups 2 and 4) and the two-sided limit for the central grouping (group 3)

behavior of stage groupings used for OIP ***w9j601*** as a function of shifting the *MMAD* of the target APSD and superimposes the limits derived from the limiting cumulative APSDs. One-sided limits were established for the groupings related to the upper and lower tail of the APSD (groups 2 and 4), as is the common current practice, while a two-sided limit was established for the central grouping (group 3).

Some immediate observations arise from closer inspection of Fig. 8.27. Firstly, this presentation supports the concept of using one-sided limits for groups 2 and 4 to detect shifts in particle size through the mass-weighted median of the APSD (*MMAD*). In principle, an increase in mass of API on stage group 2 signals an increase in *MMAD*, and conversely, an increase in mass of API collected by stage group 4 signals a decrease in *MMAD*. Secondly, the predicted behavior of stage group 3 suggests that it is unlikely to provide a practical control for changes in *MMAD*. This plot predicts only a 4.25% label claim difference corresponding to a 1 μm shift in the value of *MMAD*. Further, the relationship between stage group 3 mass content and *MMAD* is not monotonic, having its maximum value at *MMAD* = ca. 3 μm.

The final steps to generate OCCs were to compute the probability of acceptance for each metric based on the estimated precision of each metric ($s = 0.05$ for *LPM*/*SPM* and $s = 1.0$ for stage groupings) with the assumed limits as a function of the *MMAD* of the displaced APSDs.

In the case of the stage grouping strategy, the overall requirement includes the condition that the limits for all groupings are met. The final probability of acceptance could thus be computed by considering the probability that the assumed limits are simultaneously met for all stage groupings. As discussed above, the inclusion of a group 3 requirement is predicted to be problematic. This fact is borne out by the shapes of the actual OCCs generated by this scheme (Fig. 8.28). This illustration presents the contributions of the individual stage groupings to the overall OCC for one particular case and illustrates the issue of including a stage group 3 requirement with respect to monitoring movement in the value of *MMAD*. As a result of this finding, an additional set of OCCs were produced that dropped the requirement for stage group 3 and instead considered the probability of acceptance for meeting simultaneously only the stage group 2 and 4 requirements.

All resulting OCCs for *LPM*/*SPM* and stage groupings are compared and presented in Figs. 8.29, 8.30, 8.31, 8.32, 8.33, 8.34, 8.35, 8.36, and 8.37. Figure 8.29 illustrates OCCs for *LPM*/*SPM*, "stage groupings," including all three groupings, and "stage groupings," considering only groups 2 and 4, all superimposed on the ideal OCC for comparison. The reader will recall that this particular case illustrates the results for CFC suspension MDI (product ***w9j601***) where the limits were derived from setting R^2 to equal 0.9.

The issue of including group 3 is apparent in this figure where the probability of acceptance (p_{accept}) never reached 1.0 over the entire range of *MMADs* that were considered. This feature was consistently observed over all limits and OIPs that were considered. In this particular case, the same issue, though to a lesser extent, was observed for stage grouping when only groups 2 and 4 were considered. At the same time, the *LPM*/*SPM* metric demonstrated superior performance with a corresponding OCC that exhibited appropriate acceptance and rejection characteristics,

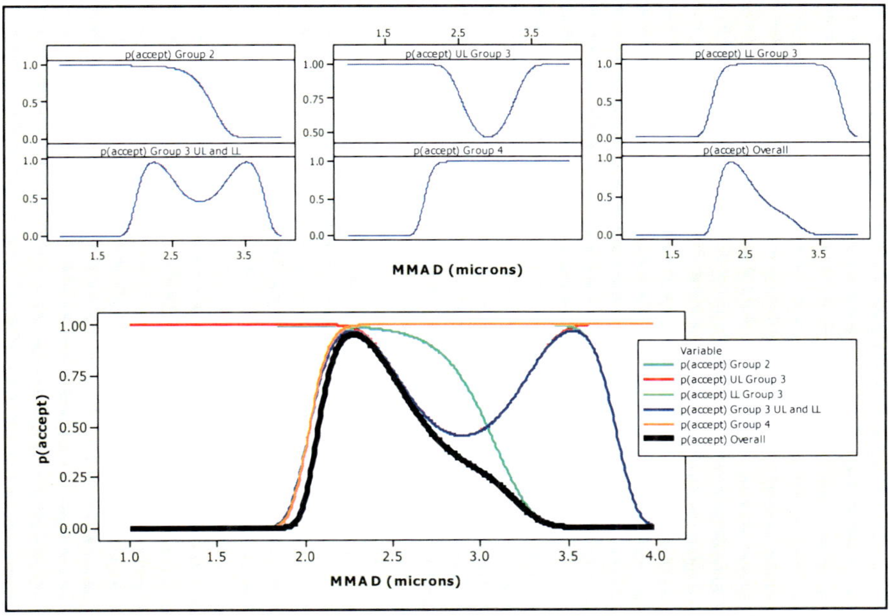

Fig. 8.28 Contributions of individual requirements for grouped stages to the overall combined probability of acceptance for OIP product ***w9j601***. *LL* lower acceptance limit, *UL* upper acceptance limit

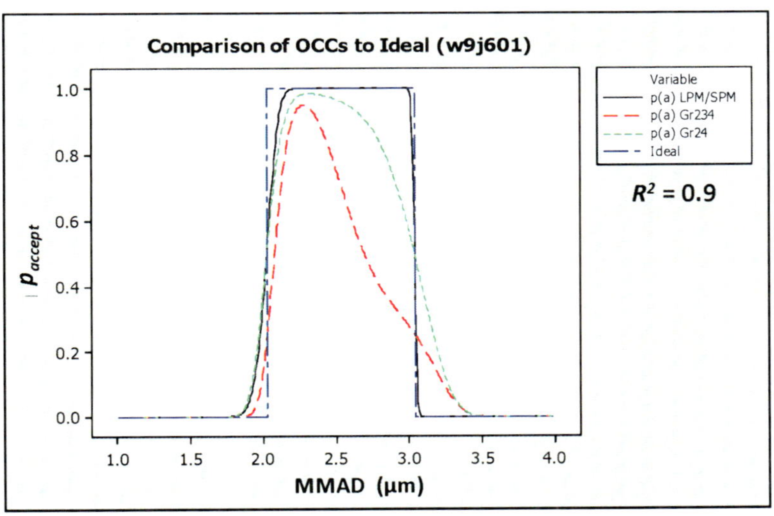

Fig. 8.29 Relative performance of *LPM/SPM* versus grouped stages based on a comparison of OCCs for OIP product ***w9j601*** at limits defined by $R^2 = 0.9$

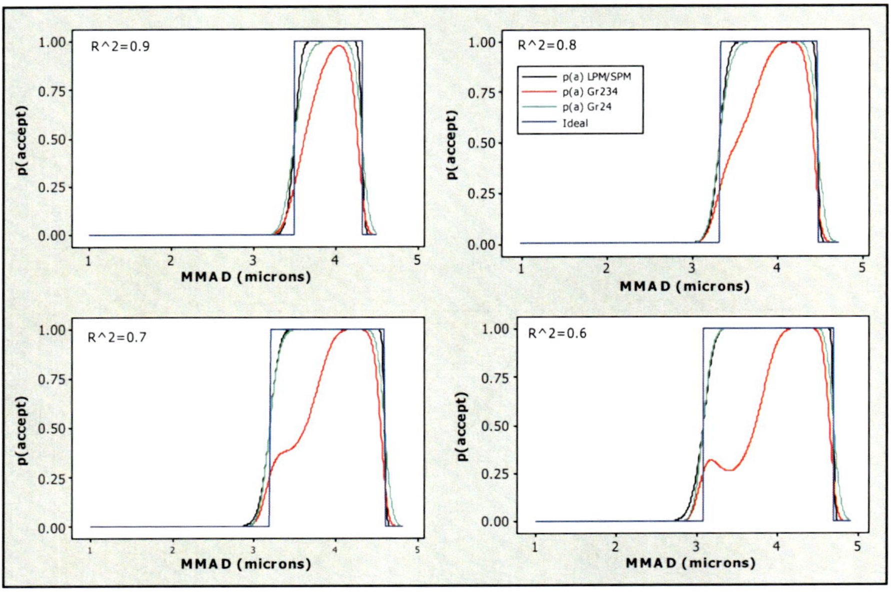

Fig. 8.30 Relative performance of *LPM/SPM* versus grouped stages based on a comparison of OCCs: HFA suspension MDI (***w9k201***)

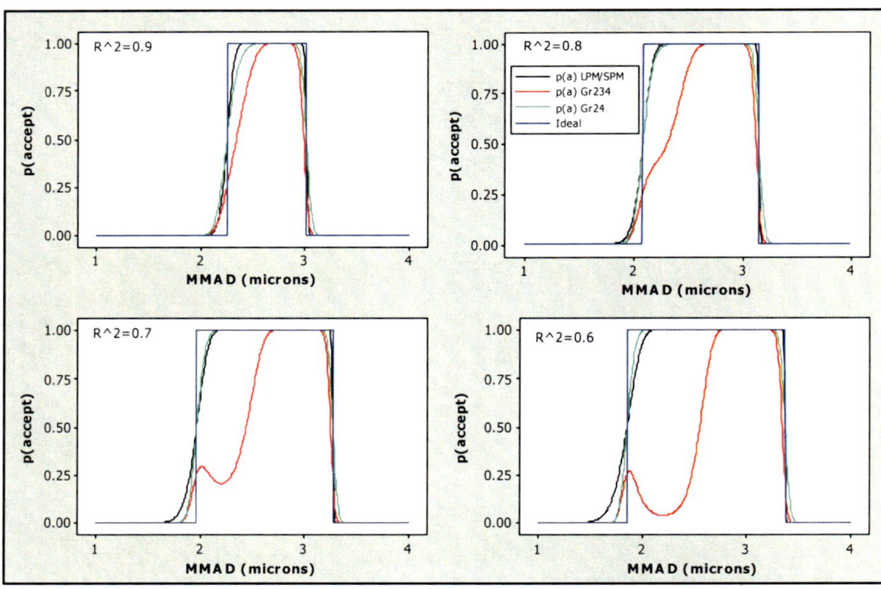

Fig. 8.31 Relative performance of *LPM/SPM* versus grouped stages based on a comparison of OCCs: HFA suspension MDI (***w9j901***)

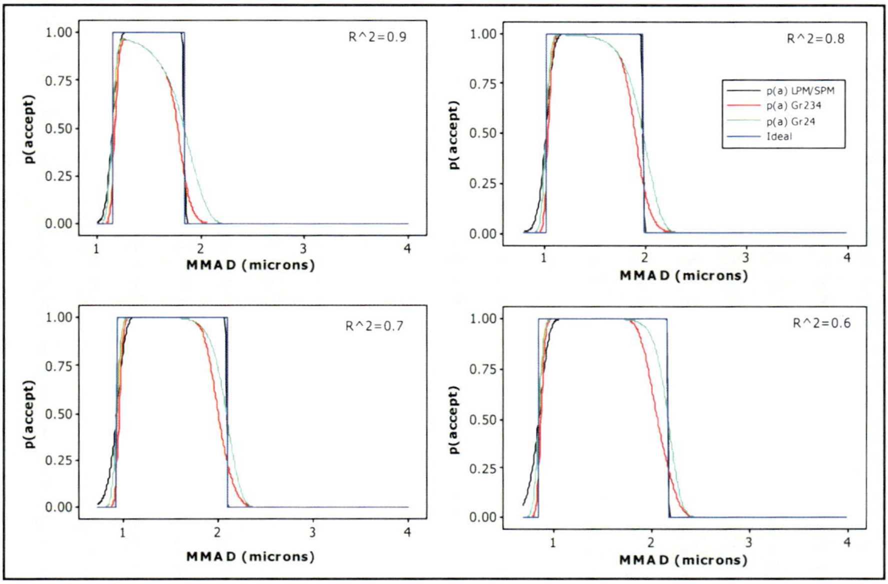

Fig. 8.32 Relative performance of *LPM/SPM* versus grouped stages based on a comparison of OCCs: HFA solution MDI (***w9j801***)

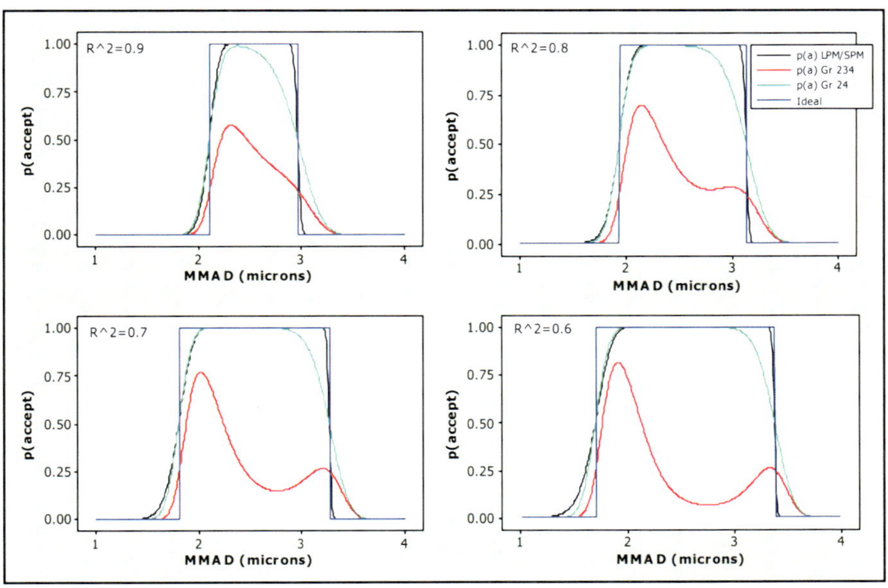

Fig. 8.33 Relative performance of *LPM/SPM* versus grouped stages based on a comparison of OCCs: DPI (***w9jk01***)

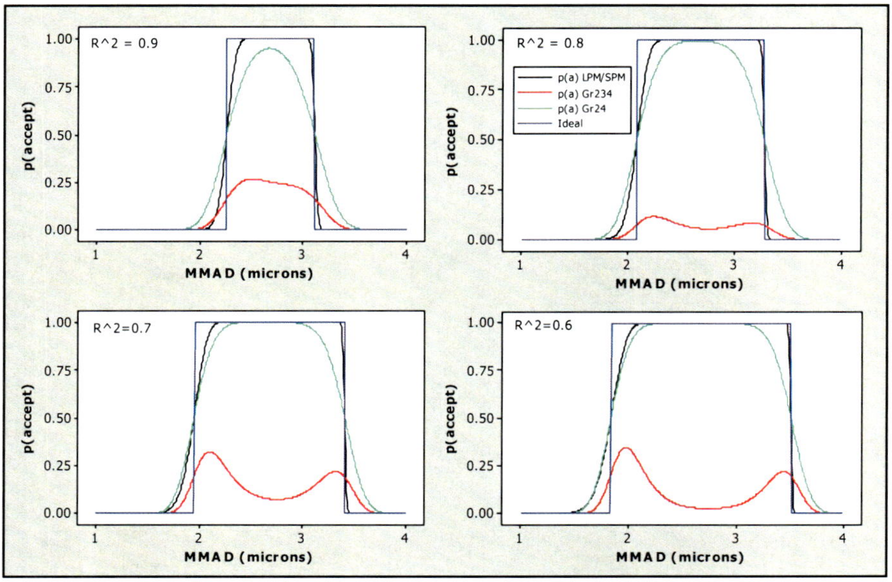

Fig. 8.34 Relative performance of *LPM/SPM* versus grouped stages based on a comparison of OCCs: DPI (***w9k901***)

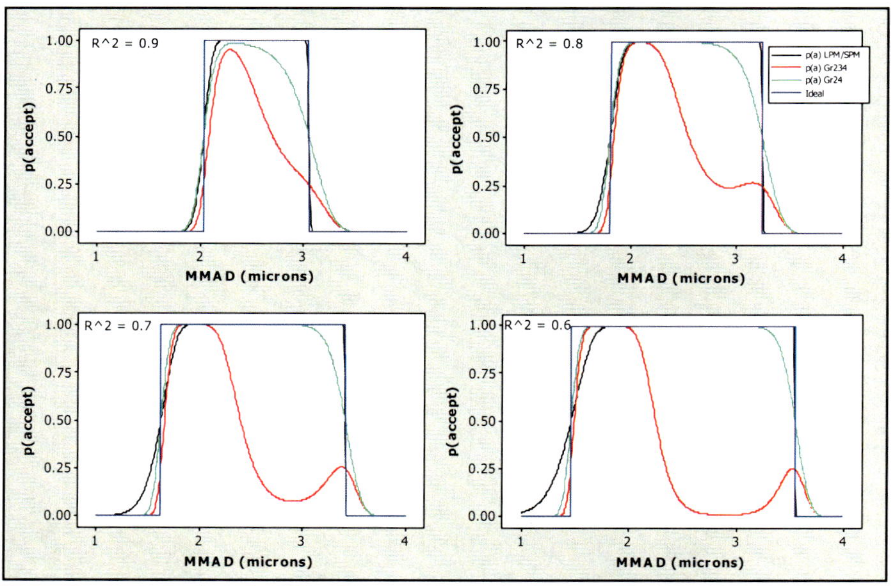

Fig. 8.35 Relative performance of *LPM/SPM* versus grouped stages based on a comparison of OCCs: CFC suspension MDI (***w9j601***)

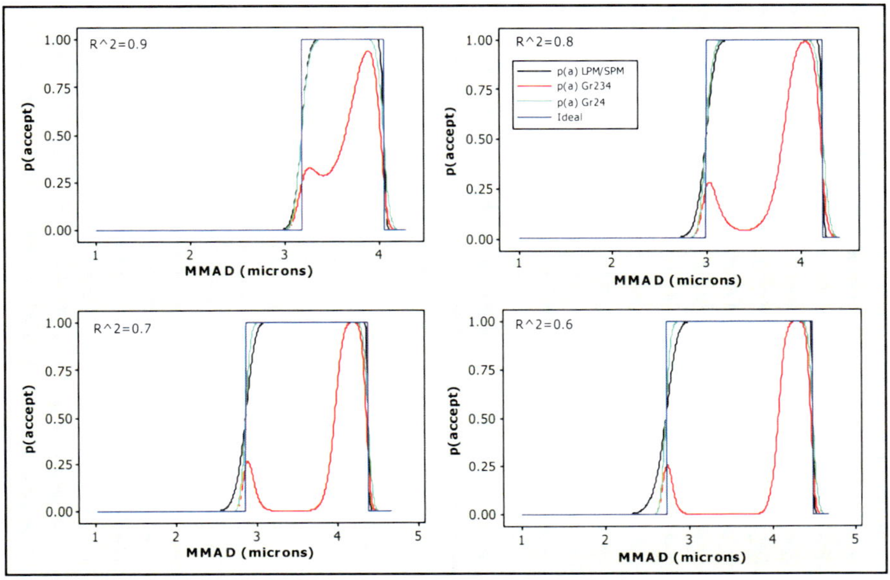

Fig. 8.36 Relative performance of *LPM/SPM* versus grouped stages based on a comparison of OCCs: CFC suspension MDI (***w9k001***)

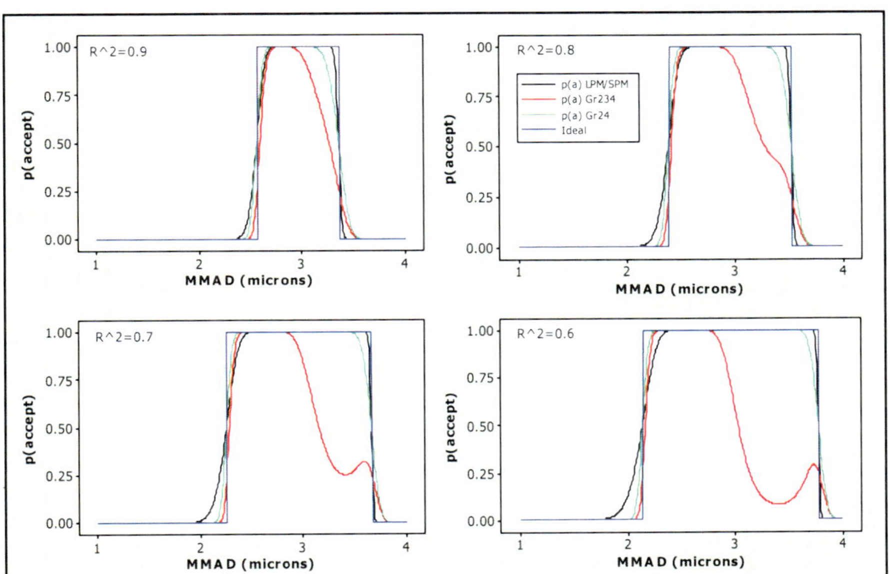

Fig. 8.37 Relative performance of *LPM/SPM* versus grouped stages based on a comparison of OCCs: CFC suspension MDI (***w9kw01***)

Table 8.7 Key to comparisons in Table 8.8

R^2 value used to establish limits	
(a) Type I region, lower limit	(b) Type I region, upper limit
(c) Type II region, lower limit	(d) Type II region, upper limit

Table 8.8 Qualitative comparison of OCCs for OIP APSDs from the IPAC-RS database in the critical regions related to type I and II errors: EDA = decision based on ratio metric (*LPM/SPM*); Gr24: decision based on stage groups 2 and 4 (excluding group 3)

HFA suspension MDI (***w9k201***)				HFA suspension MDI (***w9j901***)			
$R^2=0.9$		$R^2=0.8$		$R^2=0.9$		$R^2=0.8$	
EDA	**EDA**	EDA	EDA	EDA	EDA	≈	EDA
EDA	EDA	≈	EDA	≈	EDA	≈	EDA
$R^2=0.7$		$R^2=0.6$		$R^2=0.7$		$R^2=0.6$	
≈	EDA	≈	EDA	≈	EDA	Gr24	EDA
≈	EDA	≈	EDA	≈	EDA	Gr24	EDA
HFA solution MDI (***w9j801***)				Dry powder inhaler (***w9jk01***)			
$R^2=0.9$		$R^2=0.8$		$R^2=0.9$		$R^2=0.8$	
EDA	**EDA**	EDA	**EDA**	**EDA**	**EDA**	EDA	**EDA**
EDA	EDA	EDA	EDA	EDA	EDA	EDA	EDA
$R^2=0.7$		$R^2=0.6$		$R^2=0.7$		$R^2=0.6$	
Gr24	EDA	Gr24	EDA	≈	EDA	≈	EDA
Gr24	EDA	Gr24	EDA	≈	EDA	Gr24	EDA
Dry powder inhaler (***w9k901***)				CFC suspension MDI (***w9j601***)			
$R^2=0.9$		$R^2=0.8$		$R^2=0.9$		$R^2=0.8$	
EDA	**EDA**	**EDA**	**EDA**	**EDA**	**EDA**	Gr24	≈
EDA	EDA	EDA	EDA	≈	EDA	Gr24	EDA
$R^2=0.7$		$R^2=0.6$		$R^2=0.7$		$R^2=0.6$	
EDA	EDA	EDA	EDA	Gr24	EDA	Gr24	EDA
≈	EDA	≈	EDA	Gr24	EDA	Gr24	EDA
CFC suspension MDI (***w9k001***)				CFC suspension MDI (***w9kw01***)			
$R^2=0.9$		$R^2=0.8$		$R^2=0.9$		$R^2=0.8$	
≈	EDA	≈	EDA	≈	EDA	Gr24	EDA
≈	EDA	Gr24	EDA	Gr24	EDA	Gr24	EDA
$R^2=0.7$		$R^2=0.6$		$R^2=0.7$		$R^2=0.6$	
Gr24	EDA	Gr24	EDA	Gr24	EDA	Gr24	EDA
Gr24	EDA	Gr24	EDA	Gr24	EDA	Gr24	EDA

with the overall curve closer to the ideal OCC then to either of the stage grouping-based alternatives.

Tables 8.7 and 8.8 summarize qualitatively a comparison of these OCCs in the critical regions related to type I (false acceptance) and type II (false rejection) errors (Fig. 8.38).

The entries in the cells in Tables 8.7 and 8.8 indicate superior performance for the indicated metrics in the critical region of the OCC. Where no significant difference is observed, a character "≈" appears. Bolded values indicate that only the indicated metric is capable and that there is a serious performance issue with the alternative metric.

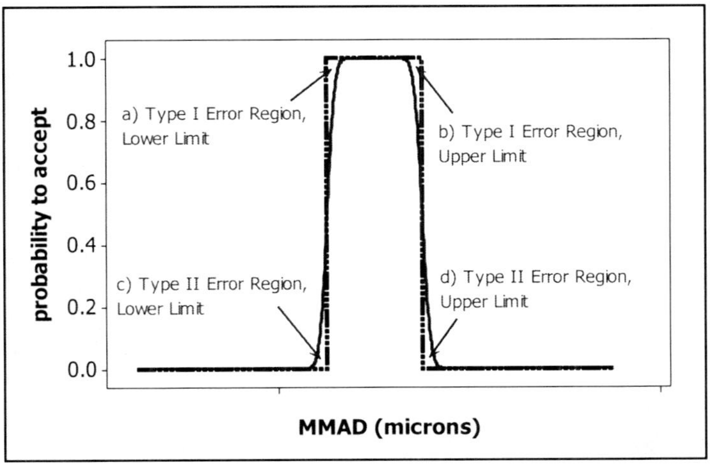

Fig. 8.38 Definitions for type I and II error regions in relation to data presented in Tables 8.7 and 8.8

8.5.2.4 Results from "Christopher–Dey" Strategy

Christopher–Dey compared the EDA and grouped-stage approaches statistically using multivariate modeling of the APSD profiles. Assessment of the relative capability of the two methods to correctly detect shifts in the APSD via shifts in *MMAD* was based on misclassification due to false rejection rates (type I errors) and false acceptance rates (type II errors) using simulations of multivariate non-normal data and OCCs.

Shifts of mass within a distribution or shifts of the entire APSD will necessarily change its *MMAD*. The ability of a given approach (EDA or grouped stages) to detect shifts in APSD can therefore be assessed by their ability to correctly determine whether *MMAD* falls within an acceptable range of *MMAD* values. The same range of acceptable *MMAD* values was used to evaluate each metric in order to ensure an equitable comparison of the two approaches.

The two sets of metrics as well as the corresponding *MMAD* value (see Fig. 8.39) were each calculated for the CI-measured APSD data from each OIP (i.e., the identical CI data values were used to calculate each metric). In the EDA approach, the focus was on the magnitude of the *LPM*/*SPM* ratio. In the grouped-stage approach, the focus was on the mass of API assigned to stage groupings 2, 3, and 4 (Table 8.1). The particular EDA and grouped-stage metrics are illustrated side by side in Fig. 8.40, together with appropriate size boundaries for the ACI, when operated at 28.3 L/min.

For an outcome indicating that *MMAD* is within the acceptable range of values, each of the grouped-stage results must fall within established acceptance limits. The ability of the grouped-stage approach to make the correct decision as to whether *MMAD* is within acceptable limits is directly related to how well the combined

Fig. 8.39 *MMAD* calculated for each APSD using a 4-parameter logistic curve fit as shown in Fig. 7.4

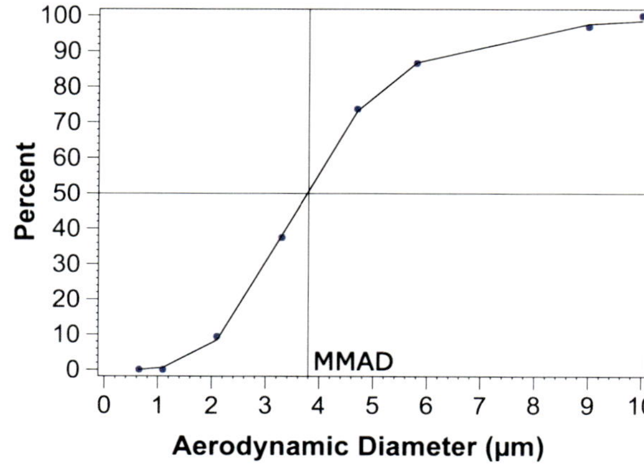

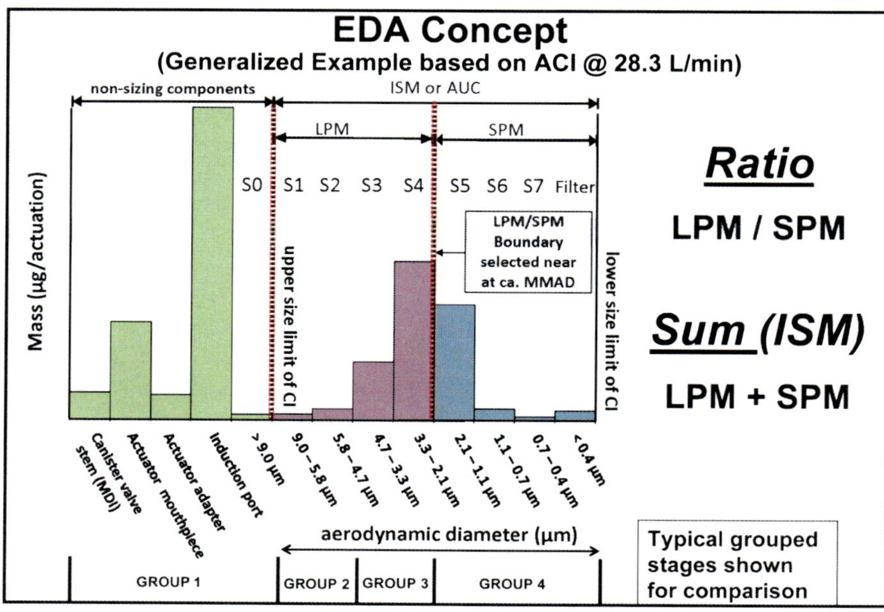

Fig. 8.40 Representative APSD histogram and corresponding EDA and grouped-stage metrics chosen by Christopher–Dey

outcome of the grouped stages predicts *MMAD*. Similarly, the ability of the EDA approach to make the correct decision is directly related to the ability of the ratio metric to predict the value of *MMAD*.

In the context of hypothesis testing, the null hypothesis presented here is that a characteristic of the CI data, *MMAD* (analogous to a population mean), lies within some range of values (upper and lower acceptance limits for *MMAD*). Three grouped-stage test statistics—group 2, group 3, and group 4 mass depositions (analogous to sample statistics such as the mean or a distribution percentile)—are calculated from the CI sample data. If all three test statistics lie within some range (upper and lower acceptance limits for the respective groupings), we fail to reject the null hypothesis (i.e., there is no evidence that the *MMAD* is outside the range of acceptable values).

If any one of the three grouped-stage test statistics falls outside acceptance limits, the null hypothesis is rejected (i.e., there is evidence that the *MMAD* is outside the range of acceptable values). In the same way, the null hypothesis is tested using the *LPM/SPM* test statistic. If the ratio falls within acceptance limits for ratio, we fail to reject the null hypothesis that the *MMAD* is outside an acceptable range of values. If the ratio falls outside its acceptance limits, the null hypothesis is rejected, indicating evidence that the *MMAD* is outside an acceptable range of values.

As described previously, for the grouped-stage approach, the value of *ISM* is completely confounded with shifts in APSD. Therefore, *ISM* is used as an additional test statistic in assessing the performance of both EDA and stage groupings. For the grouped-stage approach, all three stage grouping test statistics and *ISM must* be within their respective acceptance limits to indicate no evidence that the value of *MMAD* is outside an acceptable range. Similarly, both *LPM/SPM* ratio and *ISM must* be within acceptance limits to fail to reject the null hypothesis.

In this case it is not feasible to determine the distributions of the test statistics, so a closed-form approach, such as for a test statistic assumed to follow a t-distribution, is not used. Instead, the method used here is similar to other non-closed-form approaches such as bootstrapping.

The assessment process used to compare the two approaches is illustrated here as an example using data for a specific product in the IPAC-RS APSD database (product ***w9k001***). The basis for establishing comparable acceptance limits for EDA *LPM/SPM* ratio and grouped stages for a common range of acceptable *MMAD* values is illustrated in Fig. 8.41.

Acceptance limits for each of the stage groupings were established by calculating distribution percentiles for each of the grouped-stage results and then selecting percentiles at the extremes of the distributions (e.g., 1st and 99th percentiles), which were consistent with typical FDA limits [2]. Regression analysis was then used to determine the relationship between values of mass on each grouped stage and the corresponding *MMAD*, with a prediction interval, representing some level of uncertainty, around the regression line. The intersection of the prediction interval with the upper and lower limits for the grouped stages was used to establish, for the purpose of this assessment, a range considered to represent acceptable values of *MMAD*, as illustrated in Figs. 8.42, 8.43, and 8.44.

Group 3, which represents the middle portion of the APSD, showed poor correlation with *MMAD* ($R^2=0.009$). Groups 2 and 4, representing the "tails" of the APSD, showed similar, strong correlation with *MMAD* ($R^2=0.58$ and 0.59, respectively) and resulted in similar limits for the range of *MMAD* (3.02–4.04 μm and 3.00–3.98 μm).

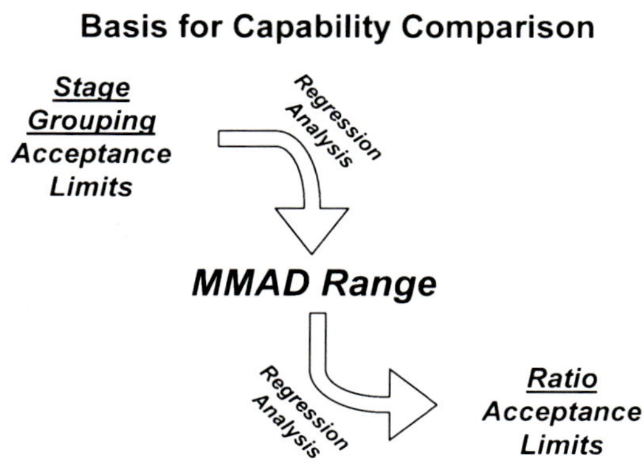

Fig. 8.41 Process diagram for establishing comparable acceptance limits for *LPM/SPM* ratio and grouped stages for a common range of acceptable *MMAD* values

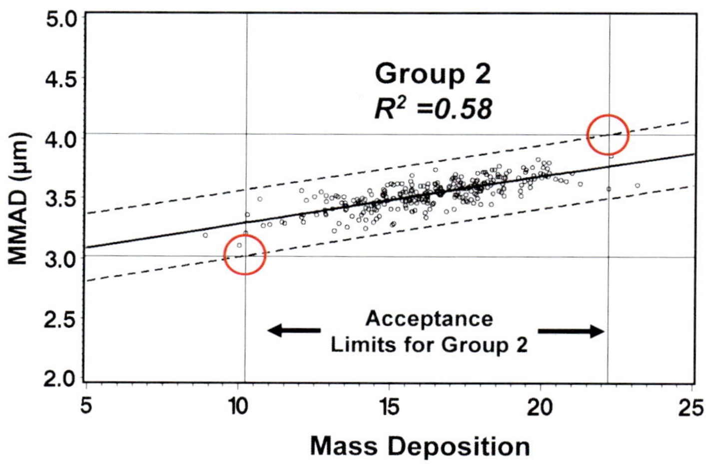

Fig. 8.42 Stage group 2 mass deposition data as a function of *MMAD*; the *dashed lines* represent the prediction interval limits with the circles defining the upper and. lower limits for *MMAD*, each defined by the intersection between the appropriate prediction interval limit and the corresponding acceptance limit for group 2

The averages of group 2 and group 4 limits were therefore subsequently used as a common range of acceptable values of *MMAD* for assessment of the two approaches.

It was found necessary to expand the range of the data using simulation techniques that maintained the characteristics of the original CI data. This procedure was undertaken in order to determine suitably robust estimates of the performance of the two approaches at the limits of the established *MMAD* range, which corresponded to the limits of the observed data. For each set of CI data, a cumulative mass-weighted APSD curve was generated using a 4-parameter curve fit (Fig. 8.39).

To shift the curve and thereby shift the *MMAD*, by some delta amount, predicted values were calculated at delta distance from each of the CI stage d_{50} sizes.

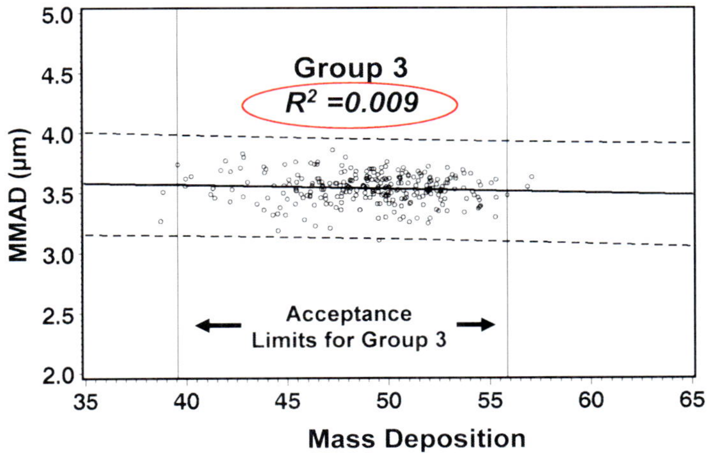

Fig. 8.43 Stage group 3 mass deposition data as a function of *MMAD*; the *dashed lines* represent the prediction interval limits, and the extremely small value of R^2 is highlighted, demonstrating the lack of a correlation

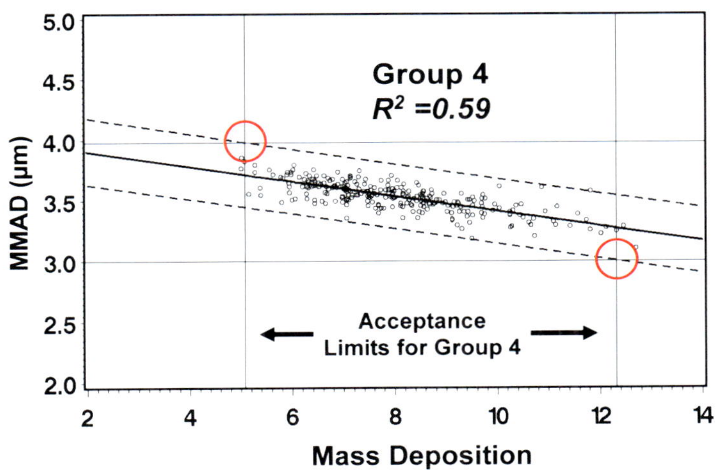

Fig. 8.44 Stage group 4 mass deposition data as a function of *MMAD*; the *dashed lines* represent the prediction interval limits with the circles defining the upper and. lower limits for *MMAD*, each defined by the intersection between the appropriate prediction interval limit and the corresponding acceptance limit for group 4

A new set of shifted data was created by matching these predicted values to the original d_{50} sizes, and by scaling the deviation from the original predicted value, assuming constant coefficient of variation (CV), to maintain variability characteristics consistent with the original data and with principles governing particle size shifts in OIPs. Shifted sets of data were generated at the appropriate deltas to provide sufficient coverage at and slightly beyond the extremes of the *MMAD* range to provide a suitably expanded data set for assessment. The process used is illustrated in Fig. 8.45.

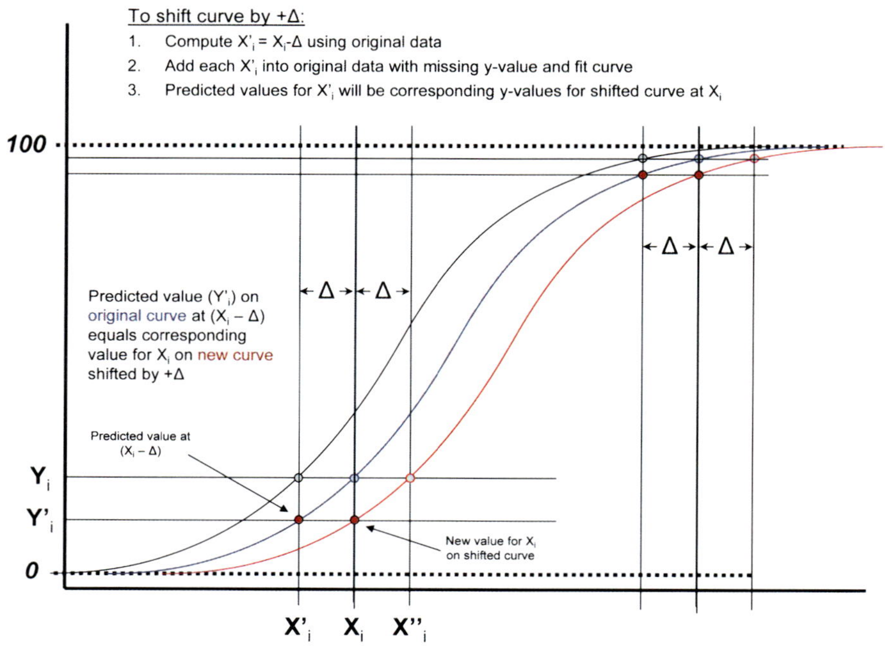

Fig. 8.45 "Christopher–Dey" expansion of the range of OIP CI-determined APSD data using simulation techniques that maintained the characteristics of the original CI data; the process involved shifting the cumulative APSD by a size amount "±Δ"

An example of this approach applied to a set of CI data is shown in Figs. 8.46 and 8.47.

Using the original data and the approach described in Tougas et al. [3], the *LPM/SPM* ratios were calculated using a boundary near the average *MMAD* to separate the *ISM* into its component parts, *LPM* and *SPM*. As with the grouped-stage data, a suitable regression model was used to determine the relationship between the *LPM/SPM* ratio and *MMAD*, with a prediction interval, representing some level of uncertainty, around the regression line. The intersection of the prediction limits with the acceptable *MMAD* range was used to determine the acceptance limits for the *LPM/SPM* ratio that corresponded to grouped-stage acceptance limits for the established range of acceptable *MMAD* values (see Fig. 8.48).

Simulated cascade impactor-generated data was generated (SAS version 9.1, SAS Institute, Cary, NC, USA) using the expanded data set to model the original CI data, incorporating not only the mean and variability of the deposited mass on each impactor stage but also the interrelationships among the stages [17, 18].

For example, the simulated data reflected the same pattern as the observed data such that in a given CI run, a higher mass deposited on one stage generally results in a lower amount on another stage, and vice versa.

Another important aspect of the simulation is that the modeling does not require that mass deposition on each stage follow a normal distribution. For each of the sets

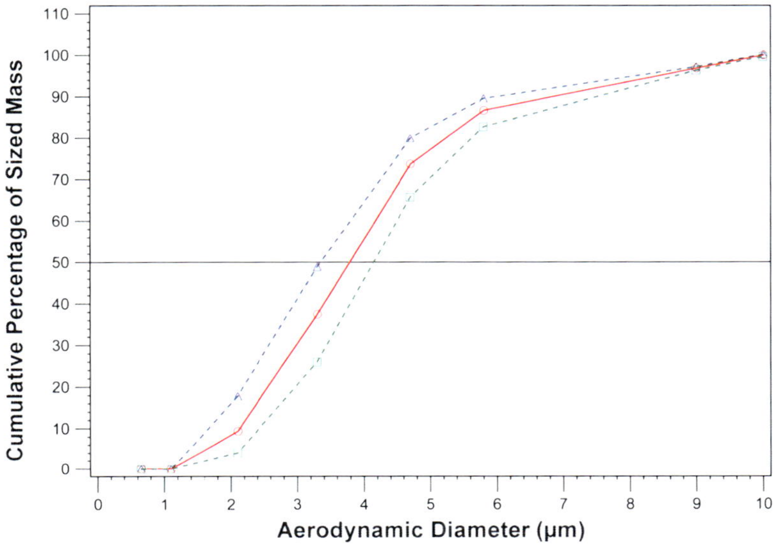

Fig. 8.46 Cumulative mass-weighted APSD curves for a set of CI data shifted both *left* and *right*, with deviations from the predicted cumulative curve adjusted to a constant CV to maintain consistent variability characteristics to the original data

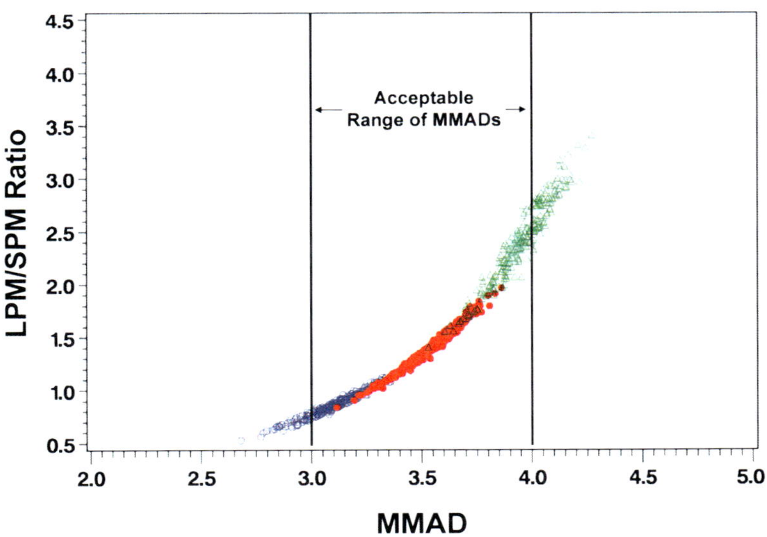

Fig. 8.47 Expanded data set showing *LPM/SPM* ratio versus *MMAD*; *red points* represent the original APSD data; left shifted are *blue*, right shifted are *green*, providing sufficient coverage at and beyond limits of the established range of acceptable *MMAD* values

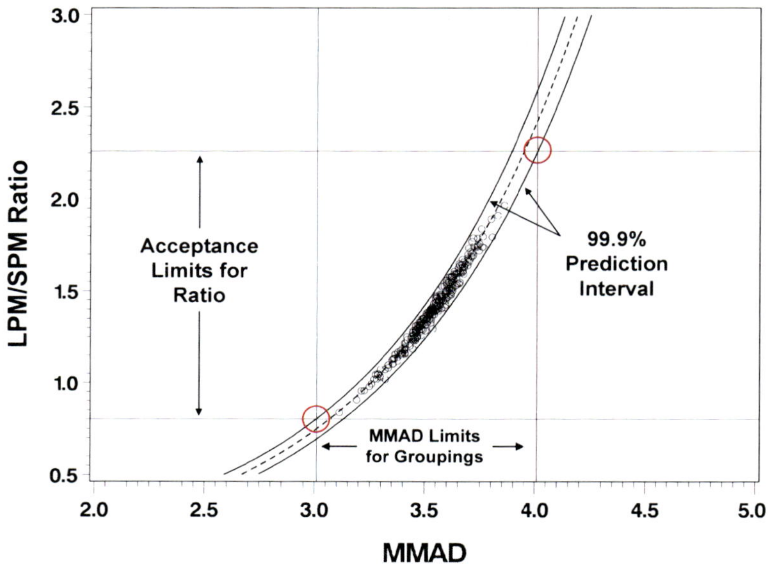

Fig. 8.48 Acceptance limits for the *LPM/SPM* ratio that corresponded to grouped-stage acceptance limits for the established range of acceptable *MMAD* values; the *red circles* represent the upper and. lower bounds for the ratio *LPM/SPM*

of original, left- and right-shifted data, 10,000 simulated APSD profiles were generated for a total of 30,000 profiles over a sufficiently wide range of *MMAD* values, with characteristics similar to the original data and consistent with the mechanisms involved in particle size shifts in OIPs.

This simulated data set was used to assess the performance characteristics of the two approaches under a variety of comparable acceptance limits.

It was possible to compare the two approaches on the same basis under different conditions, by using different combinations of distribution percentiles to set grouping acceptance limits (e.g., 5th and 95th versus 1st and 99th percentiles) as well as different prediction intervals (e.g., 95% versus 99.9% prediction intervals). For example, using the 1st and 99th percentiles to establish grouped-stage acceptance combined with a 99.9% prediction interval to establish the *MMAD* range resulted in 48% and 0% false rejection (type I) and false acceptance (type II) rates, respectively, for grouped-stage decisions based on combined group 2, 3, and 4 results. A different scenario with the same grouping distribution percentiles but with a 90% prediction interval resulted in a 0.09% type II error rate and a 21% type I error rate for the grouped-stage approach. A variety of scenarios and the corresponding error rates for the *LPM/SPM* ratio metric are shown in Table 8.9.

It is also possible to apply a scaling factor to the prediction interval to adjust acceptance limits for the *LPM/SPM* ratio to produce false acceptance rates to match that of the grouped-stage method.

The results of these simulations can be presented in the form of OCCs. For example, the scenario presented in the last row of Table 8.9 is shown in Figs. 8.49, 8.50, 8.51, and 8.52.

Table 8.9 Simulation results for various scenarios showing type I and type II error rates for the two approaches and the relative type I error rates for grouped stages compared to EDA; results from groups 2–4 are combined to yield the overall grouped-stage decision (i.e., if any of the groups 2–4 fails, the overall decision fails)

Grouping distribution percentiles	Prediction interval (%)	Error rates				Relative type I error rates: grouped stages versus EDA
		LPM/SPM ratio		Grouped stages		
		Type I	Type II	Type I	Type II	
10, 90	95	6.57	0.03	22.82	0	3.47
	99	10.86	0.01	30.51	0	2.81
	99.9	17.01	0.01	41.82	0	2.46
5, 95	95	8.48	0.04	23.08	0	2.72
	99	13.89	0.01	32.39	0	2.33
	99.9	19.66	0.01	43.09	0	2.19
1, 99	90	9.61	0.09	20.95	0.09	2.18
	95	12.24	0.08	26.05	0.02	2.13
	99	17.38	0.05	39.03	0	2.25
	99.9	**20.81**	**0.02**	**48.04**	**0**	**2.31**

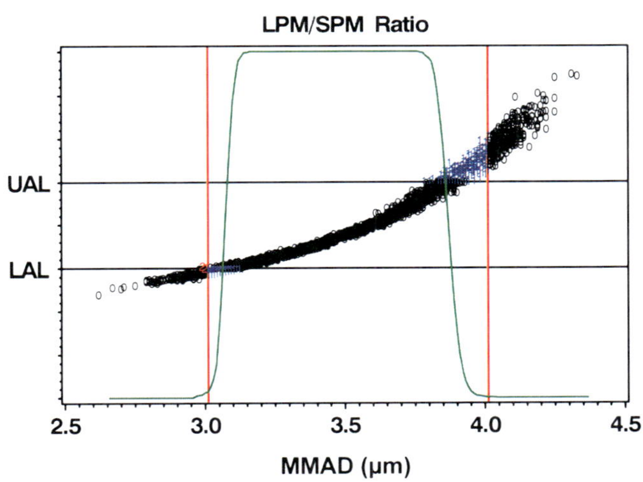

Fig. 8.49 OIP product ***w9k001***: OC curve (*solid green line*) and results for *LPM/SPM* ratio; *black symbols* are correct decisions, *blue symbols* are false rejections (type I errors), *red symbols* are false acceptance (type II errors). LAL and UAL are lower and upper acceptance limits, respectively

In addition to the relative rates of false rejections and false acceptances, the pattern of incorrect decisions is also important in assessing the decision-making capability of a method. If the errors are near acceptance limits, but a method is clearly discriminating in the majority of acceptable and unacceptable regions, that method could be considered more useful in making correct decisions than the one

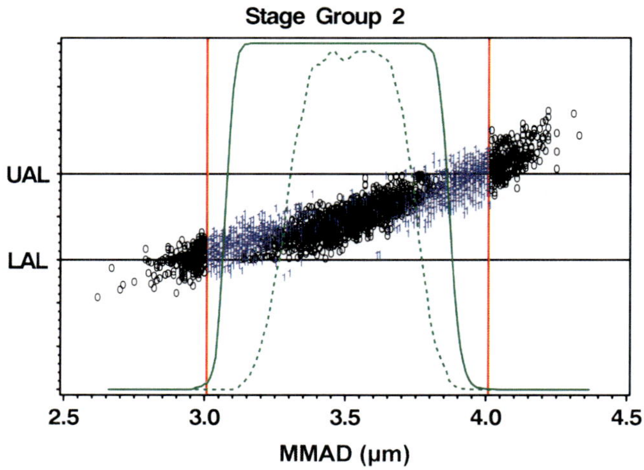

Fig. 8.50 OIP product ***w9k001***: OC curve (*dashed green line*) and results for stage group 2; *black symbols* are correct decisions; *blue symbols* are false rejections (type I errors). Outcomes and the OC curve are based on the combined results of all three groups. The OC curve for *LPM*/*SPM* ratio is superimposed for comparison

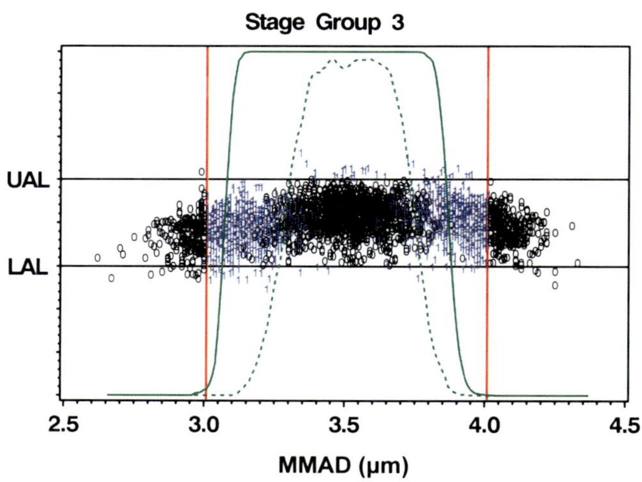

Fig. 8.51 OIP product ***w9k001***: OC curve (*dashed green line*) and results for stage group 3; *black symbols* are correct decisions; *blue symbols* are false rejections (type I errors). Outcomes and the OC curve are based on the combined results of all three groups. The OC curve for *LPM*/*SPM* ratio is superimposed for comparison

that produces errors well away from regions near the limits (and thus in the areas of clearly acceptable or clearly unacceptable quality). This can be seen in the shape of the OC curves. Incorrect decisions which only occur near the boundaries of the acceptance region produce an OC curve with relatively steep sides and a flat top across the major portion of the acceptance region. This contrast can be observed in the figures by comparing the shapes of the OC curves and the pattern of incorrect

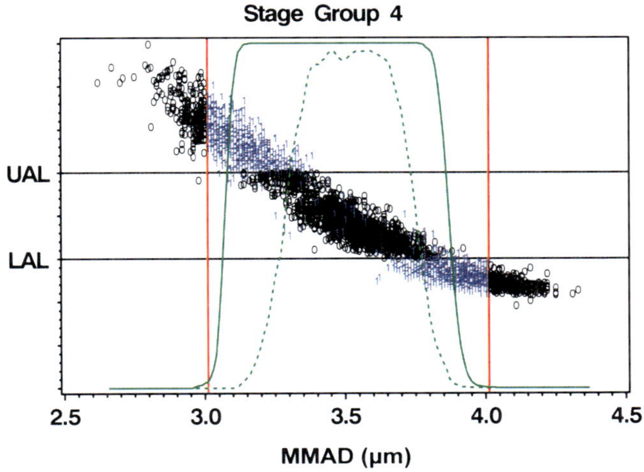

Fig. 8.52 OIP product ***w9k001***: OC curve (*dashed green line*) and results for stage group 4; *black symbols* are correct decisions; *blue symbols* are false rejections (type I errors). Outcomes and the OC curve are based on the combined results of all three groups. The OC curve for *LPM/SPM* ratio is superimposed for comparison

decisions. The *LPM/SPM* ratio exhibits a broad flat curve with relatively steep sides compared to the grouped-stage OCC, which never attains 100% probability of passing even in the center of the acceptance region. The false rejection outcomes represented by the blue plot symbols occur consistently further away from the boundaries of the acceptance region, and even through the middle of the region for the grouped-stage approach than for the *LPM/SPM* ratio approach.

Another way of comparing the patterns of outcomes between the two approaches is presented in Fig. 8.53. All nine possible pairs of outcomes for the two approaches, "G" for grouped stages and "R" for *LPM/SPM*, are plotted versus *MMAD*. Correct decisions are represented by "0," type I errors by "1" and type II errors by "2." In this case both approaches indicated the correct decision 52% of the time as shown by the "G0/R0" category. This occurred in middle of the *MMAD* acceptance region and also just outside the range and beyond. Both approaches indicated false rejection (type I errors, represented by "G1/R1") approximately 21% of the time, which occurred just inside the limits of the acceptance region. However, approximately 27% of the time, the grouped-stage approach indicated false rejection (type I errors) well inside the acceptance region, while *LPM/SPM* ratio indicated the correct decision (represented by "G1/R0"). Increased occurrence of false rejection away from the acceptance limits indicates less discrimination of the grouped-stage approach to correctly predict *MMAD*.

This difference in discrimination capability is illustrated in Figs. 8.54 and 8.55, which show the relative size of the uncertainty of the estimate of *MMAD* compared to the range of acceptable values for stage group 2 and *LPM/SPM*, respectively. A smaller amount of estimate uncertainty relative to *MMAD* acceptance limits indicates a higher capability to detect differences in *MMAD* within the acceptance region.

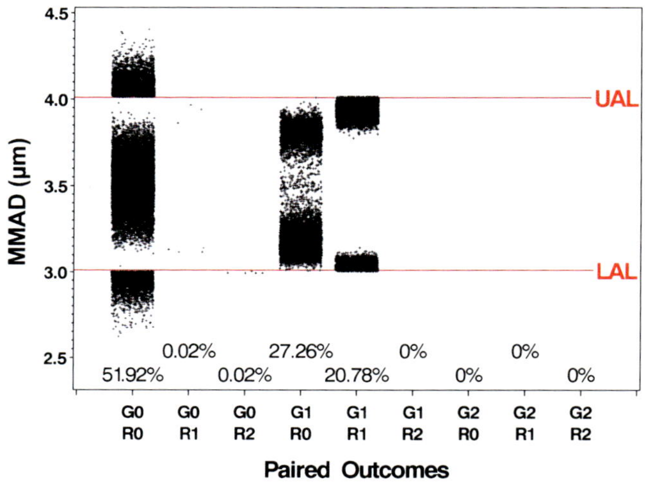

Fig. 8.53 OIP product ***w9k001***: plot showing paired outcomes for grouped stages (G) and *LPM/SPM* ratio (R) for correct decisions represented by "0," type I errors represented by "1," and type II errors represented by "2," with percent of total for each category

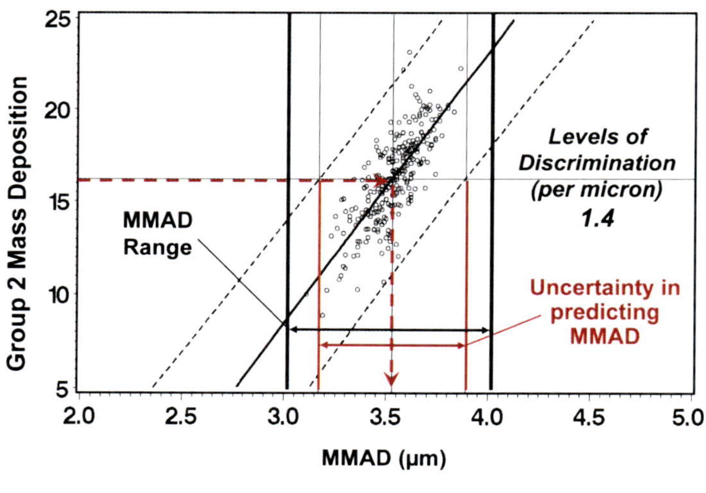

Fig. 8.54 Plot illustrating the ability of stage group 2 to predict *MMAD*; the width of the range of acceptable *MMAD* values is only slightly larger than the uncertainty interval of the prediction (*MMAD* range/uncertainty interval for prediction = 1.4 levels of discrimination per micron), indicating a relatively low capability to discriminate different values of *MMAD* compared to the range of acceptance limits

The assessment process described for IPAC-RS file code ***w9k001*** was also applied to the other products in the IPAC-RS database (Table 8.10).

The APSD characteristics of some of the products were more challenging than others to successfully model and expand the range for assessment, especially for the two products designated by file codes ***w9j901*** and ***w9j601*** for which there were only

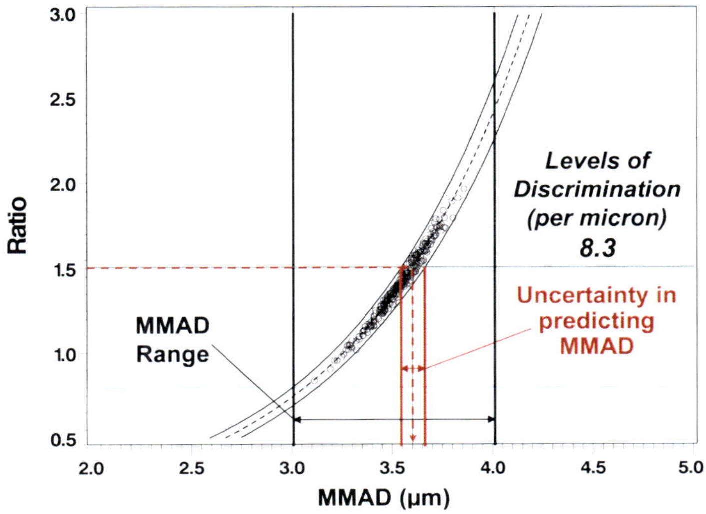

Fig. 8.55 Plot illustrating the ability of *LPM/SPM* ratio to predict *MMAD*; the width of the range of acceptable *MMAD* values is approximately eight times larger than the uncertainty interval of the prediction, indicating a relatively high capability to discriminate different values of *MMAD* with the acceptance limits

Table 8.10 IPAC-RS database file codes for OCC assessment process

File code	OIP type	Figure(s) depicting OCC assessment
w9k001	CFC suspension MDI	8.49–8.53
w9j601	CFC suspension MDI	8.56
w9j801	HFA solution MDI	8.57
w9j901	HFA suspension MDI	8.58
w9jk01	DPI	8.59
w9k201	HFA suspension MDI	8.60
w9k901	DPI	8.61
w9kw01	CFC suspension MDI	8.62

43 and 39 CI data sets, respectively. The techniques used for expanding the range of APSD shifts also allowed for potential changes in *ISM* that could be expected to occur due to physical mechanisms associated with APSD shifts in OIPs. In some cases shifts in APSD were accompanied by an increase or decrease in the mass of API entering the size-fractionating CI stages contributing to *ISM*, resulting in a corresponding increase or decrease in this metric.

These assessments show that when *ISM* is included, the decision-making process can detect changes in the amount (mass of API) as well as the shape of the APSD. The OC curve plots show instances of correct rejection, due to unacceptable *ISM*, even when the *MMAD* was still within an acceptable range. This is evident in the black plot symbols (indicating a correct decision), within the acceptable range of *MMAD*, but outside the acceptable range for grouped stages or ratio. Assessments presented in Figs. 8.56, 8.57, 8.58, 8.59, 8.60, 8.61, and 8.62 show results consistent

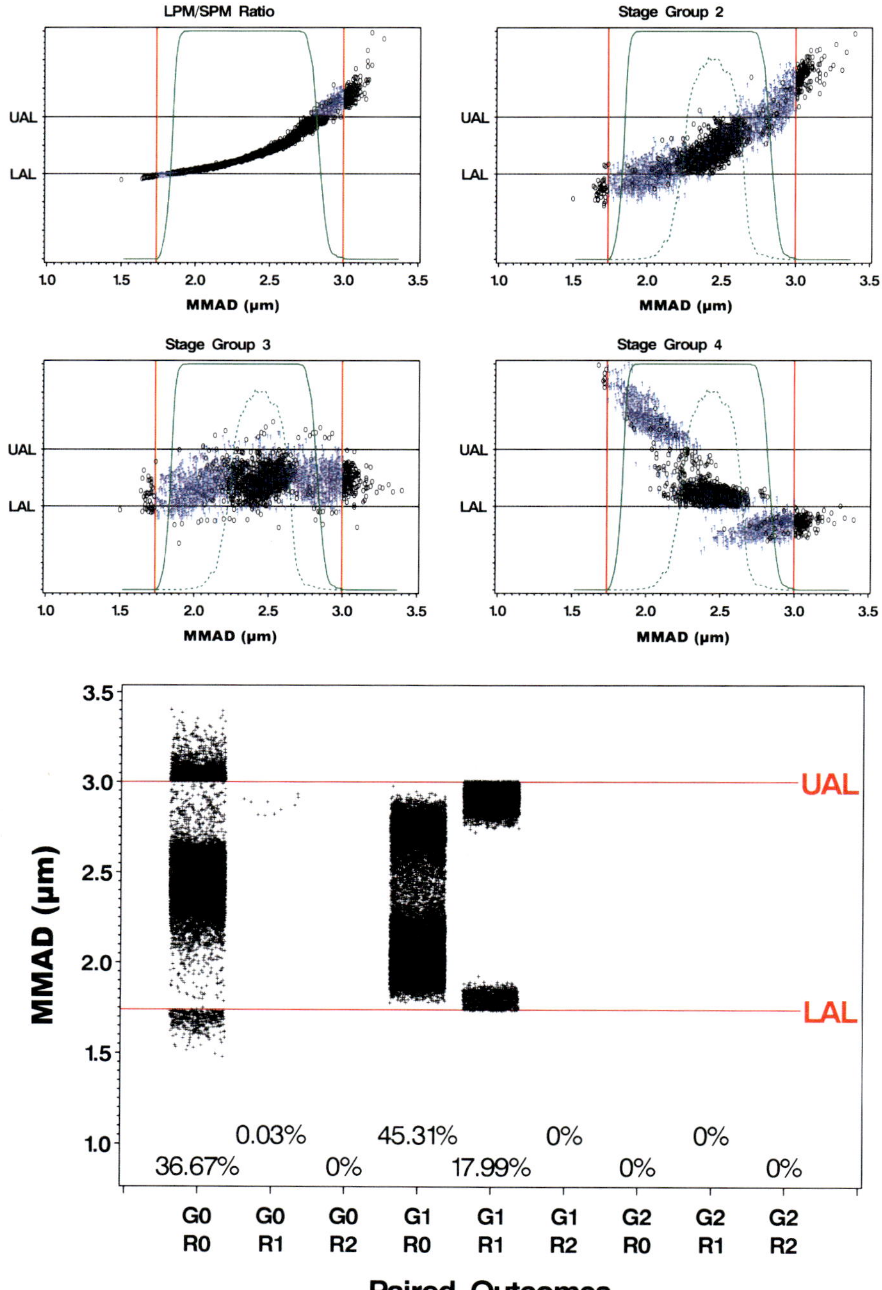

Fig. 8.56 OIP product ***w9j601***: OC curves for grouped stages (*dashed green line*) and *LPM/SPM* ratio (*solid green line*) and results for *LPM/SPM* ratio and groups 2, 3, and 4. Also shown are paired outcomes for grouped stages (*G*) and *LPM/SPM* ratio (*R*) for correct decisions represented by "0," type I errors represented by "1," and type II errors depicted by "2," with percent of total for each category. LAL and UAL are lower and upper acceptance limits, respectively

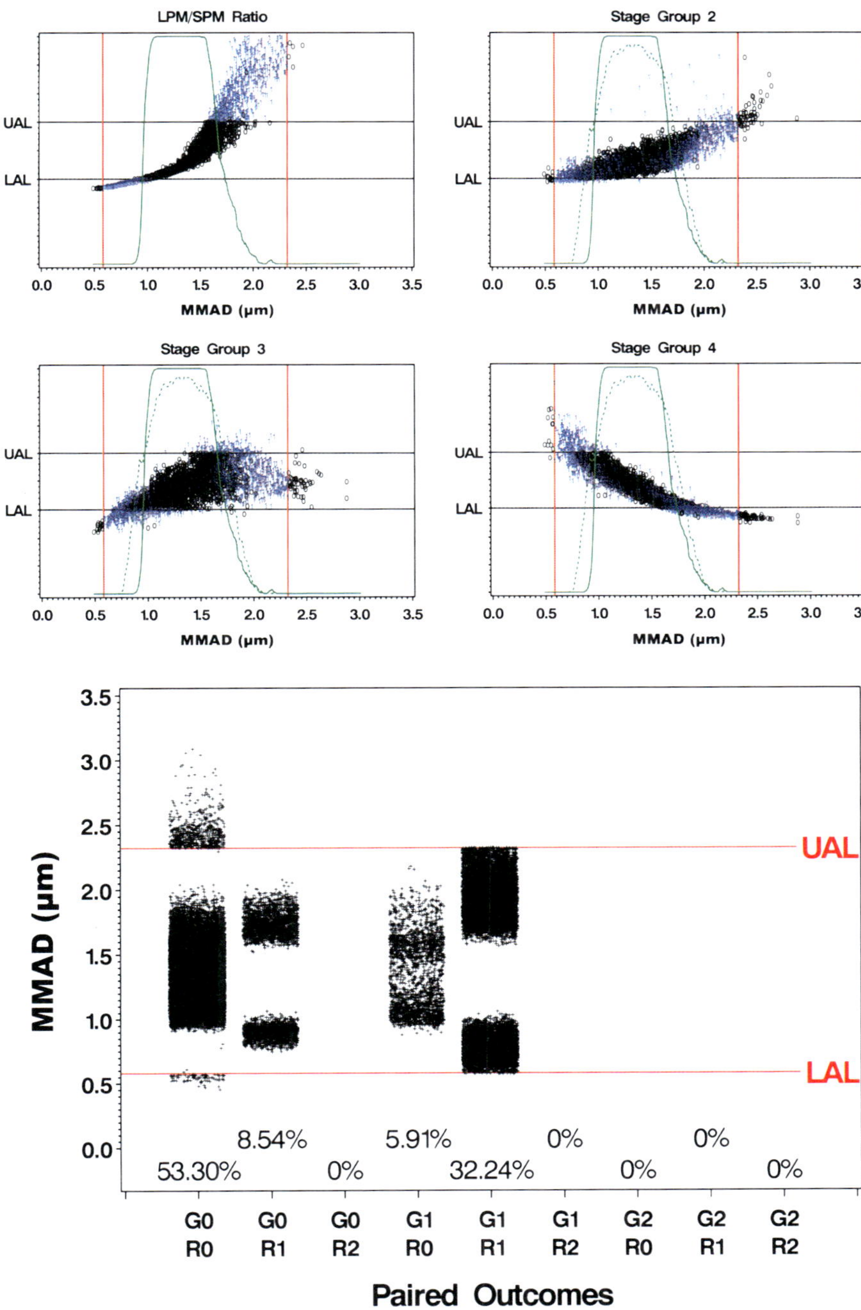

Fig. 8.57 OIP product ***w9j801***: OC curves for grouped stages (*dashed green line*) and *LPM/SPM* ratio (*solid green line*) and results for *LPM/SPM* ratio and groups 2, 3, and 4. Also shown are paired outcomes for grouped stages (*G*) and *LPM/SPM* ratio (*R*) for correct decisions represented by "0," type I errors represented by "1," and type II errors depicted by "2," with percent of total for each category. LAL and UAL are lower and upper acceptance limits, respectively

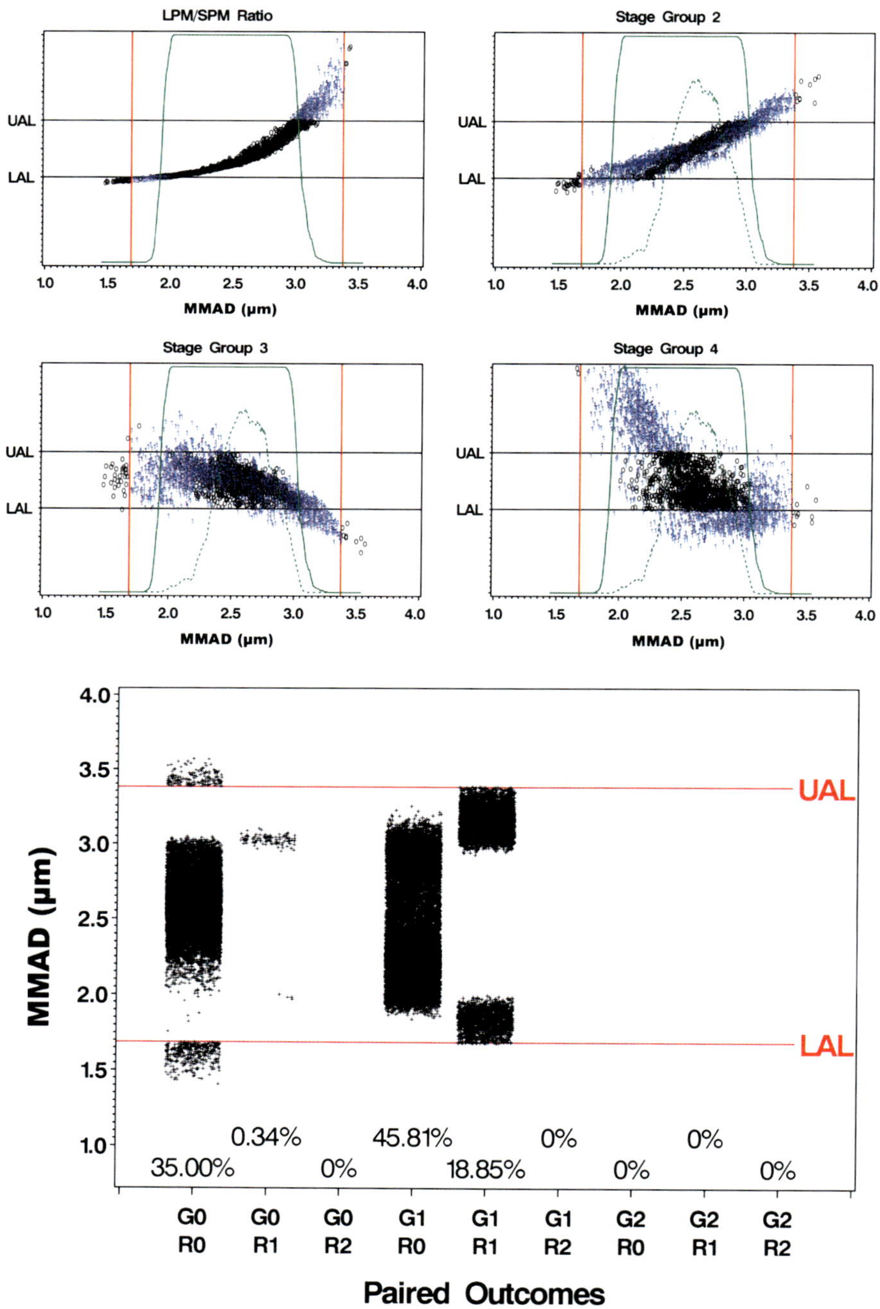

Fig. 8.58 OIP product ***w9j901***: OC curves for grouped stages (*dashed green line*) and *LPM/SPM* ratio (*solid green line*) and results for *LPM/SPM* ratio and groups 2, 3, and 4. Also shown are paired outcomes for grouped stages (*G*) and *LPM/SPM* ratio (*R*) for correct decisions represented by "0," type I errors represented by "1," and type II errors depicted by "2," with percent of total for each category. LAL and UAL are lower and upper acceptance limits, respectively

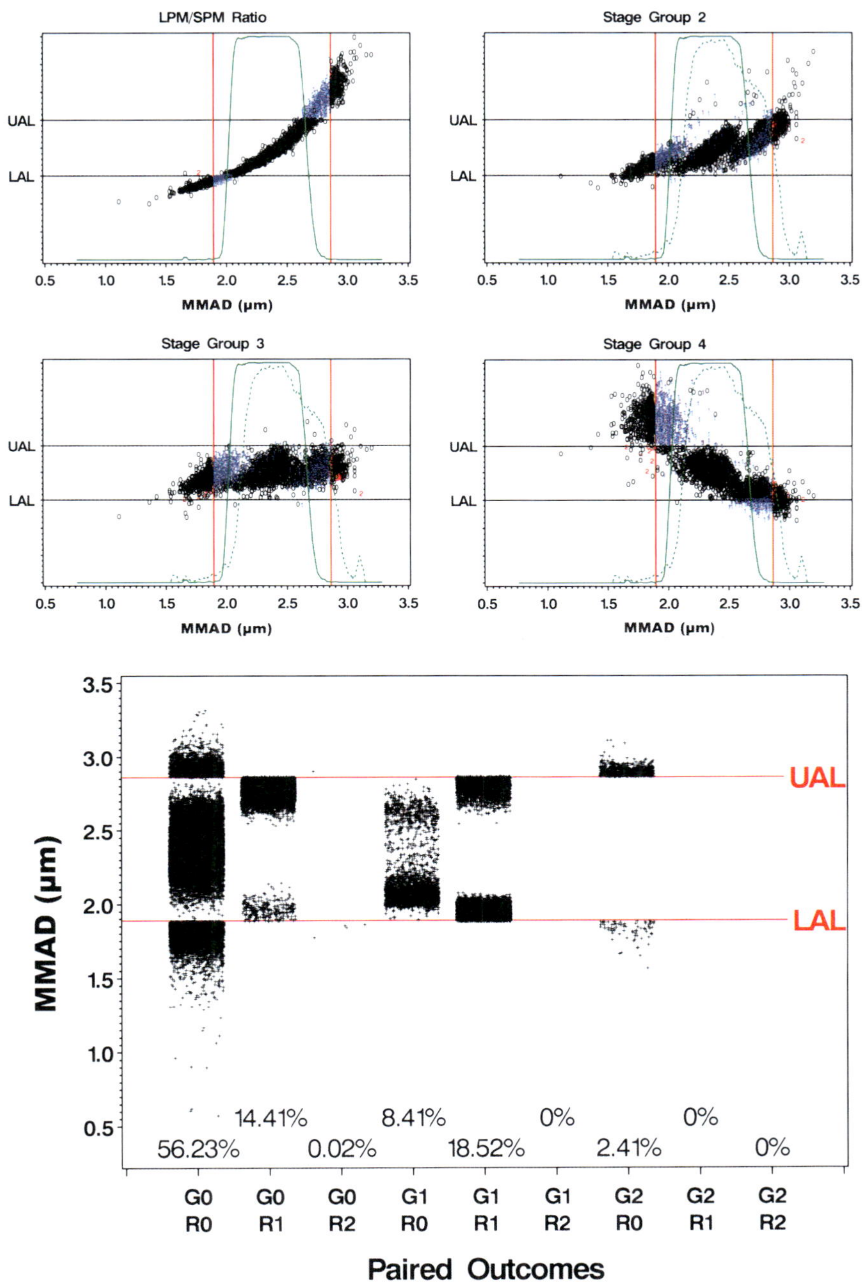

Fig. 8.59 OIP product ***w9jk01***: OC curves for grouped stages (*dashed green line*) and *LPM/SPM* ratio (*solid green line*) and results for *LPM/SPM* ratio and groups 2, 3, and 4. Also shown are paired outcomes for grouped stages (*G*) and *LPM/SPM* ratio (*R*) for correct decisions represented by "0," type I errors represented by "1," and type II errors depicted by "2," with percent of total for each category. LAL and UAL are lower and upper acceptance limits, respectively

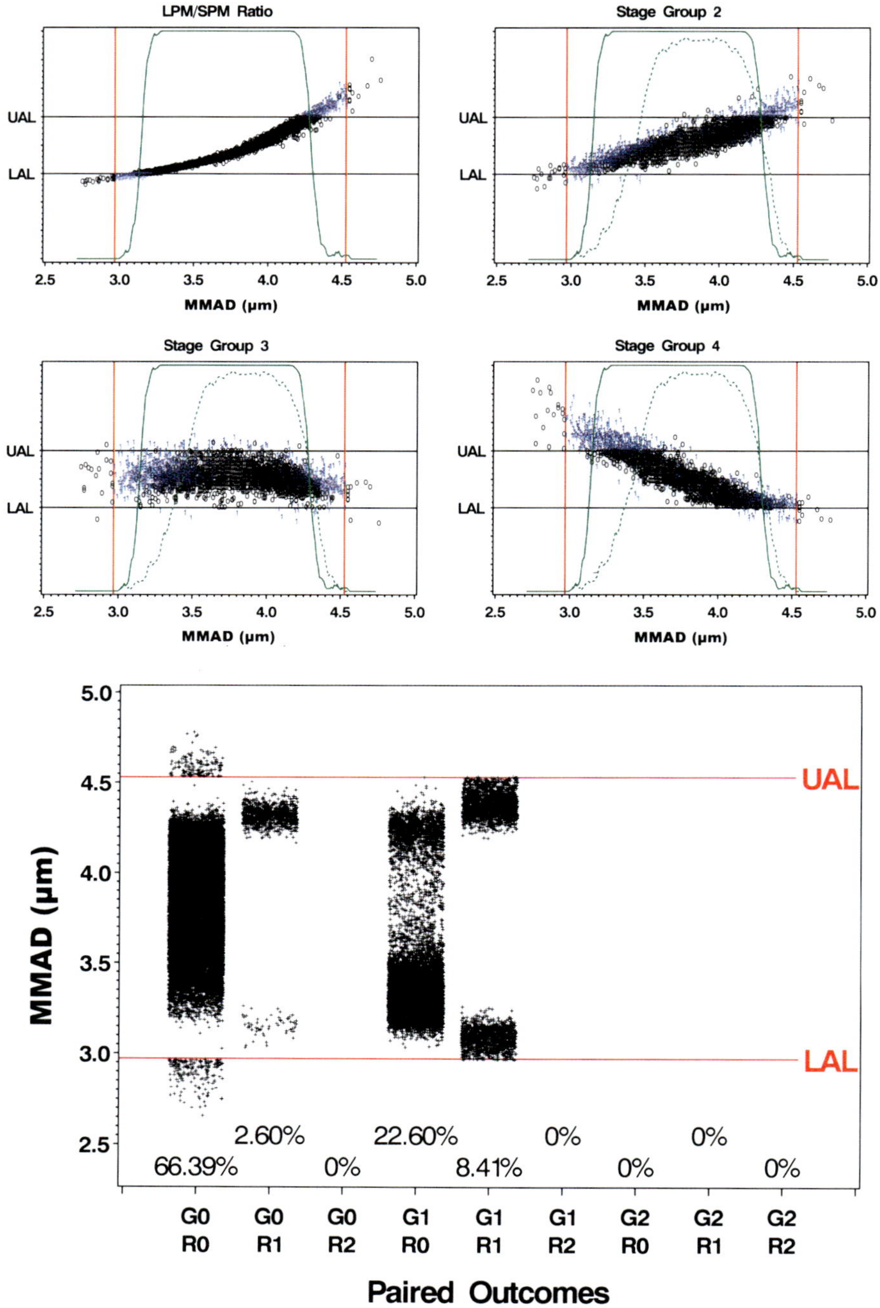

Fig. 8.60 OIP product ***w9k201***: OC curves for grouped stages (*dashed green line*) and *LPM/SPM* ratio (*solid green line*) and results for *LPM/SPM* ratio and groups 2, 3, and 4. Also shown are paired outcomes for grouped stages (*G*) and *LPM/SPM* ratio (*R*) for correct decisions represented by "0," type I errors represented by "1," and type II errors depicted by "2," with percent of total for each category. LAL and UAL are lower and upper acceptance limits, respectively

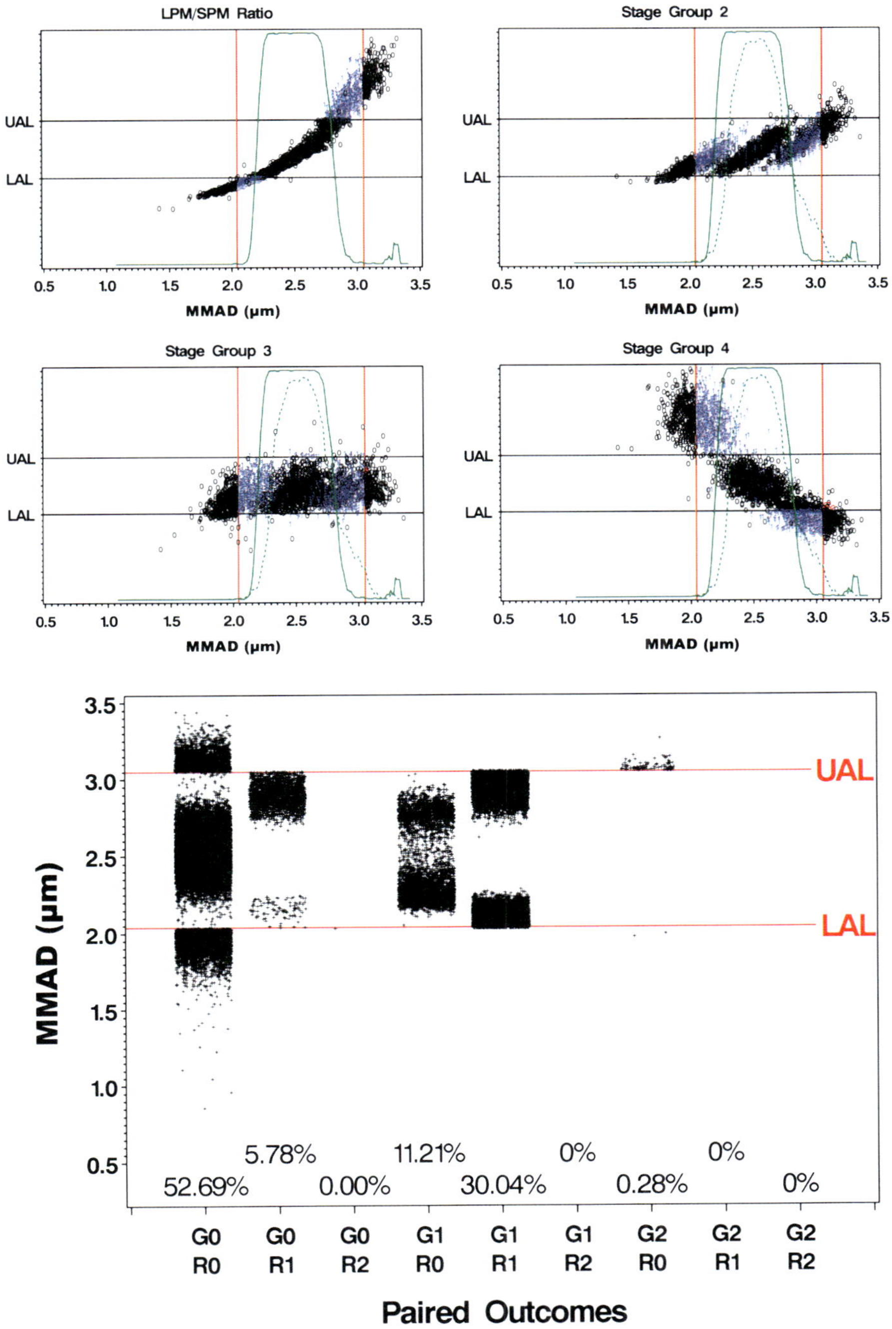

Fig. 8.61 OIP product ***w9k901***: OC curves for grouped stages (*dashed green line*) and *LPM/SPM* ratio (*solid green line*) and results for *LPM/SPM* ratio and groups 2, 3, and 4. Also shown are paired outcomes for grouped stages (*G*) and *LPM/SPM* ratio (*R*) for correct decisions represented by "0," type I errors represented by "1," and type II errors depicted by "2," with percent of total for each category. LAL and UAL are lower and upper acceptance limits, respectively

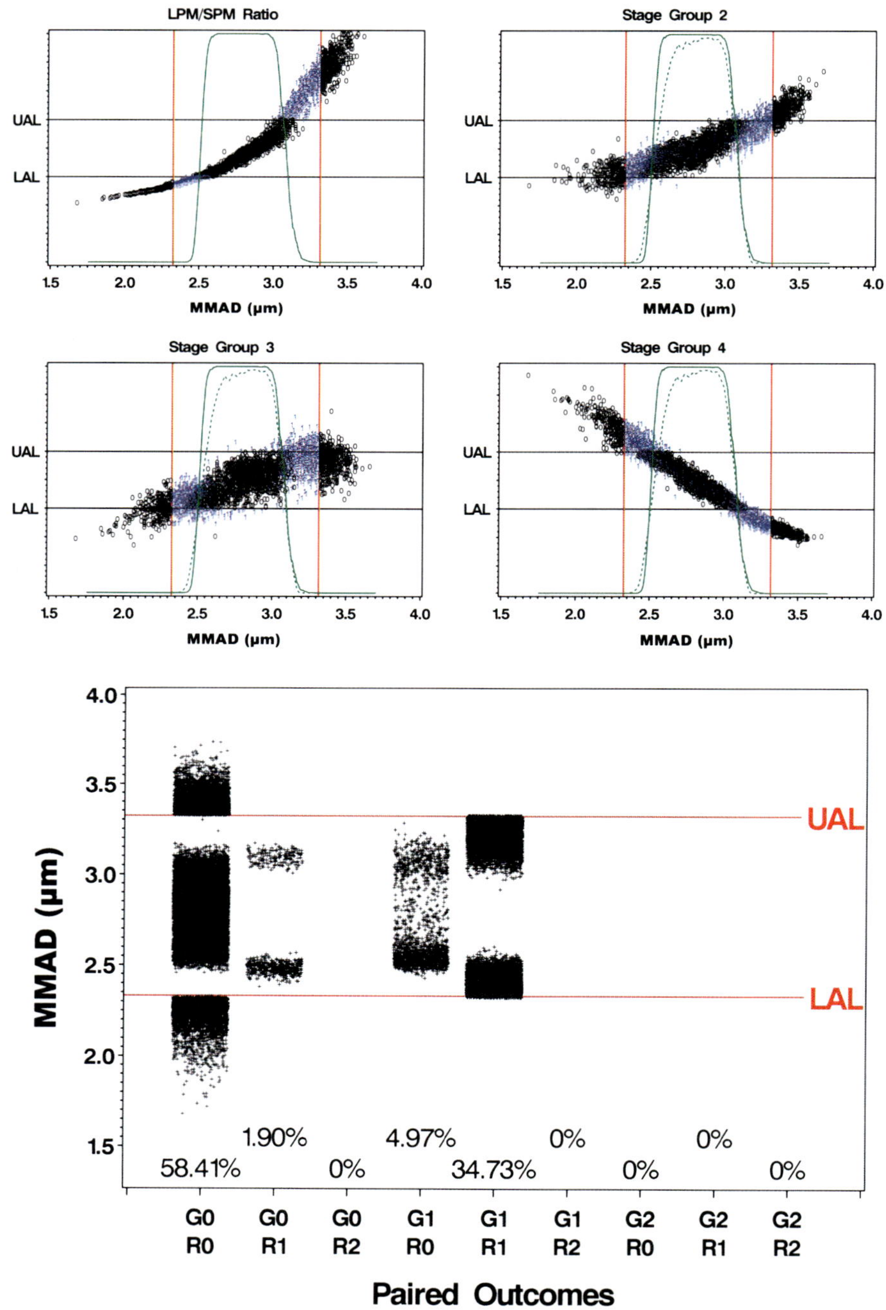

Fig. 8.62 OIP product ***w9kw01***: OC curves for grouped stages (*dashed green line*) and *LPM/SPM* ratio (*solid green line*) and results for *LPM/SPM* ratio and groups 2, 3, and 4. Also shown are paired outcomes for grouped stages (*G*) and *LPM/SPM* ratio (*R*) for correct decisions represented by "0," type I errors represented by "1," and type II errors depicted by "2," with percent of total for each category. LAL and UAL are lower and upper acceptance limits, respectively

with that for product ***w9k001***. OC curves for *LPM/SPM* ratio have steep sides with a wide, flat top compared to the grouped-stage OC curves, and the pattern of incorrect decisions for grouped stages occurs over a much broader range of the acceptable region than for the EDA approach.

8.5.3 PCA Approach

8.5.3.1 Overview

In this section, a multivariate approach to deciding whether a set of APSD profiles is similar or dissimilar to the original population is presented. In multivariate terminology, the original population is labeled the training data set, and the new measurements that are being compared to the training data are labeled the prediction set. In the case presented in this chapter, the original "training set" population consisted of 252 NGI cascade impaction measurements normalized for total impacted to account for dose differences. This population of APSD profiles reflected typical product, process, and analytical variability expected from a product in late-stage development. The approach presented here includes a principal component analysis model (PCA). The output from this multivariate technique, in terms of detecting differences or changes in a set of data, was then compared to an approach for detecting differences based on EDA and grouped stages. Summary conclusions were made from this comparison.

A PCA of the 252 NGI measurements (Sect. 8.2), used to build the model, gave very good results with >90% of the variability being captured in the first two components (Fig. 8.63).

A quantitative measure of the goodness of fit of the PCA model is given by the statistic, R^2X. The value of R^2X is a statistic that indicates how well the PCA model explains the variation in the 252 measurements. In addition to this metric, there is another measure, Q^2, which reflects the predictive ability of the PCA model to predict new data unseen by the model. These predictions are made either internally via existing data or through the use of an independent validation set of observations, a prediction set. In this case, a high value was also obtained for Q^2 from an internal validation, indicating that the model was fit for purpose.

A Hotelling T^2 ellipse [19] (a multivariate distribution analogous to the univariate t-distribution) was established from the scores plot of the data set with a 99% confidence interval, as indicated in Fig. 8.64. The influence of the individual stages in the NGI on the model can be seen from the loadings plot in Fig. 8.65.

The loadings plot is a scatter plot of the loading or weight (*p[1]* versus *p[2]*) applied to each of the individual stages to construct the principal component. The farther a data point lies from the origin, the more influence that stage has on the value of the principal component, and thus, the more overall variation in the data set is explained by changes in that factor. For example, from the loadings plot in Fig. 8.65,

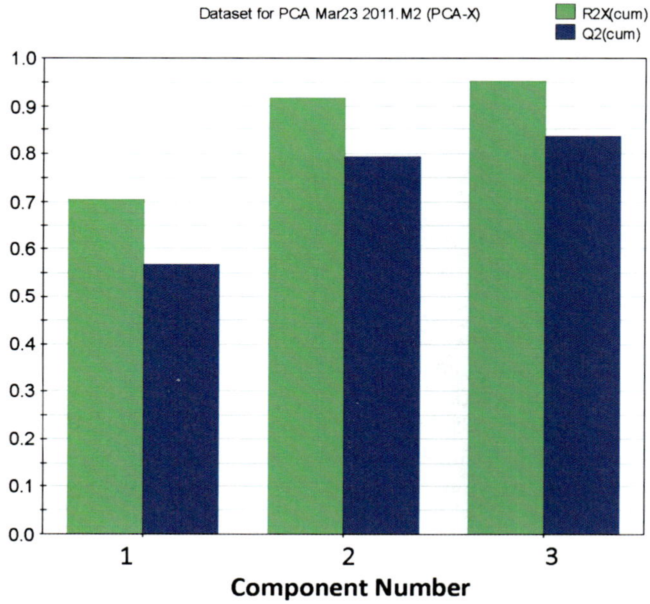

Fig. 8.63 PCA model overview: plot of the proportion of variation either explained [R^2X (*cum*)] or predicted [Q^2 (*cum*)] as a function of the number of components in the PCA model

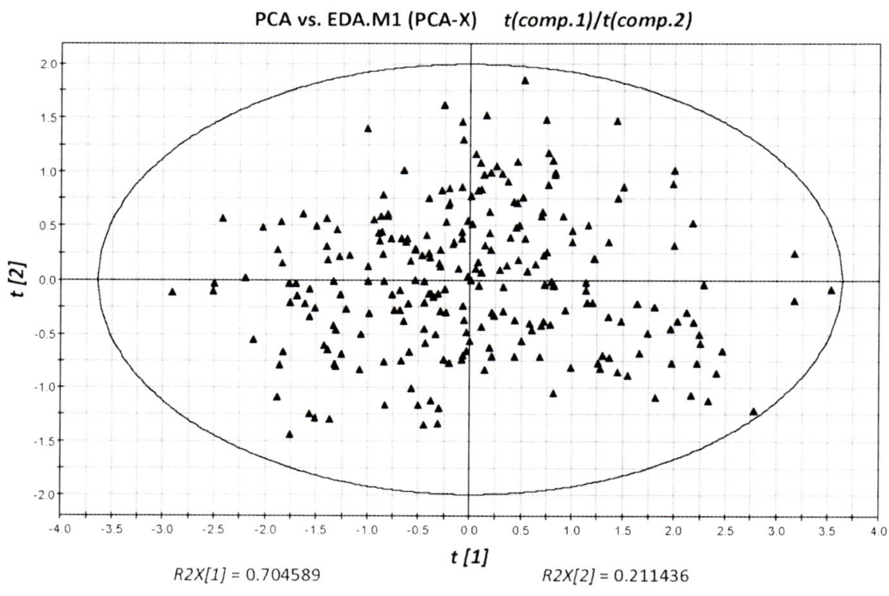

Fig. 8.64 Scores plot of clinically relevant measurements of APSD based on the 252 NGI measurements of a real (blinded) OIP

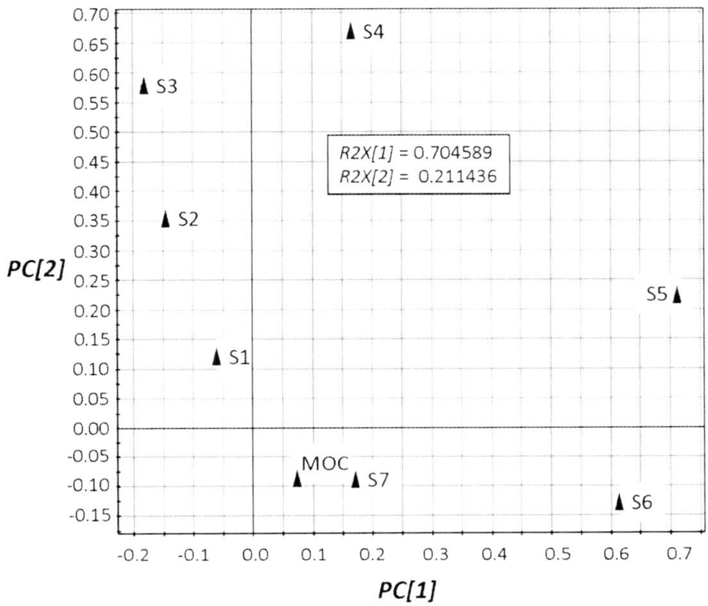

Fig. 8.65 Loadings plot of clinically relevant measurements of APSD: S1 through S7 refer to NGI stages, and MOC refers to the micro-orifice collector; *PC[1]* and *PC[2]* are principal component 1 and 2; the influence of each of the stages on *PC[1]* or *PC[2]* can be positive or negative—meaning, for example, that high deposition on S2 would increase the *PC[2]* score but decrease the *PC[1]* score

a majority of the variation or change in the APSD profiles can be explained by the first component, which applied positive loadings primarily to stages 5 and 6 and negative loadings to stages 2 and 3 so a large percentage of the variation in APSD shifts in the training data set can be explained by differences in shifts between those stages.

Using this model as the training data set, the remaining 1,738 measurements acquired throughout the product development process were overlaid, and each data point or batch indicated with green color for a "pass" (i.e., the data point lies within the Hotelling T^2 ellipse of the model) and red for a "fail" (i.e., the data point lies outside of the Hotelling T^2 ellipse of the model).

A data point in this plot represents a multistage CI APSD profile with loadings applied to the data to reduce the dimensionality down to two components. The component values or scores from these batches, the prediction set, are then plotted and presented in this scores plot. The labels, *tPS[1]* and *tPS[2]*, represent the scores or component values of the first two principal components of the prediction set. The Hotelling T^2 0.99 confidence limit ellipse represents the space of the "typical" APSD profiles from the training data set. Therefore, a profile colored as a "pass" in this plot means that the profile is "similar" to the population of APSD profiles used

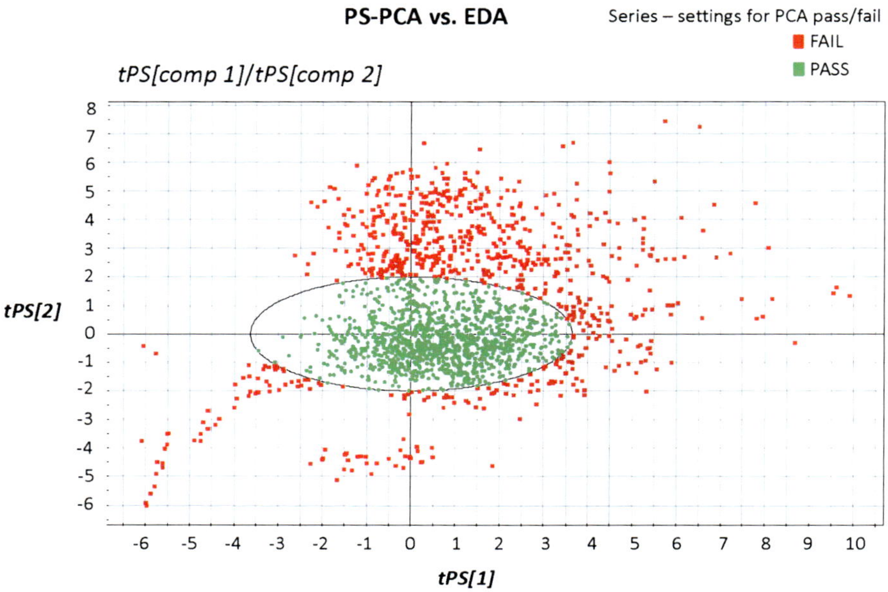

Fig. 8.66 PCA scores plot of full data set; *tPS[1]* and *tPS[2]* are predicted principal component scores (1 and 2) derived from overlaying the data onto the PCA model established using the 252 clinical measurements

to build the model, whereas a profile colored as "fail" means that the full-stage APSD profile is not similar to the 252 APSD profiles used to build the model.

The scores plot with this coloration applied is shown in Fig. 8.66. Of the 1,738 measurements, 934 fell within the Hotelling T^2 0.99 confidence limit of the clinically relevant data set and could therefore be considered as "passing," while 804 fell outside and could therefore be considered as "failing."

Next, EDA and grouped-stage approaches were evaluated using the same data set, and the data generated by each was cross-referred to the scores plot for the full-resolution NGI data.

8.5.3.2 Results for EDA Metric *LPM*/*SPM*

LPM/*SPM* ratio versus *MMAD* was compared using the same data set described in the previous section. This plot was therefore based on the 252 clinically relevant CI APSD profiles that formed the training data set of the PCA model. The *MMAD* was, again, estimated from a logistic curve fit as described in Sect. 8.4.1, in connection with data preparation for MSA. Here, the *LPM* consisted of the sum of the material deposited on S1–S3 of the NGI, and the *SPM* consisted of the sum of material

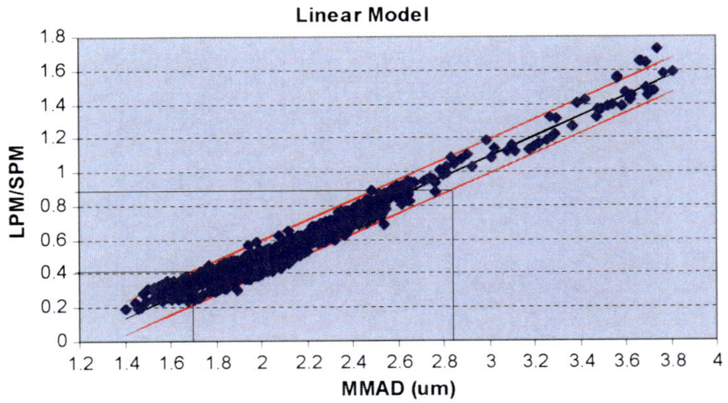

Fig. 8.67 EDA plot with linear criteria between *LPM/SPM* and *MMAD*

deposited on S4–MOC. Both linear and quadratic models (Figs. 8.67 and 8.68, respectively) were fit to the data, regressing *LPM/SPM* by *MMAD*. There was little difference between the two model fits, and therefore, the linear model was used for simplicity. A 99% prediction interval was applied [3]. This interval would be expected to contain 99% of the individual, clinically relevant APSD profiles. In this way, the prediction interval was serving the same general purpose as the Hotelling T^2 ellipse from the PCA model.

The acceptable range for the variation in *LPM/SPM* ratio was found to lie from 0.417 to 0.893, based on the linear modeling of the data (Fig. 8.67) that was comparable with quadratic modeling (Fig. 8.68). The data set generated 1,258 measurements that "passed" and 480 measurements that "failed" in the EDA approach, applying this acceptance criterion to the 1,738 points that remained outside of the clinically relevant data (i.e., the prediction set). This outcome means that 1,258 of the 1,738 APSD profiles were deemed similar, but the remaining 480 profiles were deemed dissimilar to the 252 clinically relevant profiles.

An additional criterion for *ISM*, which is the sum of deposition on S1 through to the MOC for the NGI, was included to remove data in which the *MMAD* is unchanged, but the overall impactor mass (*IM*) has increased beyond that of the clinical batch data. Thus, a range of *ISM* of 21.67–32.17% of LC was derived, which, in combination with the *LPM/SPM* ratio, reduced the number of "passed" batches to 872 and accordingly increased the number of "failed" batches to 866—again, "passing," meaning more similar to the 252 clinically relevant APSD profiles, and "failing," indicating less similar. This criterion was not necessary for the PCA model, as large changes in the *ISM* values would be reflected in changes in deposition on individual stages, and the data would then have appeared outside the Hotelling T^2 ellipse.

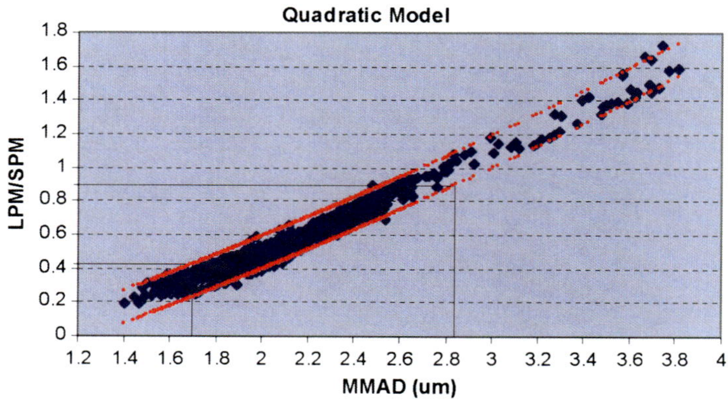

Fig. 8.68 EDA plot with quadratic criteria between *LPM/SPM* and *MMAD*

8.5.3.3 Results for Stage Grouping

Four groups of stages were used for the grouped-stage analysis, which were representative of FDA regulatory requirements. In each case, the sum of the mass obtained on each of the stages in the NGI was presented as a percent of the nominal strength of the product. Acceptance values were derived from the clinical batch data, as presented in Table 8.11.

Each of these groups was evaluated individually versus PCA as a means of detecting change in APSD, and the combination of all four stage groups was also evaluated (Table 8.11). Data for these stage groupings in combination gave 635 "passes" and 1,103 "failures."

8.5.3.4 Comparison of EDA and Stage Grouping

Comparison of the EDA and the grouped-stage analysis was made with cross-reference to the PCA as the control. PCA acts as a suitable control because it is purely a statistical representation of the data. The results of this comparison are presented in Table 8.12. In both of these analyses, there is possibility of generating an erroneous result with respect to the PCA work. These broadly fall into two categories and are labeled type I and type II errors. Type I errors represent an incorrect rejection, i.e., where the data represent a "good" batch but are rejected by the analysis. In other words, in this context, a type I error occurred when the APSD profile from a batch in the prediction set was deemed similar to the 252 clinically relevant profiles from the PCA analysis but was classified as not similar by the other methods. Type II errors represent an incorrect acceptance, i.e., where the data represent a

Table 8.11 Stage groupings and PCA acceptance limits

Group	NGI system	Acceptance range (%LC)
1	IP and PS	59.2–95.5
2	S1 to S3	7.7–12.4
3	S4 to S6	10.7–20.5
4	S7 and MOC	0.3–2.3

Table 8.12 Summary of classification of data points in test set by EDA metrics and stage grouping

Method	Pass rate	Agreement with PCA	Type I error rate	Type II error rate
PCA	934 (54%)	–	–	–
LPM/SPM ratio	1,258 (72%)	809 (46.6%)	125 (7.2%)	449 (28.8%)
LPM/SPM ratio and *ISM*	872 (50%)	783 (45.1%)	151 (8.7%)	89 (5.1%)
Grouping 1	869 (50%)	693 (39.9%)	241 (13.9%)	175 (10.1%)
Grouping 2	1,068 (61%)	777 (44.7%)	157 (9.0%)	291 (16.7%)
Grouping 3	1,366 (79%)	923 (53.1%)	11 (0.6%)	443 (25.5%)
Grouping 4	1,665 (96%)	929 (53.5%)	5 (0.3%)	736 (42.4%)
All stage groups	635 (37%)	595 (34.2%)	339 (19.5%)	40 (2.3%)

"bad" batch but are accepted by the analysis. Again, a type II error occurred when an APSD profile was deemed dissimilar from the PCA analysis but was deemed similar to the 252 clinically relevant profiles by the other methods.

A graphical representation of these errors in terms of the PCA scores plot is given in Figs. 8.69 and 8.70 for EDA and stage grouping, respectively. Batches or APSD profiles colored purple that lie inside the Hotelling T^2 ellipse were deemed similar to the training data set through the PCA analysis, but were deemed dissimilar by EDA or grouped-stage analysis.

Batches or profiles indicated by light blue coloring were deemed outside the training data set population but were deemed similar to the training data set by the EDA or grouped-stage analysis.

8.5.3.5 Outcomes

The number of type II errors was slightly lower for stage groupings (2.3%) compared to EDA metric-based criterion: *LPM/SPM* + *ISM* (5.1%), by using PCA as a reference. There was, however, a significant difference between EDA and stage grouping with regard to type 1 errors, where EDA gave considerably fewer type 1 errors than the grouped-stage approach. This analysis demonstrated the potential value of PCA as a tool in the development of APSD control strategies to allow changes to be

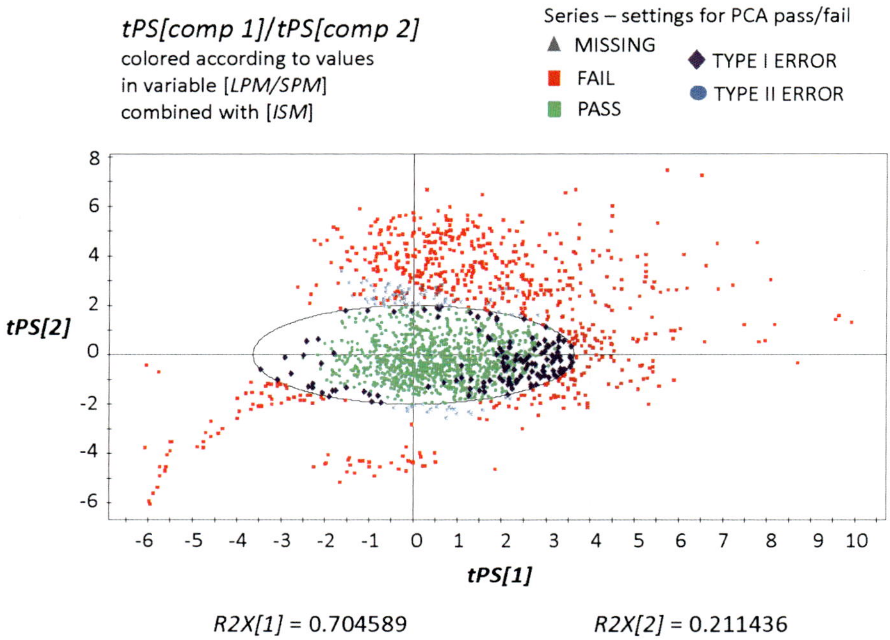

Fig. 8.69 PCA scores plot colored by *LPM/SPM* ratio combined with *ISM* criteria

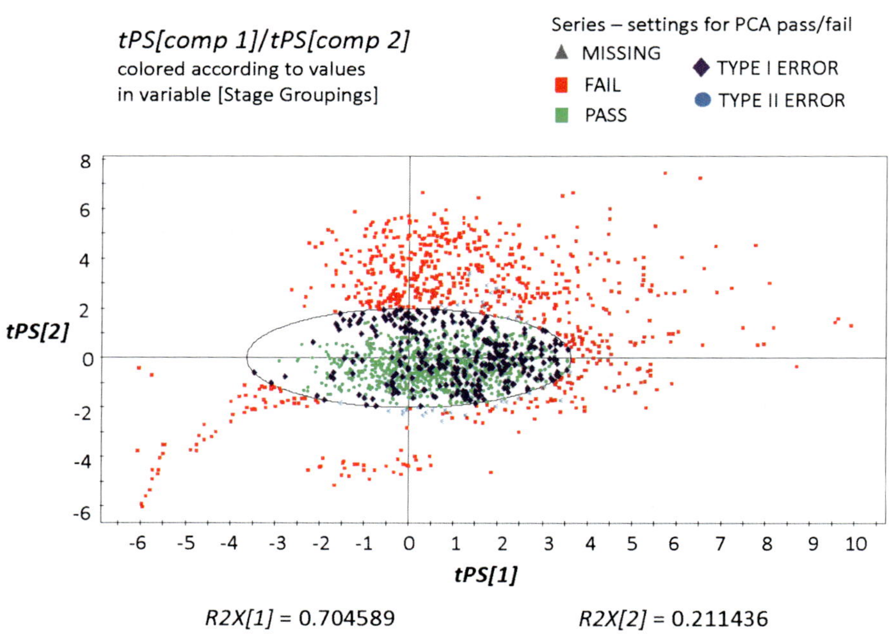

Fig. 8.70 PCA scores plot colored by stage grouping criteria

evaluated with respect to the performance of the product in an "ideal" state. It also revealed the superior sensitivity of the EDA method over the stage grouping approach, as a means of picking up incorrect rejection of batches in the OIP QC environment, i.e., as a means of detecting differences that are true differences, as defined by a multivariate approach.

In this work, batch data from dose-ranging clinical batches were used to establish the "ideal" state of the product, but in a production environment, this data set could consist of the pivotal registered stability batches or another representation of the normal-operating range of the OIP.

8.6 Outcomes from the Different Approaches and Their Relevance to the Development of Improved Methodology for OIP Aerosol Assessment

If EDA metrics are proposed to be used as an alternative compendial method for assessing compliance of an OIP, they should "possess some property for which there is an advantage" over the registered method [20]. The results of the previous sections indicate with strong evidence that the EDA metrics *LPM/SPM* and *ISM* are less variable and provide better decision-making ability than those of the grouped-stage metrics at detecting important differences in sets of individual CI-generated APSD profiles.

The comparisons made in the previous sections of this chapter are based on four distinct assessment approaches, each of which was retrospective in nature; that is, the raw data (individual APSD profiles) were not generated as part of a designed experiment with the express purpose of comparing EDA metrics to grouped-stage metrics. The first assessment (Tougas-MSA approach), the second assessment (Tougas OCC approach), and the third assessment (Christopher–Dey OCC approach) were each used as the starting basis, individual APSD profile data on various marketed products obtained from the blinded IPAC-RS database (six MDIs and two DPIs all measured by the compendial ACI technique). On the other hand, the fourth assessment (PCA) compared two independently generated data sets of a single product (252 clinically relevant APSD profiles to 1,738 APSD profiles generated throughout the development of an unspecified OIP).

Although APSD measurements for OIPs are destructive, in that the aerosol tested is consumed by the measurement system and therefore unable to be remeasured, both the EDA and the grouped-stage metrics were derived from the same physically and analytically generated APSD profiles. From a statistical perspective, any performance differences detected in the EDA metric and grouped-stage metrics would not be a result of the physical process that generated the APSD profile.

The goal of each performance evaluation was to assess the ability of the *LPM/SPM* ratio to discriminate or detect differences in APSD as it relates to changes in the central tendency. *MMAD* is not sufficient by itself to describe the entire APSD; however, once the expected distribution of *MMAD* values for an OIP is characterized, it can be used to assess atypical behavior or acceptability of the product, much like any other QC metric.

In the performance evaluations excluding PCA, the *MMAD* value for each APSD was used as the pivotal comparison parameter (i.e., *MMAD* was treated as a known variable without error). The approaches developed by Tougas to compare EDA with grouped stages involved two quite different techniques. The first methodology was essentially a MSA that focused on the predictive relationships between each of the grouped-stage metrics and *MMAD* as well as the relationship between *LPM/SPM* ratio and *MMAD*. This MSA technique is analogous to the variance equivalent approach discussed by Hauck et al. [20]. Its objective involved comparing the magnitude of uncertainty observed in the predicted *MMAD* values from an inverse regression analysis of each grouped-stage metric and *LPM/SPM*. In other words, a regression analysis was performed with *MMAD* as the known independent variable (x-space), and the metric (*LPM/SPM* or stage group mass) represents the measured response (y-space). However, instead of predicting the response/dependent variable (y = *LPM/SPM*, group 2 mass, group 3 mass, or group 4 mass) from the explanatory/independent variable (x = *MMAD*), the reverse process was put in place to predict *MMAD* from a given metric value. The uncertainty in that prediction was represented in terms of its associated 95% prediction interval. This uncertainty measurement was then used to determine performance indicators such as the discrimination index, the precision to total variability ratio. These performance indicators presented in Table 8.5 show that the *LPM/SPM* metric is more precise in determining and assessing the central tendency of an APSD for OIPs.

The second Tougas approach was similar in form to that adopted by Christopher–Dey approach, in that both groups used OCC-based techniques that are analogous to the decision equivalent approach described by Hauck et al. [20]. Both OCC-based techniques made use of the blinded IPAC-RS OIP APSD data sets to simulate different APSD populations. The APSD profiles simulated by Tougas were based on a simplistic model with a number of assumptions that have been described in Sect. 8.5.2.1. The Christopher–Dey OCC approach employed a more sophisticated modeling of the actual CI data (see Sect. 8.5.2.2), with simulations relying on fewer assumptions and driven more by characteristics of the actual CI data.

In the decision equivalent techniques, the objective was to determine how the EDA metrics, relative to the grouped-stage metrics, would declare an individual APSD profile acceptable or unacceptable when compared to specifications. Since the IPAC-RS database did not contain information regarding the regulatory specifications applicable to the particular OIP concerned, pseudo-specifications were created to judge acceptable versus unacceptable outcomes.

The results of the Tougas OCC approach were summarized qualitatively; that is, the false accept and false reject regions of the OC curves generated for all products and for each metric were visually compared. As shown in Tables 8.7 and 8.8, in the majority of the cases, the EDA metric demonstrated better performance in making the correct decision relative to the grouped-stage metrics. The results of the Christopher–Dey performance evaluation approach was summarized quantitatively in Table 8.9 by providing the estimated error rates (type I = false reject and type II = false accept). For both the Tougas and Christopher–Dey approaches, the shape of the OCC for each product based on *LPM/SPM* ratio more closely resembled the shape of the ideal "top-hat"-shaped OCC than those based on grouped stages. The conclusion from both Tougas and Christopher–Dey approaches was therefore that the ability of the *LPM/SPM* ratio metrics to correctly discriminate between acceptable and unacceptable APSDs was better than the combined-grouped-stage metrics.

The PCA multivariate approach is also analogous to a decision equivalent procedure. In this instance, the objective was to decide whether or not a set of APSD profiles were similar or dissimilar (typical versus atypical); whereas, in the Tougas and Christopher–Dey OCC-based approaches, the focus was on distinguishing acceptable versus unacceptable through the use of pseudo-product specifications. As there were two distinct PCA data sets, the original data set was used as the training or reference data set and the other as the performance evaluation set, and therefore, it was not necessary to simulate APSD data. The results of the PCA performance evaluation method indicate that the false declaration of similarity (type II error) is slightly lower for the grouped-stage metrics (overall 2.3%) than the EDA metrics (5.1%). And the false declaration of dissimilarity (type I error) is significantly higher (19.5%) for the grouped-stage metrics than the EDA metrics (8.7%).

8.7 Conclusions

Because the *LPM/SPM* ratio is more predictive of particle size changes in the CI measured APSD than API mass data derived from grouped-stages, the relative proportion of incorrect rejections (type I errors) to incorrect acceptances (type II errors) is much lower for the EDA approach than for the grouped-stages approach. As a consequence, when the *LPM/SPM* ratio acceptance limits are adjusted to control incorrect acceptance to the same rate for both approaches, the number of incorrect rejections are much lower for the EDA approach. Figure 8.71 therefore provides the answer to the question posed in Fig. 8.2 that EDA could be more discriminating at making correct decisions about changes in OIP APSD that influence specific particle size ranges, than using the grouped-stage approach currently recommended by the FDA in OIP quality assessments.

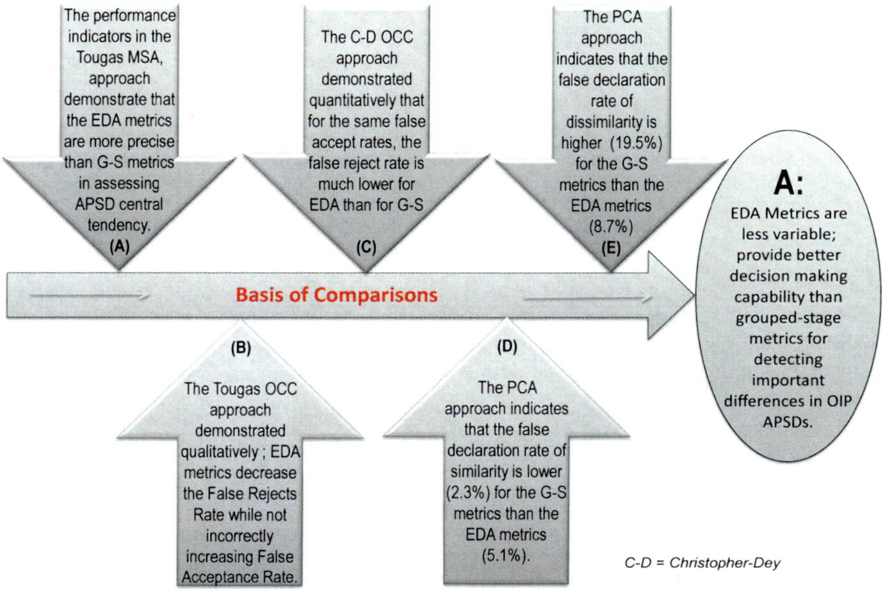

Fig. 8.71 "Road map" for APSD assessments: answering the question

References

1. US Food and Drug Administration (FDA) (1998) CDER. Draft guidance for industry metered dose inhaler (MDI) and dry powder inhaler (DPI) drug products chemistry, manufacturing, and controls documentation, Rockville, MD, USA. http://www.fda.gov/cder/guidance/2180dft.pdf. Accessed 15 July 2012
2. Peri P (2011) Assessing quality of inhaled products and links to efficacy and safety. Presentation at the IPAC-RS Conference "Bringing value to the patient in a changing world", Rockville, MD, USA. http://ipacrs.com/2011%20Conference.html. Accessed 13 July 2012
3. Tougas TP, Christopher D, Mitchell JP, Strickland H, Wyka B, Van Oort M, Lyapustina S (2009) Improved quality control metrics for cascade impaction measurements of orally inhaled drug products (OIPs). AAPS PharmSciTech 10(4):1276–1285
4. Wheeler DJ (2006) EMP (evaluating the measurement process) III: using imperfect data. SPC, Knoxville, TN
5. AIAG (Automotive Industry Action Group) (2010) Measurement system analysis, Reference Manual, 4th edn. AIAG, Southfield, MI, USA, ISBN#: 978-1-60-534211-5
6. Eisenhart C (1963) Realistic evaluation of the precision and accuracy of instrument calibration. J Res Natl Bur Stds 67C:161–187 [Reprinted, with corrections, in (1969) Precision measurement and calibration: statistical concepts and procedures In: Ku HH (ed) Natl Bur Stds Spec Publ 300 1:21–48]
7. American Society for Testing and Materials (2012) Standard terminology for relating to quality and statistics ASTM E456-12, West Conshohocken, PA, USA. http://www.astm.org/Standards/E456.htm. Accessed 15 July 2012

8. Christopher JD, Dey M, Lyapustina S, Mitchell J, Tougas T, Van Oort M, Strickland H, Wyka B (2010) Generalized simplified approaches for *MMAD* determination. Pharma Forum 36(3):812–823
9. Sandell D, Tougas T (2012) Quality considerations in the establishment of specifications for pharmaceuticals. Stat Biopharm Res 4(2):125–135
10. Jolliffe IT (2002) Principal component analysis. Springer, New York
11. Jackson JE (1959) Quality control methods for several related variables. Technometrics 1(4):359–377
12. Berntsson O, Danielsson LG, Johansson MO, Folestad S (2000) Quantitative determination of content in binary powder mixtures using diffuse reflectance near infrared spectrometry and multivariate analysis. Anal Chim Acta 419(1):45–54
13. Eriksson L, Hagberg P, Johansson E, Rännar S, Whelehan O, Åström A, Lindgren T (2001) Multivariate process monitoring of a newsprint mill: application to modeling and predicting COD load resulting from de-inking of recycled paper. J Chemometr 15(4):337–352
14. Euerby MR, Petersson P (2003) Chromatographic classification and comparison of commercially available reversed-phase liquid chromatographic columns using principal component analysis. J Chromatogr A 994(1–2):13–36
15. Intelmann D, Haseleu G, Dunkel A, Lagemann A, Stephan A, Hofmann T (2011) Comprehensive sensomics analysis of hop-derived bitter compounds during storage of beer. J Agric Food Chem 59(5):1939–1953
16. Sandler N, Wilson D (2010) Prediction of granule packing and flow behavior based on particle size and shape analysis. J Pharm Sci 99(2):958–968
17. Pan Z, Christopher JD, Lyapustina S, Chou E (2004) Statistical techniques used in simulation of cascade impactor particle size distribution profiles. In: Dalby RN, Byron PR, Peart J, Suman JD, Farr SJ (eds) Respiratory drug delivery-IX. Davis HealthCare International, River Grove, IL, pp 669–672
18. Christopher D, Pan Z, Lyapustina S (2005) Aerodynamic particle size distribution profile comparisons: considerations for assessing statistical properties of profile comparisons tests. Am Pharm Rev 8(1):68–72
19. Hotelling H (1931) The generalization of Student's ratio. Ann Math Stat 2(3):360–378
20. Hauck WW, DeStefano AJ, Cecil TL, Abernethy DR, Koch WF, Williams RL (2009) Acceptable, equivalent, or better: approaches for alternatives to official compendial procedures. Pharma Forum 35(3):772–778

Chapter 9
Verification of the EDA Concept Through an Assessment of Theoretical Failure Modes, Failure Mode Analysis, and Case Studies with Real Data

Helen Strickland, Beth Morgan, J. David Christopher, Volker Glaab, Adrian Goodey, Keyur Joshi, Lei Mao, and Jolyon P. Mitchell

Abstract Previous chapters have presented the robust theoretical case for EDA in comparison with current ways of analyzing API mass distribution profiles from OIPs. This chapter is in two distinct parts; the first part examines from the theoretical standpoint, ways in which changes in APSD could potentially go undetected by EDA; the second presents a series of case studies with a variety of OIP types that demonstrate the appropriateness of EDA as a powerful, yet simple-to-use tool for in vitro assessment of CI data. Discussion of theoretical failure modes is presented for general awareness. In a given product/method development, each sponsor would have to conduct their own analysis of potential failure modes based on their situation. Similarly, the case studies are presented as illustrations of EDA and AIM applications for several real OIPs. Each sponsor may develop a different way to implement AIM and EDA, depending on their purpose.

H. Strickland (✉)
GlaxoSmithKline, N226, 1011 North Arendell, Zebulon, NC 27597, USA
e-mail: helen.n.strickland@gsk.com

B. Morgan
GlaxoSmithKline, Zebulon Manufacturing and Supply, 1011 N. Arendell Avenue, Zebulon, NC 27597, USA
e-mail: beth.e.morgan@gsk.com

J.D. Christopher
Nonclinical and Pharmaceutical Sciences Statistics, Merck Research Laboratories, WP37C-305, 770 Sumneytown Pike, West Point, PA 19486-0004, USA
e-mail: j.david.christopher@merck.com

V. Glaab
Boehringer Ingelheim, Respiratory Drug Delivery, Binger Strasse 173, Ingelheim am Rhein 55216, Germany
e-mail: volker.glaab@boehringer-ingelheim.com

A. Goodey
Merck & Co. Inc., 556 Morris Ave, Summit, NJ 07901, USA
e-mail: adrian.goodey@merck.com

T.P. Tougas et al. (eds.), *Good Cascade Impactor Practices, AIM and EDA for Orally Inhaled Products*, DOI 10.1007/978-1-4614-6296-5_9,

9.1 Introduction

EDA, being a new concept, will require a period of time to allow for confidence building, based on the experience of individual organizations involved with the OIP life cycle. The acceptance by regulators that this approach is valid, whether approached simply using one of the existing compendial full-resolution CIs or augmented by measurements obtained with an appropriately validated AIM-based apparatus, will require a body of validated evidence in which all possible scenarios that might result in failure have been assessed. This chapter begins the process of acceptance by looking at the EDA concept from two different viewpoints:

1. Theoretical considerations, probing the EDA approach beyond conditions that are likely in association with currently marketed OIPs, by considering hypothetical scenarios in which this methodology might fail to have sufficient discriminating power to detect APSD changes that may be important in terms of product performance;
2. Practical considerations, involving an examination of several case studies involving OIP products whose brand names are blinded, that are either in production or in development.

9.2 How EDA Detects APSD Changes

Before looking at hypothetical scenarios in which EDA might fail (Sect. 9.3), a failure modes analysis for EDA (Sect. 9.4), and actual case studies (Sect. 9.5), it is worthwhile reviewing again in brief how EDA works to detect changes in APSDs of OIP-generated aerosols.

One of the two EDA metrics, *ISM*, represents the ability to determine the area under the curve of the APSD when presented in differential mass-weighted form. The other EDA metric, the ratio *LPM*/*SPM*, enables shifts in the measure of central tendency (i.e., either the mass-weighted mean diameter or more usually and conveniently the *MMAD* value) to be observed with respect to the size axis scaled as aerodynamic diameter. Taken together, the combination of these metrics enables

K. Joshi
Nasal & Inhalation Analytical R&D, Catalent Pharma Solutions,
160N Pharma Drive, Research Triangle Park, Morrisville, NC 27709, USA
e-mail: Keyur.Joshi@catalent.com

L. Mao
Catalent Pharma Solutions, 160 Pharma Dr, Morrisville, NC 27560, USA
e-mail: lei.mao@catalent.com

J.P. Mitchell
Trudell Medical International, 725 Third Street, London, ON N5V 5G4, Canada
e-mail: jmitchell@trudellmed.com

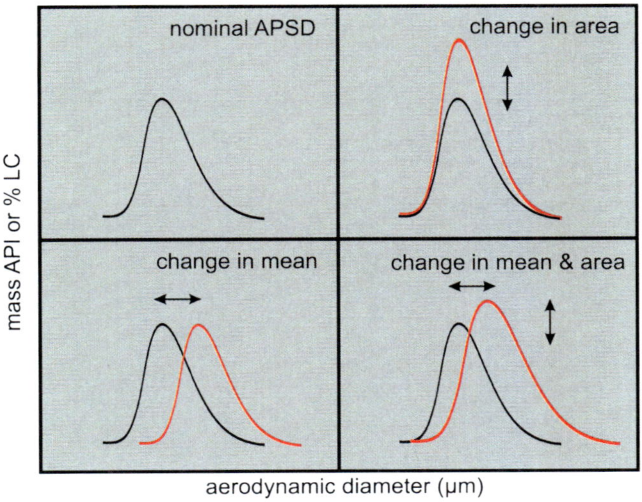

Fig. 9.1 Basic changes to hypothetical unimodal APSD; (upper left) represents a hypothetical unimodal APSD; (upper right) describes changes in APSD amplitude, reflected in AUC variations; (lower left) describes changes in APSD position on the size axis, reflected in MMAD movement; and (lower right) describes changes in both APSD amplitude and position

all the commonly encountered types of a single-mode APSD shift to be detected, as illustrated schematically by Fig. 9.1 [1].

At this stage, it is worthwhile reiterating a key message presented in Chap. 3 that the underlying physical processes that might cause a change in APSD will not lead to fine structure development, defined as the appearance of one or more separate modes encompassing size ranges discernible by one or at most two adjacent stages of a 7- or 8-stage CI. Instead, changes will take place to an observable extent over a wide portion of the size range of interest. This understanding is supported by evidence from a Product Quality Research Institute working group survey of patterns of changes observed in real products [2].

Table 9.1 is an elaboration of the APSD shifts illustrated in Fig. 9.1, defining the eight possible scenarios reported by Mitchell et al., in which the metrics *LPM* and *SPM* (and therefore the ratio metric (*R*) and *ISM*) may change with respect to each other [3].

For scenarios 1 and 2, a change in *LPM* with approximately the same magnitude but opposite directional change in *SPM* will result in a ratio metric (*R*) change but not a movement in *ISM* metric and therefore translates to an *MMAD* change without an *AUC* change in the APSD.

For scenarios 3 and 4, a similar directional but proportionate change in *LPM* and *SPM* will result in the same directional change in *ISM* while maintaining a reasonably constant *R* and *R*-value and thus translates to an *AUC* change in the APSD without a significant change in *MMAD*.

For scenarios 5 and 6, the *ISM* and *R* metrics will show an increasing or decreasing *LPM*, respectively. In each case, *ISM* and *R* will show the same directional change as *LPM*. Increased *ISM* and *R* correspond to a larger *MMAD* and an increase

Table 9.1 Eight possible scenarios associated with APSD shifts

Scenario	Summary of APSD changes	ΔMMAD[a]	ΔAUC[b]	ΔLPM	ΔSPM	Δ*R*[c]	ΔISM[d]
1	MMAD change with	↑	≈	↑	↓	↑	≈
2	no change in AUC	↓	≈	↓	↑	↓	≈
3	AUC change with no	≈	↑	↑	↑	≈	↑
4	change in MMAD	≈	↓	↓	↓	≈	↓
5	MMAD and AUC	↑	↑	↓	(≈↓)	↑	↑
6	change	↓	↓	↓	(≈↓)	↓	↓
7		↓	↑	(≈↓)	↓	↓	↑
8		↑	↓	(≈↓)	↓	↑	↓

[a]ΔMMAD = shift in MMAD

[b]ΔAUC = shift in area under differential mass-weight APSD profile

[c]$R = \text{Ratio} = \dfrac{LPM}{SPM}$

[d]ISM = *LPM* + *SPM*

in *AUC*. Likewise, decreased *ISM* and *R* correspond to a smaller *MMAD* and a decrease in *AUC*.

For scenarios 7 and 8, the *SPM* metric will show an increase or decrease, respectively, with *ISM* and *R* moving in opposite directions with respect to each other. In each case, the *ISM* change will be in the same direction as the *SPM* movement. An *ISM* decrease with an increase in *R* corresponds to an *AUC* decrease associated with a larger *MMAD*. Conversely, an *ISM* increase associated with a decrease in *R* corresponds to an increase in the *AUC* with a smaller *MMAD*.

In summary, an increase in *MMAD* without a change in the *AUC* should be readily evident when the value of *R* increases while at the same time, the *ISM* does not exhibit change. For four of the scenarios, both the *MMAD* and the *AUC* of the APSD change at the same time, while for the remaining four scenarios, two result in a change only in the median value and two result in an area only movement.

The next question to ponder is how these scenarios play out if the magnitude of total mass ex-inhaler that enters the CI system (*TM*) is allowed to vary. *TM* includes the non-sizing as well as the sized component of the aerosol emitted by the inhaler. It is conceivable that a change in the former alone might go undetected by EDA because movements in either or both *R* and *ISM* may not occur as the result of changes in the non-sized component.

The outcomes from this type of behavior are examined for various sub-scenarios. Hypothetical APSDs described in terms of TM, ratio of NISM to ISM, LPM, SPM, Ratio of LPM to SPM and ISM were compared to the baseline APSD characteristics representing conditions commonly encountered with OIP aerosols:

Total Mass (TM) = 100 mass units
Non-impactor-sized (NISM) mass component = 65 mass units,
Impactor sized mass (ISM) component = 35 mass units,

Ratio of NISM to ISM =1.857,
LPM component = 17.5 mass units
SPM component = 17.5 mass units
Ratio Metric = 1.00
ISM Metric = 35.0 mass units.

The hypothetical APSDs were generated using

1. TM values at 90%, 100%, and 110% of the baseline APSD
2. Non-impactor-sized (NISM) mass to impactor sized mass (ISM) component ratios from 0.667 (40:60 NISM to ISM) to 3.0 (75:25 NISM to ISM), and
3. LPM component to SPM component split from 0.667 (40:60 LPM to SPM) to 1.50 (75:25 LPM to SPM)

The EDA metrics for each of the hypothetical APSDs were computed and categorized according to the eight possible types of APSD shifts described in Table 9.1, a subset of the hypothetical APSDs are provided in Table 9.2.

An alternative way to look at the detectability of APSD shifts is to examine how possibilities for APSD changes might be observed by the CI method, and then consider the conditions that would have to exist for EDA to fail as a detection method (Table 9.3). It is useful to consider the small particle (*SPF*) and large particle (*LPF*) mass fractions, rather than the corresponding absolute values of *SPM* and *LPM* for this type of analysis, as the mass fractions normalize the APSD, making it easier to identify changes in shape in data comparisons.

From the data shown in Table 9.3 (equivalence conditions), several conditions can be foreseen where EDA failure would be possible (Table 9.3). However, before

Table 9.2 EDA metrics representing a subset of the previously described hypothetical APSDs as compared to the hypothetical baseline APSD characteristics and metrics, and categorized according to the eight possible scenarios associated with APSD shifts

Scenario	TM mass units	Ratio NISM to ISM (LPM to SPM)	LPM mass units	SPM mass units	Ratio (LPM to SPM)	ISM (LPM + SPM) mass units
Baseline	100	65:35 (50:50)	17.5	17.5	1.00	35.0
1. Increased ratio and equivalent ISM	90	61:39 (55:45)	19.3 (↑)	15.8 (↓)	1.22 (↑)	35.1 (≈)
		61:39 (60:40)	21.1 (↑)	14.0 (↓)	1.50 (↑)	
	100	65:35 (55:45)	19.2 (↑)	15.8 (↓)	1.22 (↑)	35.0 (≈)
		65:35 (60:40)	21.0 (↑)	14.0 (↓)	1.50 (↑)	
	110	68:32 (55:45)	19.4 (↑)	15.8 (↓)	1.22 (↑)	35.2 (≈)
		68:32 (60:40)	21.1 (↑)	14.1 (↓)	1.50 (↑)	
2. Decreased ratio and equivalent ISM	90	61:39 (40:60)	14.0 (↓)	21.1 (↑)	0.67 (↓)	35.1 (≈)
		61:39 (45:55)	15.8 (↓)	19.3 (↑)	0.82 (↓)	
	100	65:35 (40:60)	14.0 (↓)	21.0 (↑)	0.67 (↓)	35.0 (≈)
		65:35 (45:55)	15.8 (↓)	19.2 (↑)	0.82 (↓)	
	110	68:32 (40:60)	14.1 (↓)	21.1 (↑)	0.67 (↓)	35.2 (≈)
		68:32 (45:55)	15.8 (↓)	19.4 (↑)	0.82 (↓)	

(continued)

Table 9.2 (continued)

Scenario	TM mass units	Ratio NISM to ISM (LPM to SPM)	LPM mass units	SPM mass units	Ratio (LPM to SPM)	ISM (LPM + SPM) mass units
Baseline	100	65:35 (50:50)	17.5	17.5	1.00	35.0
3. Equivalent ratio and increased ISM	90	25:75 (50:50)	33.75 (↑)	33.75 (↑)	1.00 (≈)	67.5 (↑)
		60:40 (50:50)	18.0 (↑)	18.0 (↑)		36.0 (↑)
	100	25:75 (50:50)	37.5 (↑)	37.5 (↑)	1.00 (≈)	75.0 (↑)
		60:40 (50:50)	20.0 (↑)	20.0 (↑)		40.0 (↑)
	110	60:40 (50:50)	22.0 (↑)	22.0 (↑)	1.00 (≈)	44.0 (↑)
		65:35 (50:50)	19.25 (↑)	19.25 (↑)		38.5 (↑)
4. Equivalent ratio and decreased ISM	90	65:35 (50:50)	15.75 (↓)	15.75 (↓)	1.00 (≈)	31.5 (↓)
		75:25 (50:50)	11.25 (↓)	11.25 (↓)		22.5 (↓)
	100	70:30 (50:50)	15.0 (↓)	15.0 (↓)	1.00 (≈)	30.0 (↓)
		75:25 (50:50)	12.5 (↓)	12.5 (↓)		25.0 (↓)
	110	70:30 (50:50)	16.5 (↓)	16.5 (↓)	1.00 (≈)	33.0 (↓)
		75:25 (50:50)	13.75 (↓)	13.75 (↓)		27.5 (↓)
5. Increased ratio and increased ISM	90	40:60 (55:45)	29.7 (↑)	24.3 (↑)	1.22 (↑)	54.0 (↑)
		60:40 (60:40)	21.6 (↑)	14.4 (↓)	1.50 (↑)	36.0 (↑)
	100	40:60 (55:45)	33.0 (↑)	27.0 (↑)	1.22 (↑)	60.0 (↑)
		60:40 (60:40)	24.0 (↑)	16.0 (↓)	1.50 (↑)	40.0 (↑)
	110	40:60 (60:40)	39.6 (↑)	26.4 (↑)	1.50 (↑)	66.0 (↑)
		60:40 (60:40)	26.4 (↑)	17.6 (↓)	1.50 (↑)	44.0 (↑)
6. Decreased ratio and decreased ISM	90	65:35 (40:60)	12.6 (↓)	18.9 (↑)	0.7 (↓)	31.5 (↓)
		75:25 (45:55)	10.1 (↓)	12.4 (↓)	0.8 (↓)	22.5 (↓)
	100	70:30 (40:60)	12.0 (↓)	18.0 (↑)	0.7 (↓)	30.0 (↓)
		75:25 (45:55)	11.2 (↓)	13.8 (↓)	0.8 (↓)	25.0 (↓)
	110	70:30 (40:60)	13.2 (↓)	19.8 (↑)	0.7 (↓)	33.0 (↓)
		75:25 (45:55)	12.4 (↓)	15.1 (↓)	0.8 (↓)	27.5 (↓)
7. Decreased ratio and increased ISM	90	50:50 (40:60)	18.0 (↑)	27.0 (↑)	0.7 (↓)	45.0 (↑)
		60:40 (45:55)	16.2 (↓)	19.8 (↑)	0.8 (↓)	36.0 (↑)
	100	50:50 (40:60)	20.0 (↑)	30.0 (↑)	0.7 (↓)	50.0 (↑)
		60:40 (45:55)	18.0 (↓)	22.0 (↑)	0.8 (↓)	40.0 (↑)
	110	60:40 (40:60)	17.6 (≈)	26.4 (↑)	0.7 (↓)	44.0 (↑)
		65:35 (45:55)	17.3 (↓)	21.2 (↑)	0.8 (↓)	38.5 (↑)
8. Increased ratio and decreased ISM	90	65:35 (55:45)	17.3 (↓)	14.2 (↓)	1.22 (↑)	31.5 (↓)
		70:30 (60:40)	16.2 (↓)	10.8 (↓)	1.50 (↑)	22.5 (↓)
	100	70:30 (55:45)	16.5 (↓)	13.5 (↓)	1.22 (↑)	30.0 (↓)
		75:25 (60:40)	15.0 (↓)	10.0 (↓)	1.50 (↑)	25.0 (↓)
	110	70:30 (55:45)	18.2 (↓)	14.8 (↓)	1.22 (↑)	33.0 (↓)
		75:25 (60:40)	16.5 (↓)	11.0 (↓)	1.50 (↑)	27.5 (↓)

coming to a conclusion concerning the overall robustness of the EDA principle, an evaluation of their likelihood has to take place, and this step requires an in-depth analysis of the physical processes that operate on aerosol particles once they are created by operation of the OIP.

Table 9.3 Changes in APSD of aerosols from OIPs, their detection by the CI method, and potential for EDA to fail to detect such shifts

Nature of change	CI observation	Condition required for EDA failure
Increasing *MMAD*	Shift in mass from higher- to lower-numbered stages in CI	Entire APSD contained within either *LPF* or *SPF*
Decreasing *MMAD*	Shift in mass from lower- to higher-numbered stages	Entire APSD contained within either *LPF* or *SPF*
Broadening APSD (constant *MMAD*)	Decreased mass on middle stages and increased mass on peripheral stages	Either: *LPF/SPF* boundary coincident with *MMAD* Or: entire APSD contained within either *LPM* or *SPM*
Narrowing APSD (constant *MMAD*)	Increased mass on middle stages and decreased mass on peripheral stages	Either: *LPF/SPF* boundary coincident with *MMAD* Or: entire APSD contained within either *LPF* or *SPF*
Change in overall shape	Change in mass distribution across all stages accompanied by change in *MMAD*	Entire APSD contained within either *LPF* or *SPF*
Change in modality (unimodal to bimodal)	Emergence of mass at new mode, balanced by decrease mass at original mode	Entire APSD contained within either *LPF* or *SPF*

9.3 Potential Failure Modes: Theoretical Considerations

This topic is best approached by first deriving conceptual changes to the APSD in which both the sum, *ISM*, and the ratio, *LPM/SPM*, terms would remain unaltered during transport from the inhaler through to the CI-based measurement system. This strategy is similar to that which would have to be taken in the wider consideration of APSD changes to the OIP-produced aerosol, in the context of product stability evaluations, forming part of that specific OIP development process.

Mitchell et al. envisaged a series of hypothetical situations that have the potential for EDA to fail [3]. However, at the outset, it was recognized that the probability is small that such changes would occur precisely so that the portion of the APSD beyond the boundary between either *LPF* or *SPF*, depending on the scenario under consideration, is unaffected.

Figure 9.2 summarizes in simplified form the aerosol transport conditions applicable in a typical CI system.

Apart from the intentional deposition of particles to the various stages following size-fractionation on the basis of their differing inertia, nonideal deposition to the interior walls of the system may occur to a greater or lesser extent, depending on the CI design. The backup filter is assumed 100% efficient as a collector of particles that penetrate beyond the last impaction stage, but this assumption may not be true for the MOC used with the NGI. Besides mechanisms responsible for particle deposition, there is the possibility of APSD change being brought about as the result of particle–particle interactions in the aerosol phase.

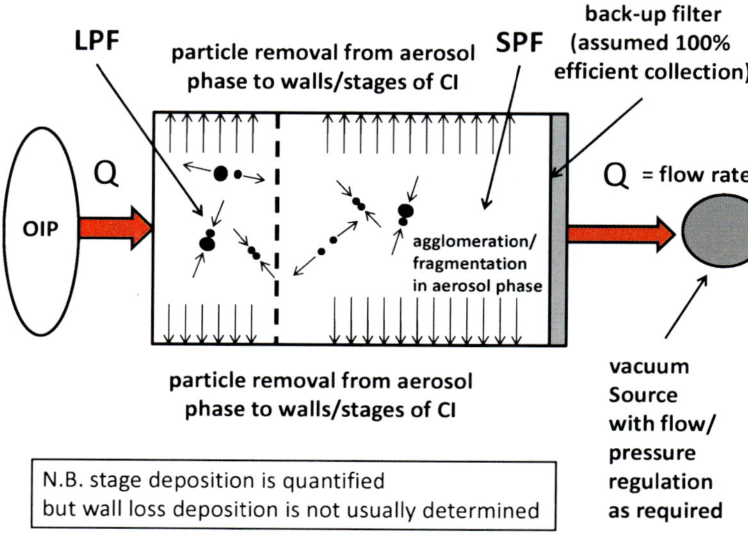

Fig. 9.2 Highly idealized model of aerosol transport from OIP through the CI system showing potential mechanisms that could influence the measured APSD

Note that the presence of an induction port and in some cases also a pre-separator is not identified here. These components would also contribute to losses of particles, primarily as the result of turbulent inertial deposition downstream of the 90° bend of the USP/Ph. Eur. configuration [4], and where the air flow direction changes take place in the highly turbulent environment associated with flow through a pre-separator [5].

The situations that Mitchell et al. considered might lead to potential EDA failure [3] are described in the following three subsections.

9.3.1 Change of Shape in the Large Particle Mass Fraction Alone But the Same Absolute LPM Before and Afterwards

In this first situation, illustrated in Fig. 9.3, the transfer of mass within the APSD takes place exclusively within the size limits defined for the *LPF*. If the underlying physical processes influencing the OIP aerosol are considered, both the mechanisms of particle-particle agglomeration (coagulation) and de-agglomeration in the aerosol phase, as well as inertial/turbulent deposition to nearby surfaces, can account for movements in mass from one size to another. However, considering the APSD as a whole, Mitchell et al. [3] considered that their effects are likely never to be so selective to the extent that the small particle fraction is entirely unaffected, while just the large particle fraction experiences the size-related shift. Instead, typically peak

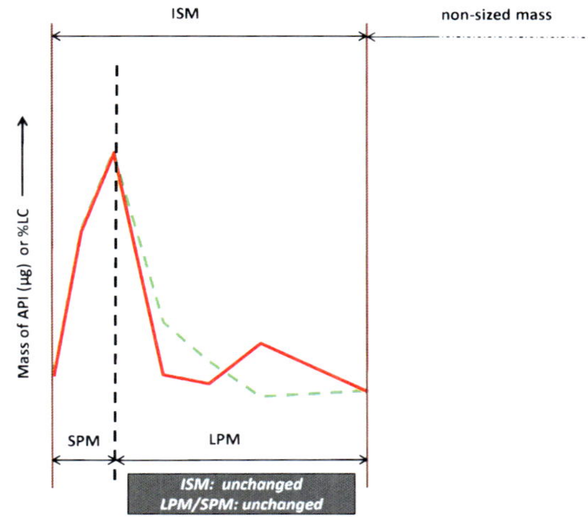

Fig. 9.3 Situation 1 for potential failure of EDA: change of shape in *LPF* alone but the same absolute *LPM* before and afterwards (*From* [3]—*used with permission*)

broadening occurs, which by changing the values of both *SPM* and *LPM* affects *LPM*/*SPM*, making the change detectable by EDA.

Particles that come into contact can agglomerate (Chap. 3). Typically, four processes that will potentially bring aerosol particles together need to be considered during the lifetime of an OIP aerosol:

1. The larger particles can sweep out smaller particles along their settling path, because the former settle more rapidly under the influence of gravity which is the dominant force affecting micron-sized particles if no other processes, in particular turbulence in the flow, are operating.
2. Brownian (molecular) motion can move particles across the streamlines of flow so that they intersect other particles. This process only becomes important for particles whose size approaches the mean-free path of the surrounding air [0.068 μm at ambient pressure (101.3 kPa)] and can therefore be largely ignored with OIP aerosols.
3. Particles can experience diffusion in random directions as the result of localized turbulence, resulting in collisions with neighboring particles.
4. Larger particles, unable to respond to acceleration of the gas phase around bends and/or obstructions, will cross the streamlines of the flow and thereby intersect other particles, resulting in inertial-induced agglomeration.

Agglomeration, being a particle number—as well as a time-dependent stochastic process, results in a smooth change to the whole APSD as particle growth occurs, most evident at the fine end as there are many more finer than coarser particles on a number-weighted basis in a typical APSD. Aerosol removal/depletion processes such as gravitational sedimentation to nearby surfaces or inertial impaction to obstructions will have most influence on the large sizes of the APSD, with a smooth steadily decreasing effect throughout the remainder of the APSD. Similarly, de-agglomeration that might occur as the result of locally induced shear stress in turbulent flow will cause mass migration to the finer end of the APSD as a smooth process in terms of its size-dependency, rather than a sudden cutoff located precisely at the *SPF*/*LPF* boundary. These shifts in APSD are therefore all detectable by EDA through changes in *LPM*/*SPM* and, in the case of removal processes, by a decrease in *ISM*.

Other processes Mitchell et al. [3] deemed worthy of consideration are the following:

1. Brownian diffusion as a removal process from the aerosol phase to the walls of the CI system would selectively reduce the concentration of the finer particles, resulting in a smooth but steadily decreasing effect as particle size increases. Such changes, if significant at the size range typical of currently marketed OIPs (from 0.5 to 10 μm aerodynamic diameter), would be detectible by EDA as an increase in *LPM*/*SPM*.
2. Phoretic processes (i.e., thermophoresis, diffusiophoresis) are generally size independent, so that an *LPM*/*SPM* change might not be anticipated. However, for thermophoresis to be effective as a removal process, a temperature gradient would need to exist along the axis perpendicular to the cooler walls of the CI system. The phenomenon can be readily avoided by operating the CI system under controlled temperature conditions, thereby preventing the presence of thermal gradients. Diffusiophoresis would require the presence of a condensing vapor to the CI walls. Since OIP aerosols are typically not present with this type of atmosphere and the water vapor content of the air containing the emitted aerosol is usually well below saturation, this type of removal process would not be expected to occur.
3. The effect of electrostatic charge on APSD is linked with electric charge distribution of the aerosol being sampled [both positive and negative charges (Chap. 3)]. It is theoretically more complex and therefore difficult to predict. Significant loss of particles by attraction to charges on the walls of the CI can be avoided with the use of conducting metal surfaces, but some charge may reside on the surface of stage-coating materials, such as silicone oil that are insulators. The use of electrically conducting tacky substrates is an obvious solution to the problem. However, in the authors' experience, particle removal from the aerosol phase due to this cause is not so size selective as to account for changes solely within the measured *LPF*.
4. Particle fragmentation (de-agglomeration) or more likely the introduction of increased numbers of larger particles into the APSD that is sized by the CI by changes to the way they are bound to larger carrier particles might conceivably result in a shift within only the *LPF* component of an APSD for DPI-based assessments. This outcome would most likely be the result of a transfer of mass retained in the non-sizing components (induction port and pre-separator (if used)) into the aerosol reaching the CI. However, such a change would be detected by an increase in both *LPM* and *ISM* and therefore be detectable by EDA. Even if in the extremely unlikely case where a simultaneous and exact amount of mass transfer also takes place from the *LPF* to the *SPF* component thereby maintaining *LPM* constant, *SPM* and therefore *ISM* would increase in response, making the change detectable.

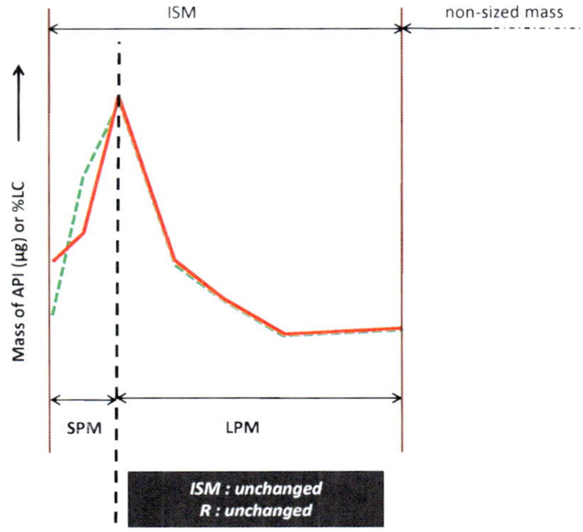

Fig. 9.4 Situation 2 for potential failure of EDA: change of shape in *SPF* alone but the same absolute *SPM* before and afterwards (*From* [3]*—used with permission*)

9.3.2 Change of Shape in Small Particle Fraction Alone But the Same Absolute SPM Before and Afterwards

In the second situation considered by Mitchell et al. [3] (Fig. 9.4), the transfer of mass within the APSD takes place exclusively within the bounds for the *SPF*, essentially making this process complementary to the first situation.

Significant removal of particles to the walls of the CI by Brownian diffusion will almost certainly be entirely confined to the *SPF*. If present at all, this effect is only likely to be observable on a mass-weighted basis in the extra-fine range <0.5 μm aerodynamic diameter. Since all *SPF* particles would be captured by the walls of the CI or on the backup filter regardless of size, it is conceivable that *SPM* might remain constant, but only if the API deposited on the walls of the CI is recovered, assayed, and assigned to the appropriate mass fraction. *LPM* would also remain constant by definition in this scenario, so that *ISM* would also be unaffected. However, in reality this scenario is a highly remote possibility, given the typically short duration of aerosol transport through the portion of the CI in which the *SPF* collects when the compendial apparatuses are operated within their recommended flow rate range.

Since none of these apparatuses operate at flow rates <15 L/min, they are not low-flow impactors, such as the QCM system (California Measurements Inc., Sierra Madre, CA, USA) that operates at only 0.24 L/min [6].

It should be noted that if the MOC is used with the NGI, it is not a perfect filter, and so any significant changes in the penetration of this stage brought about through Brownian diffusion would likely be detected as variations in the *LPM*/*SPM* ratio.

APSD changes due to Brownian diffusion are likely only to affect the tail at the fine end of the APSD (<1.0 μm aerodynamic diameter), so that they are most probably only a consideration with solution pMDI products, considering the entire range of currently marketed OIPs. Even if EDA were to fail to detect a scenario 2 shift in

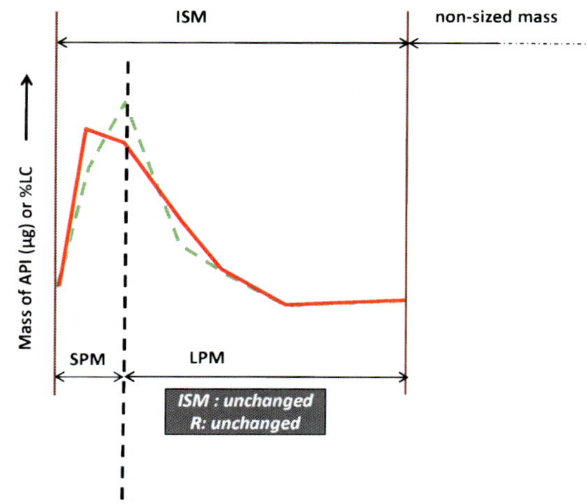

Fig. 9.5 Situation 3 for potential failure of EDA: simultaneous change of shape in *LPF* and *SPF* but the same absolute *SPM* and *LPM* before and afterwards (*From* [3]—*used with permission*)

APSD, the significance in terms of medication delivery of such a change would likely be minimal, given that most extra-fine particles do not deposit in the HRT as efficiently as larger fine particles [7], unless combined with a breath-holding maneuver [8].

Turbulent diffusion is not particularly size selective in the range of interest for the sized-portion of the OIP-generated aerosol, although this phenomenon is likely to be more important than Brownian diffusion in the associated non-sizing components of the CI system.

9.3.3 Simultaneous Change of Shape in LPF and SPF But the Same Absolute LPM and SPM Before and Afterwards

Mitchell et al. [3] considered that two different processes must happen simultaneously for the third situation to occur (Fig. 9.5). One of these must take place entirely within the *SPF*, thereby unaffecting *SPM* and the other entirely within the *LPF*, thereby unaffecting *LPM*.

Realistically, the probability of such an event is vanishingly small, given the stringency of the criteria for EDA to fail. Nevertheless, considering potential mechanisms further, particle-particle agglomeration could increase *LPM* but would also decrease *SPM*, as has been explained in connection with scenario 1, making the effect detectible by EDA. This process always gives rise to peak broadening as well as an overall shift in APSD to larger sizes. Note that the mass in the *SPF* would always decrease in proportion to the gain in the mass of *LPF*, because there is no further input of fine particles to compensate for the shift in APSD.

APSD shifts towards fine particle sizes brought about either by gravitational sedimentation and/or inertial (turbulent) deposition to the walls of the CI system would always result in a reduction in *LPM* that is larger than any corresponding fall in *SPM*, making the effects detectible by a shift in *LPM*/*SPM* and also by a decrease in *ISM*. Likewise, turbulent de-agglomeration of the aerosol particles would result in a shift to finer sizes, detectible as a decrease in *LPM*/*SPM*.

The introduction of increased numbers of larger particles by changes to the way they are bound to the carrier particles of certain DPIs might conceivably result in a shift within both *LPF* and *SPF* components of the APSD. However, such a change would be detected by an increase in *LPM*/*SPM*. In the extremely unlikely case where the exact mass transfer to *SPF* exactly balances the incoming mass to the *LPF* component, *SPM* and therefore *ISM* would have to increase, again making the change detectible by EDA.

9.3.4 *Summary*

The theoretical foundation upon which the metrics associated with EDA are constructed appears to be robust, given the outcomes from both the case studies with real OIPs and from a consideration of aerosol mechanics associated with plausible scenarios (however remote) where this approach might fail to detect changes in measured aerosol APSDs. The practical examples that follow in Sect. 9.5, in which case studies are examined, are based on full-resolution cascade impactor experience, because AIM-based measurements are not yet in routine use for product qualification.

If abbreviated systems are adopted, the underlying physical processes described in the scenarios for potential EDA failure will not be affected. However, it would be prudent to conduct an appraisal of the effect of reducing internal wall surfaces for potential particle loss by deposition in the abbreviated impactor as part of the method development process, whichever AIM system is chosen. The next chapter describes the large amount of validation work that has already been undertaken in support of a variety of AIM-based apparatuses that are available commercially, or which have been constructed from existing full-resolution CI components.

9.4 Failure Mode Analysis

Glaab et al. have looked at the robustness of the EDA concept by undertaking a failure mode analysis approach [9]. They first considered the factors that could contribute to changes in the CI-measured APSD associated with (a) the formulation, (b) the device, and (c) the raw material of a hypothetical DPI (Fig. 9.6).

They then went on to look at causes arising from the manufacturing process (Fig. 9.7).

Looking further down the chain of events leading to the final OIP, they then focused their attention on potential causes of APSD changes attributable to the analytics for the API(s) and the measurement process itself (Fig. 9.8).

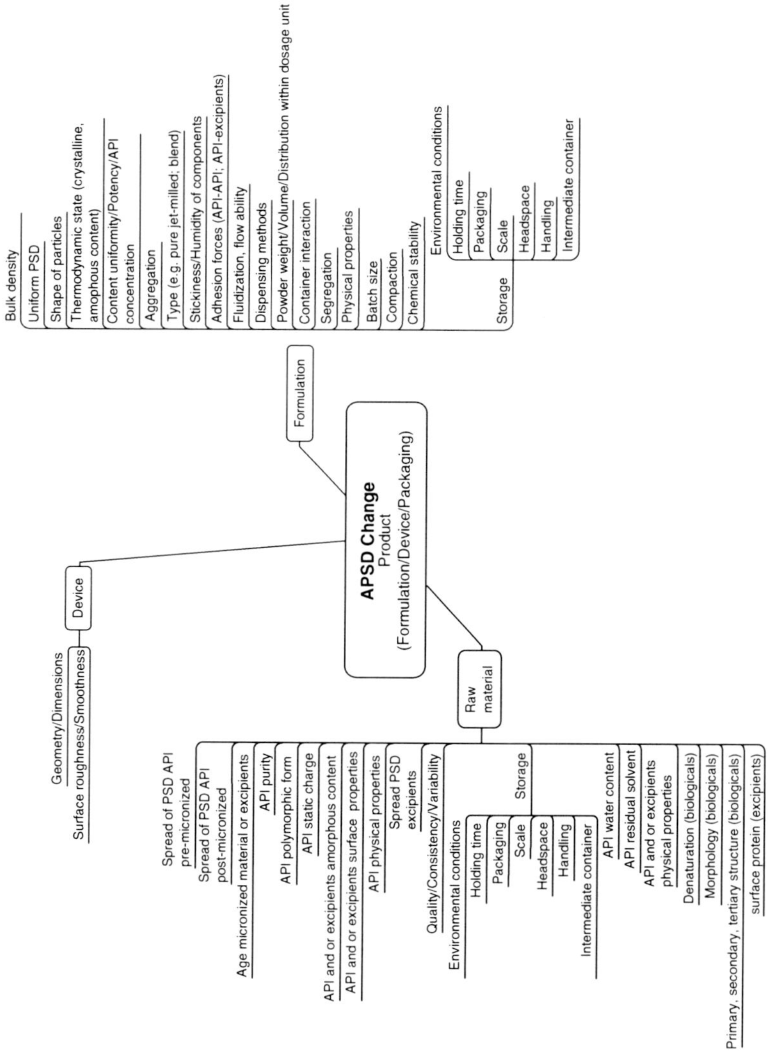

Fig. 9.6 Potential causes for APSD changes related to a DPI (*From* [9]—*used with permission*)

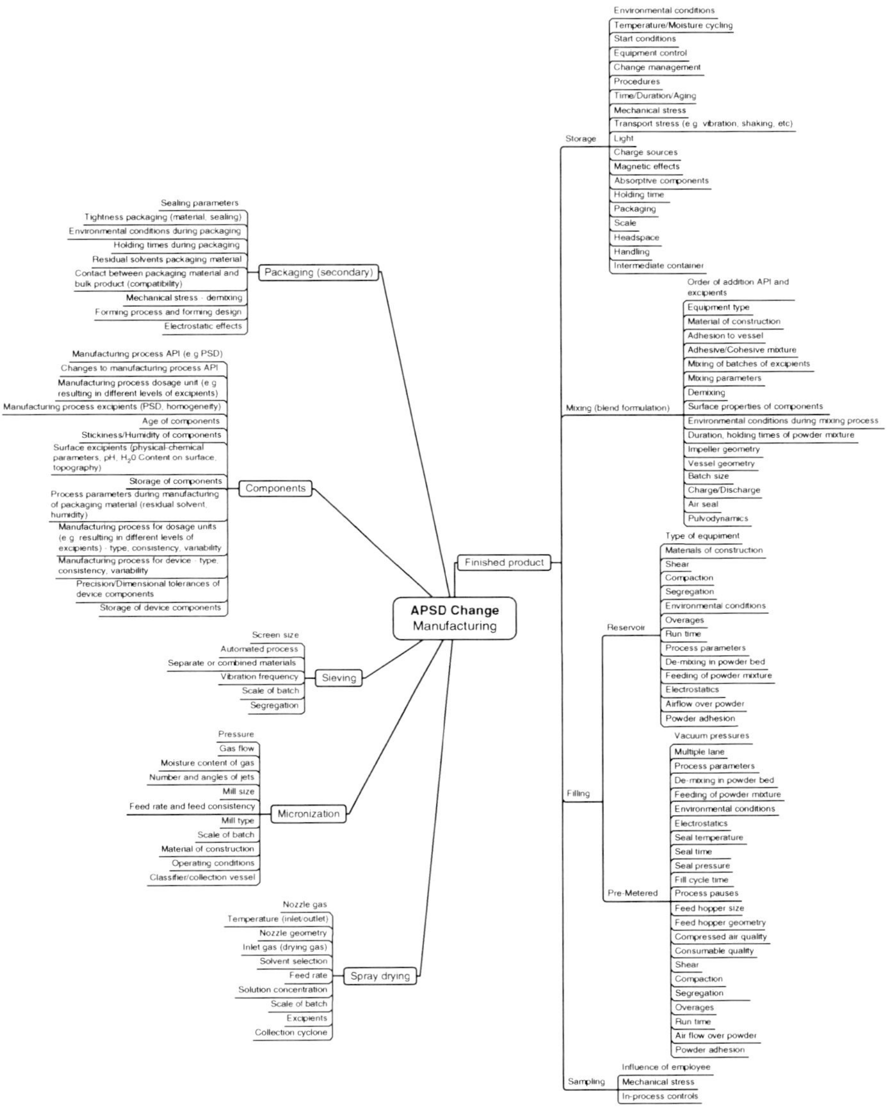

Fig. 9.7 Potential causes for APSD changes related to the DPI manufacturing process (*From* [9]*—used with permission*)

In the second part of the assessment, they repeated the same exercise for a hypothetical MDI, considering manufacturing process, formulation, device, device-formulation interactions, storage and accessories, or add-on devices such as VHCs (Fig. 9.9).

Glaab et al. [9] then went on to examine the various control strategies that could be put in place not only to detect changes in the quality of the product, excipient, or

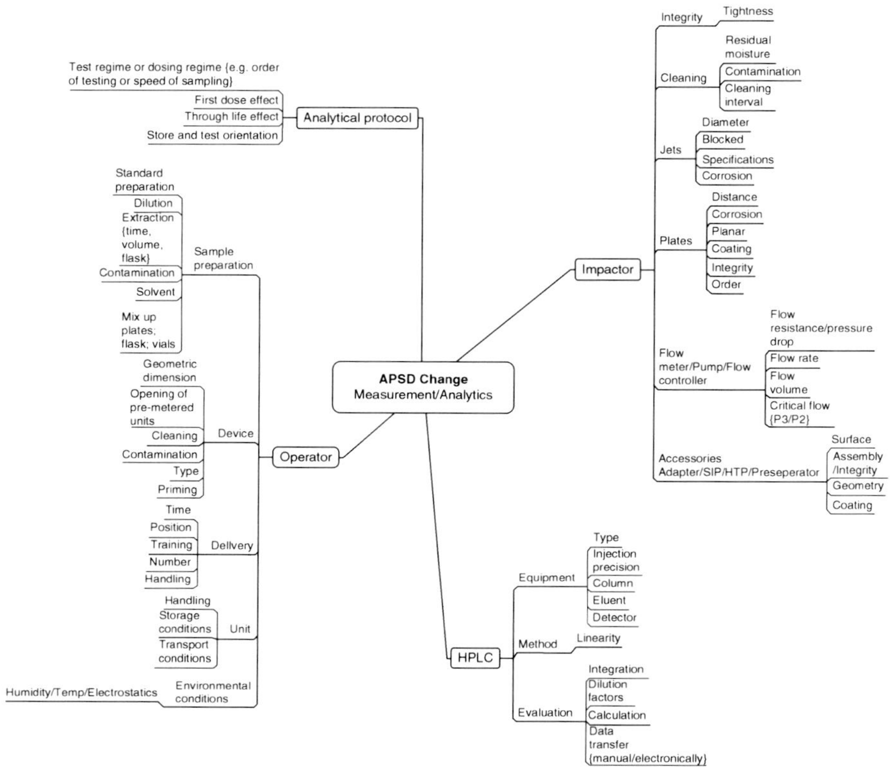

Fig 9.8 Potential causes for APSD changes related to the DPI. API analytics and measurement processes (*From* [9]*—used with permission*)

API but also to detect or prevent changes within manufacturing or analysis of the OIP. They found that these strategies can be subdivided into three different categories:

1. Quality assurance and manufacturing controls
2. Product handling controls
3. Overall quality control for the OIP itself

These operate in a continuous cycle as the OIP moves through its life cycle (Fig. 9.10).

Glaab et al. [9] continued their analysis by examining the various control strategies that can be put in place to mitigate APSD changes for both types of OIP (Fig. 9.11). Their hypothesis was that if all these strategies were to be adopted, changes that could influence the aerosol APSD at the final release testing stage could be easily detected and mitigated or eliminated at the earliest possible stage. If adopted, this degree of control would therefore make the application of EDA in the life cycle of such products easier to manage [10].

In the final stage of their evaluation, Glaab et al. [9] considered the relative magnitudes that the identified risk factors could lead to catastrophic product failure.

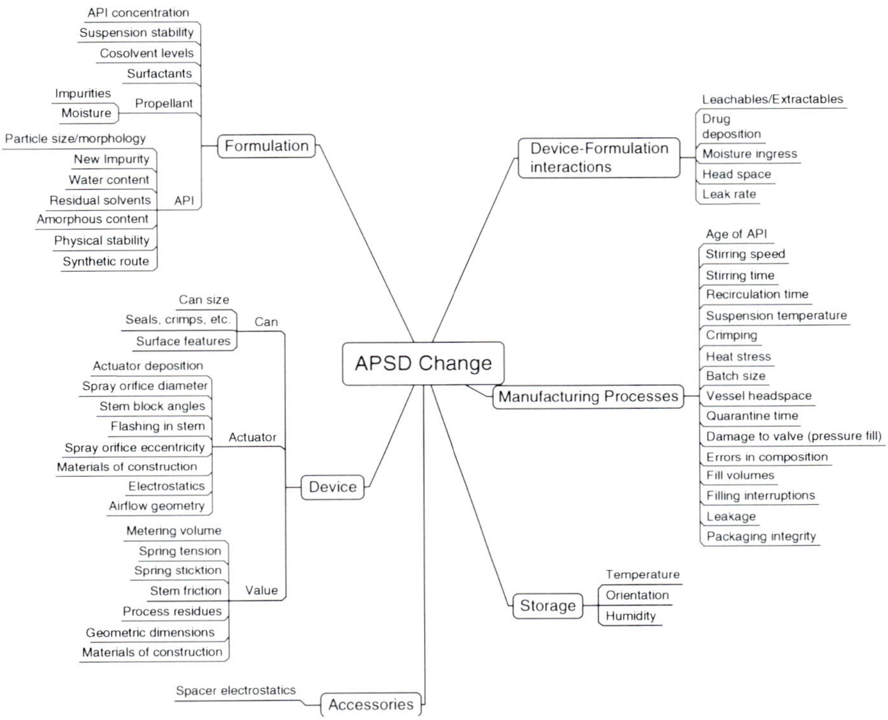

Fig. 9.9 Potential causes for APSD changes related to MDIs (*From* [9]—*used with permission*)

Fig. 9.10 Control strategies in OIP development and manufacture (*From* [9]—*used with permission*)

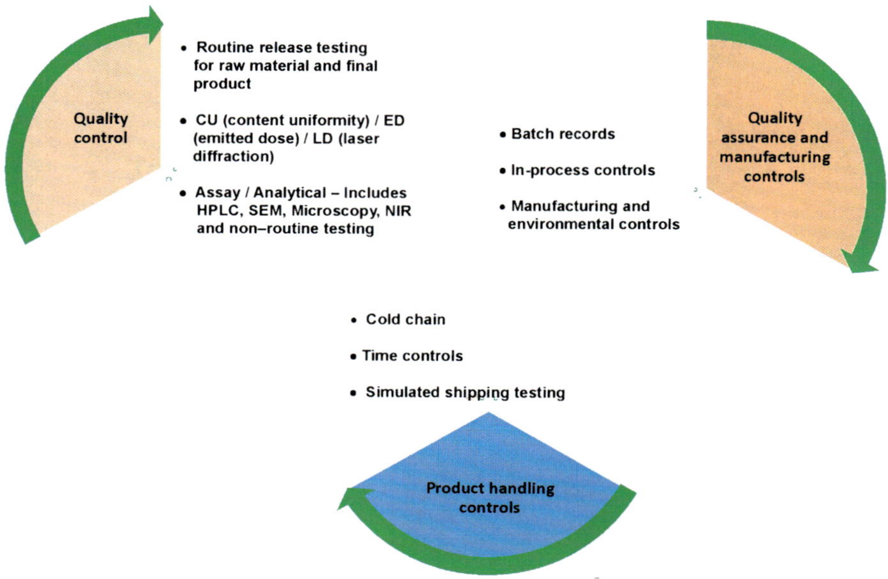

Fig. 9.11 Control strategy breakdown by process in OIP design and manufacture (*From* [9]—*used with permission*)

They appreciated that conducting a classical risk assessment of each factor in a way that was generic to all products in a particular class of OIP was not a practical proposition. They assumed that all the causal factors identified had the potential to impact the aerosol APSD. However, they recognized that it was difficult to generalize the severity of the consequences, since they were likely to be unique to the magnitude of the APSD shift related to each potential causative factor.

Glaab et al. also recognized that it would be a challenging prospect to quantify the effect on product performance with particular patient groups or even individual patients. In light of these limitations, the risk evaluation was concluded by addressing the following questions:

1. How dependent on cascade impaction is the detection of the APSD change?
2. How likely will this change lead to catastrophic product failure?

They responded to each of these questions in general terms, based upon individual experience of the authors. In a further simplification of the assessment, they evaluated the likelihood of a catastrophic product failure resulting from a given factor as either conceivable or inconceivable, with no intermediate conditions. In a similar way, they assessed factors as either likely to be detected by tests other than CI or likely to only be detected by a CI-based test.

Given this approach, it was possible to position the factors relative to quadrants defined by two orthogonal axes representing the potential for catastrophic product failure in relation to the dependence on the CI method for its detection in terms of a discernable APSD shift (Fig. 9.12).

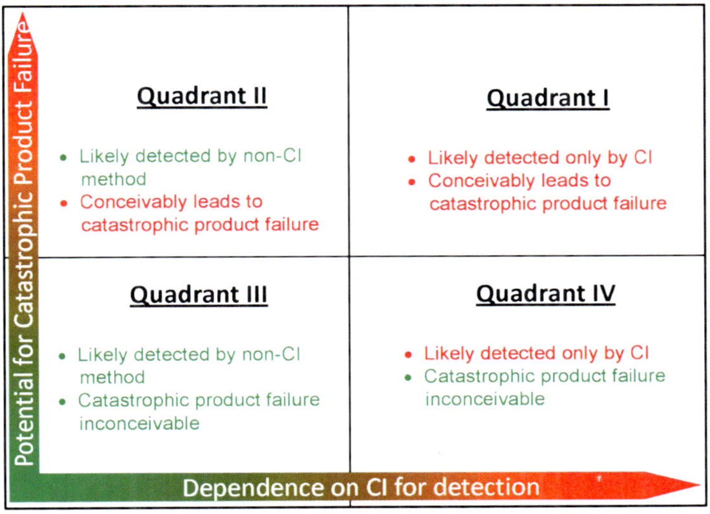

Fig. 9.12 Mapping of relative severity of factors that could influence the APSD of an OIP (*From* [9]—*used with permission*)

Once this framework was in place, Glaab et al. [9] proceeded to locate the factors associated with DPI (Fig. 9.13) and MDI products (Fig. 9.14). Items assigned to quadrant I represent the highest risk given that they are dependent upon CI for detection and could conceivably lead to catastrophic product failure. The risk associated with items placed in other quadrants was believed to be mitigated either by other means of detection method (quadrant II), by the inconceivability of resulting in catastrophic product failure (quadrant IV), or by both considerations (quadrant III).

They concluded that the likelihood that the factors identified above influencing the APSD of an aerosol emitted by an OIP for product release testing depends not only on the individual product but also on its developmental status towards commercialization. At late stage development, the majority of the factors identified in this risk assessment should be controlled within well-defined process parameters, supported by rigorous implementation of quality assurance procedures and compliance within quality control parameters.

However, there are some APSD changes which are detectable only by cascade impaction, primarily because this method, although cumbersome and exacting in terms of analyst performance, directly determines mass of API(s) present in terms of aerodynamic particle size.

This exercise forms a useful complement to the theoretical analysis of aerosol APSD changes presented earlier in the chapter, by providing a framework whereby it is possible to assess in some detail the risk of EDA failure when developing APSD-assessment methodology for a given OIP.

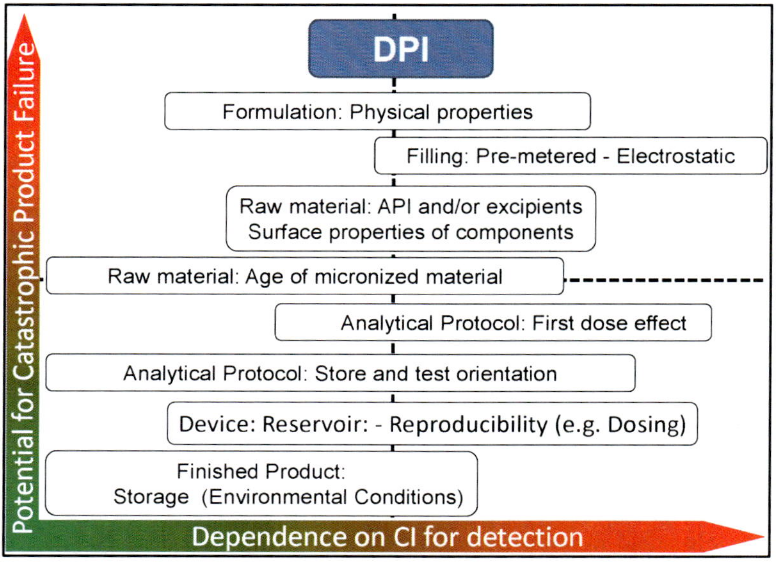

Fig. 9.13 Relative severity of the significant factors linked to possible APSD changes in DPIs (*From* [9]—*used with permission*)

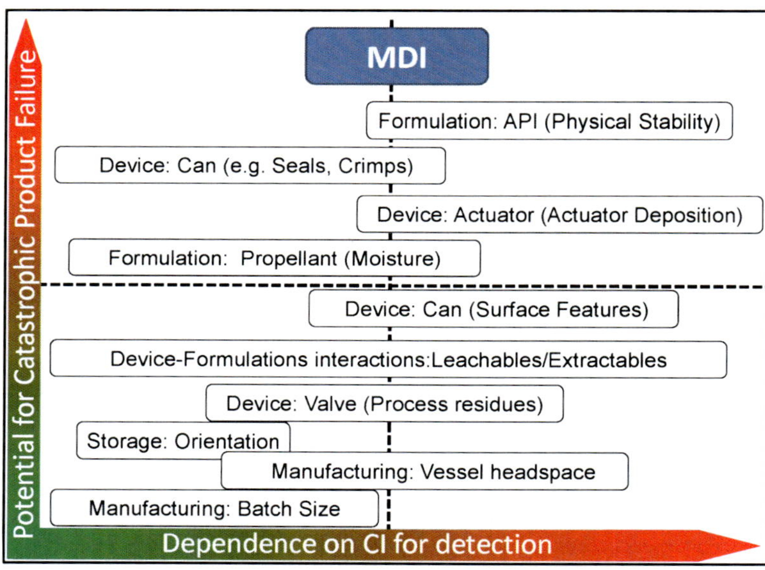

Fig. 9.14 Relative severity of the significant factors linked to possible APSD changes in MDIs (*Adapted from* [9]—*used with permission*)

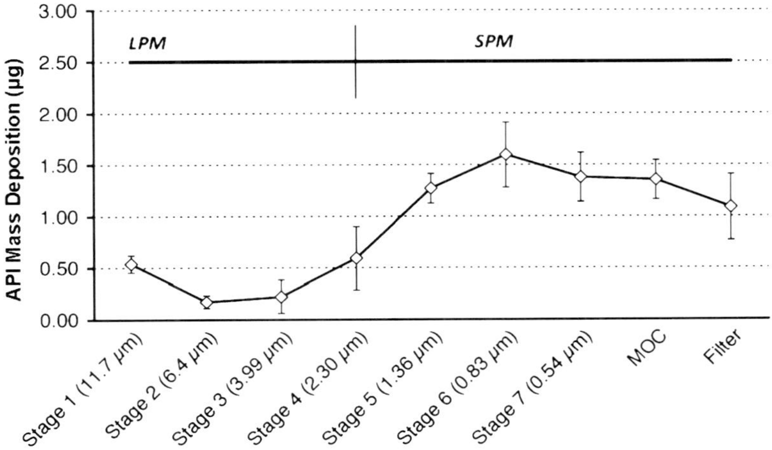

Fig. 9.15 NGI stage deposition (mean ± SD) profile in relation to *LPM* and *SPM* with bound size at 2.30 µm aerodynamic diameter (*From* [11]*—used with permission*)

9.5 Case Studies in Which EDA Has Been Successfully Applied

Perhaps the most significant way in which to demonstrate the potential of the EDA principle is through case studies with OIPs, based on real-life experience. The first such study was undertaken at Catalent Pharma Solutions and reported by Mao et al. [11]. This group applied EDA to analyze three separate studies where the NGI had been used to measure APSDs of aerosols emitted by a group of HFA-based solution MDIs. These data were pooled for the analysis.

All NGI measurements were performed at 30 L/min on the same model solution MDIs for each of the three separate series. The raw data expressed in terms of mass-per-CI component were assessed in terms of grouped stages as follows:

1. Group 2 = sum of drug deposited on stages 1–3
2. Group 3 = sum of drug deposited on stages 4–6
3. Group 4 = sum of drug deposited on stage 7 to the MOC

Group 1, which was not part of the analysis, comprised the sum of the mass of API collecting in the non-sizing components (i.e., induction port) upstream of the NGI.

MMAD values for the APSD were calculated using CITDAS® (version 2.0, Copley Scientific Ltd, Nottingham, UK) that interpolates the data points from the cumulative mass-weighted APSD closest to the 50th mass percentile. *LPM* was calculated as the sum of API mass deposited on each of the stages above and including the stage at the d_{50} value defining the bound between small and large particles (e.g., stage 3 with its d_{50} size corresponding to 3.99 µm aerodynamic diameter associated with stage 3). *SPM* was likewise calculated as the sum of API mass deposited on each of the stages below the NGI stage at the defined d_{50}, in this case from stage 4 to the MOC. It is important to note that the boundary size between *LPM* and *SPM* was varied in the subsequent analysis of the sensitivity of the EDA technique.

In the example illustrated in Fig. 9.15, the location of the boundary between *LPM* and *SPM* was set close to the *MMAD* at 2.30 μm aerodynamic diameter [i.e., $LPM = \Sigma(m_1 + m_2 + m_3 + m_4)$] for optimum EDA sensitivity [1]. Group deposition, *MMAD*, and *LPM/SPM* ratio at different values of d_{50} calculated from the component-by-component drug mass deposition (studies 1–3) are summarized in Table 9.4 and the corresponding correlations presented in Fig. 9.16.

Mao et al. [11] found that the correlations (R^2-values) between *LPM/SPM* and API mass deposition in stage groups 2 and 4 were better compared with the correlation between *LPM/SPM* and API mass in group 3, regardless of the boundary size selected for the calculation of *LPM/SPM* (Table 9.5). Such an outcome was fully consistent with predictive analyses [12, 13] and reflects the introduction of confounding associated with the movement of API mass from either group 2 or 4 into this neighboring grouping. Thus, API mass transfer from either direction (group 2 to 3 or group 4 to 3) is exactly counterbalanced by the change in mass in group 3.

This case study demonstrates the high degree of sensitivity of the *LPM/SPM* ratio to small variations in MDI-generated aerosol APSDs influencing the location of the *MMAD* values. *LPM/SPM* can be readily calculated from CI stage deposition data with no more difficulty than is the case in determining API mass-per-stage groupings.

In another series of case studies, Strickland examined real stability data from two older products no longer in the market in which probable causes for observed multistage CI-measured APSD shifts had been identified [14]. The data from multiple APSD assessments of product A by ACI over a 24-month stability trial period were initially analyzed in terms of stage groupings applying to the following components of the CI system, as would be the case in the US regulatory environment:

Group 1: non-sizing components including stage 0
Group 2: coarse mass fraction captured on stages 1 and 2
Group 3: fine mass faction captured on stages 3–5
Group 4: extra-fine mass fraction captured on stages 6, 7, and the backup filter.

When the data expressed as mass % of label claim (LC) are evaluated for the groupings involved with the APSD measurement (groups 2–4), it is evident that no trend was obtained with the group 2 data, whereas both groups 3 and 4 demonstrated a decrease in API mass as a function of duration in the trial (Fig. 9.17). However, it is important to note that the group 4 stability change would not have been identified as a specification risk as the specification for this grouping is typically a one-sided upper limit. This significant stability trend was also detected when the same data were analyzed in terms of FPM, as would be the case in a European regulatory environment (Fig. 9.18).

Table 9.4 Calculated group deposition (μg), *LPM/SPM* ratio, and *MMAD* (*From* [11]—*used with permission*)

(a) Study 1

MDI	Variant 1		Variant 2		Variant 3		Variant 4		Variant 5		Variant 6		Variant 7		Variant 8	
Can	1	2	1	2	1	2	1	2	1	2	1	2	1	2	1	2
Group 2 (S1–S3)	1.22	1.25	0.84	1.11	0.64	0.71	0.69	0.72	0.73	0.67	0.78	0.75	0.67	0.75	0.79	0.76
Group 3 (S4–S6)	3.02	3.15	3.02	3.33	3.36	3.32	3.69	3.34	3.36	3.15	3.72	3.87	2.91	3.42	3.34	3.62
Group 4 (S7–F)	3.17	3.17	3.56	3.41	4.51	4.67	5.02	4.50	4.77	5.12	4.76	5.39	3.36	3.87	3.70	3.76
MMAD (μm)	1.03	1.04	0.88	1.00	0.78	0.76	0.77	0.78	0.77	0.70	0.81	0.76	0.86	0.87	0.88	0.90
LPM/SPM at 0.83 μm	1.13	1.17	0.97	1.13	0.78	0.75	0.77	0.78	0.73	0.64	0.82	0.76	0.91	0.93	0.95	1.01
LPM/SPM at 1.36 μm	0.53	0.53	0.42	0.51	0.26	0.24	0.24	0.25	0.24	0.22	0.26	0.23	0.30	0.31	0.31	0.32
LPM/SPM at 2.30 μm	0.24	0.24	0.18	0.22	0.05	0.05	0.04	0.05	0.04	0.04	0.05	0.05	0.06	0.06	0.06	0.06
LPM/SPM at 3.99 μm	0.09	0.09	0.06	0.08	0.02	0.02	0.02	0.02	0.02	0.01	0.02	0.02	0.02	0.03	0.03	0.03

(b) Study 2

MDI	Variant 1		Variant 2		Variant 3		Variant 4		Variant 5		Variant 6		Variant 7		Variant 8	
Can	1	2	1	2	1	2	1	2	1	2	1	2	1	2	1	2
Group 2 (S1–S3)	1.25	1.37	1.26	1.14	0.74	0.94	0.74	0.74	0.85	0.89	0.90	0.89	0.82	0.81	0.93	0.82
Group 3 (S4–S6)	3.74	3.41	3.43	3.86	3.40	3.51	3.39	3.21	3.27	3.56	3.88	3.64	3.19	3.20	3.81	4.07
Group 4 (S7–F)	3.51	3.06	3.10	3.63	4.05	3.67	3.96	3.83	4.29	4.28	4.12	3.79	3.84	3.90	4.13	4.20
MMAD (μm)	1.11	1.20	1.16	1.06	0.85	0.95	0.86	0.85	0.81	0.85	0.92	0.94	0.86	0.85	0.90	0.90

(continued)

Table 9.4 (continued)

(b) Study 2

MDI	Variant 1		Variant 2		Variant 3		Variant 4		Variant 5		Variant 6		Variant 7		Variant 8	
Can	1	2	1	2	1	2	1	2	1	2	1	2	1	2	1	2
LPM/SPM at 0.83 μm	1.28	1.37	1.33	1.24	0.92	1.07	0.94	0.92	0.84	0.91	1.03	1.06	0.90	0.88	0.98	1.02
LPM/SPM at 1.36 μm	0.68	0.73	0.69	0.62	0.41	0.48	0.43	0.42	0.31	0.35	0.45	0.48	0.33	0.32	0.30	0.31
LPM/SPM at 2.30 μm	0.29	0.33	0.29	0.26	0.14	0.17	0.15	0.14	0.11	0.12	0.15	0.17	0.10	0.09	0.06	0.06
LPM/SPM at 3.99 μm	0.10	0.12	0.10	0.09	0.05	0.05	0.05	0.04	0.04	0.04	0.05	0.05	0.03	0.03	0.03	0.03

(c) Study 3

MDI	Variant 1		Variant 2		Variant 3		Variant 4		Variant 5		Variant 6		Variant 7		Variant 8	
Can	1	2	1	2	1	2	1	2	1	2	1	2	1	2	1	2
Group 2 (S1–S3)	1.63	1.41	1.20	1.36	1.07	1.05	1.01	1.25	0.70	0.86	0.99	1.07	0.84	0.84	0.60	0.60
Group 3 (S4–S6)	3.80	3.29	3.50	3.71	3.53	3.59	3.40	3.24	3.00	3.37	4.64	3.55	3.36	3.13	3.52	3.21
Group 4 (S7–F)*	2.94	2.56	3.09	2.90	3.53	3.44	3.29	3.18	4.41	3.80	4.48	3.60	3.78	3.63	3.43	3.10
MMAD (μm)	1.36	1.37	1.15	1.26	1.01	1.03	1.03	1.09	0.75	0.89	0.97	1.00	0.91	0.88	0.91	0.92
LPM/SPM at 0.83 μm	1.60	1.59	1.36	1.56	1.17	1.22	1.19	1.18	0.76	1.00	1.17	1.14	0.97	0.94	1.07	1.08
LPM/SPM at 1.36 μm	0.83	0.83	0.69	0.78	0.57	0.59	0.58	0.58	0.30	0.39	0.53	0.56	0.38	0.37	0.33	0.34
LPM/SPM at 2.30 μm	0.37	0.37	0.28	0.33	0.24	0.24	0.23	0.24	0.13	0.16	0.22	0.24	0.13	0.13	0.06	0.07
LPM/SPM at 3.99 μm	0.14	0.13	0.10	0.12	0.09	0.08	0.08	0.08	0.05	0.06	0.06	0.08	0.04	0.04	0.02	0.02

**F* backup filter

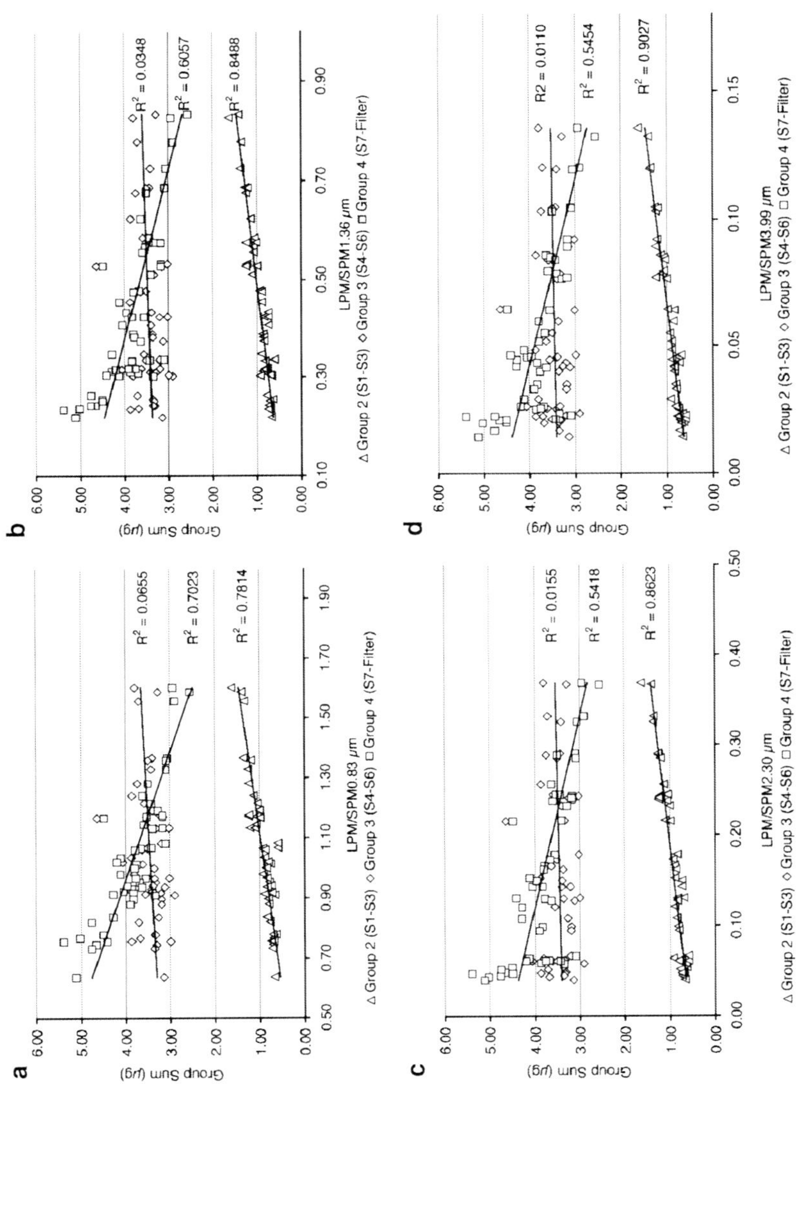

Fig. 9.16 Correlation between group deposition and *LPM/SPM* calculated based on d_{50} values of (**a**) 0.83 μm (NGI stage 6), (**b**) 1.36 μm (NGI stage 5), (**c**) 2.30 μm (NGI stage 4), and (**d**) 3.99 μm (NGI stage 3) aerodynamic diameter for the pooled data (*From* [11]—*used with permission*)

Table 9.5 Comparison between EDA and stage groupings approach to the assessment of CI-measured data from a solution-formulated MDI (*from* [9]—*used by permission*)

	Correlation (R^2 values) between mass deposition by stage grouping, *MMAD* and *LPM/SPM*			
d_{50} (μm)	*MMAD* (μm)	Group 2	Group 3	Group 4
0.83	0.98	0.7814	0.0655	0.7023
1.36	0.92	0.8488	0.0348	0.6057
2.30	0.84	0.8623	0.0155	0.5418
3.99	0.86	0.9027	0.0110	0.5454

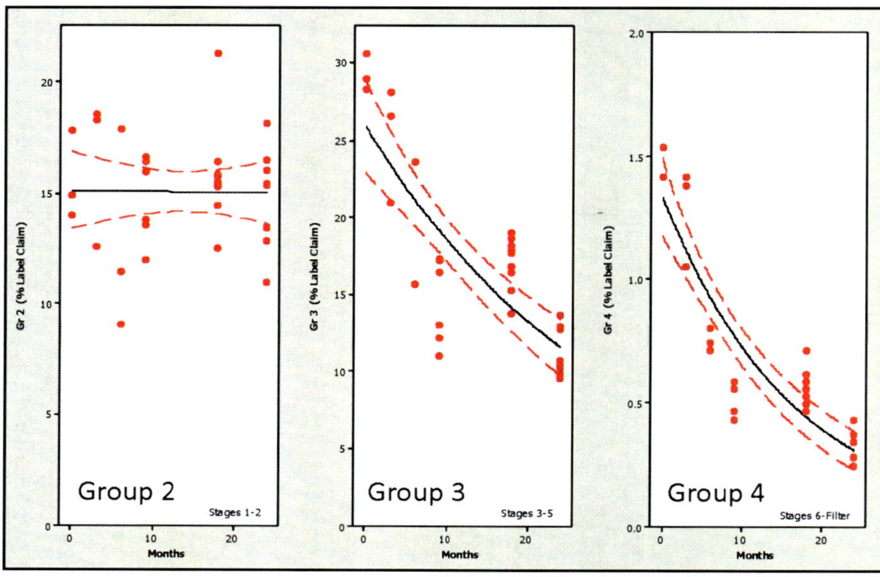

Fig. 9.17 OIP product "A"APSD changes in 24-month stability trial assessed by CI stage groupings

EDA was also able to detect the trend as a time-dependent increase in *LPM/SPM* (Fig. 9.19). However, the corresponding decrease in *ISM* pointing to a change in mass entering the CI from the OIP would not have been as obvious if viewed either from variations in individual stage groupings or *FPM*.

In the case of product B, stability trending was observed with all three groupings (Fig. 9.20); the shift to larger values for stage 2 is accompanied by decreases in both groups 3 and 4 pointing to an overall APSD movement towards larger sized particles.

This shift in APSD was also detected by a trend towards decreased values of *FPM* with time (Fig. 9.21).

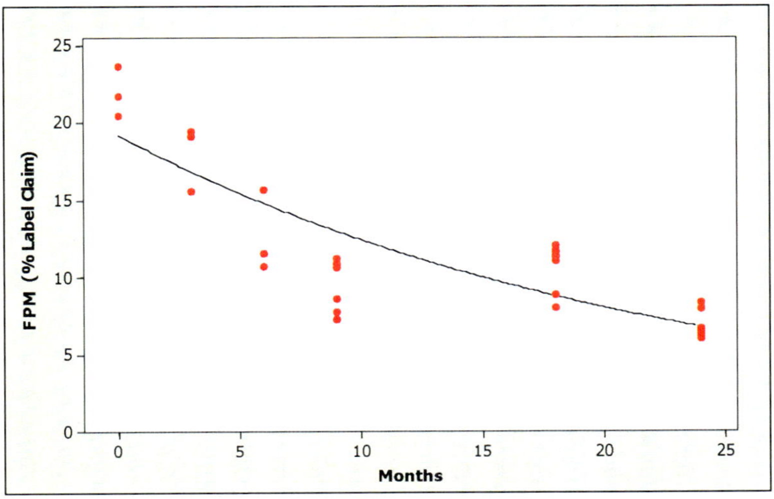

Fig. 9.18 OIP product "A" APSD changes in 24-month stability trial assessed by movement in *FPM*

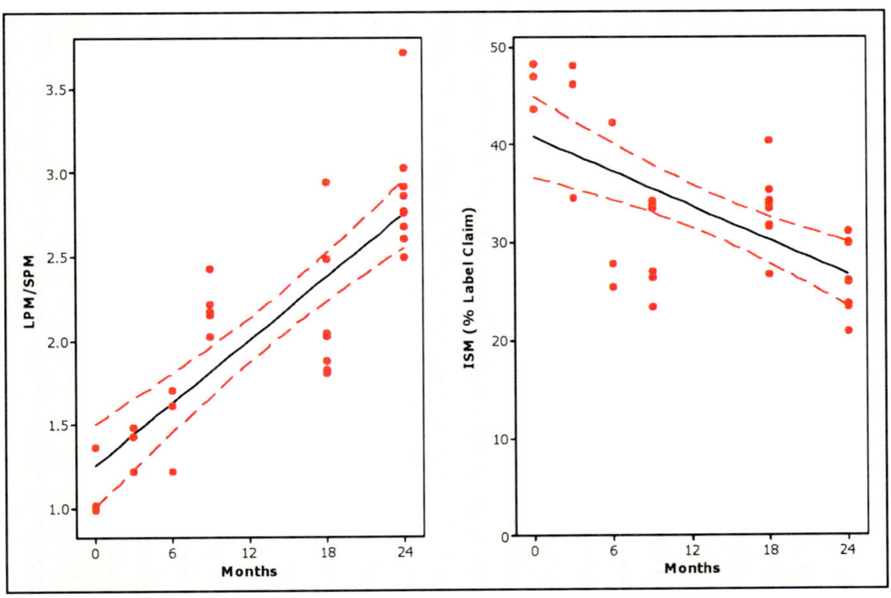

Fig. 9.19 OIP product "A" APSD changes in 24-month stability trial assessed by EDA metrics *R* and *ISM*

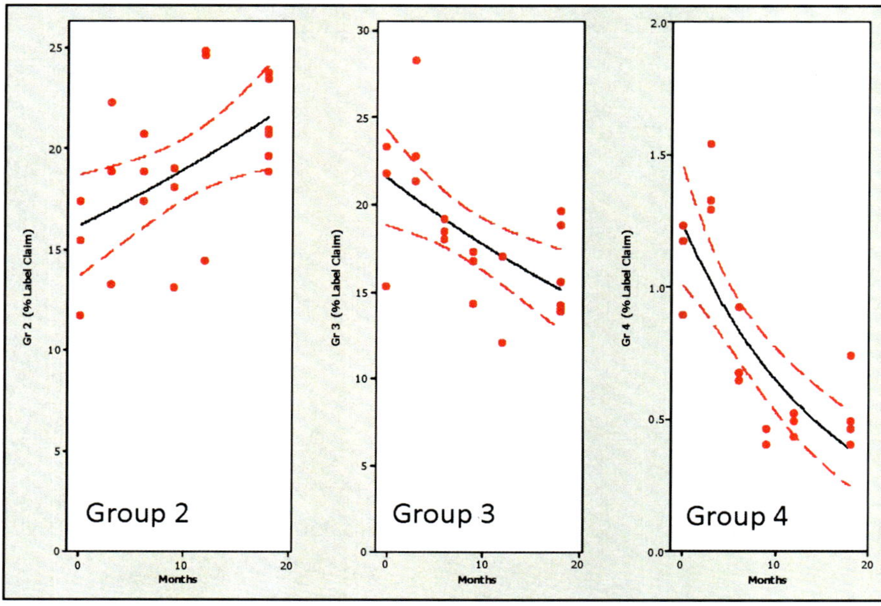

Fig. 9.20 OIP product "B"APSD changes in 24-month stability trial assessed by CI stage groupings

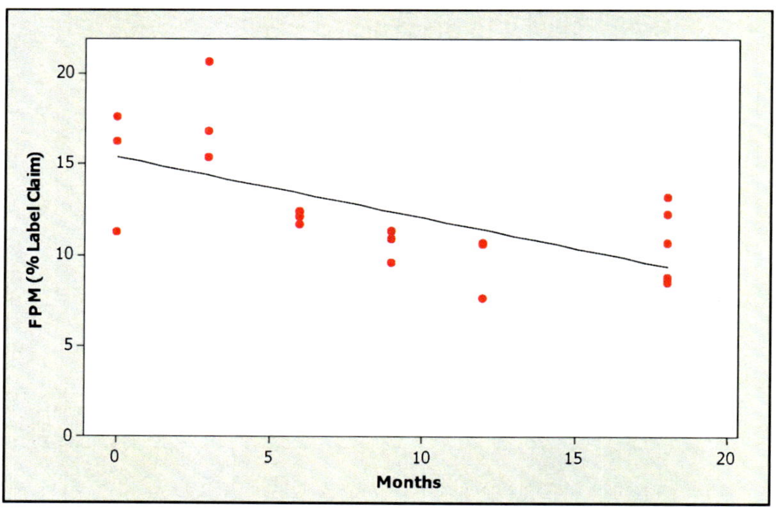

Fig. 9.21 OIP product "B"APSD changes in 24-month stability trial assessed by movement in *FPM*

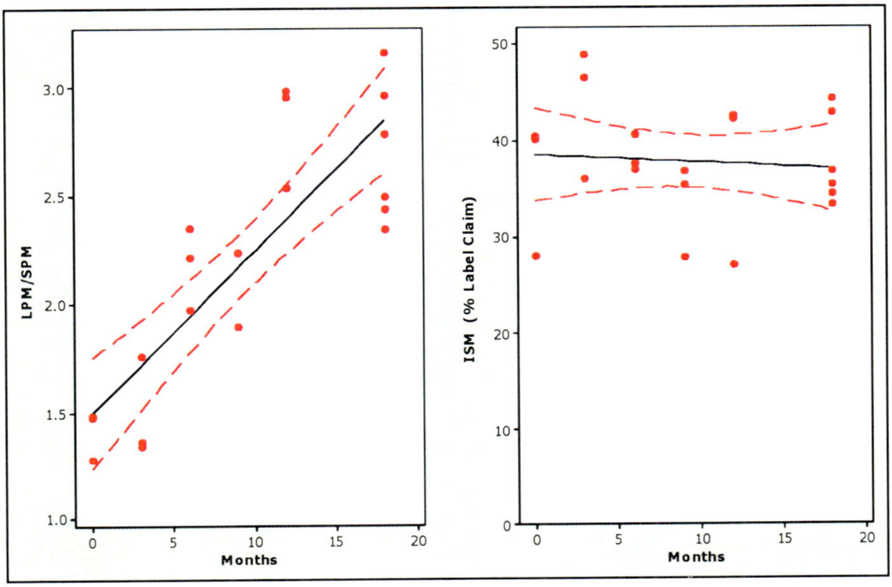

Fig. 9.22 OIP product "B"APSD changes in 24-month stability trial assessed by EDA metrics *R* and *ISM*

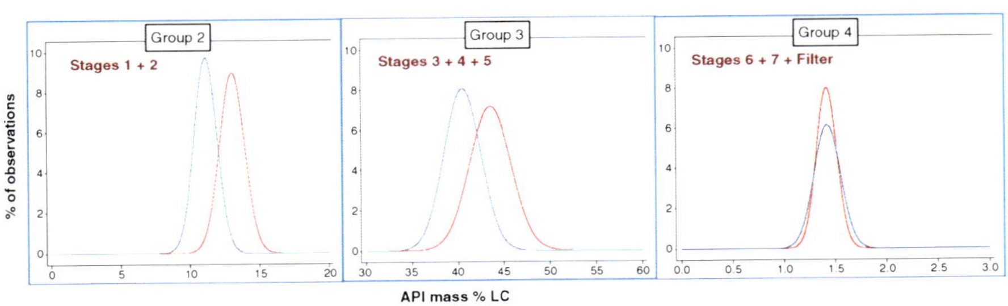

Fig. 9.23 Intentional design change to OIP product aerosol assessed by CI stage groupings

This time, EDA also detected the shift by an increase in *R* (Fig. 9.22) However, *ISM* remained largely unaffected, suggestive of a cause that was associated with the portion of the dose that entered the CI and was size fractionated.

In the third case (product C), Strickland presented APSD data associated with an intentional product design modification, in which the aerosol size properties were anticipated to change [14]. In the case of analysis by stage groupings based on normally distributed values for each metric (Fig. 9.23), the transition from design 1 to design 2 was evident with both groups 2 and 3, but not with group 4.

The same data assessed by EDA demonstrated changes in both position of central tendency and *AUC*, evident when measures of *LPM*, *SPM*, *R*, and *ISM* were considered together (Fig. 9.24).

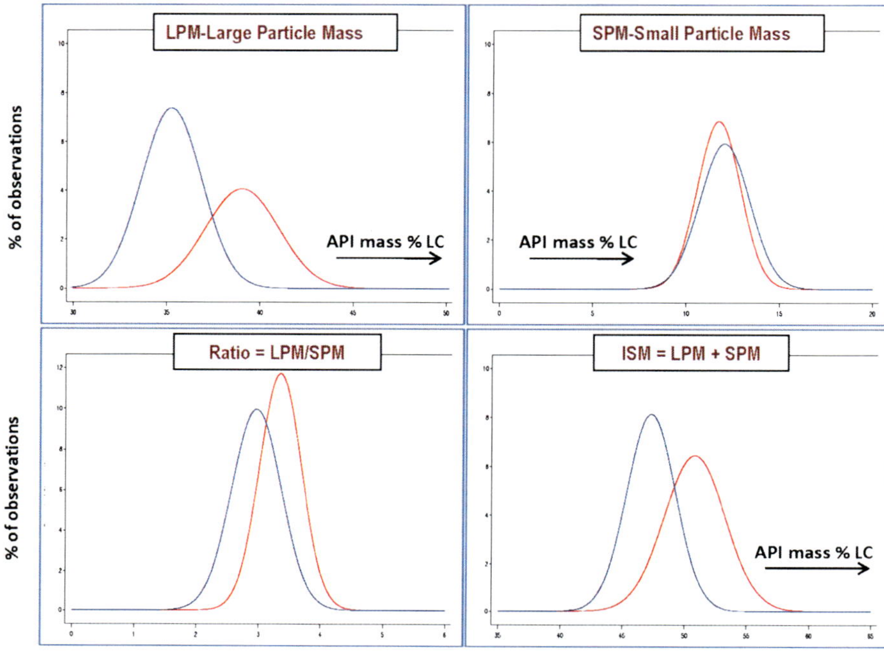

Fig. 9.24 Intentional design change to OIP aerosol assessed by EDA

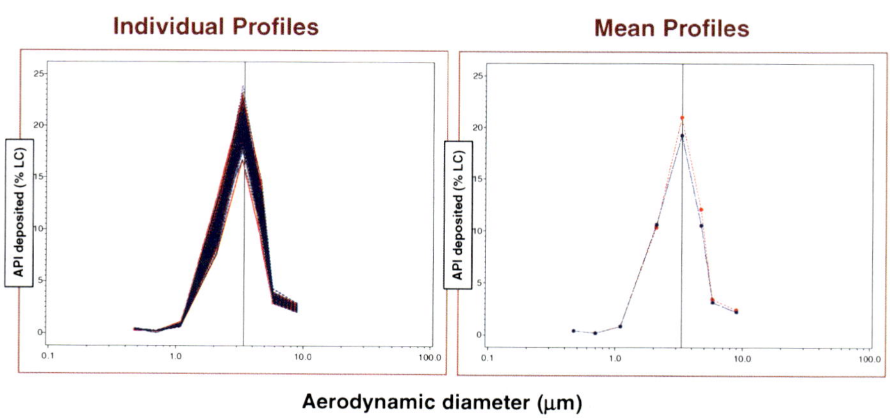

Fig. 9.25 Multistage CI-measured APSD profiles for product C; the *blue lines* represent the original design, and the *red lines* were obtained from the modified design to the OIP

These case studies [11, 14] demonstrate that far from reducing the information that is useful at diagnosing changes to OIP APSDs, EDA enriches the knowledge base available from CI data, thereby helping the analyst make correct decisions about the product. Although this information is contained in the collection of individual and averaged CI APSD profiles (Fig. 9.25), it is less accessible as a decision-making tool.

9.6 Conclusions

Strong arguments supporting the robustness of the EDA approach for the assessment of CI-determined APSD changes have been presented, both from theoretical considerations based on aerosol mechanics during the measurement process, failure mode assessments based on two of the major OIP categories, and lastly from case studies based on products that have been marketed or are in development. Although the assumption has been that the underlying measurements were made with full-resolution CI systems, it should be recalled that EDA can also be applied to AIM as well as full-resolution CI data. Such an AIM-EDA approach has the potential to combine the sensitivity of the CI data interpretation to changes in APSD described in this chapter with the simplicity and other advantages reviewed in Chap. 5 associated with abbreviating the measurement system.

References

1. Tougas TP, Christopher D, Mitchell JP, Strickland H, Wyka B, Van Oort M, Lyapustina S (2009) Improved quality control metrics for cascade impaction measurements of orally inhaled drug products (OIPs). AAPS PharmSciTechnol 10(4):1276–1285
2. Christopher D, Adams W, Amann A, Bertha C, Byron PR, Doub W, Dunbar C, Hauck W, Lyapustina S, Mitchell JP, Morgan B, Nichols S, Pan Z, Singh GPJ, Tougas T, Tsong Y, Wolff R, Wyka B (2007) Product quality research institute evaluation of cascade impactor profiles of pharmaceutical aerosols: Part 3 – Final report on a statistical procedure for determining equivalence. AAPS PharmSciTechnol 8(4):article 90. http://www.aapspharmscitech.org/view.asp?art=pt0804090. Accessed 10 Jan 2012
3. Mitchell JP, Christopher JD, Tougas T, Glaab V, Lyapustina S (2011) When could efficient data analysis (EDA) fail? Theoretical considerations. In: Dalby RN, Byron PR, Peart J, Suman JD, Young PM (eds) Respiratory drug delivery 2012. Davis Healthcare International, River Grove, IL, pp 237–245
4. Stein SW, Gabrio BJ (2000) Understanding throat deposition during cascade impactor testing. In: Dalby RN, Byron PR, Farr SJ, Peart J (eds) Respiratory drug delivery VII. Serentec, Raleigh, NC, pp 573–576
5. Sethuraman VV, Hickey AJ (2001) Evaluation of pre-separator performance for the 8-stage nonviable Andersen impactor. AAPS PharmSciTech 2(1):article 4. http://www.aapspharmscitech.org/. Accessed 11 Jan 2012
6. Mitchell JP, Nagel MW (2003) Cascade impactors for the size characterization of aerosols from medical inhalers: their uses and limitations. J Aerosol Med 16(4):341–377

7. Pedersen S, Dubus JC, Crompton G (2010) The ADMIT series—issues in inhalation therapy 5: inhaler selection in children with asthma. Prim Care Resp J 19(3):209–216
8. Roller CM, Zhang G, Troedson RG, Leach CL, Le Souëf PN, Devadason SG (2007) Spacer inhalation technique and deposition of extrafine aerosol in asthmatic children. Eur Respir J 29(2):299–306
9. Glaab V, Goodey A, Lyapustina S, Mitchell J (2011) Efficient data analysis for MDIs and DPIs: failure mode effect analysis. In: Dalby RN, Byron PR, Peart J, Suman JD, Young PM (eds) Respiratory drug delivery Europe 2011. Davis Healthcare International, River Grove, IL, pp 225–236
10. Mitchell JP, Tougas T, Christopher D, Bauer R, Church T, Dey M, Glaab V, Goodey A, Holmes S, Iley T, Lyapustina S, Patel R, Quiroz J, Russell-Graham D, Strickland H, Svensson M, Van Oort M, Wu Z (2010) Abbreviated impactor measurement (AIM) and efficient data analysis (EDA) concepts: a partnership for the assurance of oral inhaled product (OIP) quality. Inhalation 4(6):6–10
11. Mao L, Ponder D, Hughes A, White J, Glaab G, Mitchell J, Lyapustina S (2012) Efficient data analysis (EDA) case study I: pressurized metered dose inhalers (MDIs) containing a solution formulation. In: Dalby RN, Byron PR, Peart J, Suman JD, Young PM (eds) Respiratory drug delivery 2012. Davis Healthcare International, River Grove, IL, pp 441–446
12. Christopher JD, Dey M (2011) Detecting differences in APSD: efficient data analysis (EDA) vs. stage groupings. In: Dalby RN, Byron PR, Peart J, Suman JD, Young PM (eds) Respiratory drug delivery Europe 2011. Davis Healthcare International, River Grove, IL, pp 215–223
13. Tougas TP, Christopher JD, Mitchell JP, Lyapustina S (2010) Efficient data analysis and abbreviated impactor measurement concepts. In: Dalby RN, Byron PR, Peart J, Suman JD, Young PM (eds) Respiratory drug delivery 2010. Davis Healthcare International, River Grove, IL, pp 599–603
14. Strickland H (2011) EDA and stage group metrics: can these metrics detect product changes. IPAC-RS conference "Bringing value to the patient in a changing world", Rockville, MD, USA, Mar 2011, http://www.ipacrs.com/PDFs/IPAC-RS%202011%20Conference/AIM%20 Workshop/6-CI%20Workshop%202011-Strickland.pdf. Accessed 11 Jan 2012

Chapter 10
Validating AIM-Based Instrumentation and Associated Measurement Techniques

Mark Copley, Jolyon P. Mitchell, Mårten Svensson, J. David Christopher, Jorge Quiroz, Geoffrey Daniels, Melanie Hamilton, and Dave Russell-Graham

Abstract The validation of the wide variety of equipment capable of making abbreviated impactor measurements is a key component providing proof that the AIM concept works in practice. This chapter provides a comprehensive collection of validation experiments that have been provided by a variety of different laboratories, mainly through the support of the Cascade Impactor sub-team of the European Pharmaceutical Aerosol Group (EPAG), who held a Workshop on the topic in December 2010. These studies have involved the whole range of OIP formats, thereby increasing confidence in the wide applicability of the approach. A series of

M. Copley (✉)
Copley Scientific Limited, Colwick Quays Business Park,
Private Road No. 2, Colwick, Nottingham NG4 2JY, UK
e-mail: M.Copley@copleyscientific.co.uk

J.P. Mitchell
Trudell Medical International, 725 Third Street, London, ON N5V 5G4, Canada
e-mail: jmitchell@trudellmed.com

M. Svensson
Emmace Consulting AB, Vinkelhaken 1D, Sodra Sandy 247 32, Sweden
e-mail: marten@emmace.se

J.D. Christopher
Nonclinical and Pharmaceutical Sciences Statistics, Merck Research Laboratories,
WP37C-305, 770 Sumneytown Pike, West Point, PA 19486-0004, USA
e-mail: j.david.christopher@merck.com

J. Quiroz
Novartis Pharmaceuticals Corporation, One Health Plaza,
Bldg. 438/2417B, East Hanover, NJ 07936, USA
e-mail: jorge.quiroz@novartis.com

G. Daniels
Inhaled Product & Device Technology, GlaxoSmithKline,
Park Road, Ware, Hertfordshire SG12 0DP, UK
e-mail: geoffrey.e.daniels@gsk.com

T.P. Tougas et al. (eds.), *Good Cascade Impactor Practices, AIM and EDA for Orally Inhaled Products*, DOI 10.1007/978-1-4614-6296-5_10,

"learnings" are summarized at the end of the chapter as guidance for those planning on implementing an AIM-based method.

10.1 Introduction

The introduction of AIM-based apparatuses into the mainstream of OIP inhaler performance testing is a highly desirable goal, as has been demonstrated in previous chapters. Not only can the methodology for acquiring aerodynamic size-related metrics be simplified, with the attendant prospect of reducing measurement variability, but the application of EDA principles may afford the prospect of better discrimination in terms of product quality than is possible with current methods that are based on grouped stages from full-resolution CI measurements or from a single performance measure by itself, such as fine particle mass. Validation of the wide variety of AIM-based apparatuses with all of the major OIP formats is a critical component of this process.

Experimental studies were recognized from the outset as being of crucial value to the development of the AIM concept, alongside detailed theoretical rationalization. The following is a brief synopsis of key events. Following proof-of-concept studies by Trudell Medical International with commercially available suspension and solution formulated MDIs undertaken in 2007–2008, a comparative precision experiment between abbreviated and full-resolution ACI systems was undertaken at the same location by the CI Working Group of IPAC-RS the following year. Subsequently, EPAG, through their Impactor sub-team, was responsible for initiating many follow-on investigations with a variety of inhaler types and abbreviated impactor configurations. Since 2010, experimental work has also been undertaken by several organizations outside these industry groups and is included in this chapter in order to demonstrate that the AIM-EDA concept is of interest and is gaining wider acceptance in the OIP manufacturing community. Included are studies carried out using the following abbreviated impactor systems: the C-FSA (Copley Scientific Ltd., Nottingham, UK); the FPD (Westech Instruments Services, Upper Stondon, Beds., UK); other generic abbreviated Andersen CI stacks, in particular, the Trudell (Medical) fast screening Andersen impactor (T-FSA); the FSI (MSP Corporation, St. Paul, MN, USA); and differently modified versions of the NGI. In combination, these initial results provide information to support use of the AIM concept with each type of OIP (DPI, pMDI, and nebulizers).

M. Hamilton
Inhaled Delivery Sciences, Inhaled Product & Device Technology,
GSK R&D Park Road, Ware, Herts SG12 0DP, UK
e-mail: Melanie.2.hamilton@gsk.com

D. Russell-Graham
Mylan Global Respiratory Group, Mylan Inc., Discovery Park, Sandwich CT13 9NJ, UK
e-mail: david.russell-graham@mylan.co.uk

Presentations covering much of the large amount of work undertaken by many groups in the past 5 years can be found on the websites of EPAG [1] and IPAC-RS [2, 3].

10.2 AIM-Based Apparatuses: Developments Before the Present Campaigns

The "Copley-Fisons" two-stage metal impactor [4] (Fig. 10.1) and the "Glaxo" Twin Impinger [5] (Fig. 10.2) can be considered as two of the earliest physical manifestations of the AIM concept. Both were listed as apparatuses A and B, respectively, in the European Pharmacopoeia up until the 4th edition published in 2002, after which apparatus B was withdrawn on the basis that it was no longer used.

The Twin Impinger (TI) is a glassware design based on the multistage liquid impinger (MSLI) [6]. The upper stage represents the throat, stage 1 and stage 2 of an MSLI, while the final stage corresponds to stages 3 and 4 and the filter. With both stages, particles are collected by impaction on liquid surfaces, an especially suitable method for both pMDI and DPI analysis. The reported d_{50} for the upper stage of the instrument is 6.3–6.4 μm at 60 L/min [7, 8]. This limit is likely to be viewed as too large by today's standards. However, it would not be too difficult to reduce the d_{50} value closer to 5μm by decreasing the tube diameter leading into the impingement vessel.

Both instruments separate the aerosol emitted by the inhaler into just two size fractions: an upper stage captures particles above a certain stage cut-off diameter

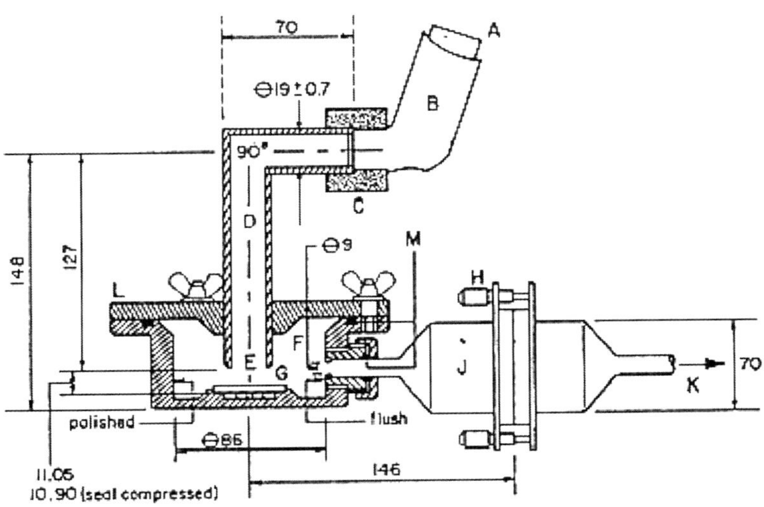

Fig. 10.1 "Copley-Fisons" two-stage metal impactor (*Courtesy of Copley Scientific Ltd., Nottingham, UK*)

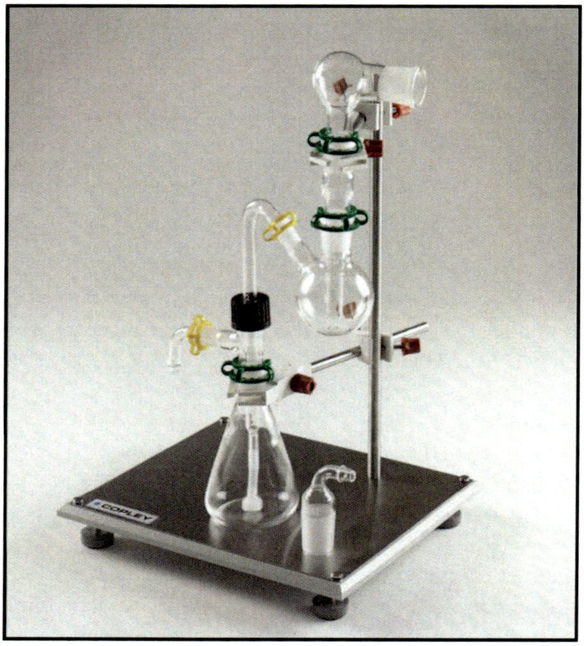

Fig. 10.2 "Glaxo" twin impinger (TI) (*Courtesy of Copley Scientific Ltd., Nottingham, UK*)

and all remaining particles pass through to a final collection media. The upper stage of the Copley-Fisons impactor has a cut-off diameter of 9.8 μm d_{ae} at 60 L/min, so the boundary between fine-coarse particle size separation would again likely be viewed as higher than acceptable for present-day OIP applications.

The introduction of the TI was supported at the time by details of its experimental use, in that the value of the device was perceived as its ability for distinguishing between "good" and "poor" aerosols, in particular its application in the detection of agglomerating pMDI formulations, during product development [5]. The lack of sensitivity relative to fuller APSD measurement by the 4-stage multistage liquid impinger was, however, identified in the mid-1990s as a potential drawback for more discriminating investigation, especially for the comparison of generic with innovator OIPs [9]. Further studies confirmed initial claims about the insensitivity of Twin Impinger measurements to variability in operational parameters such as collecting fluid composition and volume and test flow rate [7]. The use of the TI at flow rates in excess of the design value of 60 L/min for low resistance DPI testing may, however, be a practical proposition [10]. Calibration with solid monodisperse aerosols has confirmed, that like other types of inertial impactor, its cut-off size for the upper stage decreases at sampling flow rates in excess of the routinely used 60 L/min [8], as predicted by the relationship:

$$d_{ae,50,1} = d_{ae,50,ref}\left[\frac{Q_{ref}}{Q_1}\right]^{1/2} \tag{10.1}$$

where Q_1 and Q_{ref} represent the calibration flow rate and reference flow rate of 60 L/min, respectively, and $d_{ae,50,1}$ and $d_{ae,50,ref}$ are the cut-off sizes at the calibration and reference flow rates, respectively.

In retrospect, it is clear that the study carried out by Miller et al. [7] fuelled uncertainty as to the ability of the TI to differentiate between OIP aerosols with markedly different APSDs, making the simplification that they are unimodal and lognormal, and therefore using *MMAD* and *GSD* to represent measures of central tendency and spread, respectively. By concluding that because the instrument only separated into two size fractions, and, moreover gave a broad rather than sharp separation, the impression was given that it could not distinguish between aerosols that fell within the same *MMAD/GSD* "family". It was possible that a discrete range of *MMAD/GSD* combinations would give the same results when analyzed using just two size fractions, and broad separation between fine/coarse size boundary and *MMAD* appeared to exacerbate the problem. Referring to the recent work by Tougas et al. [11], it is now clear that the sensitivity of this apparatus for APSD-related shifts will vary considerably according to the location of the *MMAD* of the product. The *MMAD* of many currently marketed pMDIs and DPIs is likely to be located in the region from 1 to 3 μm, and this separation is likely to be too far from the stage cut-off size for the unmodified TI, therefore potentially impairing its precision.

A further drawback of both the TI (and also the Copley-Fisons 2-stage metal impactor) is that, as they are currently supplied, neither instrument has an easily varied stage cut-off size. However, despite these disadvantages, in the near future as AIM research progresses, there may be cause to reexamine the potential role of this apparatus, possibly with a modified stage cut-off diameter, since the impinger design is attractive from the point of view of its ability to eliminate size-related bias caused by particle bounce and re-entrainment [12]. It would be a relatively easy modification to move the cut point for the TI to 5.0 μm aerodynamic diameter in the flow rate range from 30 to 100 L/min within which most OIPs are evaluated, by modifying the diameter of the tube entering the upper stage, in accordance with the relationship (Chap. 2):

$$\sqrt{C_{c,50}}\,d_{50} = \left[\frac{9\pi\eta W^3}{4\rho_0 Q}\right]^{1/2}\sqrt{St_{50}} \qquad (10.2)$$

in which St_{50} is the dimensionless Stokes number at the size where the stage collection efficiency is 50%, W = tube diameter, d_{50} = stage cut size, Q = volumetric flow rate, η = air viscosity, ρ_0 = unit density (i.e., 1 kg/m^3), and $C_{c,50}$ is the Cunningham slip correction factor for a particle of size d_{50} [13]. However, the practicality of making this change so that this apparatus could be used at close to 30 L/min to evaluate pMDIs has not, to the authors' knowledge, yet been addressed.

The issue of flow rate sensitivity to stage d_{50} size with CIs, which is discussed in more detail in Chap. 2, highlights an important potential limitation to the AIM concept in the case of DPI testing. In contrast with MDI or nebulizer performance assessments by the compendial methods in which the flow rate is maintained at a fixed value, development of DPI a product often involves testing at multiple flow rates. It is a relatively easy process with a full-resolution CI to interpolate the mass of API in particles finer than a fixed size limit, typically 5.0 μm aerodynamic diameter, from the cumulative mass-weighted APSD obtained at each required flow rate, even though the individual stage d_{50} values change. In contrast, since interpolation is not possible when using an AIM-based CI, as an APSD is not generated, a different upper stage would be needed for testing at each required flow rate in order to maintain a stage d_{50} fixed at the appropriate value.

At this point then it is fair to say that while the AIM concept, in the physical form of the Twin Impinger, was seen as a convenient and efficient analytical tool for relatively coarse differentiation, doubts remained about its sensitivity. The theoretical work by Tougas et al. [11] on the development of EDA metrics that is described in detail in Chaps. 7 and 8, followed some time after these initial practical studies. In summary, EDA points the way to achieve better measurement precision in association with OIP APSD-related data by the following approaches: adoption of a ratio of *LPM* to *SPM* rather than individual mass fractions, simultaneous use of ISM, and the selection of an optimal boundary value for *LPM/SPM* on the basis of *MMAD* value.

Limited interest in AIM precursor concepts continued through the 1990s, a decade marked at its closing by the development of the full-resolution NGI on the basis of the very latest understanding of inertial impaction [14]. In the context of AIM-based equipment, during the mid-1990s, Van Oort and Downey [15] and Van Oort and Roberts [16] returned to the issue of reducing the analytical burden by cutting the number of size fractions, this time by simply reducing the number of stages used in an Andersen CI stack (see Chap. 5). In these works, for the first time, there was recognition of the importance of tailoring the boundary between the two size fractions used for EDA to suit the product under test. Based on full-resolution data gathered using an ACI (or NGI), analysis was focused on the stages where most of the drug collects to give size fractions that could more precisely and successfully capture changes in OIP APSD.

Unfortunately, at that time and up until the early part of the next decade, the suggestion of an abbreviated way of working failed to gain traction with the regulators [17], who favored full-resolution APSD measurements, diminishing interest in continued development of simplified systems. However, since then much has changed. In particular, the regulatory landscape has altered dramatically with the introduction of new concepts, perhaps most importantly Quality by Design, an approach designed to promote product and process development on the basis of thorough and secure knowledge [18]. Nevertheless, there are legitimate concerns that this new way of working will significantly increase the analytical burden. Hence, both the pharmaceutical industry and regulatory agencies alike have become more receptive to new approaches, based on sound science, which may help reduce the amount of testing required. Interest in the use of AIM systems based on both the ACI and NGI has therefore been renewed.

10.3 Proof-of-Concept Experiments Undertaken at Trudell Medical International: Assessing the Performance of Systems Based on the Nonviable 8-Stage ACI

In practical terms, abbreviating measurement with the vertical stack design of the ACI is relatively straightforward, as the early work by Van Oort and colleagues indicated, being simply a matter of configuring the stack with fewer stages and adjusting the length of the retaining springs to compensate for the shorter configuration. However, the issue of aerodynamic performance is a critical one; reduced stacks potentially may exhibit changed air flow patterns that can significantly affect inertial impaction behavior. Particle bounce, the re-entrainment, the distribution of active losses to internal surfaces, and the effect of impactor dead volume have all been shown to be important considerations [19, 20]. Furthermore, identifying optimal stage cut-off diameter values for product QC and also for the potential support of human respiratory tract (pHRT)-pertinent studies to develop in vitro–in vivo relationships (the latter being the focus of Chap. 12) are also topics for practical AIM implementation with this system as well as other designs of full-resolution CI [21].

In 2008, two proof-of-concept studies were undertaken by the group at Trudell Medical International (TMI) in order to validate the performance of two abbreviated systems for the purpose of improving productivity in testing add-on devices (spacers and valved holding chambers) used with pMDIs. Their work was based on the full-resolution 8-stage nonviable ACI, the C-FSA and the T-FSA abbreviated systems also based on the nonviable ACI operating principle [19, 20].

The C-FSA is a commercially available (Copley Scientific, UK) two-stage pHRT-based abbreviated stack, based on the Andersen *nonviable* CI that divides the incoming dose into coarse, fine, and extra-fine fractions (*CPF*, *FPF*, and *EPF*), respectively (Fig. 10.3a, b). In its commercially available formats, a range of stages enables configuration to give stage cut-off diameters (d_{50} values) of 4.7 and 1.1 μm at 28.3, 60, or 90 L/min, or alternatively 5.0 and 1.0 μm at 28.3 L/min, depending on the specific application.

Similar in design to the C-FSA, the T-FSA was also developed from research into AIM-based methods at TMI with MDIs delivering "dry particles" of salbutamol (albuterol) following propellant evaporation (Fig. 10.4a). The T-FSA was a hybrid C-FSA, which had an upper stage cut-off size of 4.7 μm so that data from this stage could therefore be directly compared with mass deposition of API on stage 2 of the full-resolution ACI. The d_{50} size of the lower stage was 1.0 μm, rather than 1.1 μm.

In a later modification, the T-FSA also included a non-operable (collection surface removed) ACI stage 0 to provide functional dead space before the first size separating stage, enabling closer mimicry of this potentially important aspect of the full-resolution ACI.

Two discrete investigations were carried out, each focusing on pMDI-produced aerosols, one involving dry particles (after HFA-134a propellant evaporation), the other containing low-volatile liquid ethanol excipient that was associated with the aerosol particles entering the measurement equipment. In the first study, dry fluticasone propionate (FP) particles were produced using a commercially available

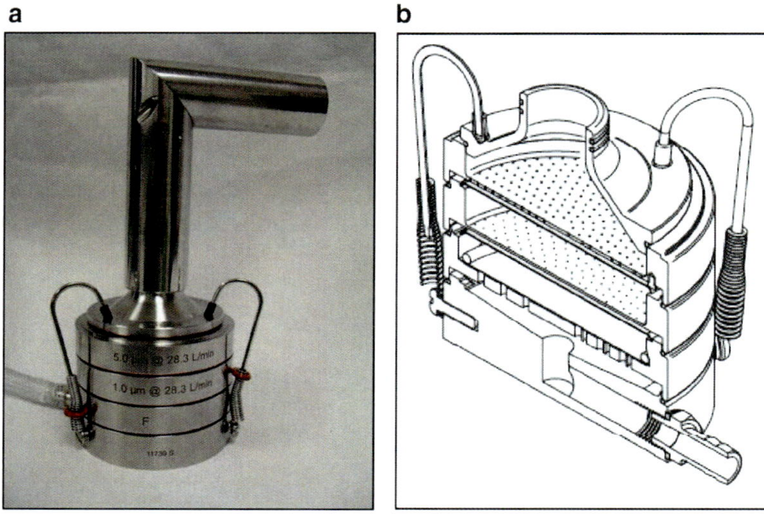

Fig. 10.3 The standard Copley C-FSA with cut-point sizes of 1.0- and 5.0-μm aerodynamic diameter—other cut-point sizes are also available. (**a**) External view showing CI with Ph. Eur./USP induction port. (**b**) Internal cross-section of CI without induction port (*Courtesy of Copley Scientific Ltd., Nottingham, UK*)

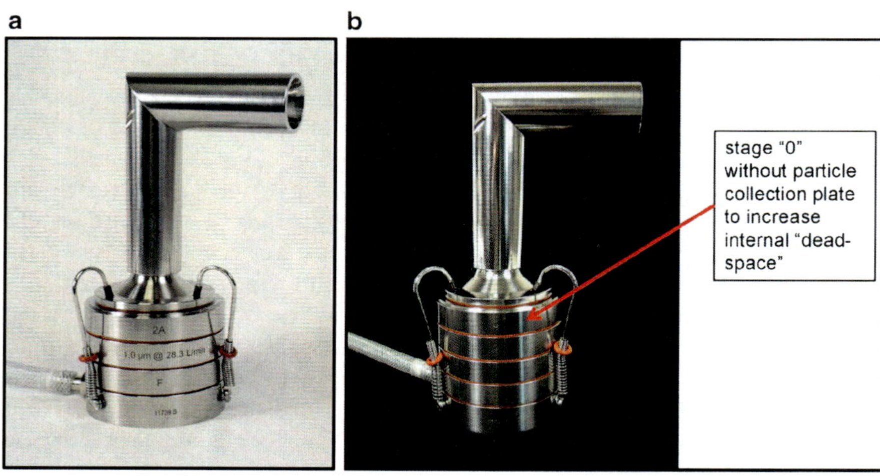

Fig. 10.4 The T-FSA system with Ph. Eur./USP inlet. (**a**) Basic T-FSA for initial experiments with "dry" aerosol particles containing salbutamol (albuterol). (**b**) Modified T-FSA containing an additional stage "0" without collection plate to increase dead space before first impaction stage to be more comparable with that in the full-resolution ACI when sampling "wet" beclomethasone dipropionate aerosol particles from formulations containing ethanol as cosolvent (*Courtesy of Trudell Medical International, London, Canada*)

Table 10.1 Cumulative mass-weighted data for Flovent®-110 measured by modified C-FSA[a] ($n=5$ replicates) with coating on collection plates (*From* [19]—*used with permission*)

Location in C-FSA	Size range (μm)	Upper size limit (μm)	Size fraction	Number of actuations per determination: 1	2	5	10
				Cumulative mass % < stated upper size limit (mean ± SD)			
Induction port	Undefined	Undefined	$CPF_{>4.7\mu m}$	60.8 ± 4.2	60.8 ± 3.3	60.1 ± 1.2	61.4 ± 2.5
Upper stage 2A	>4.7[a]						
Lower stage	1.0–4.7	4.7	$FPF_{<4.7\mu m}$	39.2 ± 4.2	39.2 ± 3.3	39.9 ± 1.2	37.9 ± 3.0
Back-up filter	<1.0	1.0	$EPF_{<1.0\mu m}$	3.1 ± 0.6	3.5 ± 0.3	3.5 ± 0.4	3.3 ± 0.2

[a] Upper stage cut size was 4.7 μm rather than 5.0 μm aerodynamic diameter

pMDI, Flovent®-HFA (GSK plc, UK); 110 μg/actuation ex-actuator mouthpiece. APSD measurements were made using the T-FSA and C-FSA with uncoated collection surfaces, and then with collection surfaces coated with polyoxyethylene lauryl ether (Brij-35) surfactant. These data were compared with analogous results generated using a full-resolution ACI (Table 10.1).

In this and other tables in this Chapter, unless otherwise stated, $CPF_{>4.7\mu m}$, $FPF_{<4.7\mu m}$, and $EPF_{<1.0\mu m}$ represent coarse, fine, and extra-fine mass fractions with the subscripts indicating the pertinent size limit. The cut-point sizes are based on the manufacturer's nominal values with Q at 28.3 L/min, and the numbering in the left-most column is based on the stage numbering sequence of the full-resolution ACI, with the "A" indicating that stage 2 was not the standard C-FSA stage with 5.0 μm cut-point size.

The impact of the number of actuations used during testing was directly investigated. In all the experiments, the mass recovery with each system was found to be broadly equivalent and well within the specification set down by the FDA [22] (±15% label claim/actuation). Furthermore, with the abbreviated systems, the mass of fine particles recovered per actuation was acceptable even with a single actuation. This is an important result since the FDA recommends minimizing the number of actuations to the clinical dose (typically 2-actuations), within the constraint of reaching a detectable limit on each stage, to improve impactor performance.

With uncoated collection surfaces (Table 10.2), the amount of material in the extra-fine fraction decreased with increasing number of actuations; from 9.4 ± 0.7 μg with a single actuation to 5.3 ± 0.4 μg with ten actuations (modified C-FSA data). This is consistent with previously reported observations suggesting that the deposition of material on an uncoated collection surface makes it progressively "stickier," potentially reducing the extent of particle bounce [23–25]. The use of surfactant-coated collection plates removed this dependence on actuation number and improved accuracy for both the T-FSA and C-FSA relative to the benchmark results generated with the full-resolution ACI (Table 10.3 and Fig. 10.5).

These measurements with either of the reduced impactors with collection surface coating were found to be substantially equivalent to the full-resolution ACI (Table 10.3). This outcome occurred despite the fact that relative API mass deposition per stage in the AIM-based systems was higher than in the full-resolution

Table 10.2 Cumulative mass-weighted data for Flovent®-110 measured by modified C-FSA[a] ($n=5$ replicates) without coating on collection plates (*From* [19]—*used with permission*)

				Number of actuations per determination			
				1	2	5	10
Location in C-FSA	Size range (μm)	Upper size limit (μm)	Size fraction	Cumulative mass % < stated upper size limit (mean ± SD)			
Induction port	Undefined	Undefined	$CPF_{>4.7\mu m}$	59.3 ± 2.3	61.7 ± 2.6	60.5 ± 1.7	61.1 ± 3.4
Upper stage 2A	>4.7						
Lower stage	1.0–4.7	4.7	$FPF_{<4.7\mu m}$	40.7 ± 2.3	38.3 ± 2.6	39.5 ± 1.7	37.8 ± 4.0
Back-up filter	<1.0	1.0	$EPF_{<1.0\mu m}$	9.4 ± 0.7	7.5 ± 0.6	6.4 ± 0.4	5.3 ± 0.4

[a] First stage cut size was 4.7 $_{\mu m}$ rather than 5.0 μm aerodynamic diameter

Table 10.3 Key size fraction metrics determined for 5-actuations of Flovent®-110 into the T-FSA ($n=5$ replicates/CI system): comparison with equivalent data from a modified[a] C-FSA and ACI (*From* [19]—*used with permission*)

Location	Size range (μm)	Upper size limit (μm)	Size fraction	Cumulative mass % < stated upper size limit (mean ± SD) T-FSA	C-FSA	ACI
Induction port	Undefined	Undefined	$CPF_{>4.7\mu m}$	57.6 ± 3.5	60.1 ± 1.2	57.7 ± 2.2
Upper stage[a]: 2A—C-FSA; 2—T-FSA	>4.7					
Lower stage	1.0–4.7: C-FSA; 1.1–4.7: T-FSA, ACI	4.7	$FPF_{<4.7\mu m}$	42.4 ± 3.5	39.9 ± 1.2	42.3 ± 2.2
Back-up filter	<1.0: C-FSA; <1.1: T-FSA, ACI	1.0: C-FSA; 1.1: T-FSA, ACI	$EPF_{<1.0\mu m}$	3.8 ± 0.5	3.5 ± 0.4	1.2 ± 0.2

[a] First stage cut size of modified C-FSA was 4.7 μm rather than 5.0 μm aerodynamic diameter

CI for the same number of actuations of the inhaler, resulting in the potential for earlier overloading of stages. Interestingly, the small but measurable wall losses associated with those stages in the full ACI that were removed to create the abbreviated designs were believed to have been transferred to the lower stage in the abbreviated systems. Fortunately this resulted in an increase in extra-fine particle mass of only ca. 2%.

It is perhaps to be expected that collection surface coating will be especially critical in abbreviated systems since any tendency toward non-ideal behavior is magnified as a consequence of the increased inertia of particles that would otherwise be collected by previous stages in the full-resolution configuration.

In the follow-on investigation [20], measurements were made with a formulation containing 8% w/v ethanol as cosolvent (Qvar™; 80 μg/actuation beclomethasone dipropionate (BDP) ex-actuator mouthpiece), using surfactant-coated collection surfaces with both the C-FSA and T-FSA. Tests with liquid ethanol-sensitive paper confirmed that the ethanol evaporated inside the impactor during measurement, penetrating only to the first stage (Fig. 10.6a, b).

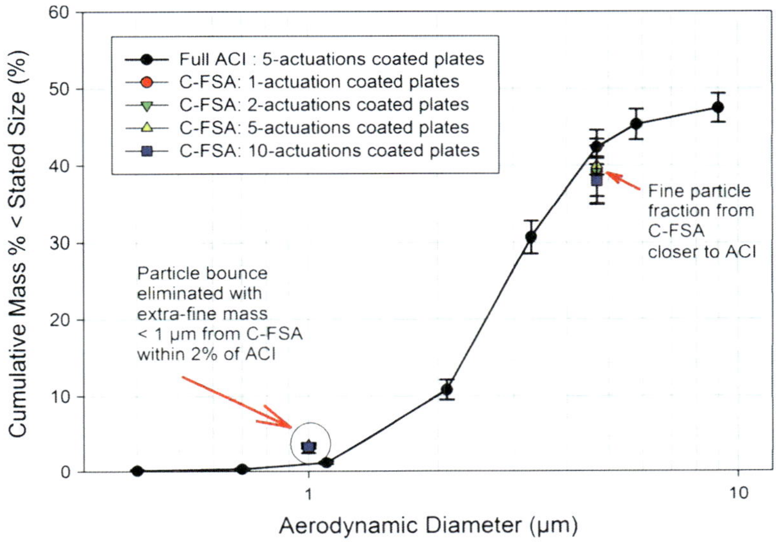

Fig. 10.5 Effect of number of actuations of Flovent-110® on C-FSA-measured data when used with surfactant-coated collection plates (*From* [19]—*used with permission*)

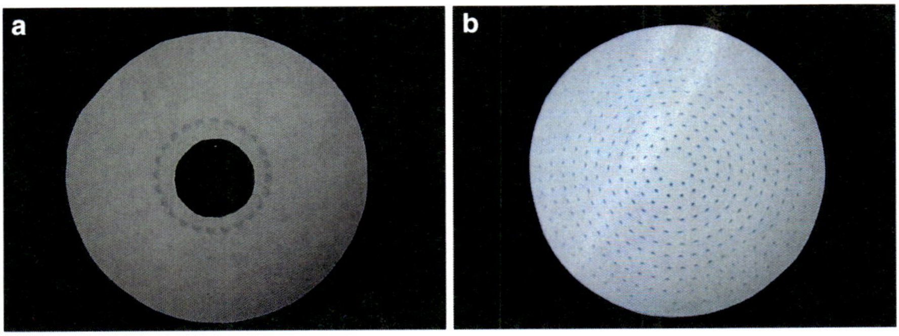

Fig. 10.6 Use of liquid ethanol-sensitive paper to diagnose presence of liquid droplets within the upper stages of the C-FSA. (**a**) Traces confirming liquid ethanol presence with incoming aerosol to C-FSA from sampling 10-actuations of Qvar™-80, using liquid ethanol-sensitive filter paper located on collection plate below stage. (**b**) Traces confirming liquid ethanol presence with incoming aerosol to C-FSA from sampling 10-actuations of Qvar™-80, using liquid ethanol-sensitive filter paper located on collection plate below stage 1 (*From* [20]—*used with permission*)

The introduction of additional dead space in the T-FSA, compared with the C-FSA, was found to improve agreement with the ACI in terms of fine particle mass, a result attributed to the provision of more similar conditions for ethanol evaporation within the former configuration (Fig. 10.7). Overall, the difference between the data using the slightly modified C-FSA and T-FSA was so small that

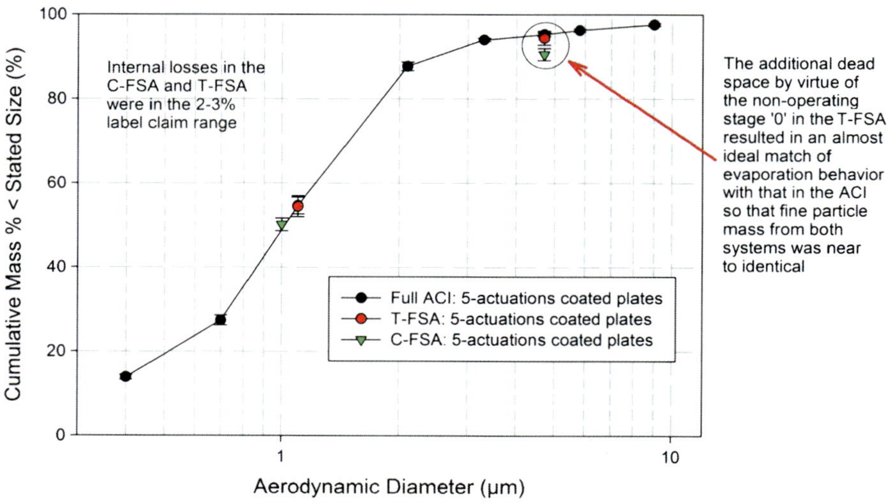

Fig. 10.7 Comparative 5-actuations per measurement data for the T-FSA, C-FSA, and ACI with Qvar™-80 (*From* [20]—*used with permission*)

for almost all practical purposes, either of these abbreviated systems could be used for this particular formulation. However, care would need to be taken to reevaluate the situation for formulations containing higher levels of low-volatile cosolvent.

10.4 The IPAC-RS Impactor Precision Comparison: Comparing the Performance of an AIM ACI-Based System Configured for pHRT Studies with a Similar System Tailored to QC Applications

One of the central issues for AIM implementation is setting the boundary value(s) for size-related metrics appropriately given the limited number of size fractions produced by abbreviated systems. To a large extent, in the OIP QC environment, this decision will likely be taken on a product-by-product basis, depending upon the product APSD obtained in early development (see Chap. 6).

Outside of the product QC environment, an alternative strategy is to set boundaries to reflect areas of potential clinical interest—the sub-5 µm fraction being an obvious target for *FPF*, being the size limit defined for the fine particle dose in the European Pharmacopoeia [4]. The stage cut-off diameter of stage 2 of the full-resolution Andersen CI instrument is slightly finer at 4.7 µm aerodynamic diameter at 28.3 L/min, and as a result, this size is often used as the limiting value for convenience during the assessment of pMDIs. In addition to the differentiation between

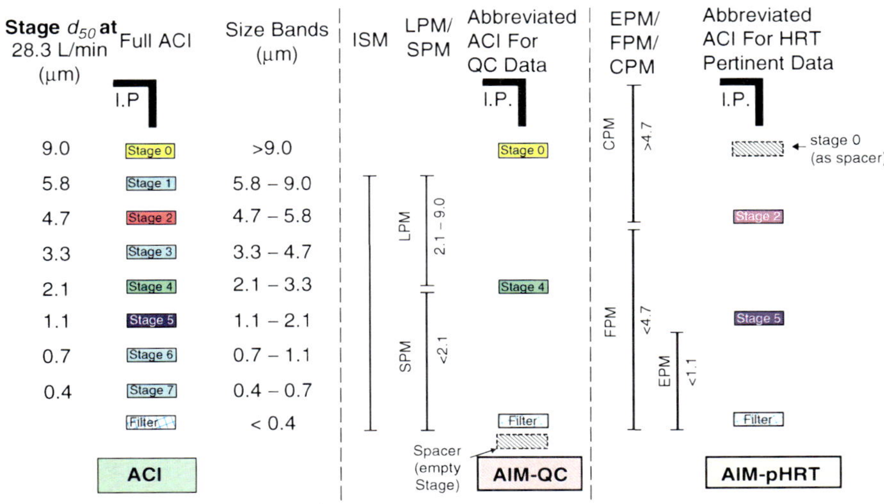

Fig. 10.8 CI configurations used in the IPAC-RS AIM precision experiment; the additional "spacer" stage used after filter stage for the AIM-QC system has no measurement function and does not interfere with analysis (*From* [26]—*used with permission*)

fine and coarse mass fractions, there may be a need to evaluate the portion of the emitted dose from some OIPs, particularly pMDI solution formulations such as Qvar™ and Alvesco® in terms of extra-fine fraction < ca. 1 µm, comprising a sub-fraction of the fine particle mass. In such instances, a second impaction stage provides added flexibility to the abbreviated impactor, with cut-off size exactly at this limit or in the case of systems derived by removing stages from a full-resolution nonviable ACI, retaining stage 5, whose cut-off size is 1.1 µm at 28.3 L/min. This was the rationale underlying the development of the so-called potential-HRT configuration, referred to from now onward as the "AIM-pHRT"-abbreviated impactor that was evaluated in the IPAC-RS precision comparison study at Trudell Medical International in 2009 [26]. This apparatus is included here for the sake of completeness, as it formed one arm of the precision comparison study. However, the underlying reasons for the development of AIM-pHRT configurations are explored in more detail in Chap. 12. Like the T-FSA described earlier, the "pHRT-FSA" configuration had a non-operable stage 0 inserted prior to the first separation stage to give dead space equivalence to the full-resolution impactor. The "pHRT-FSA" system was evaluated together with a so-called AIM-QC configuration having a single stage with cut-off size at 2.1 µm (stage 4 from the full-resolution ACI operated at 28.3 L/min) chosen to be close to the MMAD of the product used in the evaluation (AIM-QC apparatus).

The two abbreviated configurations compared with that of the benchmark full-resolution system are illustrated schematically in Fig. 10.8. C1, C2, and C3 are the configurations for the full-resolution nonviable ACI, AIM-QC, and AIM-pHRT

Table 10.4 Experiment design with three cascade impactor configurations, six inhalers, three dosing sets by inhalers, and three replicates of the same inhalers; the values in square parentheses indicate the order of test on a particular day (*From* [26]—*used with permission*)

Inhaler	Replicate	Cascade impactor configuration [dosing set]		
1	1	ACI [1]	AIM-QC [2]	AIM-pHRT [3]
	2	AIM-pHRT [5]	ACI [6]	AIM-QC [4]
	3	AIM-QC [9]	AIM-pHRT [7]	ACI [8]
2	1	AIM-QC [2]	ACI [1]	AIM-pHRT [3]
	2	ACI [4]	AIM-pHRT [6]	AIM-QC [5]
	3	AIM-pHRT [9]	AIM-QC [8]	ACI [7]
3	1	ACI [1]	AIM-pHRT [3]	AIM-QC [2]
	2	AIM-pHRT [6]	AIM-QC [5]	ACI [4]
	3	AIM-QC [8]	ACI [7]	AIM-pHRT [9]
4	1	AIM-QC [3]	AIM-pHRT [1]	ACI [2]
	2	AIM-pHRT [5]	ACI [6]	AIM-QC [4]
	3	ACI [7]	AIM-QC [8]	AIM-pHRT [9]
5	1	ACI [1]	AIM-pHRT [3]	AIM-QC [2]
	2	AIM-QC [5]	ACI [4]	AIM-pHRT [6]
	3	AIM-pHRT [9]	AIM-QC [8]	ACI [7]
6	1	AIM-pHRT [2]	ACI [3]	AIM-QC [1]
	2	ACI [4]	AIM-QC [5]	AIM-pHRT [6]
	3	AIM-QC [9]	AIM-pHRT [7]	ACI [8]

systems, respectively. A commercially available HFA-salbutamol pMDI was used as the test product, and careful attention was made to the design of the experiment to minimize the influence of possible confounding sources of error (e.g., operator, environment, inhaler, etc.).

The experiment was conducted using a design with the three impactor configurations already described, six inhalers, with three measurements made per inhaler per impactor configuration (Table 10.4). The specific objective of the study was to assess the repeatability of the impactor configurations (not to assess inhaler product performance); consequently, the experiment was conducted so that the following sources of variability were reasonably controlled:

1. Inhaler-to-inhaler variability over manufacturing run of OIP evaluated
2. Through-life trends of pMDI canister
3. Inter-operator variability
4. Inter-impactor system variability

Another feature of the study design was the inclusion of recurrent testing of each of the six inhalers on all three impactors. These precautions were taken to enable estimation of the intrinsic variability (precision) of each impactor system in isolation from other potentially confounding effects.

The statistical analysis estimated and compared the repeatability of three impactor configurations side-by-side, based on first quantifying the following

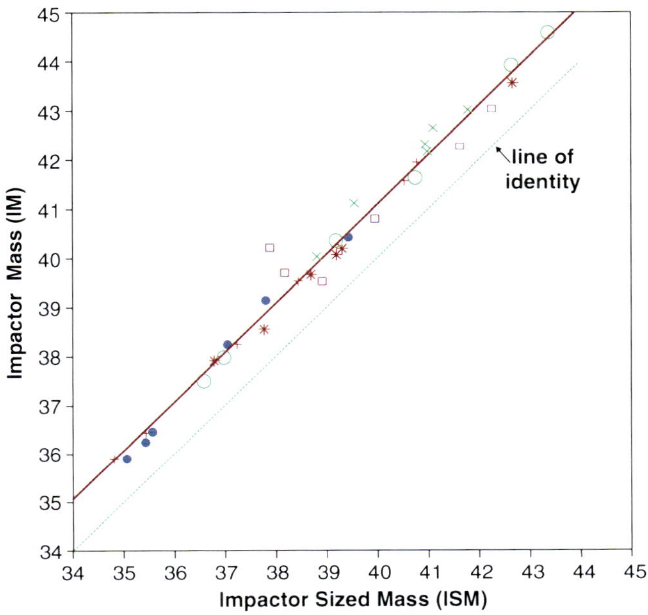

Fig. 10.9 Relationship between *IM* versus *ISM*, showing measurements with a different symbol for each of the six inhalers evaluated (*From* [26]—*used with permission*)

three principal metrics related to the total mass of salbutamol emitted from the inhaler, normalized on a per actuation basis:

1. Impactor mass (*IM*), defined as the total mass of API recovered from all components of the measurement system downstream of the induction port (i.e., mass on all stages including stage 0 for the full-resolution Andersen CI).
2. Ex-actuator mass (*Ex-ActM*), defined as the total mass of API recovered from all components of the measurement system including the induction port.
3. Ex-metering valve mass (*Ex-MVM*), defined as the total mass of API recovered from the measurement system together with that from the actuator mouthpiece of the inhaler.

Values of the subfractions of the mass/actuation were subsequently established and the relationship between *IM* (capable of being measured by all systems) and impactor-sized mass (*ISM*), determined only by the AIM-QC and full-resolution ACI, quantified (Fig. 10.9).

The consistent ca. 1 μg/actuation offset between *IM* and *ISM*, representing a fixed mass of API retained by stage 0 of the full-resolution ACI, enabled the precision comparison to take place between the two abbreviated configurations and the full-resolution CI, based on IM (Fig. 10.10).

On this basis, both abbreviated impactors had comparable precision with that of the ACI (Table 10.5).

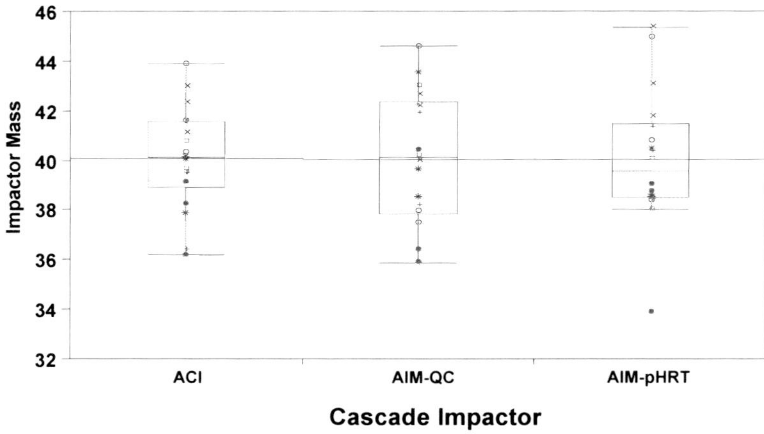

Fig. 10.10 Box-whisker plots of impactor mass by inhaler and cascade impactor configuration. The mean is represented by the tick mark at the center of the line (*From* [26]—*used with permission*)

Table 10.5 Summary statistics and 95% confidence intervals on ratio of standard deviations of abbreviated to full configurations for metrics related to inhaler APSD (*From* [26]—*used with permission*)

Metric	Impactor configuration	Mean (μg)	SD [repeatability] (μg)	Coefficient of variation (%)	95% CI[a] on ratio of SDs of abbreviated to full configuration
ISM	ACI	38.97	1.57	4.07	–
	AIM-QC	38.96	2.68	6.87	[0.93; 3.05]
LPM/SPM	ACI	2.70	0.28	9.98	–
	AIM-QC	2.69	0.35	12.83	[0.68; 2.24]

[a]Inclusion of 1.00 in confidence interval (CI) indicates no statistically significant difference

Separate estimates of variability were made for each metric (*ISM*, *LPM/SPM* with the AIM-QC and full-resolution ACI) and *FPM* (identical with *CPM* in terms of precision) for the AIM-pHRT and ACI. The estimates of precision associated with all these metrics obtained by the appropriate abbreviated impactor were substantially equivalent to the corresponding metrics with the full-resolution system (Fig. 10.11 and Tables 10.5 and 10.6).

Interestingly, all these metrics consistently tracked minor differences between the six inhalers that were used in the investigation (Fig. 10.11b, d, f). When the size fractions from either abbreviated impactor were compared with the corresponding cumulative mass-weighted APSD data from the ACI (Fig. 10.12), excellent agreement was apparent in almost all instances. However, an unexpected outcome was the magnitude of the positive bias associated with *EPF* measured by the AIM-pHRT system, which was almost 8% greater than the corresponding full-resolution data (Table 10.6), illustrated in the comparison of this metric with the expected value from the ACI (Fig. 10.12).

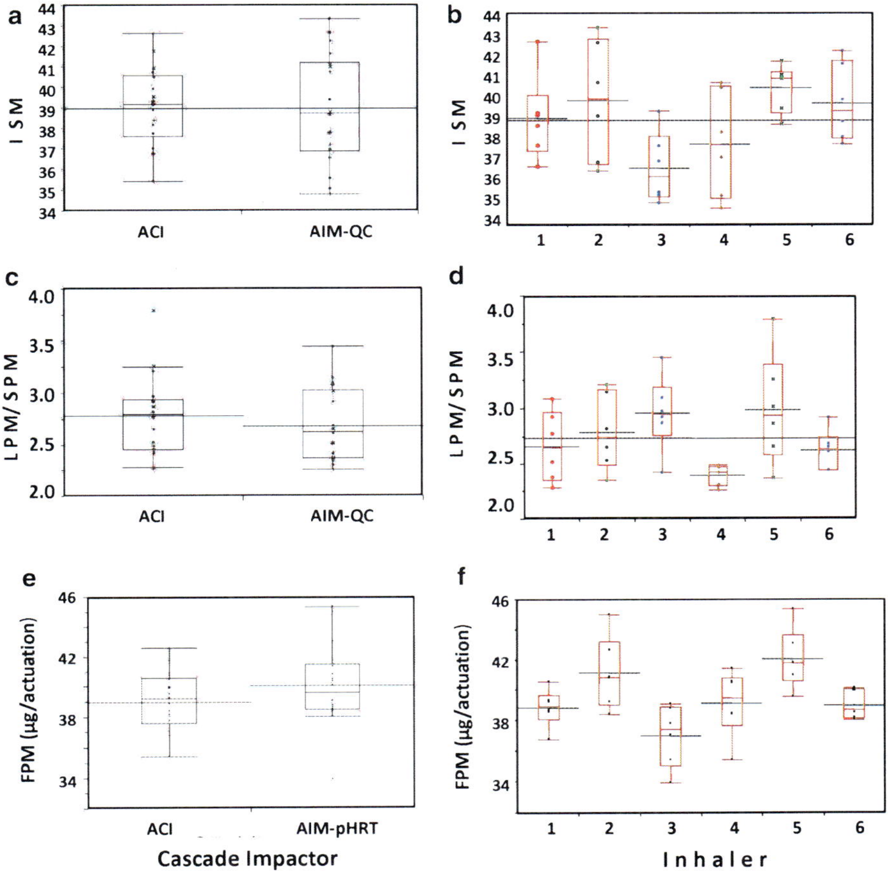

Fig. 10.11 Comparison of performance metrics in precision study: (**a**) ISM by impactor type; (**b**) *ISM* by inhaler number; (**c**) ratio LPM/SPM by impactor type; (**d**) *LPM/SPM* by inhaler number; (**e**) *FPM* by impactor type; and (**f**) *FPM* by inhaler number (*From* [26]—*used with permission*)

Table 10.6 Summary statistics and 95% Confidence Intervals (CIs) on ratio of standard deviations of AIM-pHRT system to nonviable ACI for metrics relating to inhaler APSD (*From* [26]—*used with permission*)

Metric	Impactor configuration	Mean (µg)	SD [repeatability] (µg)	Coefficient of variation (%)	95% CI on ratio of SDs of abbreviated to full configuration
CPM	ACI	44.08	2.87	6.52	[0.57; 1.89]
	AIM-pHRT	45.17	3.00	6.64	
FPM	ACI	35.43	1.40	3.88	[0.69; 2.25]
	AIM-pHRT	35.00	1.74	4.97	
EPM	ACI	2.21	0.74	33.72	[0.73; 2.40]
	AIM-pHRT	7.93	0.99	12.44	

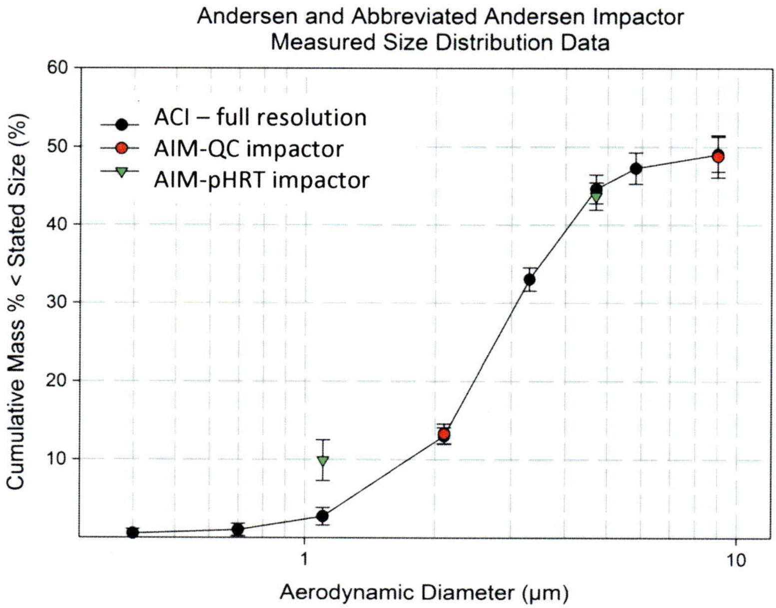

Fig. 10.12 Comparative measures of impactor-sized subfractions by AIM-QC and AIM-pHRT abbreviated systems with ACI (original AIM-pHRT data set) (*From* [26]—*used with permission*)

This observation prompted a follow-on investigation to eliminate the cause of the bias [27]. Microscopic inspection of the coated surface below the second stage (cut-off diameter 1.1 µm) revealed depressions in the coating directly beneath the nozzles, where impaction of particles would be expected (Fig. 10.13).

Interestingly, no other collection surfaces, in either of the abbreviated systems, exhibited the same problem. It was concluded therefore that the relatively high Reynolds number associated with flow through the nozzles of this lower stage ($Re_f = 292$ at 28.3 L/min) leads to displacement of the coating surface, inhibiting its ability to trap particles effectively. Particle bounce is therefore relatively high, re-entrainment carrying over material that should be efficiently collected into the *EPM*.

This problem was successfully resolved by floating a filter coated in surfactant on top of the collection plate (Fig. 10.14) that provided a surface that was both energy absorbent but at the same time resisted relocation by the incoming flow (Fig. 10.15). This unexpected outcome adds weight to the argument for a very careful consideration of particle bounce for all abbreviated systems, regardless of OIP class.

Rather surprisingly, increased precision, one of the potential benefits initially claimed for AIM systems by virtue of eliminating variability arising from stages that collected API close to the lower limit of detection, was not observed with either abbreviated system. In explanation, it was hypothesized that precision gains achieved by eliminating the analysis of material from stages where little sample collects are offset by other factors, possibly related to the flow of aerosol in the

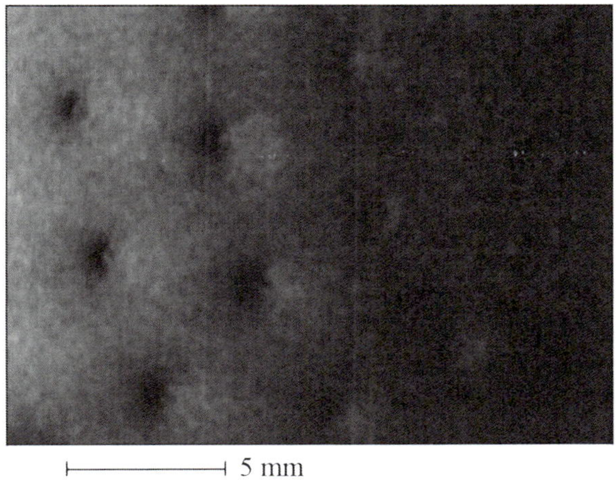

Fig. 10.13 Photomicrograph of displaced Brij 35 surfactant on collection plate for the second impaction stage of the AIM-pHRT impactor (*From* [27]*—used with permission*)

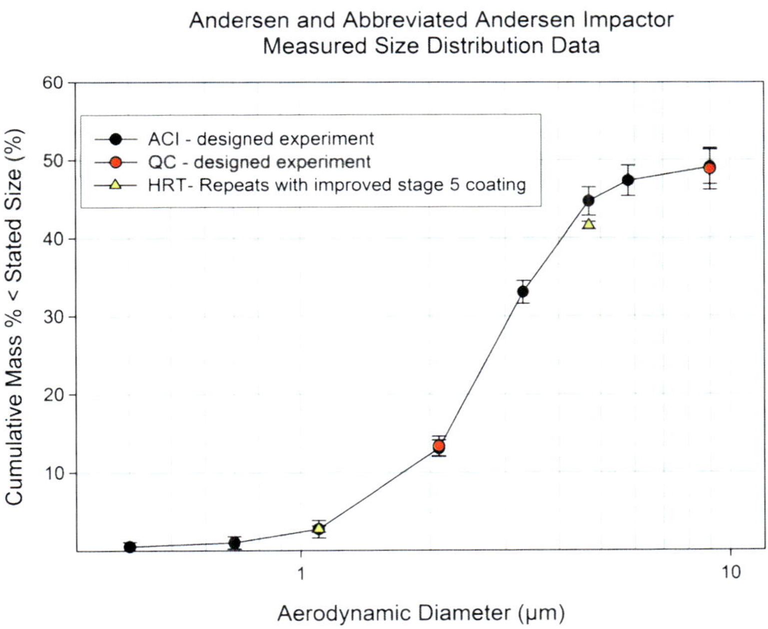

Fig. 10.14 Comparative measures of impactor-sized subfractions by ACI with measures of CPF, FPF, and EPF after modifying second impaction stage of AIM-pHRT CI with Brij-35-soaked filter (*From* [27]*—used with permission*)

reduced systems. Despite this finding, measurement time gains were found to be significant; analysis with the abbreviated systems taking around 30% of the time consumed using the ACI. This finding of more rapid determinations is a consistent trend across other reported studies to be discussed later in this chapter.

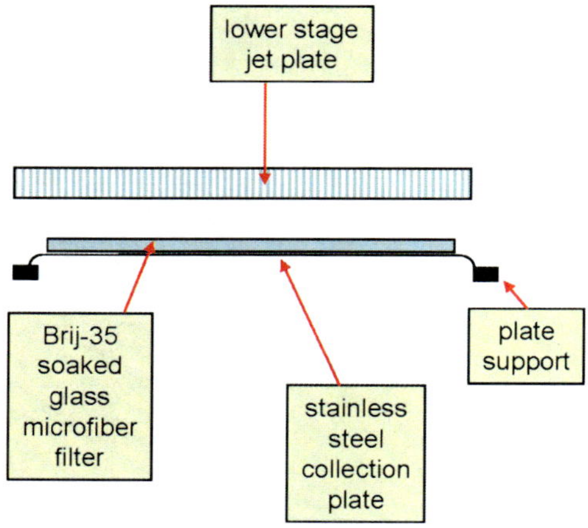

Fig. 10.15 Schematic of Brij-35-soaked glass microfiber filter, showing filter located on top of stainless steel collection plate (*From* [27]*—used with permission*)

10.5 Other Studies with ACI-Based AIM Systems

More recently, the commercialized version of the abbreviated ACI (FSA, Copley Scientific Ltd.) has been extensively evaluated by Keegan and Lewis in connection with the rapid screening of prototype pMDI actuators in early-stage product development [28, 29]. In a rapid prototyping environment, it is often desirable to optimize a device or system based upon the delivered mass and fine particle mass (≤5 μm) that are obtained through the cascade impactor method. A number of prototypes with slightly altered configurations may be screened to provide an optimized embodiment.

In their first study [28], MDIs containing beclomethasone dipropionate (BDP) (100 μg/50 μL) with 13% w/w ethanol cosolvent in HFA134a propellant were manufactured for use as the model product, equipped with a Bespak 630 series actuator with a 0.22 mm orifice diameter (Bespak, UK). A standard FSA with cut-point sizes of <5 μm and <1 μm d_{ae} was evaluated as the abbreviated CI configuration. BDP was recovered only from the impaction plates in the so-called rapid (rFSA) procedure, whereas this API was recovered from all CI surfaces following the standard method (FSA). The rFSA method therefore permitted as few as three separate actuations from each prototype formulation to be analyzed in terms of $FPM_{<5.0}$μm, compared with 4-actuations that were required to achieve the required sensitivity with the standard (FSA) procedure. The impaction plates were coated after FSA assembly (using either method) using aerosolized 1% w/w glycerol delivered in HFA134a propellant by a proprietary process. The recovery solvent was an 85:15 methanol–water mixture.

Following each actuation into the FSA, samples were collected from the USP/Ph. Eur. induction port and impaction plates (including back-up filter), but interstage drug loss was not determined. The stack was then reassembled with clean components. Following the final actuation, a sample was collected from the actuator and an average deposition over the appropriate number of pMDI actuations was reported.

Table 10.7 Comparison of particle distribution metrics for BDP (100 μg per actuation/50 μL metered volume) using three impactor methods (*From* [28]—*used with permission*). $n=3$; mean ± SD

	TEM (μg)		$FPM_{<5.0\mu m}$ (μg)		$EFM_{<1.0\mu m}$ (μg)	
Method	Mean	SD	Mean	SD	Mean	SD
ACI	88.2	2.5	50.8	3.1	18.4	1.1
FSA	88.2	3.3	45.9	3.7	20.9	1.9
rFSA	85.9	3.2	46.0	0.5	21.3	1.4
p-Value[a]	*0.59*		*0.13*		*0.12*	

[a]One-way ANOVA

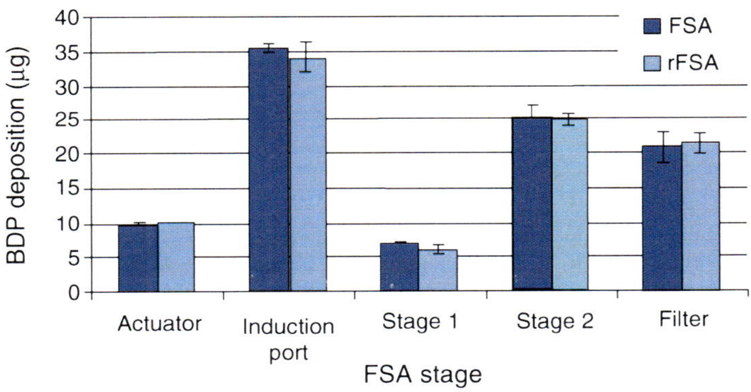

Fig. 10.16 Deposition profiles for BDP (100 μg per actuation/50 μL metered volume) within the abbreviated impactor both with (FSA) and without (rFSA) recovery of interstage drug loss ($n=3$; mean ± SD for each data series) (*From* [28]—*used with permission*)

Assay for BDP was undertaken by UPLC-MS in both standard FSA and rFSA procedures.

In addition to the abbreviated CI-based measurements, benchmark full-resolution APSD determinations based upon two actuations of the model product were made using the ACI with collection plates coated using 1% w/w glycerol and also equipped with the same induction port. All measurements were undertaken at a flow rate of 28.3 L/min.

Comparison of the key metrics, *TEM*, $FPM_{<5.0\mu m}$, and $EPM_{<1.0\mu m}$, are summarized in Table 10.7. Note that in this work, delivered dose is equivalent to *TEM*.

No statistical difference ($p>0.05$; ANOVA) between the reported metrics was evident. However, both FSA methods showed a tendency to slightly underestimate $FPM_{<5.0\mu m}$ and marginally overestimate $EPM_{<1.0\mu m}$, when compared to equivalent measures derived from the benchmark ACI. This finding is consistent with that discussed earlier for pMDIs containing ethanol as low-volatile cosolvent [20]. Importantly, this study showed no significant differences ($p>0.05$; ANOVA) between either rFSA or FSA methods, despite the omission of the interstage drug deposition in the latter (Fig. 10.16).

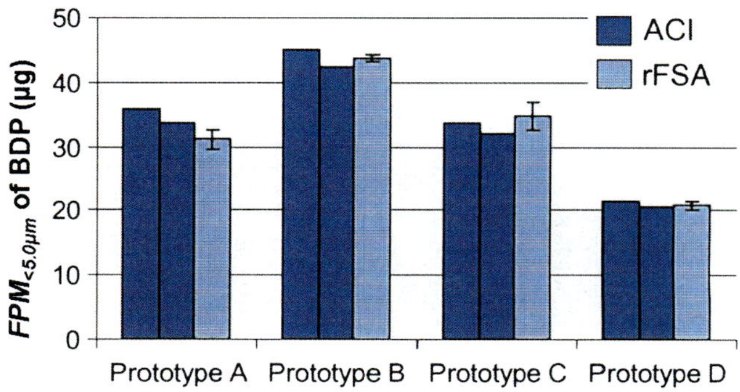

Fig. 10.17 Comparison of values of $FPM_{<5.0\mu m}$ achieved by aerosolizing BDP (100 μg per actuation/50 μL metered volume) with novel prototype actuators A–D: series 1 and 2 refers to the two ACI measurements ($n=2$) shown separately: series 3 refers to the rFSA measurements ($n=3$), in which mean ± SD is illustrated for group (*From* [28]—*used with permission*)

In the second phase of their investigation, four prototype actuators were assessed using a standard procedure ($n=2$; ACI) and rFSA method ($n=3$) in order to determine the screening capability of the rapid method. The trend in $FPM_{<5.0\mu m}$ observed between prototypes A–D was found to be independent of the impactor method, with prototype B delivering the optimal dose (Fig. 10.17).

Keegan and Lewis observed that their rFSA method reduced the time taken to obtain replicate ACI measurements by 50%. This time saving was in addition to other reported benefits such as reduced solvent consumption, analysis time, and data processing. They concluded that their AIM-based methodology represents a sufficiently precise method for determining key particle distribution metrics such as $FPM_{<5.0\mu m}$ for the purpose of screening and design optimization studies.

In their second investigation [29], Keegan and Lewis turned their attention to the use of their FSA as a tool for the rapid screening of solution pMDIs containing increasing concentrations of ethanol. The effect of adding 1.3% w/v glycerol was also investigated with the preparation containing 13% ethanol. The general test procedure used by Keegan and Lewis was similar to that for the full FSA method described in their first study [28].

An additional spacing stage was included above the first impaction stage in the FSA, as recommended by Mitchell et al. (see Fig. 10.4) [20] for some measurements. In this modified FSA (mFSA) configuration, more time was allowed for ethanol to evaporate, with the aim of achieving closer agreement with full-resolution benchmark ACI measurements for these MDI products. However, this annular spacer stage differed from the approach adopted by Mitchell et al., in that it consisted of a metal ring without the nozzle plate (Fig. 10.18).

The MDIs, again containing BDP as API (100 μg per actuation/50 μL metered volume), were manufactured containing 8%, 13%, and 26% w/w ethanol in HFA134a propellant. Each MDI was equipped with the same actuator as used in their first study. Values of $CPM_{>5.0\mu m}$, $FPM_{<5.0\mu m}$, and $EPM_{<1.0\mu m}$ obtained from each of the

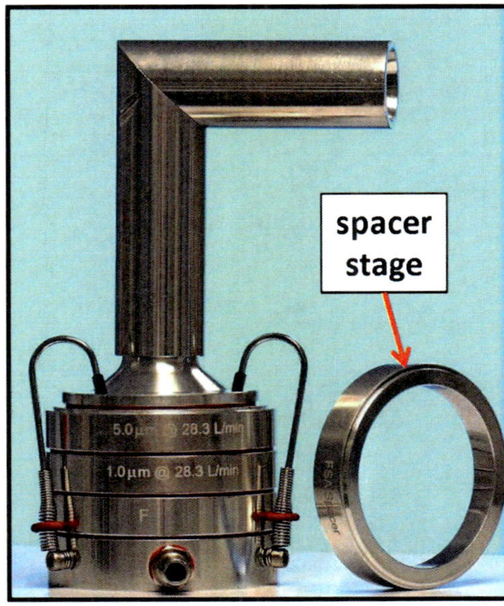

Fig. 10.18 FSA with annular spacer stage (mFSA) used by Keegan and Lewis (*From* [29]—*used with permission*)

impactor configurations are summarized in Fig. 10.19. There was no difference between the reported values for the glycerol-containing formulation regardless of the impactor used ($p>0.05$). The addition of glycerol to a solution pMDI formulation modulates the MMAD, increasing it from 1.3 µm (13% ethanol) to 2.8 µm. This effect may reduce impaction due to incomplete ethanol evaporation since the residual droplets are larger. Significant differences ($p<0.01$) in the metrics obtained when the ethanol concentration was at its highest (26%w/w) became evident between the impactor systems.

The difference between the average BDP masses at each particle size fraction for the ethanol-containing MDIs is reported in Table 10.8. The residual values in this table represent the absolute magnitude of the difference between FSA- and ACI-measured values of each metric, expressed as a percentage and in micrograms. Keegan and Lewis observed a consistent increase in the magnitude of the difference between the FSA value and that calculated from ACI stage deposition as the ethanol concentration in the formulation increased. Importantly, however, the inclusion of the additional "spacer" stage in the FSA attenuated these divergence between FSA and ACI-measured values of both $CPM_{>5.0\mu m}$ and $FPM_{<5.0\mu m}$, as would be expected from the earlier observations of Mitchell et al. [20]. However, this behavior was offset by increases in divergence with values of $EPM_{<1.0\mu m}$ when the "spacer" stage was included.

Figure 10.20 shows the effect of the "spacer" stage on deposition of BDP on the collection plate of the first FSA stage in comparison with the calculated equivalent from the full-resolution ACI (impaction stage mass less $FPM_{<5.0\mu m}$).

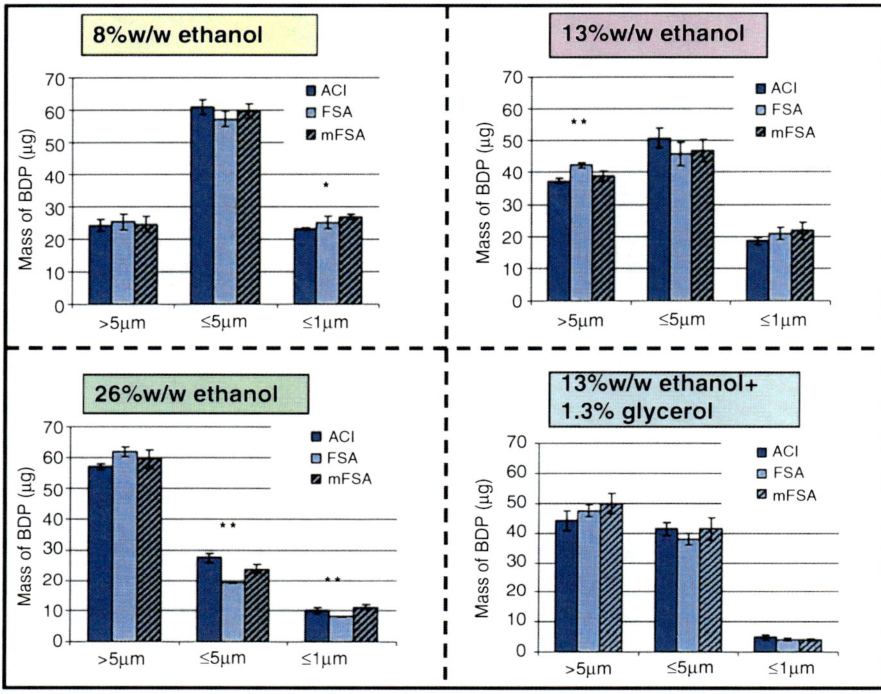

Fig. 10.19 Comparison of impactor size fractions (ANOVA; $*p<0.05$; $**p<0.01$) from solution formulations containing 100 µg BDP per 50 µL dose ($n=3$; ±SD). *mFSA* modified FSA as described above (*From* [28]—*used with permission*)

Table 10.8 Residual FSA-measured particle dose values relative to associated ACI values[a] (*From* [28]—*used with permission*)

		Ethanol content (%w/w)					
		8%		13%		26%	
Metric	FSA type	µg	%	µg	%	µg	%
$CPM_{>5.0\mu m}$	FSA	−1.1	4.5	−4.9	**13.1**	−4.8	8.4
	mFSA	−0.3	1.2	−1.4	4.1	−2.5	4.4
$FPM_{<5.0\mu m}$	FSA	3.8	6.2	4.9	9.6	8.4	**30.4**
	mFSA	1.3	2.0	3.8	7.5	3.6	**13.0**
$EPM_{<1.0\mu m}$	FSA	−2.0	8.6	−2.4	**13.0**	2.3	**22.8**
	mFSA	−3.6	**15.5**	−3.2	**17.3**	−0.9	8.9

[a]Bold typeface indicates a difference of >10% of the ACI reported value

The outcome depicted confirms the previously reported observation of impaction of partially evaporated droplets from solution MDI formulations containing ethanol [20]. However, Keegan and Lewis noted that the addition of the "spacer" stage to the FSA did not make any significant difference to the BDP deposition observed for

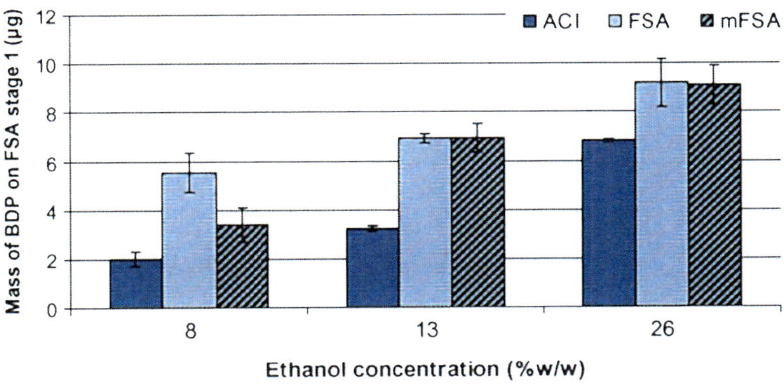

Fig. 10.20 Mass BDP deposited on stage 1 of the FSA configurations compared with the equivalent mass deposited in the ACI ($n=3$; ±SD). *mFSA* modified FSA as described above (*From* [28]—*used with permission*)

formulations with ethanol concentrations >8% w/w. They suggested that the inclusion of a "spacer" stage may help to attenuate some differences between the FSA and ACI. However, it is likely that the magnitude of the effect will be formulation/device specific.

Keegan and Lewis concluded that, in general, the FSA may provide representative values for key particle metrics that are not significantly different from the full-resolution ACI for solution pMDI formulations. If it can be used, this abbreviated impactor offers a tool that eliminates the need for post-analysis data processing to obtain key metrics when screening formulations.

The outcomes from all of the work reported in this section reinforce the recommendation that the implementation of an AIM-based method should always be preceded by some form of validation study with the particular products of interest, using an appropriate full-resolution CI as the benchmark technique.

10.6 Assessing the Performance of AIM Systems Based on the Andersen Viable Cascade Impactor

The Andersen viable cascade impactor (AVCI) is the earliest version of the Andersen multistage CIs to be developed [30]. It is similar in operating principle to the nonviable ACI, with the important exception that the stage wells are larger so that they can each accommodate a Petri dish instead of a collection plate [31]. The Westech fine particle dose impactor (Westech Instrument Services, Upper Stondon, Beds., UK) was designed as a simple two-impaction stage and filter sampler for the rapid determination of fine particle mass from pMDIs at 28.3 L/min (Smurthwaite MJ (2012) Westech instrument services, UK, personal communication). The design is based on an abbreviated AVCI (Fig. 10.21) but incorporates some new design features

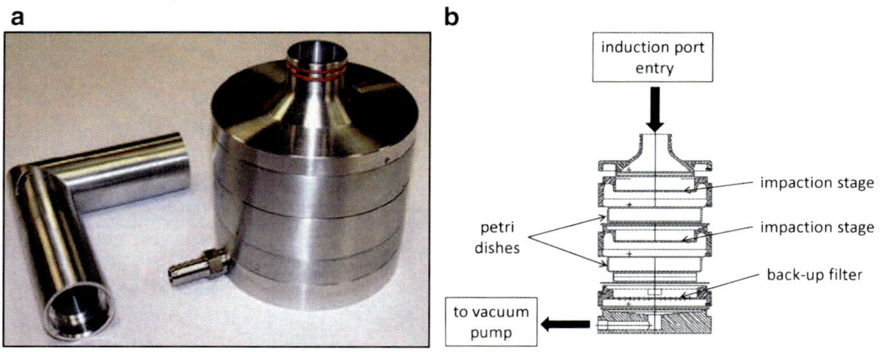

Fig. 10.21 Westech Instrument Services fine particle dose impactor (FPD–AVCI). (**a**) External appearance. (**b**) Cross-section through interior (*Courtesy of Westech Instrument Services Ltd.*)

to improve ease of use, notably a bayonet fixing between stages to allow rapid assembly/disassembly of the CI. The replacement of the standard ACI collection plates with glass or metal Petri dishes facilitates in situ recovery of API. The size-separation stages themselves are based on stages 2 and 6 of the AVCI, having nominal jet diameters of 0.914 mm and 0.254 mm, respectively, with corresponding cut points of 4.7 and 1.1 μm aerodynamic diameter.

In 2010, Chambers and colleagues at AstraZeneca (AZ), UK undertook a performance evaluation study of the FPD apparatus, comparing it against a 6-stage nonviable ACI, and also a 2-stage abbreviated nonviable ACI (sACI) that utilized ACI stages 2 and 5 with a blank stage 0 present (and was therefore equivalent to the T-FSA/AIM-pHRT design), as benchmark systems [32]. Here the prefix "s" stands for standard (i.e., nonviable ACI) components. The performance of the CIs was assessed against the following metrics:

1. Total mass [dose collected by impactor equivalent to impactor mass (*IM*)]
2. Mass of API collected by induction port (USP/Ph. Eur.) [equivalent to nonimpactor-sized mass (*NISM*)]
3. Coarse particle mass >4.7 μm aerodynamic diameter ($CPM_{>4.7\mu m}$)
4. Fine particle mass <4.7 μm aerodynamic diameter ($FPM_{<4.7\mu m}$)
5. Extra-fine mass collected on back-up filter ($EPM_{<1.1\mu m}$)

Three pMDIs containing a single component API were used to evaluate the performance of the FPD–AVCI and sACI relative to data generated from the 6-stage ACI. The collection plates of the 6-stage ACI were uncoated in accordance with normal practice for working with this class of OIP. However, the respective plates and Petri dishes of the sACI and FPD–AVCI were coated with a Brij 35-coating solution based on the findings from the earlier work reported in Sect. 10.3.

An additional experiment was carried out in order to investigate the extent of possible particle re-entrainment whereby Brij-coated Westech filter papers were placed on the plates and Petri dishes of both stages of the sACI and FPD–AVCI,

Table 10.9 Experiment order in the AstraZeneca (UK) 2011 FPD–AVCI evaluation study

Experiment	pMDI no.	pMDI actuation numbers	Impactor
6-Stage ACI (control)	1	1–6	A
	2		B
	3		C
sACI with Brij-coated plates	1	9–11	A
	2		C
	3		B
FPD–AVCI with Brij-coated petri dishes	1	14–16	A
	2		B
	3		C
FPD–AVCI with Brij-coated filter on petri dishes	1	19–21	B
	2		A
	3		C
sACI with Brij-coated filter on plates	1	24–26	C
	2		A
	3		B

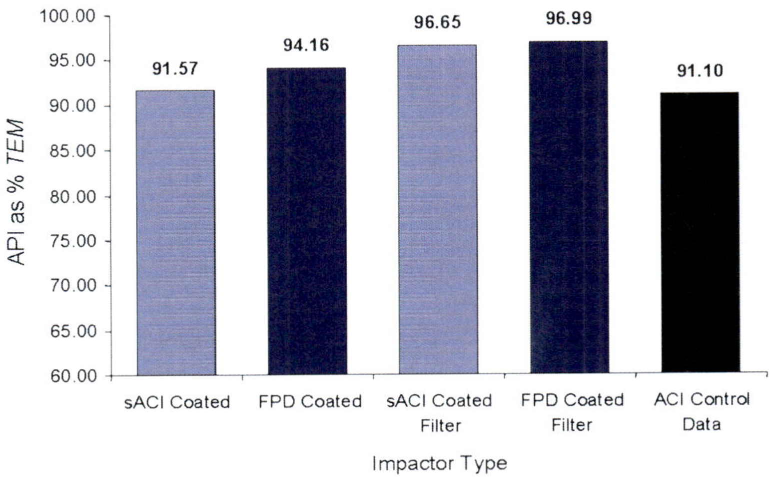

Fig. 10.22 Impactor mass (*IM*) of API expressed as % of nominal dose recovered from the various systems evaluated by Chambers et al. (*From* [32]—*used with permission*)

following the practice described in Sect. 10.4 for the second stage of the AIM-pHRT system evaluated at TMI [27]. Three impactors of each type were evaluated, but in order to guard against systematic bias in the different experiments, the impactor order was swapped between experiments (Table 10.9). The measurement of *IM* as percentage of the nominal dose by sACI with Brij-coated Petri dishes was in closest agreement with that obtained from the control ACI (Fig. 10.22).

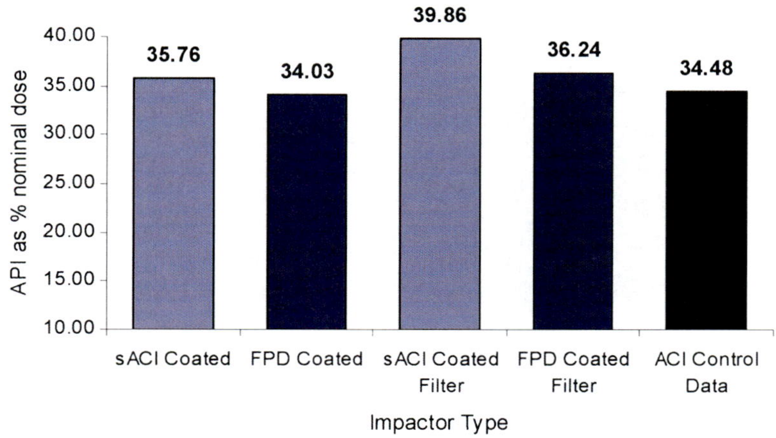

Fig. 10.23 $FPM_{<4.7\mu m}$ as percentage of *TEM* (mass/actuation) ex-MDI for the various systems evaluated by Chambers et al. (*From* [32]—*used with permission*)

IM, determined by the FPD–AVCI, was greater than with the control ACI by +3.4% for the surfactant (Brij)-coated FPD–AVCI, and +6.5% for the same CI when Brij-coated filters were used. Importantly, the addition of Brij-coated filters to the sACI resulted in a similar increase in *IM* relative to the ACI, in this instance being +6.7%.

These data are suggestive of decreased internal losses when the Brij-soaked filters were used to mitigate particle bounce further than could be achieved by simply coating the collection plates with the same surfactant. The measure of agreement between sACI with coated plates and the ACI control is probably to be expected, given the similarity in internal geometry and dead space.

All abbreviated systems were closely correlated to the 6-stage ACI in terms of either $CPM_{>4.7\mu m}$ or $FPM_{<4.7\mu m}$ (Fig. 10.23). The sACI slightly underestimated $CPM_{>4.7\mu m}$ relative to that measured with the reference ACI, but that introduction of the Brij-coated filter papers on the sACI stages resulted in an increase in $CPM_{>4.70\mu m}$, providing additional support for an argument that bounce and re-entrainment were not entirely eliminated when the plates were coated with surfactant, without the means to stabilize the coating when flow passed through these CIs.

$FPM_{<4.7\mu m}$ determined by the Brij-coated sACI and the FPD data both agreed closely with the same measure obtained using the reference ACI. The addition of Brij-coated filters increased $FPM_{<4.7\mu m}$ in line with similar effects associated with both $CPM_{>4.7\mu m}$ and *IM*.

When both $FPM_{<4.7\mu m}$ and $CPM_{>4.7\mu m}$ were calculated as a percentage of *IM* (Fig. 10.24), this form of data presentation highlighted more clearly that both abbreviated CIs correlated well with the ACI; although, $CPM_{>4.7\mu m}$ was slightly underestimated by the sACI. The increase in upper stage deposition with this formulation was confirmed by the addition of Brij-coated filters paper to the sACI. However, these observed differences are slight and comparable with true variability associated with the product itself.

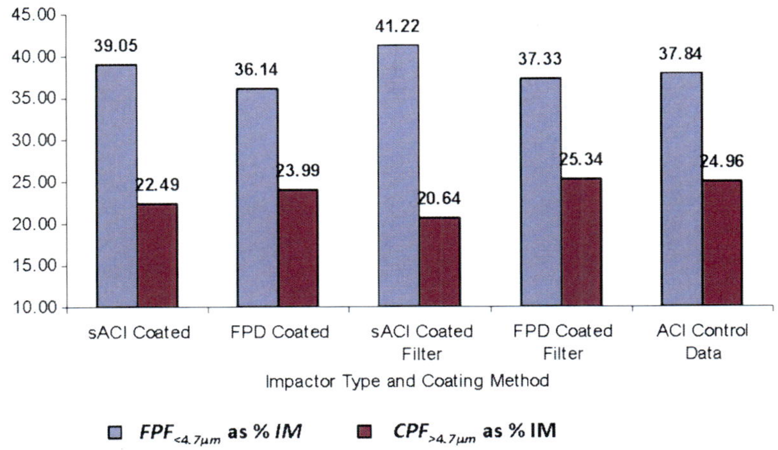

Fig. 10.24 $FPF_{<4.7\mu m}$ and $CPF_{>4.7\mu m}$ as a percentage of *IM* for the various systems evaluated by Chambers et al. (*From* [32]—*used with permission*)

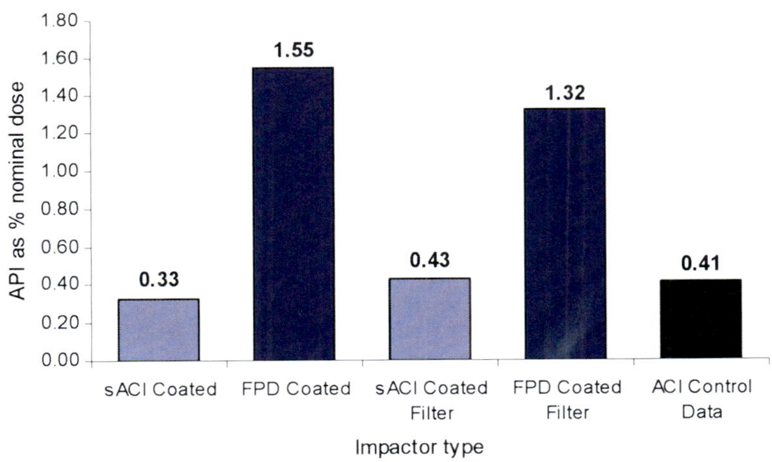

Fig. 10.25 Measures of $EPM_{<1.1\mu m}$ as percentage of *TEM* (mass/actuation) ex-MDI for the various systems evaluated by Chambers et al. (*From* [32]—*used with permission*)

The largest discrepancy seen between the impactors (Fig. 10.25) was in filter deposition ($EPM_{<1.1\mu m}$), where the magnitude obtained with the FPD–AVCI (1.55% of nominal dose with Brij-coated surfaces; 1.32% with added Brij-coated filters) was more than three times greater than equivalent values obtained using either the sACI (0.33% Brij-coated plates; 0.43% with added Brij-coated filters) or the ACI control (0.41%).

Table 10.10 pMDI products evaluated by Guo et al. with the FPD–AVCI [33]

Product name	API	Propellant	Excipients	Formulation type
Aerobid®	Flunisolide	CFC	Sorbitan trioleate	Suspension
Combivent®	Salbutamol (AS)/ipratropium bromide (IB)	CFC	Soya lecithin	Suspension
MaxAir™	Pirbuterol	CFC	Sorbitan trioleate	Suspension
Advair®	Fluticasone propionate (FP)/ salmeterol xinafoate (SX)	HFA	None	Suspension
Flovent-110®	Fluticasone propionate	HFA	None	Suspension
Proair®	Salbutamol	HFA	Ethanol	Suspension
Proventil™	Salbutamol	HFA	Oleic acid, ethanol	Suspension
Atrovent®	Ipratropium bromide	HFA	Water, citric acid, ethanol	Solution

This pattern of divergence between the various abbreviated options was comparable in absolute terms with the behavior observed with measures of $FPM_{<4.7\mu m}$. The fact that the addition of Brij-coated filters to the collection dishes of the FPD–AVCI only slightly reduced $EPM_{<1.1\mu m}$ suggests that particle re-entrainment is an unlikely cause. However, it should be emphasized that these differences between the abbreviated systems with or without surfactant-saturated filters to control particle bounce and re-entrainment and the control ACI were small, and therefore unlikely to prevent any of these options being used as an abbreviated impactor of choice with this particular formulation. This type of detailed study illustrates well the approach that should be taken when validating a potential AIM-based system for any OIP.

Guo et al. have also recently presented measurements undertaken with the Westech FPD–AVCI impactor, evaluating eight different suspension and solution pMDIs (Table 10.10) [33]. $CPF_{>5.0\mu m}$, $FPF_{<5.0\mu m}$, and $EPF_{<1.0\mu m}$ determined by the FPD–AVCI were compared to the same metrics obtained by means of an 8-stage nonviable ACI. Ten actuations from the pMDI-on-test were delivered into the FPD–AVCI or ACI, and API recovery proceeded afterward by validated quantitative chemical analysis. Since the ACI does not have stages with cut-off diameters precisely at 5 μm and 1 μm, the three measures of interest were interpolated from the cumulative APSD data obtained with the full-resolution CI.

Equivalent total API recovery was observed for all pMDI products between FPD and ACI (Fig. 10.26). Although agreement between FPD–AVCI and ACI data was generally good, Guo et al. found small but significant differences in all three subfractions with some of the formulations, especially with $CPF_{>5.0\mu m}$, with the values obtained using the FPD–AVCI higher than their ACI-determined counterparts (Fig. 10.27).

Guo et al. concluded that, whether or not the pMDI is a solution or suspension, formulation was not the only deciding factor on whether or not there is a divergence between abbreviated and full-resolution data [33].

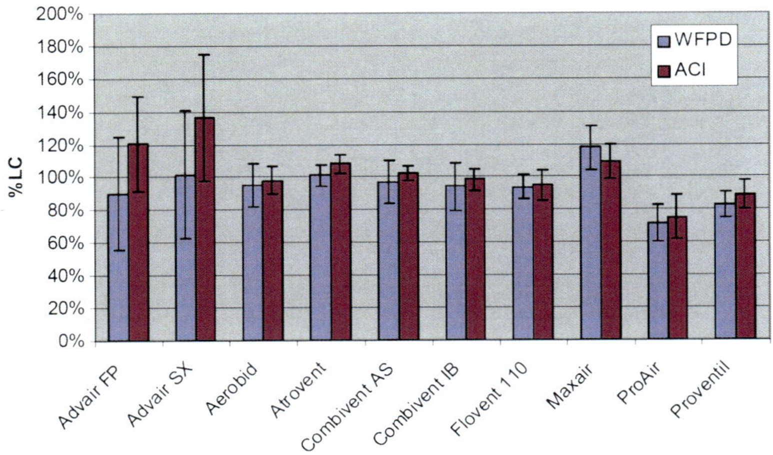

Fig. 10.26 Total API recovery for eight different pMDI products comparing FPD–AVCI and non-viable 8-stage ACI. *%LC* percent label claim (*From* [33]—*courtesy of W. Doub*)

In retrospect, the internal dead space associated with the FPD–AVCI is significantly greater than that for the abbreviated nonviable ACI systems, due to the need to accommodate the three dimensional structure of the Petri dish rather than near-to-flat collection plates associated with the nonviable ACI internal configuration (compare Fig. 10.28a, b). The extra dead space in the FPD–AVCI might therefore be expected to result in increased $FPF_{<5.0\mu m}$ through increased time for particle shrinkage due to cosolvent evaporation. However, this outcome was not seen to a marked extent with formulations containing cosolvent except with Proventil™.

Furthermore, cosolvent evaporation would not explain the observed bias toward larger values of $CPF_{>5.0\mu m}$, that was apparent with almost all the products, whether or not cosolvent was present in the formulation. It therefore appears that another explanation is needed to explain these results in a more satisfying way. One possibility could be the potential for increased impaction of coarse particulate at the first stage of the FPD–AVCI, again brought about by differences in internal geometry between this abbreviated impactor and the nonviable ACI.

Further investigation is therefore warranted, this time, ideally using both AVCI and nonviable ACI as control impactors, given the similarity of this impactor to the interior of the FPD–AVCI (compare Fig. 10.16 with Fig. 10.28a).

10.7 Assessing the Performance of AIM Systems Based on the Fast Screening Impactor

Up until this section, the focus has been on abbreviated impactors that are based on either the nonviable or viable ACI internal configurations. Such systems were the first to be evaluated, as in the case of abbreviated nonviable ACI systems, they can be

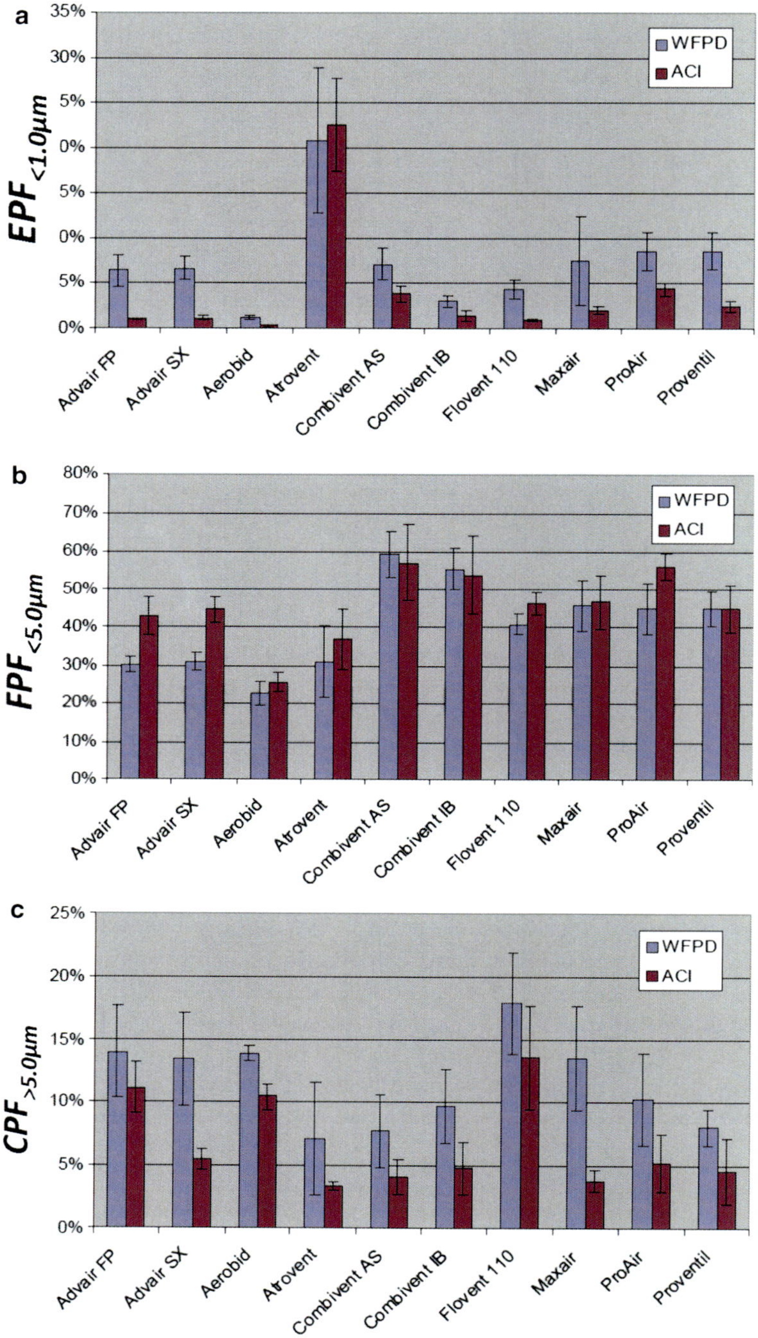

Fig. 10.27 Comparisons of measures of $CPF_{>5.0\mu m}$, $FPF_{<5.0\mu m}$, and $EPF_{<1.0\mu m}$ for 8-different pMDI products comparing FPD–AVCI and nonviable 8-stage ACI. *%LC* percent label claim (*From* [33]—*courtesy of W. Doub*)

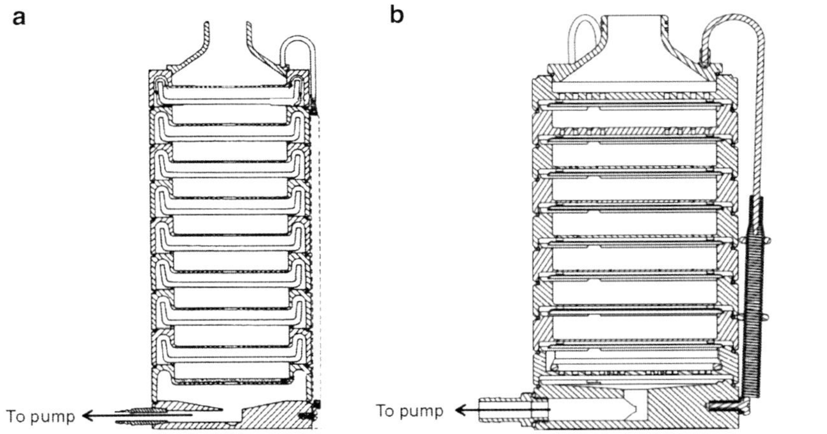

Fig. 10.28 Andersen 8-stage impactors (**a**) viable (AVCI); (**b**) nonviable (ACI) [*Courtesy of Westech Instrument Services Ltd. (**a**) and Copley Scientific Ltd. (**b**)*]

constructed easily from existing components of the full-resolution system. In the past 10 years since commercialization, the NGI has become increasingly used to characterize OIPs, especially since its acceptance as a pharmacopeial apparatus for APSD determinations (Apparatus 5 and 6 in Chap. 601 of the US Pharmacopeia, and apparatus E in Chap. 2.9.18 of the European Pharmacopoeia—see Chap. 2). Abbreviated systems based on this impactor are either newly introduced units or involve modification of a full-resolution impactor, which is discussed later in this chapter.

The fast screening impactor (FSI) is a newly introduced one-stage impactor (Fig. 10.29), importantly with no "parent" full-resolution apparatus; although its design is based on a modified NGI pre-separator [34]. With this abbreviated impactor, large non-inhalable liquid boluses from nebulizing systems and powder boluses in the case of DPIs are captured in a liquid trap, and the sample is then separated by a fine-cut impaction stage whose d_{50} is 5.0 µm. A filter collector below the pre-separator body collects the fine fraction.

The stage collection efficiency characteristics of four slightly different designs of FSI insert having a nominal cut point of 5 µm aerodynamic diameter have been reported by Roberts and Romay [34] (Table 10.11), permitting the FSI to be operated in the flow rate range from 30 to 90 L/min, and therefore making it suitable for DPI aerosol characterization, as well as with the other forms of OIP.

The sharpness of cut, given by the closeness of the geometric standard deviation (GSD_{stage}) to unity (see Chap. 2), was excellent in all the cases for which data were presented (the GSD_{stage} of the insert developed to operate at 35 L/min was subsequently also confirmed to be close to 1.10). These values compare with values of GSD_{stage} in the range 1.1–1.4 for the NGI [35, 36] intentionally designed to have optimum performance in terms of size-fractionation and resolution capability [14]. There are now inserts manufactured with 5 µm cut point for use at 5 L/min intervals

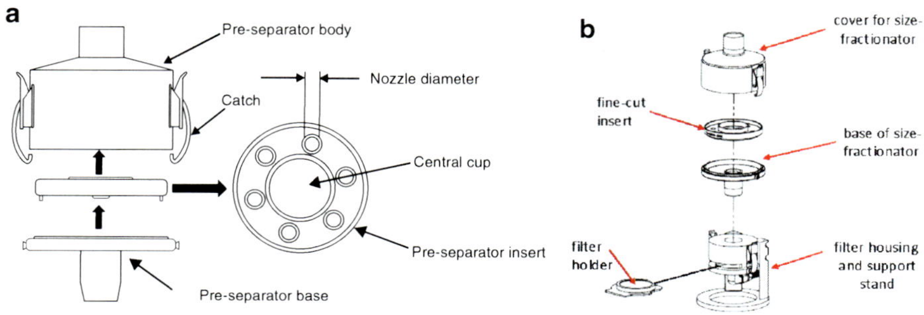

Fig. 10.29 Fast screening impactor (FSI). (**a**) Components. (**b**) Assembly (*Courtesy of MSP Corp., St. Paul, MN*)

Table 10.11 Design and calibration data for the FSI at different flow rates with inserts having a nominal stage d_{50} of 5.0 $_{\mu m}$ (*From* [34]—*courtesy of D.L. Roberts*)

Flow rate (L/min)	Number of nozzles	Size of nozzles (mm)	Nozzle-to-nozzle circumferential distance/nozzle diameter (dimensionless)	$STK^{1/2}$ (critical Stokes number for impaction)	GSD_{stage}
30	6	4.08	11.0	0.49	1.07
35	7	4.08	9.4	0.49	1.10[a]
60	12	3.94	5.7	0.51	1.12
90	9	5.00	5.9	0.51	1.12

[a]Measured subsequent to publication of [34]

from 30 to 100 L/min [37], primarily for DPI testing in accordance with the method given in monograph 2.9.18 of the Ph. Eur.

If greater control of cut point is required and the sampling flow rate from the inhaler can be varied, the use of a fixed insert at different volumetric flow rates (Q) can allow fine adjustment of the boundary for the coarse/fine fraction, as cut-off diameters (d_{50}) shift in accordance with Marple-Liu theory [Eq. (10.2)]. Other inserts can therefore be manufactured to order, providing any desired cut point in the range 1.0–10 μm aerodynamic diameter for measurements at a given flow rate (Roberts DL (2012) MSP Corporation, St. Paul, MN, USA, personal communication).

The first reports issued in late 2009 by a group from Pfizer (Sandwich, UK) [38, 39] described the use of an FSI for DPI studies. To begin with, a commercially available DPI was measured at 70 L/min using a FSI with a 5 μm insert. The NGI APSD measurements were made in accordance with the compendial DPI methodology, which was adapted, in the case of the FSI evaluation, to reflect the simplified information obtainable from this abbreviated system. Results were gathered for the uncoated FSI and again following collection surface coating with a very fine layer of silicone fluid applied as a 1% v/v solution in cyclohexane. Their metrics, $CPF_{>5.0\mu m}$, $FPF_{<5.0\mu m}$, and total impactor recovery (equivalent to IM) were compared with data from a full-resolution NGI, with data from the full-resolution APSD

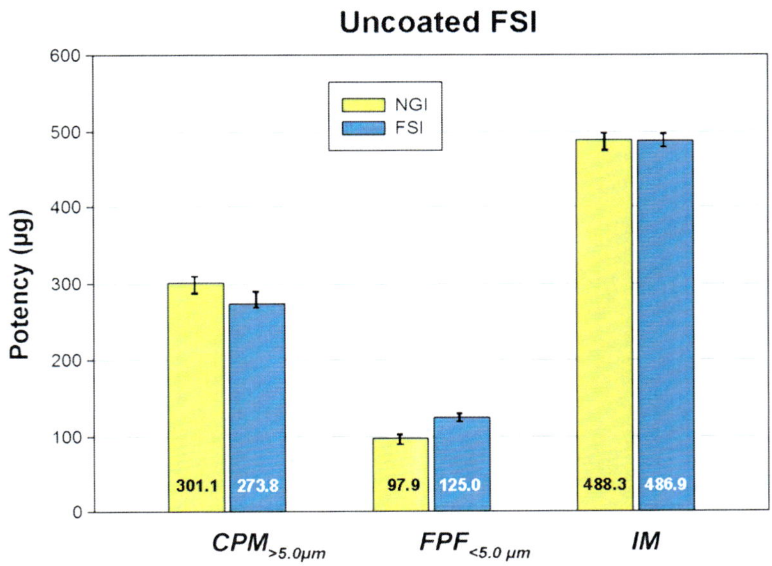

Fig. 10.30 Size-related metrics from FSI with uncoated collection surfaces with equivalent NGI-generated metrics (*From* [38]—*courtesy of D. Russell-Graham*)

interpolated to give an equivalent $FPF_{<5.0\mu m}$ assuming a unimodal and lognormal APSD; since the NGI has no stage with a cut-off diameter located at precisely 5.0 µm aerodynamic diameter. Their metrics obtained with the particle collection surfaces of the FSI uncoated were in reasonably close agreement with those measured using their full-resolution NGI (Fig. 10.30), but there was a small but systemic bias toward higher values of $FPF_{<5.0\mu m}$ with the FSI.

These findings are consistent with particle bounce and re-entrainment and were largely eliminated, using the silicone fluid coating (Fig. 10.31).

Similar trials were carried out with other DPIs at flow rates determined via the standard pharmacopeial method for DPI testing, using an appropriate insert to maintain the 5 µm stage cut-off at each flow rate. These results confirm the initial conclusion that a coated FSI produced $FPF_{<5.0\mu m}$ values that were closely comparable to those obtained using the full-resolution NGI. However, the magnitude of minor discrepancies between the NGI and FSI varied from product to product, serving to highlight the necessity of justifying the use of AIM techniques through suitable method development and comparative testing against a full-resolution instrument, for each inhaler product.

A further study was investigated by this group to assess the value of AIM as a screening tool during early-stage formulation screening [38]. This investigation was structured to simulate a Design of Experiments (DoE) of the type routinely applied during early-stage product development. Here, the goal was to investigate the ability of the FSI to correctly identify trends in FPF resulting from changes in percentage

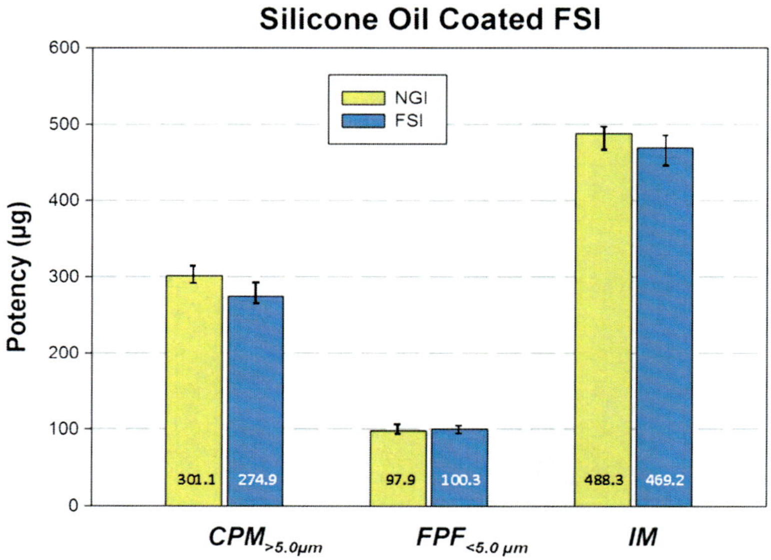

Fig. 10.31 Size-related metrics from FSI with silicone-oil-coated collection surfaces with equivalent NGI-generated metrics (*From* [38]—*courtesy of D. Russell-Graham*)

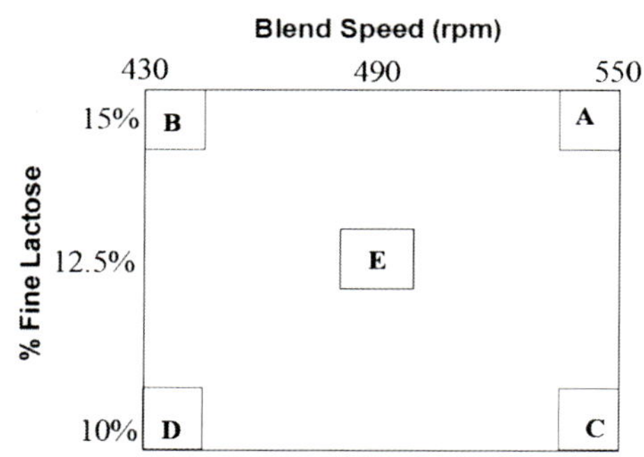

Fig. 10.32 DoE for dry powder blending parameters investigated by FSI (*From* [38]—*courtesy of D. Russell-Graham*)

fine lactose content and blend speed. Five DPI formulations were prepared: A, B, C, and D mark the four extremes of the experimental range defined by high (15%) and low (10%) fine lactose content and high (550 rpm) and low (430 rpm) blend speed, while E lies in the center of the defined space (Fig. 10.32). Theoretically, $FPM_{<5.0\mu m}$ would be expected to increase with increasing fine lactose content and increasing blend speed, excipient content being the dominating effect, giving a ranking for the formulations of A, B, E, C, and D, with A expected to deliver superior performance.

The FSI results correctly identified the influence of excipient fine content (Table 10.12) but failed to indicate any statistical difference between formulations

Table 10.12 Comparison between FSI and NGI for screening of DPI lactose blends based on $FPM_{<5\mu m}$ (*From* [38]—*courtesy of D. Russell-Graham*)

	NGI		FSI	
Blend	Mean (μg)	RSD (%)	Mean (μg)	RSD (%)
A	58.4	4.9	66.6	3.8
B	59.2	4.5	66.2	4.0
C	50.4	13.4	60.9	3.1
D	55.4	3.7	60.6	3.6
E	63.2	6.8	62.8	6.0

produced at different blend speeds. The values of FSI-determined $FPM_{<5.0\mu m}$ for the blends were as expected, according to the predicted order in relation to percentage of fine lactose in each blend.

However, surprisingly, blend speed appeared to have had no effect on this metric. In contrast, corresponding values of $FPM_{<5.0\mu m}$ determined by NGI showed the anticipated differentiation between both blend speed and % lactose fines, except for a higher than expected value of $FPM_{<5.0\mu m}$ for blend E.

In a supplementary study, a commercial DPI device was loaded with capsules of different fill weights to investigate in greater detail the tracking capability of the FSI over a wider variation in fine particle mass ($FPM_{<5.0\mu m}$) than was evident in the preceding study. The tracking ability for this metric was assessed over a far greater magnitude (~90 μg) than expected in product development. A new marketed DPI was filled with capsules of four different fill weights—and therefore different values of $FPM_{<5.0\mu m}$—using a different formulation to that evaluated in the previous studies. Three actuations of each fill weight were delivered into each CI system in accordance with the compendial methodology as previously described.

This head-on comparison for $FPM_{<5.0\mu m}$ by both measurement techniques revealed a near 99% correlation within a wide range of potency (Fig. 10.33). On that basis, it appears that if precautions are taken to eliminate bias from particle bounce, the FSI can track changes in this performance metric as well as the NGI. However, these authors acknowledged at the time that more work would need to be done to compare the tracking ability of their FSI compared with that for the NGI in cases where differences in blend performance are as small as typically found in product development. While $FPM_{<5.0\mu m}$ tracking between the FSI and NGI was the main objective of this study, such an approach would also be useful to test a new inhaler and drug blend in early product development using also $CPF_{>5.0\mu m}$ and total impactor recovery (*TIR*) of API as a supplementary performance metric and system suitability check, respectively.

Aside from the technical contribution made by the Pfizer work, their investigations also provided useful practical information relating to the potential time savings associated with AIM (Table 10.13).

The total time quoted for six FSI measurements was estimated to be about one-third that required for six equivalent analyses with a fully optimized NGI set-up utilizing multiple sets of equipment and analysts. Given that FSI techniques were

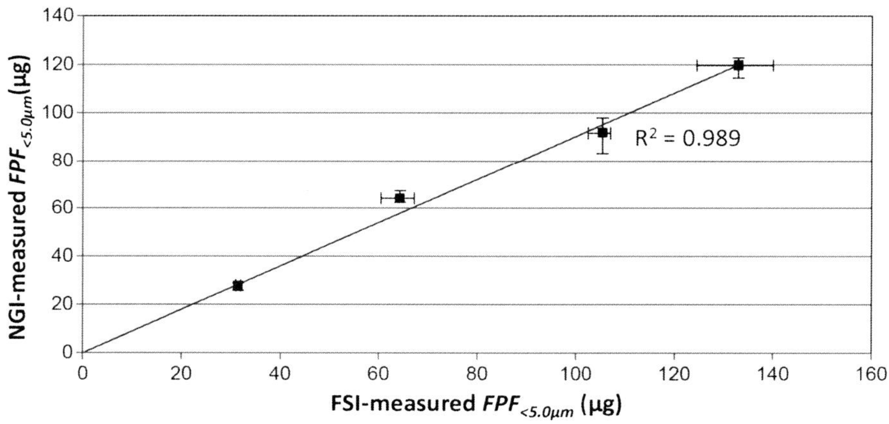

Fig. 10.33 $FPF_{<5.0\mu m}$ comparison by NGI and FSI from supplementary study investigating effect of dry powder blending properties (*From* [38]—*courtesy of D. Russell-Graham*)

Table 10.13 Comparative measurement times (h) between FSI and NGI use in aerosol assessment from 6-DPI replicate measurements (*From* [38]—*courtesy of D. Russell-Graham*)

Impactor	Experimental	Analysis	Data processing	Total
FSI	2.0	0.67	0.33	3.0
NGI	6.0	2.5	0.5	9.0

not yet fully optimized during this assessment, these findings would suggest that, conservatively, time savings of more than 50% are easily achievable using abbreviated systems, a potentially massive gain in productivity.

This group has since extended their investigation of the FSI by comparing it with the NGI across a range of drug products dispensed from different DPIs, exploring the potential impact of differences in pressure drop profile experienced by the device on the data obtained from the abbreviated and full-resolution impactors [40]. The following conditions were evaluated:

1. Product (1) contained two dosage strengths of the same API, each evaluated at 60 L/min (Fig. 10.34a).
2. Products (2) and (3) were each two-component combination DPIs, tested at 60 and 70 L/min, respectively (Fig. 10.34b, c).
3. Product (4) was a DPI having a greater flow rate dependency on API APSD, evaluated at 60 L/min by FSI and at 60 and 90 L/min by NGI (Fig. 10.34d).

It was postulated that the difference in internal volume between the FSI and NGI could cause differences in the pressure drop profile experienced by the product, by varying the characteristics of the time-dependent flow which passes through the impactor when the flow is initiated at the controller. The internal volume of the FSI (estimated to be approx. 960 mL) is substantially smaller than that of the NGI which is close to 1,940 mL, when the pre-separator is included [41].

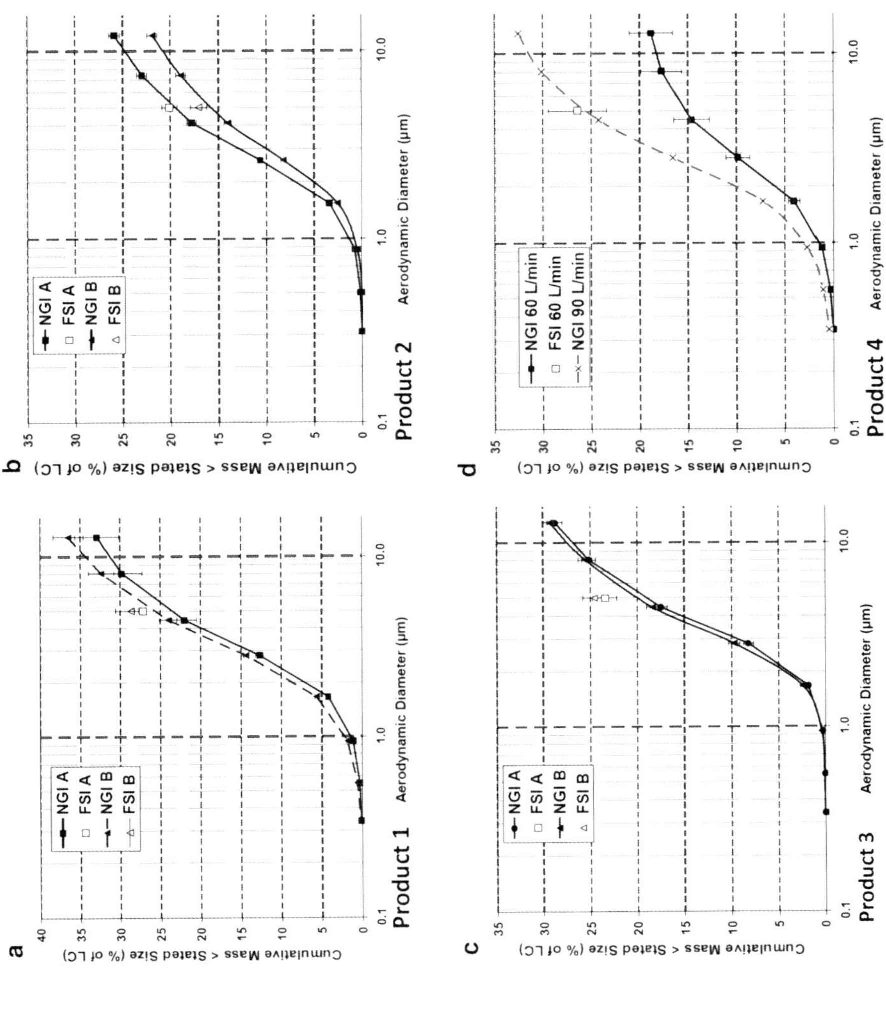

Fig. 10.34 Comparisons between FSI and NGI with different DPI products in follow-on phase of studies by Russell-Graham et al.; the terms A and B in the legends for each plot refer to different (undisclosed) dosage strengths of API (*From* [40]—*courtesy of D. Russell-Graham*)

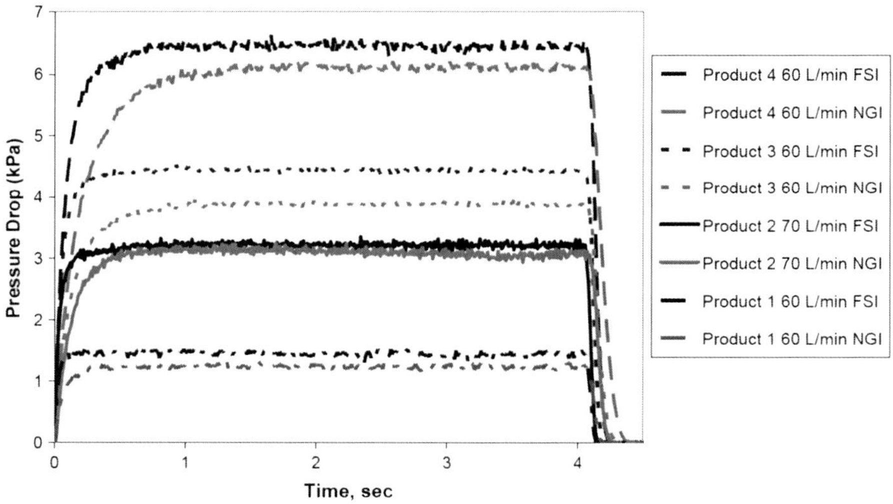

Fig. 10.35 Pressure drop-elapsed time profiles for DPI combinations studied by Russell-Graham et al. (*From* [40]—*courtesy of D. Russell-Graham*)

Their start-up kinetics for each DPI, shown by the time-dependent variation in pressure drop at the mouthpiece for each device-impactor combination (Fig. 10.35), show that the set flow (as measured via pressure drop) is not experienced instantaneously by the product and that the pressure drop profile varied depending on both the device resistance and choice of abbreviated or full-resolution impactor.

The plateau (equilibrium) pressure drop attained, which would be expected to depend on the device resistance, was also observed to differ between the two measurement apparatuses, particularly for Product 3. This was an unexpected observation, but was not thought to be large enough to account for the differences in FPF observed, particularly for Products 1–3 which are not known to be highly flow dependent. An alternative explanation provided by this group involved the observation that the initial acceleration of the pressure drop experienced by each DPI product varied substantially comparing abbreviated with full-resolution apparatus (Fig. 10.36).

In the initial 0.1–0.2 s period, which may be critical for the de-agglomeration of the powder to generate the fine particle fraction delivered from the DPI, the pressure drop measured with the FSI was almost double that measured with the NGI. This effect appeared most pronounced for the product with the highest device resistance (Product 4) which was, perhaps coincidentally, the most flow dependent of the group of devices. This work lends support to the hypothesis that the different internal volume of the FSI may be significant in considering its comparability to the NGI for the testing of DPI products, but further work, including the introduction of additional extra-impactor volume to the FSI apparatus, needs to be done to provide confirmatory data.

More recently, Pantelides et al. have reported further assessments of pressure drop-elapsed time profiles for the FSI, exploring the influence of dead space, systematically increasing the internal volume of the FSI in 500 mL steps by the addition of glass vessels of varying volume and shape configurations (Fig. 10.37), in

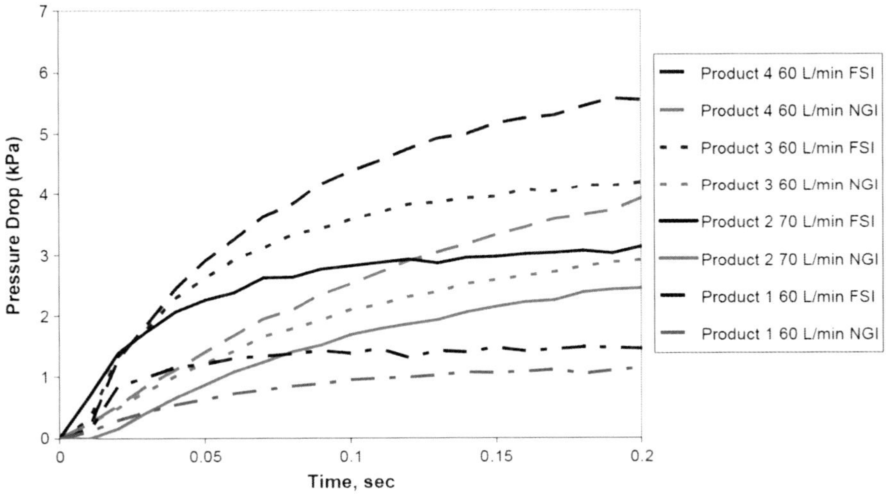

Fig. 10.36 Expanded pressure drop time profiles at the start of measurement for DPI combinations studied by Russell-Graham et al. (*From* [40]—*courtesy of D. Russell-Graham*)

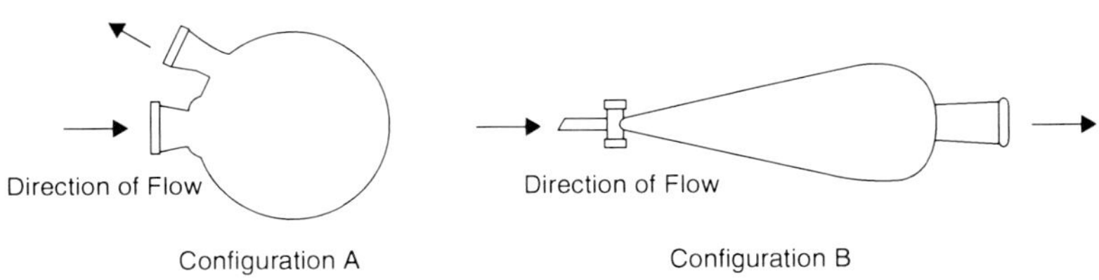

Fig. 10.37 Dead volume configurations used by Pantelides et al.: (A) rapid expansion of flow and (B) more gradual expansion of flow (*From* [42]—*used with permission*)

order to simulate rapid expansion of flow (configuration A) and more gradual expansion (configuration B) [42]. These vessels were located between the FSI and switching valve (Fig. 10.38).

Additionally, the FSI was tested locating a standard NGI pre-separator prior to the modified fine-cut pre-separator stage. Pressure drop profiles were recorded via a pressure tap at the DPI mouthpiece using an inhalation profile recorder similar to the method described by Burnell et al. [43].

The base of the coarse particle collector of the FSI was coated with silicone oil to mitigate particle bounce and re-entrainment as described previously for the work performed with the FSI by this group. The pressure drop profiles for the NGI, FSI, and modified FSI configurations are shown in Fig. 10.39.

The flow initiation portions are superficially similar to those illustrated in Fig. 10.35, taken from the previous study by this Pfizer (UK) group. The new data showed a consistent trend of slowed pressure ramp (acceleration) rate with increasing impactor dead volume in the order FSI (1,045 mL with USP/Ph. Eur. induction port), FSI +500 mL Configuration A, NGI (2,025 mL with pre-separator and USP/

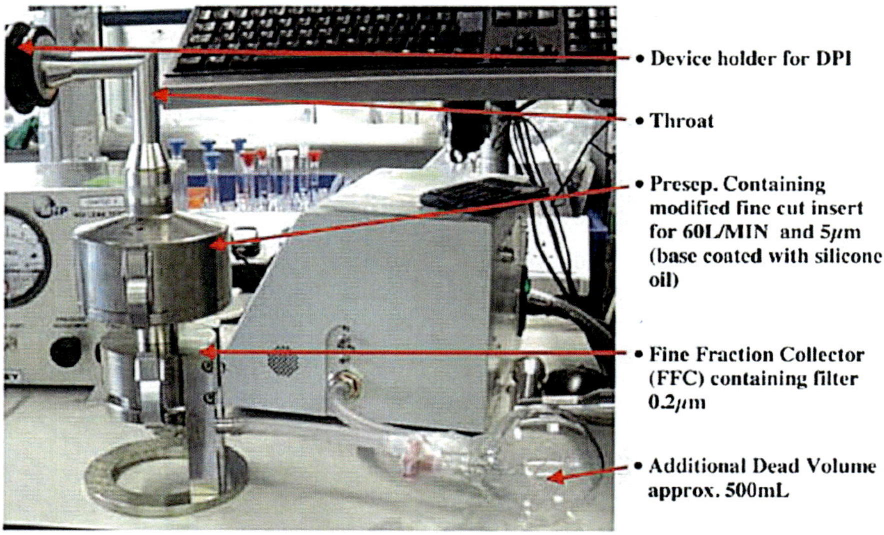

Fig. 10.38 Experimental set-up for the FSI with 500 mL additional dead volume Configuration A investigated by Pantelides et al. (*From* [42]—*used with permission*)

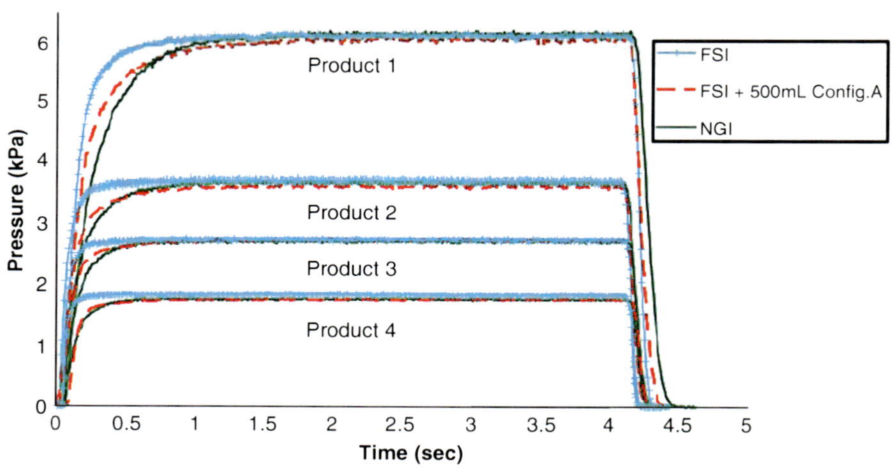

Fig. 10.39 Pressure drop-elapsed time profiles at a nominal flow rate of 60 L/min for DPI products 1–4 (*From* [42]—*used with permission*)

Ph. Eur. induction port [41]) with all of the DPIs. Pressure ramp rate is related to the volume of air which is removed from the impactor before peak pressure drop is attained during the stable flow rate-time portion of the measurement. Importantly, changing the configuration, shape, and size of the extra dead volume added to the FSI made a clear difference to the pressure ramp rate, as illustrated with a series of measurements made with DPI product 1 (Fig. 10.40). Thus Configuration B, that

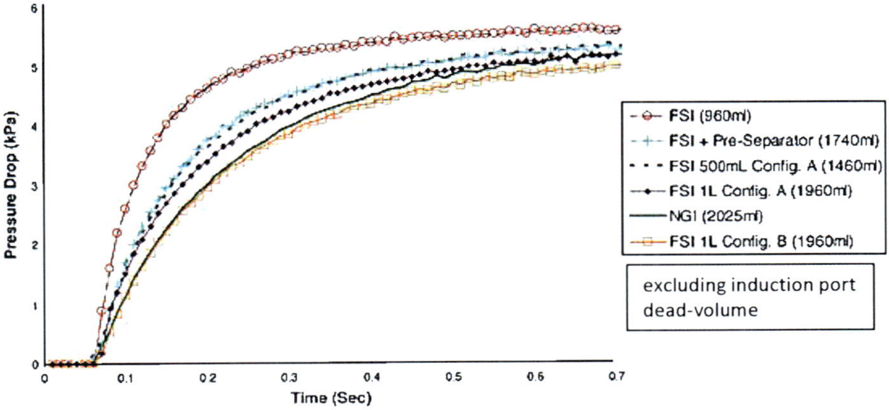

Fig. 10.40 Pressure drop profiles at a nominal flow rate of 60 L/min for DPI product 1 evaluated by Pantelides et al. (*From* [42]—*used with permission*)

allowed a gradual flow expansion on start-up created a FSI-generated pressure drop profile that closely matched that of the NGI. In contrast, when Configuration A that permitted rapid flow expansion was used with the same 1 L volume, the FSI-generated pressure drop profile was observably steeper, reflecting an increased pressure ramp rate, but not as steep as that for the FSI without any added dead volume.

This study showed that if the FSI is equipped with a standard NGI pre-separator (780 mL [41]) before the fine-cut stage, the pressure drop profile was closer to that for the NGI, suggesting that this simple modification to the FSI may be all that is required to achieve comparability for the fluid dynamics for the two systems. However, there was still a significant difference in the pressure drop values (around 1 kPa at about 0.25 s), and this disparity may have been the cause of the significantly higher FPF% shown in Fig. 10.40 for the combined system.

Pantelides et al. completed the investigation by comparing $FPM_{<5.0\mu m}$ per actuation obtained by all the various FSI configurations with reference data (full APSD) obtained for DPI product 1 by the NGI (Fig. 10.41). Increasing the volume of the FSI reduced the magnitude of fine particle mass closer to the value interpolated from the reference APSD. However, dead volume shape and flow resistance of the measurement system [44] likely also had an influence. Thus fine particle mass obtained with the 1-L "Configuration B" added to the FSI was closer to the reference NGI value than the corresponding measure obtained by the FSI when "Configuration A" was used, also adding 1 L to the internal volume of the system. Importantly, such behavior would have been predicted from the relative positions of the pressure drop-elapsed time profiles in Fig. 10.40. These validation measurements added further confirmation that despite the pressure differences mentioned above, the addition of the NGI pre-separator to the FSI improves the correlation between this apparatus and the NGI in terms of fine particle mass.

The small discrepancy in the data shown in Fig. 10.41 between the FSI with pre-separator and NGI was the subject of a follow-on study in which the internal volume of the FSI was augmented to 1,740 mL by the arrangement shown in Fig. 10.42 [45].

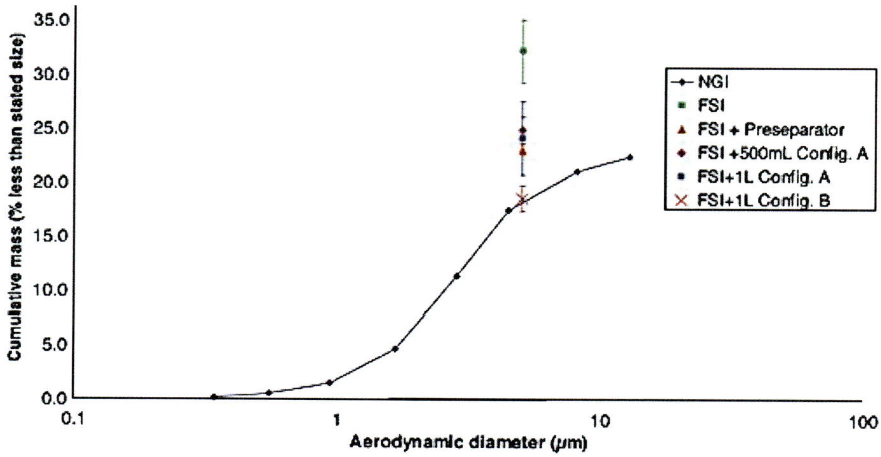

Fig. 10.41 $FPM_{<5.0\mu m}$ for DPI product 1 determined by various FSI configurations in comparison with NGI-measured APSD from measurements of Pantelides et al. (error bars represent ±1 SD for three replicates) (*From* [42]*—used with permission*)

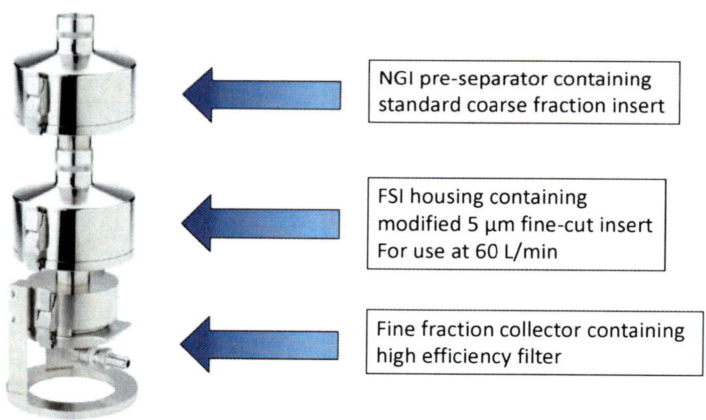

Fig. 10.42 Configuration of FSI for follow-on study by Pantelides et al. matching internal volume more closely to that for the NGI (*From* [45]*—courtesy of D. Russell-Graham*)

The resulting pressure drop-elapsed time profiles for the same DPI product 1 as was evaluated in the previous study are illustrated in Fig. 10.43.

As expected from a comparison of internal dead volumes, the pressure drop-time profile for the volume-enhanced FSI was much closer to that for the NGI. This closer match of dead space resulted in a closer agreement in fine particle mass/actuation between the two systems. Figure 10.44 illustrates the near-to-linear correlation between fine particle mass and air acceleration rate for the FSI, volume-enhanced

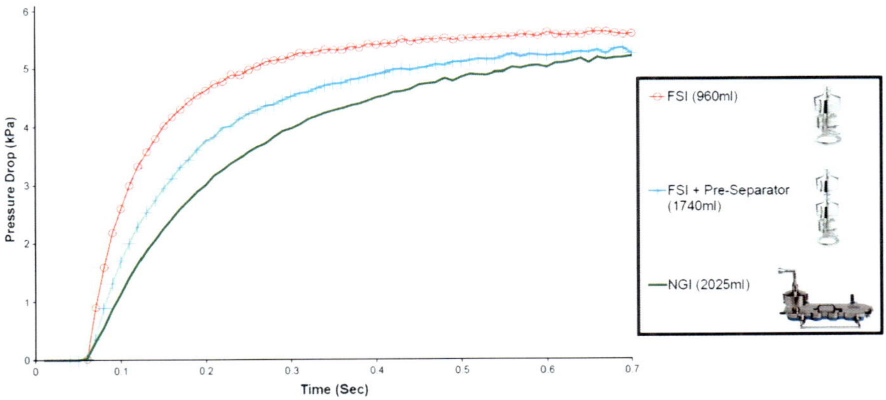

Fig. 10.43 Pressure drop-elapsed time profiles for FSI, FSI with enhanced internal volume and NGI for DPI Product 1 determined by Pantelides et al. at a nominal flow rate of 60 L/min (*From* [45]—*courtesy of D. Russell-Graham*)

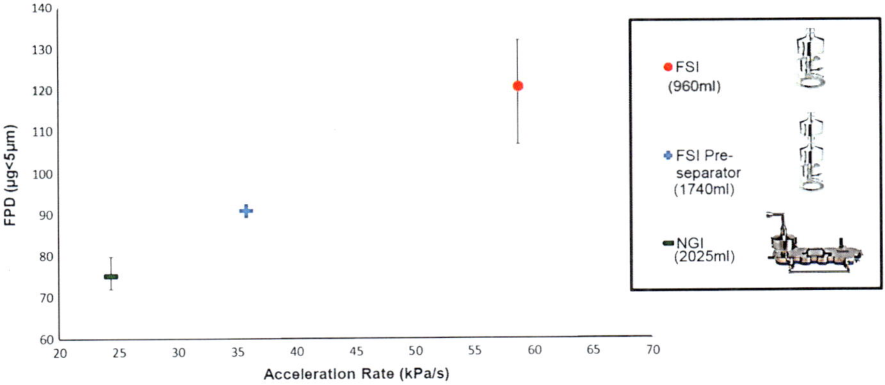

Fig. 10.44 Correlation between $FPM_{<5.0\mu m}$ and air acceleration rate for FSI alone, FSI with enhanced internal volume and NGI for DPI Product 1 determined by Pantelides et al. at a nominal flow rate of 60 L/min (*From* [45]—*courtesy of D. Russell-Graham*)

FSI, and NGI, demonstrating that it is necessary to match internal dead space for the abbreviated CI as closely as practical with that for the corresponding full-resolution impactor for the most accurate work. However, at this early stage in the application of the FSI to the measurement of aerosols from DPIs, it is possible that the effects observed by Pantelides et al. may be DPI type specific. It follows that the addition of dead volume may only be necessary in certain cases following a suitable comparison to a full-resolution impactor.

The basic FSI configuration with 5 µm aerodynamic diameter cut-off between coarse and fine particle fractions has been used by Daniels and Hamilton as part of

Fig. 10.45 Reduced NGI GSK configuration (*From* [46]—*used with permission*)

a comparison between this abbreviated impactor system and that produced by reducing the NGI (rNGI) to an abbreviated system [46].

This modification was undertaken by eliminating collection stages in the NGI so that following the inlet, the coarse-fine size separation took place at the stage 2 location as normal. However, the filter collection stage containing a bespoke filter was relocated to a position directly above stage 3 (Fig. 10.45), and the remaining stages of the NGI were not used (see Sect. 10.7). API was recovered by rinsing Stages 1 and 2 together (representing the *LPM*) and separately from the recovery of API from the Stage 3 filter (representing the *SPM*, i.e., the fine particle mass <4.46 μm aerodynamic diameter).

A DPI simultaneously delivering two components with different *MMAD* values was used for this study and was tested at a nominal flow rate of 60 L/min with a 4-s "inhalation" time. The FSI was operated with a 5 μm insert. All three systems incorporated the same Copley vacuum pump with TPK flow controller. Similar to the Pfizer group studies, a multichannel Inhalation Profile Recorder (GSK) was used to record the pressure drop-elapsed time profiles via the connection of a pressure transducer to a pressure tap located in the throat of the apparatus-on-test.

Values of *SPM* (here equivalent to $FPM_{<4.51\mu m}$) and *LPM* (equivalent to $FPM_{>4.51\mu m}$ ex NGI and rNGI), each expressed as a percentage of the label-claim dose, from the systems are summarized in Fig. 10.46. SPM for the FSI was observed to be between 5% and 10% greater than the equivalent values obtained when either the standard or reduced NGI systems were used. As expected, the converse was observed for the $CPM_{>5.0\mu m}$.

The pressure drop-elapsed time profiles for the three configurations (Fig. 10.47) were almost identical for both NGI and rNGI, as might be anticipated based on their similar internal dead volumes. However, the air acceleration rate during start-up of flow with the FSI was noticeably faster than for the other systems. Daniels and

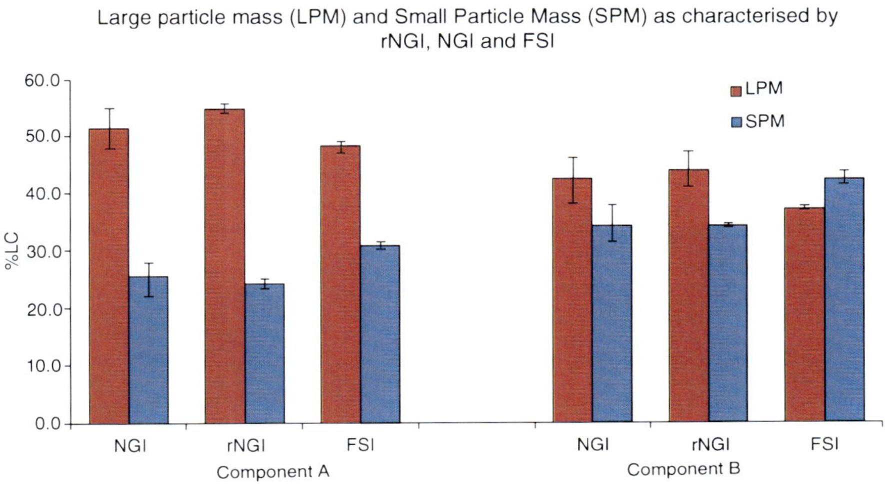

Fig. 10.46 *SPM* (equivalent to $FPM_{<5.0\mu m}$) and *LPM* (equivalent to $FPM_{>5.0\mu m}$) determined for a DPI simultaneously delivering two components with different *MMAD* values evaluated by Daniels and Hamilton at a nominal flow rate of 60 L/min (*From* [46]—*used with permission*)

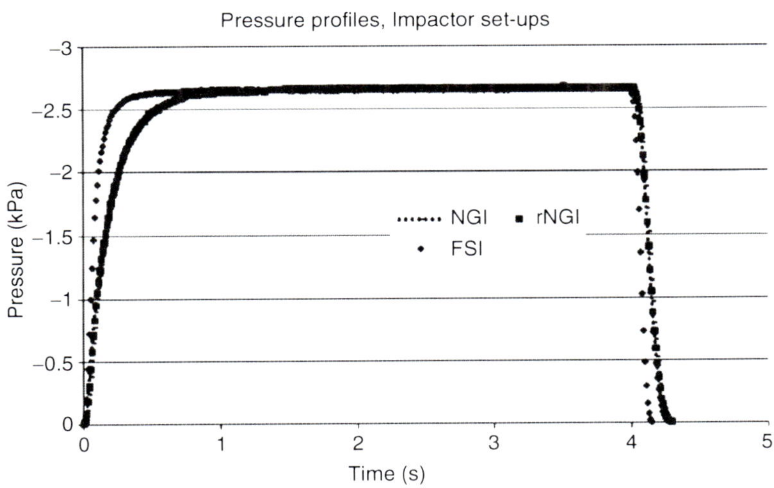

Fig. 10.47 Pressure drop-elapsed time profiles for a DPI simultaneously delivering two components with different *MMAD* values evaluated by Daniels and Hamilton at a nominal flow rate of 60 L/min (*From* [46]—*used with permission*)

Hamilton made the important observation that these differences are likely to be significant in determining the efficiency of the size-fractionation process; and therefore, the ratio of small (fine) to large (coarse) particle mass, because they occur during the critical period of the flow-time profile when powder aerosolization, is

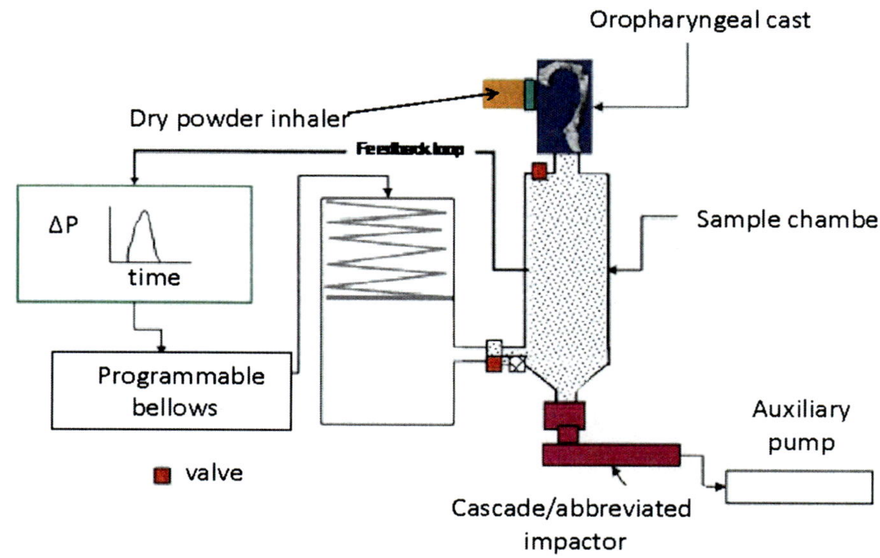

Fig. 10.48 Electronic Lung™ (eLung) set-up for DPI testing by Daniels and Hamilton (*From* [47]*—used with permission*)

most likely to be occurring. This group also noted that the testing time to perform an rNGI measurement and analysis sequence was approximately 2–3 min longer than that to perform similar operations with the FSI.

To confirm the important observations described earlier, in a further study [47], Daniels and Hamilton investigated the significance of the difference in the ramp-up phase of the profiles through replication of two patient representative (asthmatic and COPD) Inhalation profiles using the Electronic Lung™ (eLung) (Fig. 10.48).

In this work, they sampled a proprietary DPI simultaneously delivering two components A and B (Fig. 10.49) having different *MMAD* values, enabling discrimination on the basis of *LPM/SPM* ratio with the boundary still fixed at 5.0 μm. Values of *SPM* and *LPM* for either component (Fig. 10.49) were found to be comparable between the FSI and full-resolution NGI (note their rNGI was not included in this comparison).

This outcome was believed to be a result of eliminating any potential for the flow into the cascade impactor (abbreviated or full resolution) to influence the ramp-up profile of the DPI and therefore its dose emission characteristics. Separation of the flow rate profile controlling dose emission to that of dose characterization confirmed that the data previously reported by this group [46] had arisen as the result of the difference in ramp-up kinetics between the FSI and NGI.

Both the Pfizer and the initial GSK studies demonstrated the importance of matching the internal dead volume of an abbreviated CI with that of the full-resolution CI reference technique for the most accurate results when used in DPI performance evaluations.

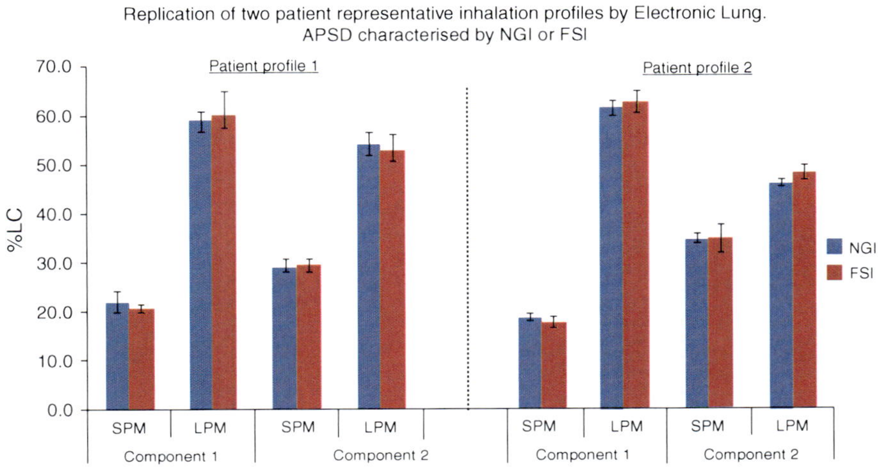

Fig. 10.49 Comparison of NGI and FSI data post replication of human patient profiles by eLung (*From* [47])

Aptar Pharma (Le Vaudreuil, France) also recently reported feasibility studies for the FSI detailing its use for the measurement of another DPI (Prohaler™), featuring their proprietary OBIC™ (open, breathe-in, close) technology [48]. The objectives of their study were to look at several aspects of impaction testing comprising:

1. The performance of the FSI versus the NGI
2. The effect of coating the collection surface of the insert in the FSI
3. The comparison of emitted dose data from the FSI versus a dose unit sampling apparatus (DUSA)—standard equipment recommended by the regulators for the measurement of delivered dose uniformity
4. A comparison of testing parameters for the NGI and FSI
5. Estimation of time and cost advantages of FSI testing versus NGI testing

Comparative tests were carried out at a flow rate of 35 L/min with a 2 L sample volume using the FSI equipped with a 5 μm stage cut-off and an NGI, with three actuations of the device in each case. With three actuations of the DPI the FSI determined slightly higher values of both total emitted dose (equivalent to *TEM*) and fine particle dose <5 μm (equivalent to $FPM_{<5\mu m}$) than did the NGI (Fig. 10.50).

Mean values of *TEM* and $FPM_{<5.0\mu m}$ with a single actuation of the DPI were both around 10% (*TEM*) to 30% ($FPM_{<5.0\mu m}$) higher by FSI compared with the corresponding data from the NGI. However, the 3-dose data for the FSI were in much closer agreement with the corresponding benchmark NGI metrics, being only 1% (*TEM*) and 15% ($FPM_{<5.0\mu m}$) higher than the corresponding NGI values. It was hypothesized that the observed differences were due to increased particle bounce and entrainment when single doses were measured, but collection surface coating

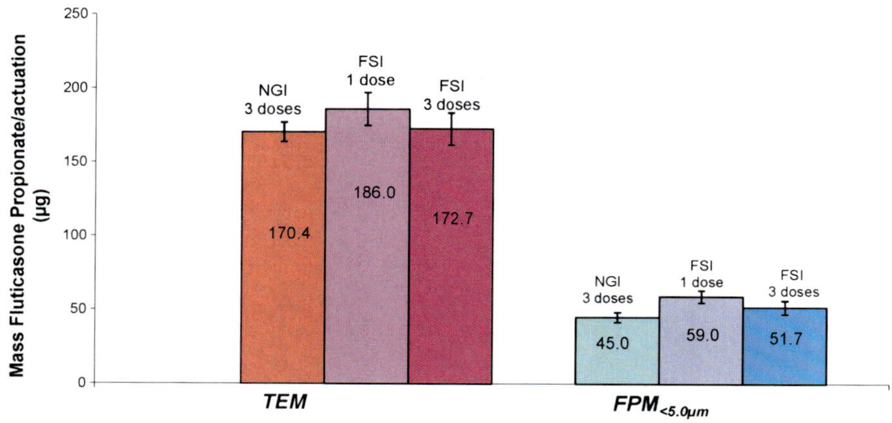

Fig. 10.50 Comparison between NGI and FSI for measures of *TEM* and $FPM_{<5.0\mu m}$ reported by Després-Gnis and Williams (error bars correspond to ±1 SD) (*From* [48]—*courtesy of G. Williams*)

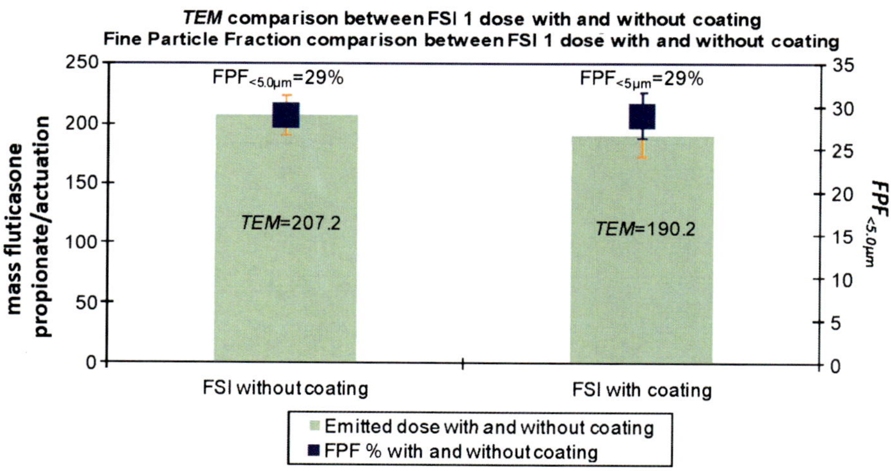

Fig. 10.51 FSI performance with and without stage coating reported by Després-Gnis and Williams (error bars correspond to ±1 SD) (*From* [48]—*courtesy of G. Williams*)

proved unnecessary with a single actuation of the DPI, based on comparative tests showing no statistical difference between *TEM* and $FPF_{<5.0\mu m}$ measured with and without the application of a standard stage coating prepared from a solution of 1% v/v glycerol in ethanol (Fig. 10.51).

Nevertheless, even though these differences were not statistically significant, mean *TEM* with coating (207.2 µg/actuation) was numerically greater than the corresponding value uncoated (190.2 µg/actuation) suggesting that further examination of the stage coating methodology might still be beneficial, possibly resorting

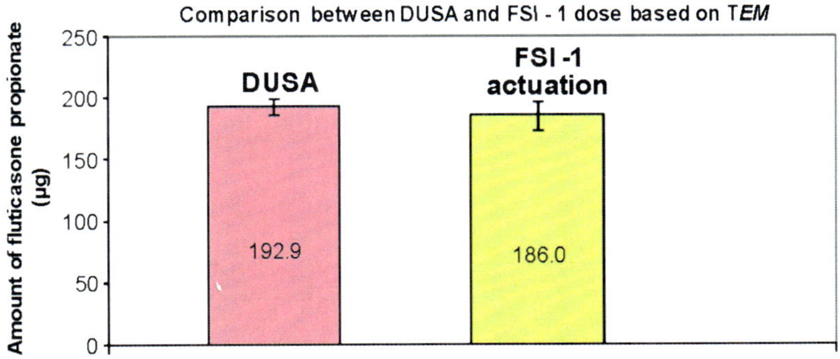

Fig. 10.52 Comparison between FSI and DUSA reported by Després-Gnis and Williams (single dose data) (*From* [48]—*courtesy of G. Williams*)

Table 10.14 Comparison between FSI and NGI in terms of savings in time per measurement, HPLC-based analysis time, and API recovery solvent use (*From* [48]—*courtesy of G. Williams*)

Apparatus	Test duration (min)	API assay duration (min)	Solvent quantity (mL)
NGI	60	140	300
FSI	25	63	200
Gain (%)	**58**	**55**	**33**

to a filter saturated with coating medium, as was done in the follow-on IPAC-RS precision study to eliminate entirely bias from this source [27].

In a follow-on study, a dose uniformity sampling apparatus (DUSA, Copley Scientific Ltd., Nottingham, UK) sampling at 35 L/min with a 4 kPa pressure differential and inhalation volume of 2 L was used to compare with single dose FSI measurements for the same DPI. There was close agreement between the single actuation *TEM* results and those measured with the DUSA (Fig. 10.52).

Taken together, these results suggest that using the FSI with just a single actuation is a suitable approach for *TEM* and $FPM_{<5.0\mu m}$ measurement in support of screening during development work, provided precautions are taken to mitigate bias due to particle bounce and re-entrainment by coating the collection surface of the insert.

The findings of Després-Gnis and Williams support the outcome reported by Stobbs et al. [38], in terms of time savings, suggesting gains in excess of 50% may be possible in both test duration and HPLC analysis time. Reductions in solvent usage were also quantified at 33%, important in terms of the increasing attention being paid to so-called green chemistry initiatives (Table 10.14).

Finally, additional data relating specifically to the test conditions they employed, demonstrated that the sealing integrity of the FSI is comparable to that of the NGI, while the overall pressure drop was considerably lower (Table 10.15). Interstage wall losses were also reduced with the FSI, possibly as the result of needing fewer inhaler actuations with this system. It is important to note, however, that the lower pressure drop of the FSI merits careful consideration in the context of achieving

Table 10.15 Sealing integrity, apparatus air resistance and interstage API loss for the FSI compared with the NGI (*From* [48]—*courtesy of G. Williams*)

(a) Sealing integrity ($\Delta_{\Delta P}$, kPa)

FSI ($n=5$ replicates)		NGI ($n=4$ replicates)	
Mean	0.3	Mean	0.4
SD	0.07	SD	0.04

(b) Equipment air resistance ($\text{cm}_{H_2O}{}^{1/2}$ L / min)[a]

FSI ($n=3$ replicates)		NGI ($n=3$ replicates)	
Mean	0.12	Mean	0.19
SD	0.00	SD	0.00

(c) Interstage API loss [wall losses] (μg)

FSI: 1 actuation ($n=3$ replicates)		NGI: 3 actuations ($n=3$ replicates)	
Mean	0.6	Mean	1.7
SD	0.33	SD	1.03

n number of replicates
[a]Sealing integrity verified prior to use

comparable start-up kinetics (pressure drop versus elapsed time) for DPI testing in general, reinforcing the advice that such considerations should be a key part of AIM-based method development with this class of OIP.

Rogueda et al., from Novartis Pharma, also presented a comparison of an FSI with insert providing a cut point of 5 μm aerodynamic diameter at 90 L/min with the NGI as full-resolution reference impactor [49]. Their data for various low resistance DPI products used with lactose-blended powders also confirmed that fine particle fraction determined by the FSI could be up to 20% higher $FPF_{<5.0\mu m}$ than equivalent measures using the NGI (see Fig. 10.53 for DPI "system" C) when the pre-separator of both devices were used without coating the interior surfaces with an adhesive agent (i.e., as it would normally be used with the NGI).

Coating the bottom plate of the pre-separator with surfactant in order to reduce particle bounce resulted in a discernible decrease in $FPF_{<5.0\mu m}$ with DPI product "C" (Fig. 10.53), with this value almost comparable to that obtained by the reference NGI (Fig. 10.54). However, the FSI-based data for other products "A" and "B" evaluated were still marginally lower and higher, respectively, than corresponding values determined from the NGI (Fig. 10.54). Nevertheless, these differences compared with equivalent values using the NGI were less than the 10% limit that they set as their acceptance criterion for these measurements.

This group also reported between 35% and 42% savings in time using the FSI compared with the NGI, based on timed measurements with other similar DPIs, corresponding to between 22% and 37% time saved on the experimental work (HPLC analysis excluded).

Confirmatory evidence of the value of the FSI for measuring particles delivered by pMDIs and nebulizers was also provided in early 2009 by Sheng et al. from MAP Pharmaceuticals Inc. (California, USA) [50]. In the first part of their study, two

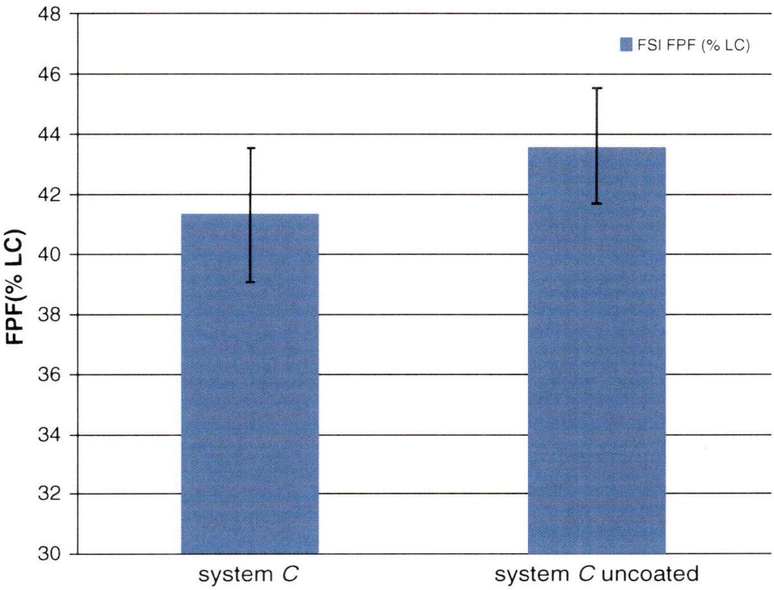

Fig. 10.53 $FPF_{<5.0\mu m}$ measured by FSI with uncoated and surfactant-coated pre-separator bottom plate for DPI product (system) "C"; error bars represent ±1 SD (*From* [49]—*courtesy of P. Rogueda*)

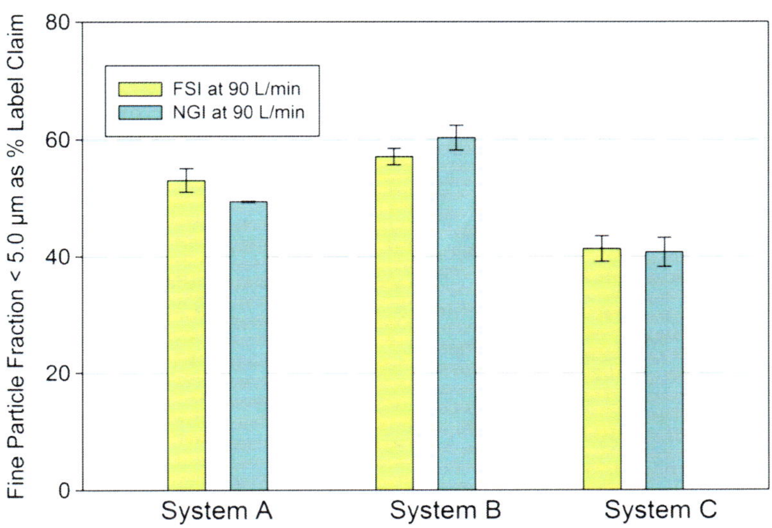

Fig. 10.54 Measurement time comparison between FSI and NGI with coated FSI pre-separator (90 L/min flow rate, 4 L total volume, $n = 5$ replicate measurements by one operator); error bars represent ±1 SD (*From* [49]—*courtesy of P. Rogueda*)

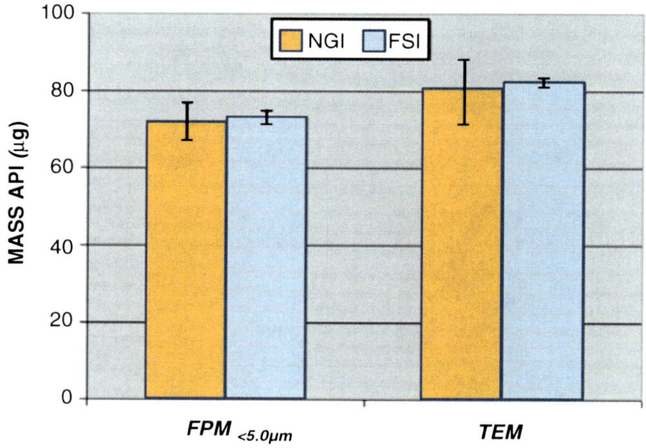

Fig. 10.55 Comparison of MDI formulation using NGI and FSI reported by Sheng et al. based on $FPM_{<5.0\mu m}$ and *TEM*; error bars represent ±1 SD (*From* [50]—*used with permission*)

proprietary budesonide HFA suspension pMDI formulations were measured at a flow rate of 30 L/min, using the FSI, also with 5 µm cut-off for *FPM*, using the NGI as the reference apparatus. The results (Fig. 10.55—one product) were compared in terms of fine particle dose ($FPM_{<5.0\mu m}$) and total emitted mass (*TEM*) using a two-side *t*-test on the basis of the null hypothesis $FPM_{5.0\mu m\text{-}NGI} = FPM_{5.0\mu m\text{-}FSI}$ and $TEM_{NGI} = TEM_{FSI}$. No statistical difference was observable between these two sets of data ($p > 0.05$).

In the second part of their investigation, four proprietary budesonide liquid suspension formulations (A, B, C, and D) containing particles of different morphologies and concentrations were nebulized with an Aeroneb Go® (Aerogen Ltd., Ireland) vibrating membrane system ($n = 3$ replicates at each condition). The aerosol passed through an Aeroneb Go®/NGI induction port mouthpiece adaptor prior to reaching the impactor.

A flow rate of 15 L/min has been recommended for nebulizer testing [51]. However, since at that time, there was no commercially available option to operate the FSI at flow rates <30 L/min (*but see method adopted by Tservistas et al.* [52] *later in this section*), Sheng et al. operated their FSI at this lower limit in this investigation. The FSI yielded similar results for $FPM_{<5.0\mu m}$ and *TEM* to those of NGI for all formulations (Fig. 10.56). Importantly, the values of *TEM* encompassed a wide dosage range of ca. 80 µg, but as with the pMDI-aerosol measurements, the difference between the two measurement techniques for each formulation was not statistically significant for each formulation ($p > 0.05$).

In 2010, Sheng and Watanabe extended the original work to evaluate the FSI as a tool for the rapid screening of pMDI-based formulations in early-stage OIP development [53]. They confirmed agreement ($p > 0.05$) between measures of $FPM_{<5.0\mu m}$ determined by both FSI and NGI during the screening of pMDI formulations (Fig. 10.57).

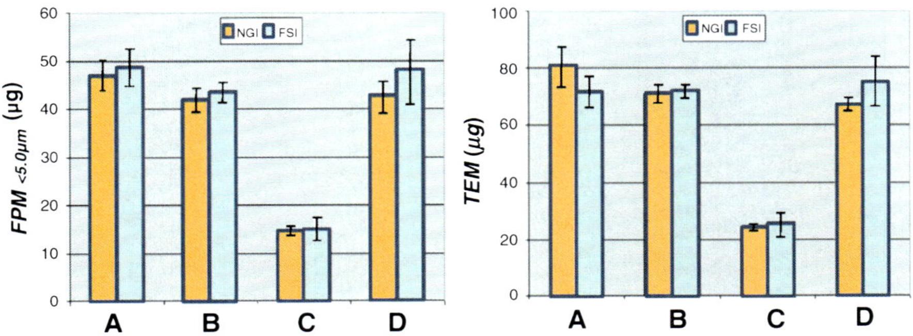

Fig. 10.56 Comparison of four nebulizer-generated budesonide aerosols based on $FPM_{<5.0\mu m}$ and *TEM* reported by Sheng et al. using NGI and FSI systems; error bars represent ±1 SD (*From* [50]*—used with permission*)

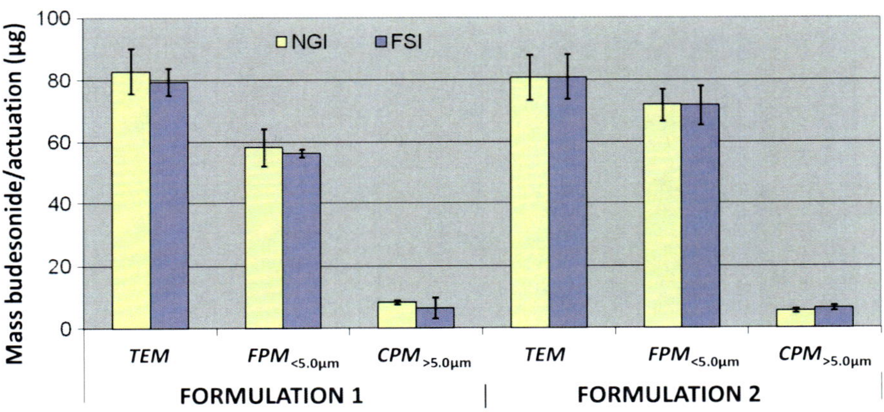

Fig. 10.57 Comparison of *TEM*, $FPM_{<5.0\mu m}$ and $CPM_{>5.0\mu m}$ from two different pMDI formulations measured by Sheng and Watanabe for NGI and FSI operated at 30 L/min; error bars represent ±1 SD ($n=3$ replicates for each impactor) (*From* [53]*—courtesy of G. Sheng*)

In several instances, they also noted slightly lower variability in measurements by their FSI, although further replicate experiments would be needed to confirm this finding.

Interestingly, in a comparison of their pMDI data measured at 30, 45, and 65 L/min [private communication post Workshop], Sheng and Watanabe also included the ACI as a reference impactor. Values of $FPF_{<5.0\mu m}$ by FSI were generally higher than their equivalent from the ACI, with the greatest divergence at the highest flow rate (Fig. 10.58).

In the second part of their study, they compared $FPF_{<5.0\mu m}$ for eight preparations for nebulization containing suspended submicron and super-micron-sized budesonide particles having different morphologies, formulated with and without

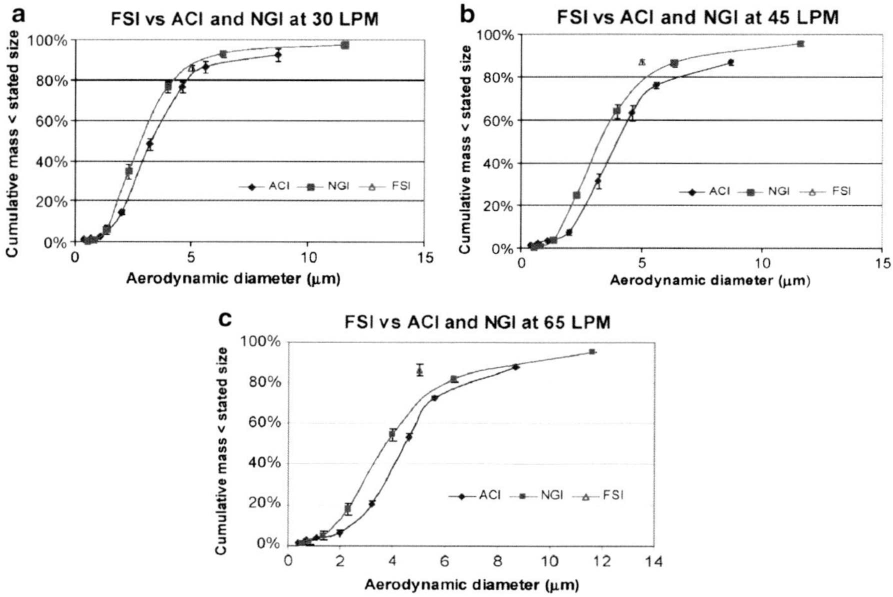

Fig. 10.58 Comparisons of $FPF_{<5.0\mu m}$ reported by Sheng and Watanabe from a pMDI-produced aerosol measured by FSI, NGI, and ACI at different flow rates; error bars represent ±1 SD. (**a**) 30 L/min. (**b**) 45 L/min. (**c**) 65 L/min (*From* [53]—*courtesy of G. Sheng*)

Table 10.16 Nebulizer formulations tested in the study of Sheng and Watanabe; "+" high, "0" medium, "–" low value relative to series average (*From* [53]—*courtesy of G. Sheng*)

Formulation code	Particle morphology based on aspect ratio (–, +)	Surfactant type (0, I, II)/ concentration (+, –)	API concentration (–, 0, +)	Size of particles in formulation
A	–	I/+	+	Submicron
B	–	I/+	+	Submicron
C	–	I/– and II/–	–	Submicron
D	–	I/0 and II/–	0	Submicron
E	–	I/+	+	Submicron
F	+	I/+	+	Micron
G	–	I/+	+	Micron
H	–	I/+	+	Micron
I	+	I/+	+	Micron

surfactant (Table 10.16). Their results (Fig. 10.59), obtained at higher flow rates than recommended (15 L/min) for evaluating this class of inhaler (28–30 L/min), nevertheless demonstrated an excellent correlation ($r^2 \sim 0.97$) between these abbreviated and full-resolution apparatuses.

Sheng and Watanabe also extended their comparison of the FSI to the evaluation of a vibrating membrane nebulizer (Aeroneb® Go, Aerogen Ltd., Galway, Ireland)

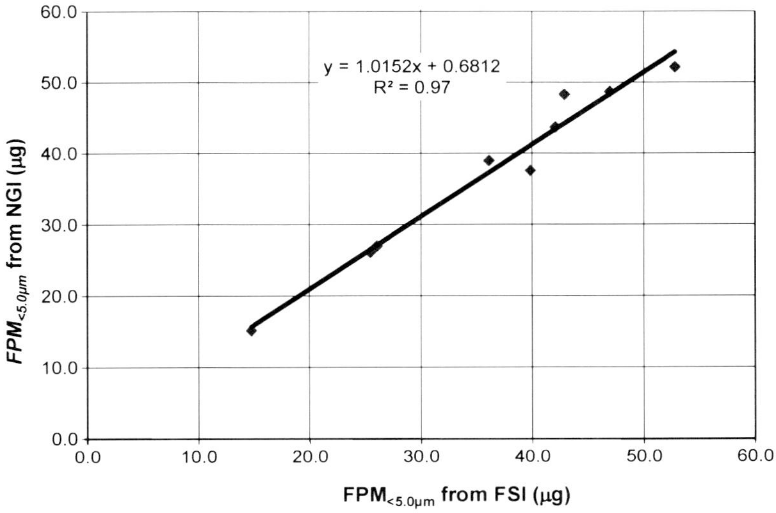

Fig. 10.59 Correlation between $FPM_{<5.0\mu m}$ obtained from NGI and FSI reported by Sheng and Watanabe from 0.5 mL aliquots of different nebulized budesonide preparations (Table 10.16) (*From* [53]—*courtesy of G. Sheng*)

that was operated with a 0.5 mL fill of a proprietary corticosteroid formulation in aqueous suspension for inhalation.

Comparative measurements with an NGI ($n=6$ replicates) and FSI ($n=5$ replicates) undertaken at both 28.3 L/min (Fig. 10.60) and the preferred flow rate of 15 L/min for nebulizer testing (Fig. 10.61) revealed similar values of $FPF_{<5.0\mu m}$, with slightly less variability associated with the FSI data compared with the NGI at the higher flow rate. However, it should be noted that neither the NGI nor the FSI were chilled for this work, as their preliminary studies had shown this precaution to prevent heat transfer-related evaporative loss with the aerosol droplets was not warranted with the Aeroneb® Go system. Furthermore, Dennis et al. had shown in a comparative study with the Aeroneb Go® and a jet nebulizer (MistyMax™, Cardinal Health, USA) that bias in measures of MMAD, GSD, and fine droplet fraction from not chilling the NGI were relatively small (<10% difference between measurements made with this CI at room ambient and chilled to +5 °C) [54]. It is possible, therefore, that NGI chilling may only be needed for the most accurate measurement, perhaps depending upon the formulation concerned [55, 56], and the much lower thermal mass of the FSI is likely to make this precaution even less necessary in routine work.

However, droplet evaporation can be a significant concern when applying the CI measurement technique to the measurement of aerosols produced by jet nebulizers, especially those devices that do not entrain ambient air into the nebulized droplet stream. The considerable heat capacity of the impactor, especially when the NGI is

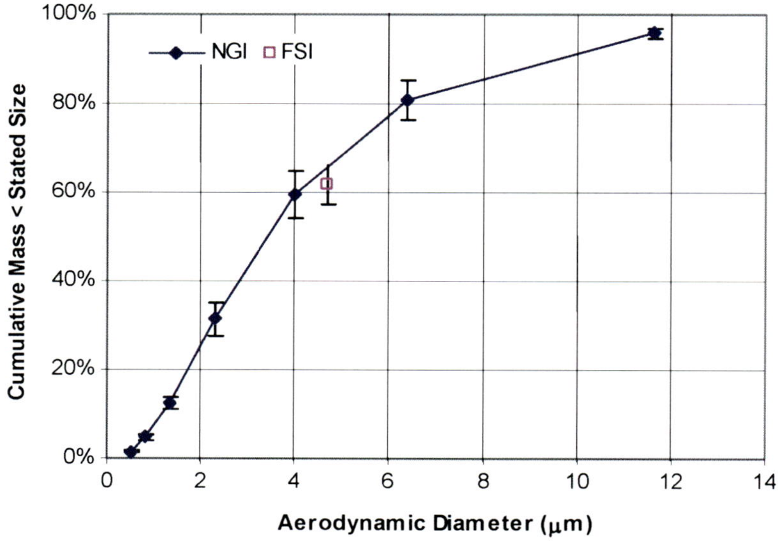

Fig. 10.60 Comparison between FSI and NGI for the assessment of vibrating mesh and jet nebulizer-generated droplets of a 0.5 mL sample of an aqueous suspension product measured at 28.3 L/min by Sheng and Watanabe (*From* [53]—*courtesy of G. Sheng*)

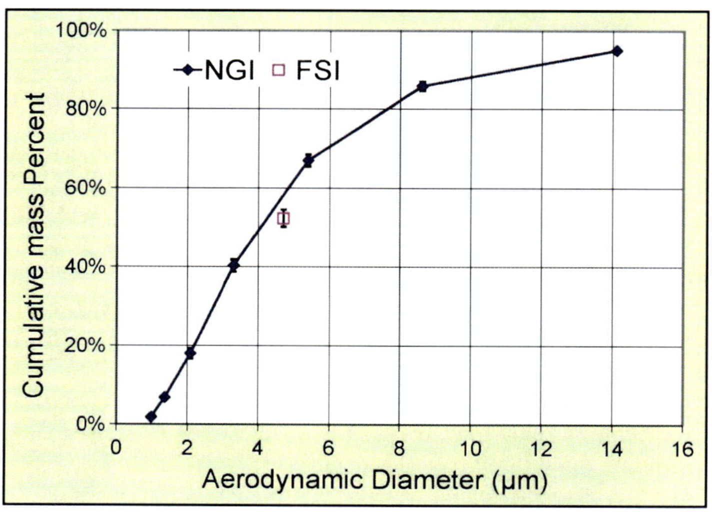

Fig. 10.61 Comparison between FSI and NGI for the assessment of vibrating mesh and jet nebulizer-generated droplets of an aqueous suspension product measured at 15 L/min by Sheng and Watanabe (*From* [53]—*used with permission*)

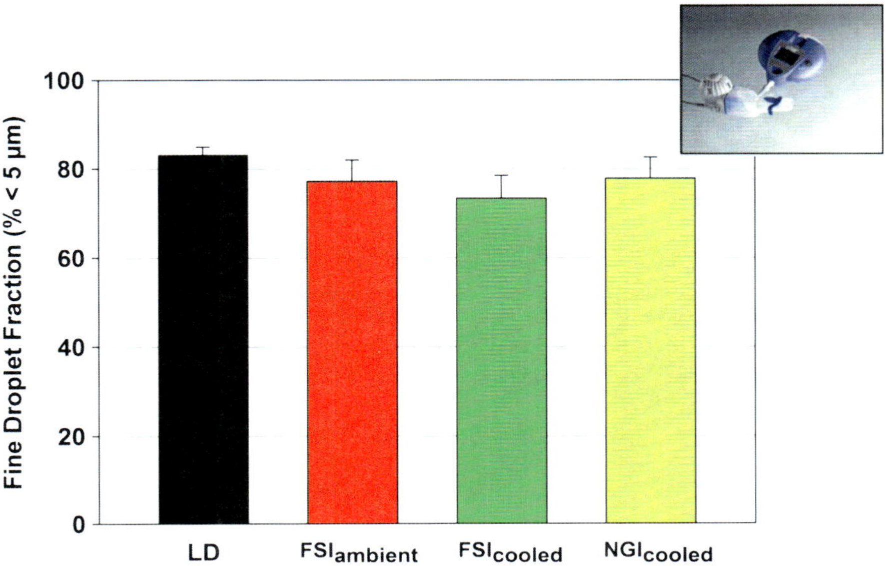

Fig. 10.62 $FPF_{<5.0\mu m}$ (reported as fine droplet fraction, $FDF_{<5.0\mu m}$) by Tservistas et al. for an aqueous aerosol generated by a vibrating membrane nebulizer using FSI, NGI, and laser diffractometry (LD) as measurement methods (*From* [52]—*courtesy of M. Tservistas*)

used, promotes this undesirable effect. The precise mechanism of heat transfer to the aerosol is uncertain; however, in one explanation, water evaporation takes place by heat transfer from the mainly metallic CI as the aerosol passes through the equipment, resulting in undersizing of the formulation [57]. Evaporative changes originating from this cause are routinely avoided by cooling the CI before and/or during measurement [54].

In a feasibility study designed to assess the use of the FSI for nebulizers, Tservistas et al. [52] took a similar approach, comparing fine droplet fraction ($FDF_{<5.0\mu m}$) measured with a cooled FSI (down to 18 °C) from an aqueous formulation delivered by an investigational e-Flow® vibrating mesh nebulizer (PARI GmbH, Starnberg, Germany) with those recorded under ambient conditions at 22 °C (Fig. 10.62). Interestingly, they extended the lower flow rate range of their FSI to 15 L/min by blocking three of the six nozzles on the FSI insert to retain the stage cut-off diameter of 5 µm aerodynamic diameter at 50% of the design flow rate for this AIM apparatus (Fig. 10.63).

Under ambient conditions, the FSI-generated $FDF_{<5.0\mu m}$ values at 15 L/min were substantially equivalent to those produced using a cooled NGI. The cooled FSI produced a lower $FDF_{<5.0\mu m}$, although the difference was relatively small, less than 5 per cent. The highest $FDF_{<5.0\mu m}$ was obtained by laser diffractometry (LD).

These results suggest that the lower thermal capacity of the FSI, a function of its much reduced mass relative to the NGI, is advantageous in terms of accuracy

Fig. 10.63 Modified FSI insert used by Tservistas et al. for nebulizer aerosol measurements at 15 L/min (*From* [52]—*courtesy of M. Tservistas*)

Table 10.17 Comparative times for 6-replicate abbreviated and full-resolution impactor measurements compared with laser diffractometry reported by Tservistas et al. in the context of nebulizer aerosol assessment (*From* [52]—*courtesy of M. Tservistas*)

	Time allocation (h)			
Assessment method	Measurement	API analysis	Data processing	Total
LD	1.5	N/A	1.0	2.5
FSI (cooled)	7.0	6.5	1.0	14.5
FSI (ambient)	5.0	6.5	1.0	12.5
NGI	8.0	18.0	2.0	28.0

associated with nebulizer-generated aqueous droplet size measurements, as the result of less pronounced heat transfer-related evaporative bias.

Comparative data for time savings provided by Tservistas et al. (Table 10.17) add further evidence of approximately 50% gain in productivity with the cooled FSI compared with the NGI.

As might be expected, LD was by far the most rapid technique; however, this measurement method is not directly related to mass of API and is therefore limited to the assessment of solutions, rather than more complex formulations (i.e., suspensions, emulsions, liposomal forms, etc.) that may be encountered with this class of OIP [58].

10.8 Assessing the Performance of AIM Systems Based on the NGI

Since the NGI in its standard design does not have movable stages [14], the development of an abbreviated system based on NGI geometry at first sight seems to be unattractive and possibly a difficult process. However, this CI has a horizontal configuration that lends itself to semi- or full automation more readily than the vertical stack configuration associated with ACI designs, and the process of abbreviating the full-resolution configuration is not as daunting as might be considered at first sight.

Two distinctly different approaches are feasible; however, there is relatively limited data on either reported thus far in the open literature. The most straightforward method involves the use of deep cups to make certain stages inoperable; particles fail to impact on the collection surface and simply pass to the next stage. In addition, an insert can be fitted to the stage 1 nozzle to reduce jet diameter to give a desirable stage cut point. However, since the internal flow pathway through the NGI is not reduced, the so-called deep-cup approach has the potential drawback that losses to the internal surfaces may increase. Future studies validating this option will therefore need to address this concern.

In the second, more radical approach to the abbreviation of the full-resolution NGI (Fig. 10.64a) first adopted by Svensson and Berg [59], the NGI itself was abbreviated by moving "active" stages followed by the back-up filter so that the flow passed through these components before being returned to the NGI body. Daniels and Hamilton also adopted this arrangement for their reduced NGI (abbreviated to rNGI)-based studies [46], and their data have already been discussed in the previous section in connection with understanding how internal dead space affects start-up flow characteristics in DPI testing with the FSI. Their particular rNGI set-up involved moving the filter collection stage containing a bespoke filter to follow directly before full-resolution NGI stage 3, where separation of fine from coarse subfractions takes place (Fig. 10.45). This change could be made without altering the type of collection cup used or making penetrations through the body of the NGI itself to remove flow from unused stages. Svensson and Berg termed this apparatus the "internal filter" configuration (Fig. 10.64b).

The rNGI approach adopted by Daniels and Hamilton [46] overcomes the need for specialized collection cups, as would be the case if external connections had needed to be made between components to achieve an abbreviated design, as originally proposed by Svensson and Berg in one of their configurations (Fig. 10.64c). Daniels and Hamilton [46] recovered the deposited API by rinsing stages 1 and 2 together (representing the *LPM*) and separately from the recovery of API from the bespoke filter located before stage 3 (representing the *SPM*, i.e., $FPM_{<4.46\mu m}$). The comparative data for their particular DPI obtained with this rNGI configuration compared favorably with measurements by full NGI. However, the comparison of data from either of the two NGI configurations and an FSI was less good (Fig. 10.46). After their follow-up study using the eLung™ to replicate patient inhalation profiles (Fig. 10.48), this relatively poor agreement was attributed to the possibility that the original FSI configuration in its simpler set-up (constant Q of 60 L/min with a 4-s "inhalation" time) may have affected the ramp-up profile of the DPI.

The ability to move the cut size between large and small particle fractions by insertion of the filter stage in different stage positions in the rNGI, so that it is close to the MMAD of the OIP of interest, is a significant advantage for product QC applications. Svensson and Berg evaluated three different stage positions of the filter in

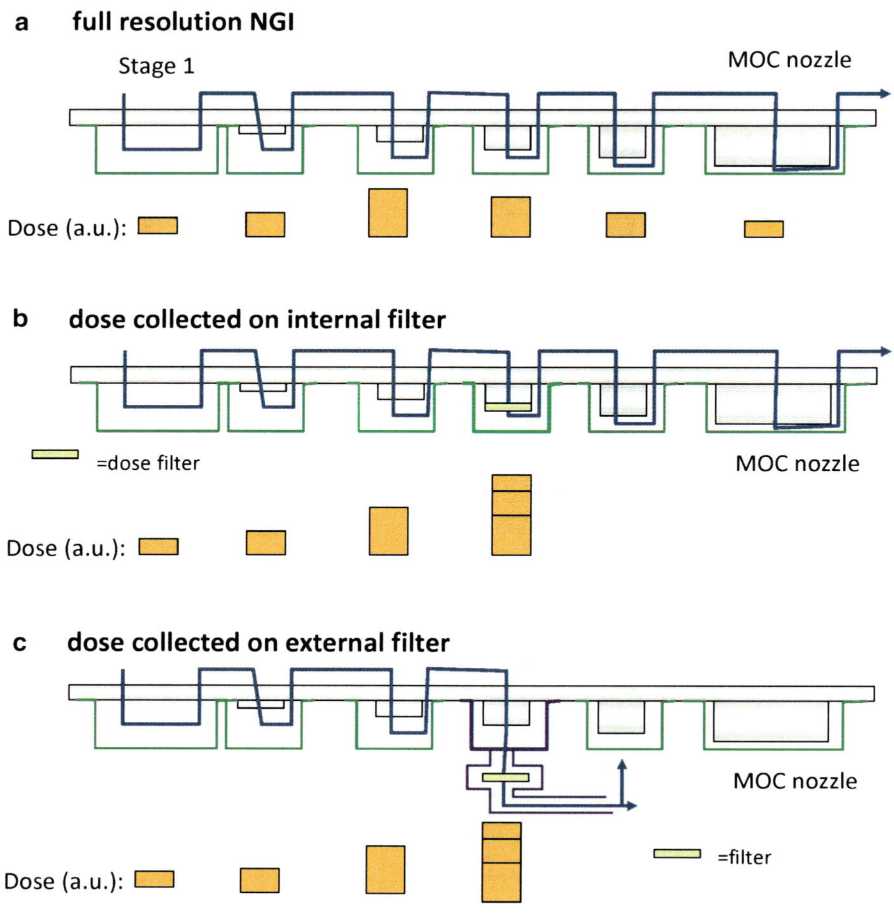

Fig. 10.64 Three concepts developed by Svensson and Berg for creating abbreviated versions of the NGI (rNGI) (*From* [59]—*courtesy of M. Svensson*)

their investigations of the rNGI [59]. Their "internal filter" or "nozzle-filter" design (Fig. 10.64b), already mentioned in connection with the evaluation undertaken by Daniels and Hamilton (Fig. 10.45), was developed and introduced immediately after stage of interest in the seal body to be able to capture the *SPM*. It should be noted that the remaining stages (after the filter) to the micro-orifice collector (MOC) were still present in the seal body during the measurements with this abbreviated configuration. This arrangement guarantees an identical flow-time profile if a filter with sufficiently low air flow resistance is used, which is important in the context of DPI testing.

In the alternative so-called external filter or outlet filter design (Fig. 10.64c), Svensson and Berg reconfigured their rNGI so that flow exiting the abbreviated

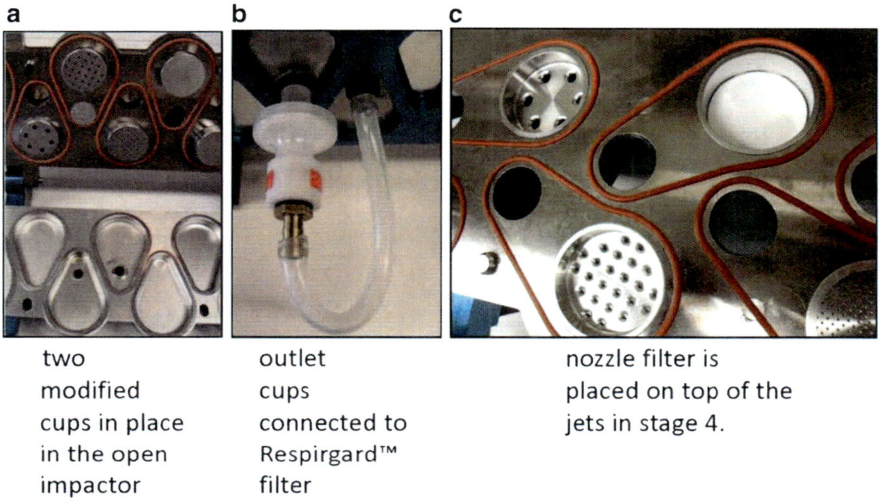

Fig. 10.65 Modified NGI by Svensson and Berg showing "O"-cup option (**a** and **b**) and internal filter option (**c**) for abbreviated testing of OIPs (*From* [59]—*courtesy of M. Svensson*)

section of the apparatus was directed through an externally mounted filter and then redirected into the NGI again after passage through the filter.

Both configurations can also be seen in Fig. 10.65; the modified outlet stage cups (necessary for the external filter approach) are shown in (a), the external and internal filter options are depicted in (b) and (c), respectively.

The main aim of their study was to judge whether these so-called filter methods can be regarded equivalent to the standard impactor method in which all stages are individually analyzed. Therefore, parallel experiments using full NGI set-up were performed and used for equivalence testing versus the proposed filter methods. Three different DPIs with three different formulations and two different pMDI products (CFC and HFA propellant) were used in this study. The comparison of the two methods was performed by comparing a total of 91 mean values from the full NGI and the corresponding filter method (Fig. 10.66).

Every mean value comprised data from either two or three impactor tests (full NGI) and 5–10 filter samples for the filter method, so that in total, around 650 individual tests were conducted. Except for the pMDI, each DPI were tested at two different flows corresponding to 2 and 4 kPa differential pressure over the device, spanning a flow range from 40 to 77 L/min. A strong linear relationship was obtained between the filter measure and the full NGI measures of *FPM* obtained at the same selected size limits. The correlation coefficients (r^2) were >0.95 for nozzle filter and >0.98 for the outlet cup method. The slopes for the fitted lines were close to the line of identity, varying from 0.89 to 1.03. However, the authors noted that the linear relationship, determined from the slope and associated r^2 value, was weaker for

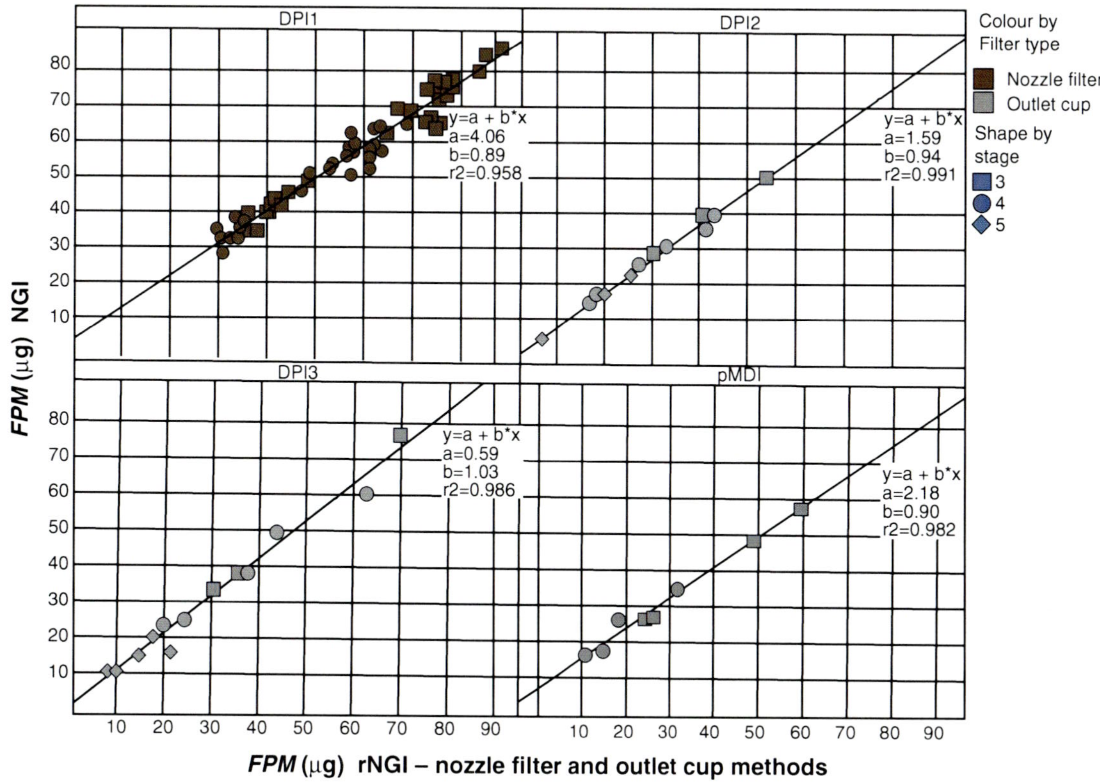

Fig. 10.66 Full-resolution versus abbreviated NGI measures of *FPM* reported by Svensson and Berg for various pMDI and DPI products; data from each inhaler category are shown in the separate panels, and the different stages are indicated by *shape* and *colors* on the markers corresponding to the filter type used (*From* [59]—*courtesy of M. Svensson*)

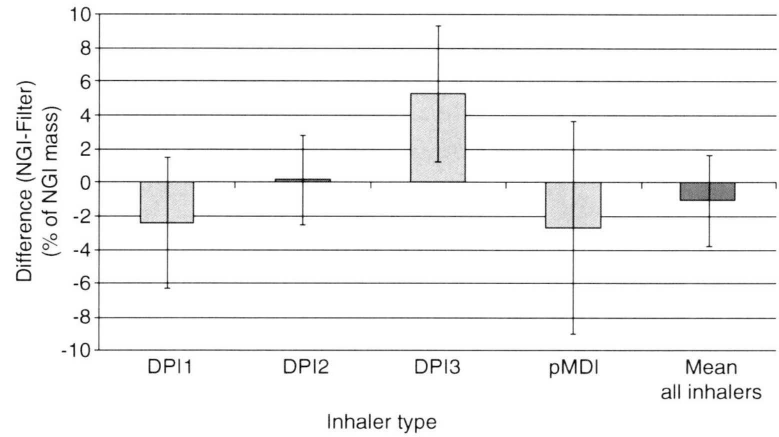

Fig. 10.67 Differences in *FPM* between the full NGI and the filter method for the inhaler types used in the study reported by Svensson and Berg; error bars are 95% confidence intervals (*From* [59]—*courtesy of M. Svensson*)

DPI-1 compared to the other inhaler types, and they provided the following rationale from their findings:

1. The nozzle-filter methodology is a less mature methodology in their laboratory, whereas the outlet "O"-cup has been established and used in development studies for several years.
2. DPI-1 as a device and the three formulations used in this device were both concepts in early development phase at the time when the work was undertaken, so the dosing properties may be more variable than the other inhalers that are products that are in late stage development.

Importantly, they did not observe that any particular stage was better (or worse) suited to be used in these filter methodologies (see marker shape in Fig. 10.66) and that no outliers were obtained in the data set.

Svensson and Berg went on to calculate the mean differences (presented in percentage terms of the found mass in the NGI test) between the NGI and the filter methods for each of the inhaler types (Fig. 10.67). Both negative and positive differences were observed, ranging from +5% to −2.5%. DPI-3 displayed significant difference between NGI and filter method, but when all 91 mean values were regarded as one data set, the difference became minimal (around 1%) and importantly, no significant difference was obtained between the two methods (standard NGI versus Filter concepts).

From a practical perspective, Svensson and Berg made the judgment that the nozzle-filter-abbreviated approach is about two times faster than the external filter method [59]. Moreover, since the external filter method, in its current design, includes a disposable plastic fixture, the nozzle filter is preferable from an environmental perspective. The authors concluded that the filter method is also very feasible for determination of the fine particle dose uniformity (FPDU) of a product; a parameter that is very difficult and laborious to retrieve from full NGI measurements. They also observed that it is necessary to size separate the dose below the first size-fractionating stage in two parts in order to achieve greater sensitivity in terms of changes in ASPD and dose passing beyond the impactor inlet. As a next step toward realization of two size fractions (and samples) in the abbreviated NGI platform, experiments are currently in progress using a modified NGI in which the stages have been physically interchanged in combination with the nozzle-filter approach. This approach should enable capture of the LPM and SPM in a cup and a filter, respectively, thereby resulting in an elegant abbreviated methodology to implement.

10.9 Short Stack ACI Systems Created by Rearranging Location of Back-Up Filter

In 2012, Horodnik et al. demonstrated an important alternative arrangement to the reduced ACI configurations previously described [60], as their approach avoided the removal of stages, therefore preserving the internal dead space of the full-resolution ACI. In their particular configuration, they relocated the back-up filter stage immediately downstream of the second impaction stage. The equipment was operated at 60 L/min and active measuring components therefore consisted of stages 1, 0, and filter.

The normal length spring-loaded clamps supplied with the full-resolution ACI could be used to ensure a tight seal between stages of their configuration, as an additional benefit [60]. The physical appearance of their abbreviated ACI was therefore comparable with that of the full-resolution system (Fig. 10.68).

Horodonik et al. went on to use this arrangement to evaluate a new DPI blend delivery system containing mometasone furoate intended for use with patients undergoing mechanical ventilation. The entry to the CI therefore comprised a spacer designed for use in such an environment with its distal end consisting of a short length of 22 mm diameter tubing representing part of a ventilator circuit, rather than the Ph. Eur./USP induction port. The focus of their study was on the proof of concept for a new in vitro method to evaluate how their DPI might perform in the clinic. They therefore did not present comparative data with the full-resolution ACI. However values of $FPM_{<ca.6.5\mu m}$ ($n = 5$ replicates at each condition) were consistent over a wide range of values of recovered mass of this particular API from the filter stage (Fig. 10.69).

This simple-to-configure arrangement may avoid both the need to take measures to match the flow rate-time profile in DPI testing, already discussed. Horodonik et al. retained the same number of stages in their short stack ACI by locating redundant stages below the filter collection stage, so that the internal dead space was the same as that for a full-resolution ACI. However, although suitable

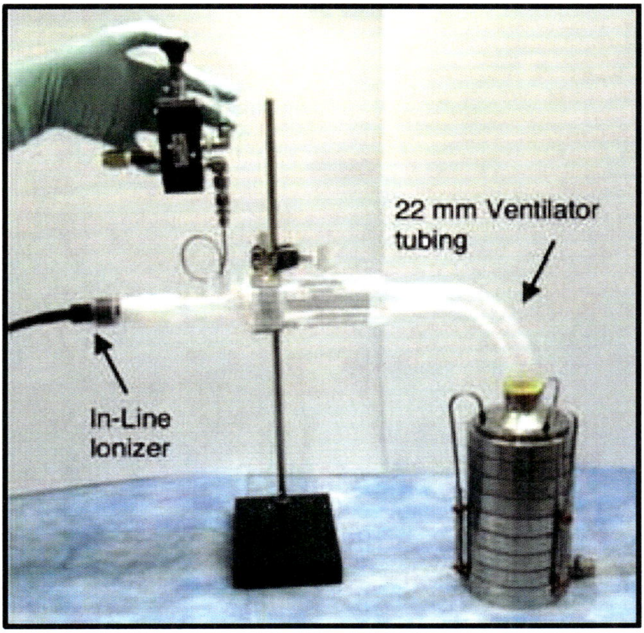

Fig. 10.68 "Short stack" arrangement of Horodnik et al. [60] in which they reconfigured a full-resolution ACI such that the back-up filter is located immediately below stage 1, and the other stages are retained (*From* [60]—*used with permission*)

for DPI testing where the total dead space in the measurement system is important, such an arrangement may not be effective if used in connection with the evaluation of pMDIs containing low-volatile cosolvent. Under such circumstances, it would be more appropriate to locate one or more redundant stages (i.e., not containing a collection surface) before the size-fractionating stage, as was done by Mitchell et al. [20], to ensure that cosolvent evaporation in the reduced system matched that of the full-resolution CI.

10.10 AIM-Based Measurement Equipment: Learning from Validation Studies, Current Status, and Future Needs

In December 2010 a workshop was organized by EPAG to act as a forum for the discussion of ways to develop AIM-based apparatus toward maturity, given the large amount of experimental data presented in the preceding sections of this chapter in support of their adoption [61]. In the panel discussion that followed,

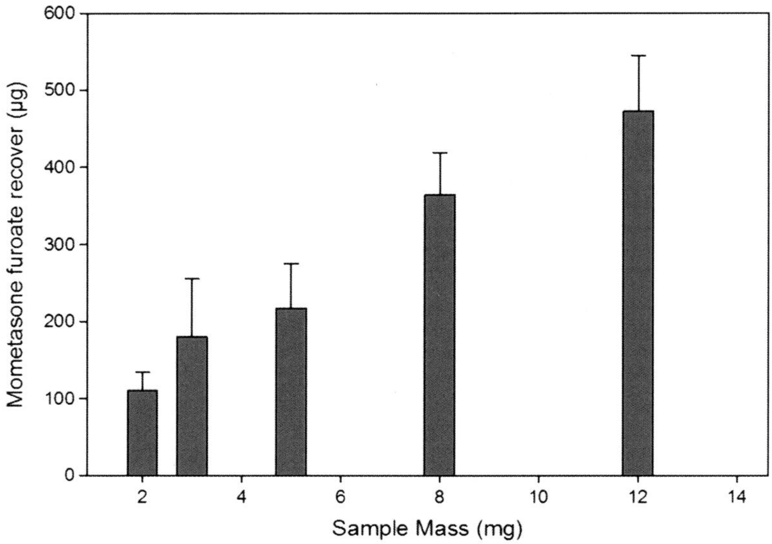

Fig. 10.69 $FPM_{<ca.6.5\mu m}$ of mometasone furoate [$n = 5$ replicates at each condition (mean ± SD)] delivered to the short stack ACI over a range of sample weights which, with a 12 mg sample load (15% blend mometasone furoate), will deliver approximately 500 µg to the filter stage (*From* [60]—*used with permission*)

the current status of the measurement technology was established and following issues were identified [62]:

1. Measurements made by AIM-based equipment for pMDIs and nebulizers provide measures of fine particle fraction that are in substantial agreement with the equivalent metric from the corresponding full-resolution impactor (either the ACI or NGI).
2. Measures of *FPF* by FSI were frequently higher than the corresponding full-resolution data for DPI testing. In contrast with the evaluation of pMDIs and nebulizers, where the impactor is operated at a fixed flow rate throughout the determination, the DPI test is more complex in that the flow rate at initiation of the measurement is zero and rapidly rises to a stable value as the pressure field within the DPI and measurement system stabilizes. Two possible causes were identified that need further investigation:

    a. The start-up kinetics of both abbreviated and full-resolution impactor systems appear to be important, since the compendial method necessitates initiating flow from the DPI at the start of measurement, so that the flow through the system is developing during the initial few hundred milliseconds of the determination.
    b. Differences in sharpness of cut for the insert in the FSI compared with both NGI and perhaps more so with the ACI whose stage collection efficiency curves are noticeably less steep than those of the NGI may also be responsible

for small upward shifts in fine particle fraction observed at 5 μm aerodynamic diameter with the FSI.

Further work is needed to understand the relative importance of both causes, as well as to determine how much of the divergence between abbreviated and full-resolution DPI-based measurements is formulation based and therefore productspecific.

3. The Twin Impinger (Apparatus A in monograph 2.9.18 of the European Pharmacopoeia) may become a suitable candidate AIM apparatus. It already has only a single cut-point size of 6.4 μm at 60 L/min [5]. Being an impingement device, the potential for bias from bounce and re-entrainment are eliminated by virtue of collecting the particles in the impingement fluid, as well as having the intrinsic advantage that recovery of active pharmaceutical ingredient from the impingement fluid can be achieved without further work-up in some cases. Modifications to achieve a slight reduction in the cut point to 5 μm at a defined flow rate in the range within which MDI and DPI testing, respectively, takes place (30–100 L/min) appear to be feasible.

   A better alternative might be to develop a reduced (say 2 or 3-stage) version of the MSLI (Fig. 10.70), which also achieves avoidance of particle bounce and

Fig. 10.70 Multistage liquid impinger: a candidate for abbreviation? (*Courtesy of Copley Scientific Ltd.*)

re-entrainment and in which the aerodynamically critical parts are manufactured from metal [63]. Both the Twin Impinger (Fig. 10.2) and MSLI systems benefit from having no interstage losses, but the MSLI has the distinct advantage, particularly for work with DPIs, of being calibrated from 30 to 100 L/min [12]. However, the possibility of using a reduced stack version of the MSLI throughout this flow rate range will be limited, without a range of flow rate dependent stages being designed and manufactured to give the desired cut-off diameter at the intended test flow rate (as in the case of the FSI).

4. The desire to develop AIM-based apparatuses should also consider designs that are potentially automatable. However, at this stage, more work is needed in understanding the role that AIM has to play in the life-cycle management of OIPs before the scope for partial or full automation will become clear. In the near future, semiautomated fixtures may have more prospects of being adopted, given the substantial financial investment required to automate AIM-based systems, despite their relative simplicity compared with their full-resolution counterparts.
5. There was a consensus that AIM-based measurements are unlikely to be allowed by themselves in regulatory submissions, given the need to have full-resolution aerodynamic particle size distribution data in order to interpret safety and efficacy data from the clinical batches. However, once relationships are established and appropriately validated, AIM-based measurements could be considered especially in a product QC environment. It is important to note that the full-resolution CI would always be available to support the process in the event that an out-of-specification investigation is needed. A role may also exist for an AIM-based approach in the speeding up of early development formulation screening, but a convincing case will likely need to be made on a company-by-company basis, given the reduction in the data relating to aerosol aerodynamic size that results with an AIM-based methodology.
6. These observations and suggestions for future work are still current, although at a recent workshop organized by the US Pharmacopeia, at which AIM was discussed together with EDA there was more understanding concerning the importance of matching dead volume of the abbreviated system to that for the full-resolution reference CI, especially in the evaluation of DPIs [64].
7. The large body of data that has been presented in this chapter illustrates the degree of commitment from stakeholders to understand both the limitations as well as the perhaps more obvious advantages that AIM-based CI measurements have to offer in the assessment of aerosols from all types of OIP. Table 10.18 summarizes the salient points in connection with the extension of good cascade impactor practice (GCIP—see Chap. 4) to include AIM-based approaches.

At the present time, there are no hard-and-fast rules for choosing the most appropriate full-resolution CI with which to match data from an AIM-based method. This situation has arisen because there is still insufficient experience with all of the various options available for AIM-based measurements. Nevertheless, given the apparent importance of matching internal dead space between AIM and full-resolution apparatuses where volatile species are present or to match start-up kinetics for DPI testing.

Table 10.18 Points to consider (PtC) when implementing an AIM-based CI method; "√" important; "≈" unimportant

PtC	Description	OIP class applicability		
		MDI	DPI	Nebulizer/SMI
1	Select appropriate full-resolution CI as benchmark method (Table 10.19)	√		
2	Eliminate particle bounce and re-entrainment	√		≈
3	Match pressure drop rise time of AIM system to benchmark CI[a]	√		
4	Chill AIM and benchmark CI	≈		√
5	Refer to benchmark CI data as first step in resolving questionable AIM data	√		

[a]In the case of DPI testing, adding a first impaction stage to match the full-resolution impactor will not by itself be sufficient to mirror the volume and flow resistance of the full impactor. However, in the case of the ACI, it would be a practical alternative to retain the redundant stages in the abbreviated impactor, but locate them beneath the filter stage (e.g., in the stage order: 0, 2, 5, Filter, 1, 2, 3, 5, 7) to achieve this goal

Table 10.19 Matching an AIM-based CI to a full-resolution benchmark CI

AIM system	Suggested full-resolution CI
FSA, T-FSA variants	Andersen 8-stage nonviable impactor (ACI)
FPD	Andersen 8-stage viable impactor (ACVI)
rNGI variants	Next Generation Pharmaceutical Impactor (NGI)
FSI Other systems including abbreviated MSLI	No obvious matching full-resolution CI, therefore use with any CI and evaluate need to match dead space for OIP being tested. Using NGI as reference full-resolution CI with FSI could result in some components (e.g., inlet and pre-separator housing) being shared, allowing reduction in cost for additional equipment. Furthermore, familiarity of test recovery procedures for these components is very similar between FSI and NGI. Validation work with appropriate benchmark CI needed with other systems

Table 10.19 provides guidance concerning this matter. The interested reader is urged to keep up-to-date with the continually developing literature concerning good AIM practices, as newer approaches to resolving compatibility concerns are published.

References

1. European Pharmaceutical Aerosol Group (2010) Workshop on AIM-Based Techniques, Edinburgh, UK. http://www.epag.co.uk/Library/Default2.asp. Accessed 12 Jan 2012
2. International Pharmaceutical Consortium on Regulation and Science (IPAC-RS) (2008) IPAC-RS conference "Bringing value to the patient in a changing world", Rockville, MD, USA. http://www.ipacrs.com/ipac2008.html. Accessed 12 Jan 2012
3. International Pharmaceutical Consortium on Regulation and Science (IPAC-RS) (2010) IPAC-RS conference "Bringing value to the patient in a changing world", Rockville, MD, USA. http://www.ipacrs.com/2011%20Conference.html. Accessed 12 Jan 2012

4. European Directorate for the Quality of Medicines and Healthcare (EDQM) (2002) Preparations for inhalation: aerodynamic assessment of fine particles. Section 2.9.18 – European Pharmacopeia – Apparatus B in versions up to 4th edition, Council of Europe, Strasbourg, France
5. Hallworth GW, Westmoreland DG (1987) The Twin Impinger: a simple device for assessing the delivery of drugs from metered dose pressurized aerosol inhalers. J Pharm Pharmacol 39(12):966–972
6. May KR (1966) Multi-stage liquid impinger. Bact Rev 30:559–570
7. Miller N, Marple VA, Schultz RK, Poon WS (1992) Assessment of the Twin Impinger for size measurement of metered dose inhaler sprays. Pharm Res 9(9):1123–1127
8. Onyechi JO, Martin GP, Marriott C, Murphy L (1994) Deposition of dry powder aerosols in cascade impactors at different flow rates. J Aerosol Med 7(2):181–184
9. Watson JP, Lewis RA (1995) Generic salbutamol metered dose inhalers. Thorax 50(5):590–592
10. Mendes PJ, Pinto JF, Sousa JMM (2007) A non-dimensional functional relationship for the fine particle fraction produced by dry powder inhalers. J Aerosol Sci 38(6):612–624
11. Tougas TP, Christopher D, Mitchell JP, Strickland H, Wyka B, Van Oort M, Lyapustina S (2009) Improved quality control metrics for cascade impaction measurements of orally inhaled drug products (OIPs). AAPS PharmSciTech 10(4):1276–1285
12. Asking L, Olsson B (1997) Calibration at different flow rates of a multistage liquid impinger. Aerosol Sci Technol 27(1):39–49
13. Marple VA, Rubow KL, Olson BA (2001) Inertial, gravitational, centrifugal, and thermal collection techniques. In: Baron PA, Willeke K (eds) Aerosol measurement: principles, techniques and applications, 2nd edn. Wiley Interscience, New York, pp 229–260
14. Marple VA, Roberts DL, Romay FJ, Miller NC, Truman KG, Van Oort M, Olsson B, Holroyd MJ, Mitchell JP, Hochrainer D (2003) Next generation pharmaceutical impactor. Part I: Design. J Aerosol Med 16(3):283–299
15. Van Oort M, Downey B (1996) Cascade impaction of MDIs and DPIs: Induction port, inlet cone, and pre-separator lid designs recommended for inclusion in the general test chapter Aerosols <601> Pharm Forum 22(2):2204–2210
16. Van Oort M, Roberts W (1996) Variable stage-variable volume strategy for cascade impaction. In: Dalby RN, Byron PR, Farr SJ (eds) Respiratory drug delivery-V. Interpharm, Buffalo Grove, IL, pp 418–421
17. Poochikian G, Bertha CM (2002) Regulatory view on current issues pertaining to inhalation drug products. In: Dalby RN, Byron PR, Peart J, Farr SJ (eds) Respiratory drug delivery-VIII. Davis Horwood International, Raleigh, NC, pp 159–164
18. Bowles N, Cahill E, Haeberlin B, Jones C, Mett I, Mitchell J, Müller-Walz R, Musa R, Nichols S, Parkins D, Petterssen G, Preissmann A, Purewal T, Schmelzer C (2007) Application of quality-by-design to inhalation products. In: Dalby RN, Byron PR, Peart J, Suman JD (eds) Respiratory drug delivery–Europe 2007. Davis Healthcare, River Grove, IL, pp 61–69
19. Mitchell JP, Nagel MW, Avvakoumova V, MacKay H, Ali R (2009) The abbreviated impactor measurement (AIM) concept: Part 1—influence of particle bounce and re-entrainment—evaluation with a "dry" pressurized metered dose inhaler (pMDI)-based formulation. AAPS PharmSciTech 10(1):243–251
20. Mitchell JP, Nagel MW, Avvakoumova V, MacKay H, Ali R (2009) The abbreviated impactor measurement (AIM) concept: Part 2—influence of evaporation of a volatile component—evaluation with a "droplet producing" pressurized metered dose inhaler (pMDI)-based formulation containing ethanol as co-solvent. AAPS PharmSciTech 10(1):252–257
21. Mitchell JP, Copley M (2010) Accelerated inhaled product testing. Pharma Mag Jan–Feb: 14–17. http://www.pharma-mag.com/pharma/index.html. Accessed 12 Jan 2012
22. United States Federal Drug Administration (FDA) (1998) Draft guidance: metered dose inhaler (MDI) and dry powder inhaler (DPI) drug products chemistry, manufacturing and controls documentation. United States Federal Drug Administration, Rockville, MD, USA. Docket 98D-0997, http://www.fda.gov/downloads/Drugs/GuidanceComplianceRegulatoryInformation/Guidances/ucm070573.pdf. Accessed 22 Aug 2011

23. Graham SJ, Lawrence RC, Ormsby ED, Pike RK (1995) Particle size distribution of single and multiple sprays of salbutamol metered-dose inhalers (pMDIs). Pharm Res 12(9):1380–1384
24. Kamiya A, Sakagami M, Hindle M, Byron P (2004) Aerodynamic sizing of metered dose inhalers: an evaluation of the Andersen and next generation pharmaceutical impactors and their USP methods. J Pharm Sci 93(7):1828–1837
25. Kamiya A, Sakagami M, Hindle M, Byron PR (2003) Particle sizing with the next generation impactor: a study of Vanceril™ metered dose inhaler. Proc 14th ISAM Congress, Baltimore, MD, USA, J Aerosol Med 16(2):216 (abstract)
26. Mitchell JP, Nagel MW, Doyle C, Ali RS, Avvakoumova V, Christopher D, Quiroz J, Strickland H, Tougas T, Lyapustina S (2010) Relative precision of inhaler aerodynamic particle size distribution (APSD) metrics by full resolution and abbreviated Andersen Cascade Impactors (ACIs): Part 1. AAPS PharmSciTech 11(2):843–851
27. Mitchell JP, Nagel MW, Doyle C, Ali RS, Avvakoumova V, Christopher D, Quiroz J, Strickland H, Tougas T, Lyapustina S (2010) Relative precision of inhaler aerodynamic particle size distribution (APSD) metrics by full resolution and abbreviated Andersen Cascade Impactors (ACIs): Part 2–Investigation of bias in extra-fine mass fraction with AIM-HRT impactor. AAPS PharmSciTech 11(3):1115–1118
28. Keegan GM, Lewis DA (2012) Rapid prototype screening with the Copley fast screening Andersen (FSA). In: Dalby RN, Byron PR, Peart J, Suman JD, Young PM (eds) Respiratory drug delivery 2012. Davis HealthCare, River Grove, IL, pp 469–472
29. Keegan GM, Lewis DA (2012) Formulation-dependent effects on aerodynamic particle size measurements using the fast screening Andersen (FSA). In: Dalby RN, Byron PR, Peart J, Suman JD, Young PM (eds) Respiratory drug delivery 2012. Davis HealthCare, River Grove, IL, pp 465–468
30. Andersen A (1966) A sampler for respiratory health assessment. Am Ind Hyg Assoc J 27(2):160–165
31. Yao MS, Mainelis G (2007) Analysis of portable impactor performance for enumeration of viable bioaerosols. J Occup Environ Hyg 4(7):514–524
32. Chambers FE, Smurthwaite M (2012) Comparative performance evaluation of the Westech fine particle dose (FPD) impactor. In: Dalby RN, Byron PR, Peart J, Suman JD, Young PM (eds) Respiratory drug delivery-2012. Davis HealthCare, River Grove, IL, pp 553–557
33. Guo C, Ngo D, Ahadi S, Doub WH (2011) Evaluation of an abbreviated impactor for fine particle fraction (FPF) determination of inhalation drugs. AAPS Annual Meeting, Washington, DC
34. Roberts DL, Romay F (2009) Design of the fast screening impactor based on the NGI preseparator. Drug delivery to the lungs-20, The Aerosol Society, Edinburgh, UK, 20:206–209. http://ddl-conference.org.uk/index.php?q=previous_conferences. Accessed 4 Aug 2012
35. Marple VA, Olson BA, Santhanakrishnan K, Mitchell JP, Murray SC, Hudson-Curtis BL (2003) Next generation pharmaceutical impactor (a new impactor for pharmaceutical inhaler testing). Part II: Archival calibration. J Aerosol Med 16(3):301–324
36. Marple VA, Olson BA, Santhanakrishnan K, Mitchell JP, Murray SC, Hudson-Curtis BL (2004) Next generation pharmaceutical impactor: a new impactor for pharmaceutical inhaler testing. Part III. Extension of archival calibration to 15 L/min. J Aerosol Med 17(4):335–343
37. MSP Corporation (2009) Fast Screening Impactor™ for quantifying "large" and "small" particles emitted by inhalable drug devices: user guide. St. Paul, MN, USA. FSI-0185-6002, Revision A, available at: www.mspcorp.com. Accessed 14 Jan 2012
38. Stobbs B, McAuley E, Bogard H, Monsallier E (2009) Evaluation of the fast screening impactor for determining dine particle fraction of dry powder inhalers. Drug delivery to the lungs-20, The Aerosol Society, Edinburgh, UK, 20:158–161. http://ddl-conference.org.uk/index.php?q=previous_conferences. Accessed 4 Aug 2012
39. Copley M, Mitchell J, McAuley E, Russell-Graham D (2010) Implementing the AIM concept. Inhalation 4(1):7–11
40. Russell-Graham D, Cooper A, Stobbs B, McAulay E, Bogard H, Heath V, Monsallier E (2010) Further evaluation of the fast-screening impactor for determining fine particle fraction of dry

powder inhalers. Drug delivery to the lungs-21, The Aerosol Society, Edinburgh, UK, 21:374–377. http://ddl-conference.org.uk/index.php?q=previous_conferences. Accessed 4 Aug 2012
41. Copley M, Smurthwaite M, Roberts DL, Mitchell JP (2005) Revised internal volumes to those provided by Mitchell JP and Nagel MW in "Cascade impactors for the size characterization of aerosols from medical inhalers: their uses and limitations". J Aerosol Med 18(3):364–366
42. Pantelides PN, Bogard H, Russell-Graham D, Cooper AD, Pitcairn GR (2011) Investigation into the use of the fast screening impactor as an abbreviated impactor measurement (AIM) tool for dry powder inhalers. In: Dalby RN, Byron PR, Peart J, Suman JD, Young PM (eds) Respiratory drug delivery–Europe 2011. Davis Healthcare, River Grove, IL, pp 391–395
43. Burnell PKP, Small T, Doig S, Johal B, Jenkins R, Gibson GJ (2001) Ex-vivo product performance of Diskus™ and Turbuhaler™ inhalers using inhalation profiles from patients with severe chronic obstructive pulmonary disease. Respir Med 95(5):324–330
44. Roberts DL, Chiruta M (2007) Transient impactor behavior during the testing of dry-powder inhalers via compendial methods. Drug delivery to the lungs-18, The Aerosol Society, Edinburgh, UK, 18:202–205
45. Pantelides PN, Bogard H, Russell-Graham D, Cooper AD, Pitcairn GR (2011) An evaluation of a fast screening impactor (FSI) set-up for abbreviated impactor measurement: quality control (AIM-QC) of dry powder inhalers. UK Academy of Pharmaceutical Sciences (APSGB) Inhalation 2011 meeting, University of Bath, UK, July (abstract)
46. Daniels GE, Hamilton M (2011) Assessment of early screening methodology using the reduced next generation and fast screening impactor systems. In: Dalby RN, Byron PR, Peart J, Suman JD, Young PM (eds) Respiratory drug delivery–Europe 2011. Davis Healthcare, River Grove, IL, pp 327–330
47. Hamilton M, Daniels G (2011) Assessment of early screening methodology using the Next Generation and Fast Screen Impactor systems. Drug delivery to the lungs-22, The Aerosol Society, Edinburgh, UK, 22:355–358. http://ddl-conference.org.uk/index.php?q=previous_conferences. Accessed 4 Aug 2012
48. Després-Gnis F, Williams G (2010) Comparison of next generation impactor and fast-screening impactor for determining fine particle fraction of dry powder inhalers. Drug delivery to the lungs-21, The Aerosol Society, Edinburgh, UK, 21:386–389. http://ddl-conference.org.uk/index.php?q=previous_conferences. Accessed 4 Aug 2012
49. Rogueda P, Morrical B, Chew YD (2010) Comparison of NGI and the fast screening impactor (FSI) for suitability for analytical drug development. Drug delivery to the lungs-21, The Aerosol Society, Edinburgh, UK, 21:394–397. http://ddl-conference.org.uk/index.php?q=previous_conferences. Accessed 4 Aug 2012
50. Sheng G, Zhang J, Simmons R, Watanabe W (2010) Fast screening impactor (FSI) as a prescreening tool for MDIs and nebulizers. In: Dalby RN, Byron PR, Peart J, Suman JD, Farr SJ, Young PM (eds) Respiratory drug delivery-2010. Davis Healthcare, River Grove, IL, pp 637–640
51. European Directorate for the Quality of Medicines and Healthcare (EDQM) (2012) Preparations for nebulisation: characterisation, general chapter 2.9.44, Council of Europe, Strasbourg, France
52. Tservistas M, Uhlig M, Mitchell J (2010) Assessment of abbreviated impactor measurement (AIM) methods for nebulizer characterization. Drug delivery to the lungs-21, The Aerosol Society, Edinburgh, UK, 21:378–381. http://ddl-conference.org.uk/index.php?q=previous_conferences. Accessed 4 Aug 2012
53. Sheng G, Watanabe W (2010) Feasibility of fast screening impactor as a screening tool. Drug delivery to the lungs-21, The Aerosol Society, Edinburgh, UK, 21:390–393. http://ddl-conference.org.uk/index.php?q=previous_conferences. Accessed 4 Aug 2012
54. Dennis J, Berg E, Sandell D, Ali A, Lamb P, Tservistas M, Karlsson M, Mitchell J (2008) Cooling the NGI: an approach to size a nebulised aerosol more accurately. Pharm Europa Sci Notes 1:27–30
55. Fowdar N, Hammond M, Solomon D (2010) A comparison of the effect of continuous cooling of an NGI on a solution and suspension based nebulized product. Drug delivery to the lungs-21,

The Aerosol Society, Edinburgh, UK, 21:275–279. http://ddl-conference.org.uk/index.php?q=previous_conferences. Accessed 4 Aug 2012
56. Williams K (2010) The influence of ambient relative humidity during cooled NGI Testing. Drug delivery to the lungs-21, The Aerosol Society, Edinburgh, UK, 21:292–294. http://ddl-conference.org.uk/index.php?q=previous_conferences. Accessed 4 Aug 2012
57. Finlay WH, Stapleton K (1999) Undersizing of droplets from a vented nebulizer caused by aerosol heating during transit through an Andersen impactor. J Aerosol Sci 30(1):105–109
58. Mitchell JP, Bauer R, Lyapustina S, Tougas T, Glaab V (2011) Non-impactor-based methods for sizing of aerosols emitted from orally inhaled and nasal drug products (OINDPs). AAPS PharmSciTech 12(3):965–988
59. Svensson M, Berg E (2010) Measuring the fine particle dose using inter-stage filters in the NGI – an overview of two methods. Drug delivery to the lungs-21, The Aerosol Society, Edinburgh, UK, 21:382–385. http://ddl-conference.org.uk/index.php?q=previous_confernces. Accessed 4 Aug 2012
60. Horodnik W, Garber N, Ewing G, Donovan B (2012) The *in vitro* delivery of a dry powder blend using pressurized air and a pMDI designed spacer and actuator combination product (AeroChamber MV) into a ventilator circuit tubing and "short stack" Andersen cascade impactor (ACI) for particle size characterization. In: Dalby RN, Byron PR, Peart J, Suman JD, Young PM (eds) Respiratory drug delivery 2012. Davis HealthCare, River Grove, IL, pp 581–584
61. Mitchell JP, Nichols SC (2011) Drug delivery to the lungs 21 – European Pharmaceutical Aerosol Group Abbreviated Impactor Measurement workshop summary. Therapeut Deliv 2(3):301–305. http://ddl-conference.org.uk/index.php?q=previous_conferences. Accessed 4 Aug 2012
62. Mitchell JP, Nichols SC (2011) European Pharmaceutical Aerosol Group (EPAG): Abbreviated Impactor Measurement (AIM) workshop – December 8th 2010. In: Dalby RN, Byron PR, Peart J, Suman JD, Young PM (eds) Respiratory drug delivery–Europe 2011. Davis Healthcare, River Grove, IL, pp 469–472
63. Bell JH, Brown K, Glasby J (1973) Variation in delivery of isoprenaline from various pressurized inhalers. J Pharm Pharmacol 25(S):32P–36P
64. United States Pharmacopeia (2011) A co-sponsored workshop by USP and AAPS on aerosols–inhalation and nasal drug products, Rockville, MD, USA, December 12–13. http://www.usp.org/meetings/asMeetingIntl/aerosols.html. Accessed 18 Jan 2012

Chapter 11
The Regulatory and Compendial Pathways to Acceptance for AIM and EDA Concepts

Steven C. Nichols, Jolyon P. Mitchell, Terrence P. Tougas, J. David Christopher, and Susan Holmes

Abstract The acceptance by the regulatory and compendial authorities of AIM and EDA concepts is a critical stage before their widespread adoption by the pharmaceutical industry will likely take place. A prerequisite for such acceptance will be the assemblage of a body of data in which close attention has been placed not only on the methods themselves but also on their applicability across as wide a portion of the spectrum of OIP classes as possible. In the case of AIM-based apparatuses, it will be important to provide experimental data from many inhalers of each type, so that the variability of the method can be assessed in the context of typical variability associated with the corresponding full-resolution method. Robust statistical analysis of either full-resolution APSDs or AIM CI-generated APSD metrics by EDA, in comparison with existing data reduction methods (i.e., stage grouping, fine/total particle

S.C. Nichols (✉)
c/o Glebe Farm, Main Street, Willey, Rugby, Warwickshire CV23 0SH, UK
e-mail: Jack_snipe@btinternet.com

J.P. Mitchell
Trudell Medical International, 725 Third Street, London, ON N5V 5G4, Canada
e-mail: jmitchell@trudellmed.com

T.P. Tougas
Boehringer Ingelheim Pharmaceuticals Inc., 900 Ridgebury Road, Ridgefield, CT 06877-0368, USA
e-mail: terrence.tougas@boehringer-ingelheim.com

J.D. Christopher
Nonclinical and Pharmaceutical Sciences Statistics, Merck Research Laboratories, WP37C-305, 770 Sumneytown Pike, West Point, PA 19486-0004, USA
e-mail: j.david.christopher@merck.com

S. Holmes
GlaxoSmithKline, CMC Regulatory Affairs, Five Moore Dr, 3 Main 443B, PO Box 13398, Research Triangle Park, NC 27709, USA
e-mail: susan.m.holmes@gsk.com

T.P. Tougas et al. (eds.), *Good Cascade Impactor Practices, AIM and EDA for Orally Inhaled Products*, DOI 10.1007/978-1-4614-6296-5_11,

mass), will be needed to justify replacing current methods. A vital part of this process will be demonstrating that EDA is capable of at least matching current methods and preferably is shown to have better decision-making capability, in the context of quality control for most OIPs. AIM-based apparatuses may have roles in the development of testing capacity to work with improved material and process understanding, implementation of in-process controls prior to final product, and minimization of capacity spent on end product controls. Given precedents for the pace of change, compendial and regulatory acceptance will likely take several years to be realized. This chapter begins by assessing the current regulatory guidance and compendial requirements with respect to the use of cascade impaction or when APSD data is required. The second part of the chapter discusses strategy and requirements that the pharmaceutical industry involved with OIP assessment is likely to have to provide to demonstrate that AIM and EDA are "fit for purpose" and thereby gain acceptance by the regulators and become a pharmacopeia-approved method.

11.1 Current Guidance

The use of multistage cascade impaction in the context of OIP-generated aerosol APSD assessment has three principal objectives:

1. As a measure of the product aerosol aerodynamic performance
2. To provide a tool for assessing product quality in commercial product batch release
3. To establish relationships with potential in vivo performance, through either an IVIVR or IVIVC

There are issues associated with meeting each of these objectives that limit applicability to the clinical reality of OIP use by the patient requiring therapy. These issues are discussed further in Chap. 12. However, as examples, the use of a single constant air flow rate through the CI system cannot represent the inhalation profile of a patient in more than a rudimentary way. Furthermore, the APSD of the aerosol cloud produced under the constant flow rate testing conditions may not necessarily be identical with that of the aerosol which the patient inhales at a variable flow rate-time profile. Inhaler dosage form is a further complicating factor affecting inhaled aerosol characteristics, as the testing methods described in the pharmacopeial compendia differ whether the OIP is an MDI, DPI, or nebulizer. Even if complicating factors on the clinical side such as patient-to-patient variability, disease condition, and relatively insensitive clinical measures of efficacy are ignored, it is not unexpected that current laboratory testing methodology for OIPs falls short by failing to provide unambiguous IVIVR/IVIVCs [1–3].

When using laboratory methodology, whether AIM or full-resolution CI based, there are therefore many compromises that of necessity have to be made and their consequences understood. As the community involved with inhaler in vitro performance assessment procedures gains a better understanding of the underlying aerosol physics as well as the limitations of their apparatuses, it should be possible to develop a wider array of methodology than exists at present. However, a goal for new methods to achieve is that they may be both universally applicable across OIP categories, and at least equal or better, improving the quality of the data for the intended purpose.

Looking at the first part of this goal in more detail, it is self-evident that it is desirable that the compendial methodologies in place to assess OIP aerosol properties have common utility across as many categories of inhaler (i.e., pMDIs, DPIs, etc.) as possible. However, it has to be recognized that minimizing the number of compendial apparatuses for AIM-based measurements, though attractive, should be balanced against the fact that OIPs are a very diverse group of products, probably requiring a degree of specialization in the AIM-based apparatuses that are ultimately developed for their assessment. Such a situation is not unique within the scope of the compendia; for example, several systems, each with their own specific purpose, characteristics, and limitations, have been developed for the assessment of tablet dissolution. Notwithstanding the above considerations, minimization of AIM-based options should certainly be considered at the outset of pharmacopeial method development, in order to avoid the development of a plethora of procedures having only minor differences that are customized for each inhaler variant within a given class of OIP. That said, it has to be recognized that the therapeutic areas treated by inhalers are quite varied, in that both local and systemic action can be targeted, so that OIPs within a given class may vary substantially both in mode of operation and formulation type(s) delivered. These characteristics are likely to facilitate development of many methods differing only between inhaler types but essentially determining the same performance attribute. In the more flexible environment created by the adoption of strategies such as Quality by Design [4], it is likely that regulatory agencies looking to provide product market approval will have different performance measurement needs than those focused on ensuring consistent "in-market" quality, e.g., post-market product surveillance. It follows that despite the overarching desire to limit method proliferation, different approaches to in vitro testing methodology for aerosol properties, by either full-resolution or abbreviated CI, may have to be accepted as appropriate during the life cycle of a particular OIP (see Chap. 6).

At the present time, many regulatory agencies require the use of multistage cascade impaction for the determination of OIP APSD based on the mass of drug substance as one of the principal means for determining likely clinical performance in terms of dose delivery to the receptors in the lungs. In the USA, the 1998 draft FDA Guidance for Industry covering chemistry, manufacturing, and controls of pMDIs and DPIs indicates that the choice of CI should allow determination of an APSD of the whole dose including the small particle size fraction [5]. This requirement is echoed in the equivalent documentation produced by CDRH in relation to the 510(k) premarket approval process for medical devices, such as spacers, valved holding chambers, and general-purpose nebulizers [6]. The draft FDA-CDER guidance further states that the number of actuations (*of the inhaler*) needed to determine APSD by multistage CI should be kept to the minimum, justified by the sensitivity of the analytical method used to quantitate the deposited drug substance. The amount of drug substance deposited on the critical stages of the cascade impactor should be sufficient for reliable assay, but not so excessive as to bias the results by masking individual actuation variation. Although the CI size resolution is not stated explicitly in this guidance document, for practical reasons a 7- or 8-stage system is needed to provide five stages having cut-point sizes in the critical range from 0.5 to 5.0 μm aerodynamic diameter if information about the coarse fraction >5 μm is to be obtained as well [7]. For post-approval release and stability testing, but not for characterization of the drug product in the initial submission to the regulatory agency, drug deposition on individual stages may be grouped (so-called stage grouping), with separate requirements placed on each of the groupings [8].

In Europe and Canada, the 2006 jointly approved Guideline on the Pharmaceutical Quality of Inhalation and Nasal Products also makes reference to the use of a multistage CI for the measurement of APSD, with the implication that sufficient stages will be used to enable fine-particle dose <5 μm ($FPM_{<5.0\mu m}$) aerodynamic diameter to be obtained, together with the *MMAD* and *GSD* of the *ISM*, if appropriate [9, 10]. Since the statement is also made to the effect that control of the portion of the APSD >5 μm may be necessary depending on the relevance of this fraction for the therapeutic index of the drug product, a 7- or 8-stage CI will most likely be used, although the 5-stage multistage liquid impinger (Apparatus C in the European Pharmacopoeia monograph 2.9.18 [11]) has been used in regulatory submissions. Stage pooling (grouping) is also permitted, as illustrated by the following example for a generic pMDI-delivered salbutamol [12]:

1. Induction port to represent the oropharyngeal deposition and hence swallowed dose
2. Pooling 1: Stage 0, 1, and 2 to represent the large nonrespirable particles deposited in upper airway
3. Pooling 2: Stage 3, 4, and 5 to represent the fine-particle dose (*FPD*) between 1.1 and 4.5 μm ($\equiv FPM_{1.1\text{–}4.5\mu m}$), deposited on bronchi and predictive of the in vivo bronchodilator efficacy and of C_{max} (early lung bioavailability)
4. Pooling 3: Stage 6, 7, and filter to represent the extrafine particles ($EPM_{<1.1\mu m}$) likely to deposit in the alveoli

Acceptance of the AIM concept as a direct substitute for multistage cascade impaction in regulatory submissions relating to OIP quality control will likely prove to be a challenging process for industry, requiring much evidence across a wide range of OIP platforms and abbreviated systems to support its adoption. From the regulatory perspective, robust demonstrations of the following advantages will be central in gaining acceptance by meeting the following criteria:

1. Comparable or possibly better precision for AIM systems compared with their full-resolution counterparts as well as freedom from measurement bias (see Chap. 10)
2. Applicability throughout the life cycle of the OIP (see Chap. 6)
3. Capability of EDA (where proposed as the AIM metric) to provide equivalent or improved sensitivity to small changes in APSD compared with stage grouping, leading to better decision making with respect to batch disposition in the QC environment (see Chaps. 7, 8, and 9)

Currently, the subsequent investigation can be difficult when using a full-resolution CI and an out-of-specification result is obtained [13], as there are many sources of potential error. However, if an AIM approach is used as a front line batch release method, then an important additional consideration is the availability of using the full-resolution CI as the primary diagnostic aid in the event of an out-of-specification AIM-EDA determination. This consideration is in addition to the application of a logical fault-tree approach for anomalous CI measurements, similar to that described for CI mass balance determination and discussed in Chap. 4 [14]. However, care would be needed in the setting of specifications to avoid different requirements pertaining to metrics determined by abbreviated or full-resolution CI systems, unless such a situation was unavoidable, but a demonstrable benefit was nonetheless available using an AIM-based approach.

11.2 Pharmacopeia Requirements

An important focus of the pharmacopeias is the provision of authoritative monographs for the assessment of drug product quality. European Union directives 2001/82/EC, 2001/83/EC, and 2003/63/EC (amended), on medicines for human and for veterinary use, maintain the mandatory character of European Pharmacopoeia (Ph. Eur.) specifications on medicines for marketing authorization applications for OIPs. In contrast, the US Pharmacopeial (USP) Convention is a nongovernmental body responsible for the USP-NF. However, the FDA, although working closely with the USP, is not obliged to use USP specifications or test methodologies in developing or application of the FDA guidance documents.

Both European and US pharmacopeial monographs associated with the measurement of particle size from OIPs describe a range of full-resolution CI apparatuses, with the exception of Apparatus A of the Ph. Eur., which is the Twin (Glass) Impinger (TI) (Table 11.1). This apparatus could possibly be used as an AIM-based

Table 11.1 Current compendial apparatuses for the assessment of aerosols from OIPs

Apparatus	US Pharmacopeia	European Pharmacopoeia
Glass "Twin Impinger" (TI)	Not used	Apparatus A for pMDIs, DPIs, and nebulizers
Andersen 8-stage without pre-separator (ACI)	Apparatus 1 for pMDIs	Apparatus D for pMDIs
Marple Miller Model 160 (MMI)	Apparatus 2 for DPIs	Not used
Andersen 8-stage with pre-separator (ACI)	Apparatus 3 for DPIs	Apparatus D for DPIs
Multi-stage liquid impinger (MSLI)	Apparatus 4 for DPIs	Apparatus C for pMDIs and DPIs
Next generation pharmaceutical impactor (NGI)	Apparatus 5 for DPIs Apparatus 6 for pMDIs	Apparatus E for pMDIs and DPIs

system, as discussed in Chap. 10, with some refinements to make its cutoff size between coarse and fine fractions more flexible at its operating flow rate of 60 L/min [15]. However, its status has been under review in recent years as the result of the exclusive requirement for higher resolution CI measurements in product license submissions, e.g., NDAs in the USA or MAAs in the EU [16]. The EMA guideline covering the pharmaceutical quality of inhalation and nasal products allows the use of alternative impactors if justified and validated [9]. An alternative strategy, also presented in Chap. 10, might therefore be to develop a reduced (say 2- or 3-stage) version of the MSLI (Apparatus 4 of the Ph. Eur. and C of the USP). However, such a strategy will require someone to undertake the necessary development work and associated business risk that the reduced MSLI may not get adopted as an AIM apparatus.

11.3 Gaining Acceptance for AIM and EDA

11.3.1 Framework for Acceptance

AIM apparatuses can provide aerodynamic particle size data with the following properties:

(a) A measure of API mass that is equivalent to the "fine"-particle dose "(mass)," as defined at a cut point of 5 μm aerodynamic diameter. This size limit would be required within the framework of the current Ph. Eur. monograph 2.9.18. The "fine"-particle "dose" using a full CI is currently derived by interpolation using the cumulative under size log-normal plot. If the AIM apparatus did not have a cut point precisely at 5 μm, then an alternate interpolation approach may be possible or justification that the AIM data provides equivalent APSD at its near 5 μm cut point when compared to the current Ph. Eur. approach for full-resolution CI data interpolation.

(b) A cut point close to or precisely at the *MMAD* of the OIP-emitted aerosol. If such data were to become accepted in a quality control environment, then the EDA approach would be pertinent to define the size-pertinent metrics (i.e., *LPM/SPM* ratio and *ISM*) and set subsequent specifications (Chaps. 5 and 6). This option is incompatible with the current EMA requirement to report fine-particle dose <5 μm aerodynamic diameter [9].

(c) Cut-point sizes other than 5 μm aerodynamic diameter (i.e., 4.7 μm, based on the cut-point size of stage 2 of the ACI when operated at 28.3 L/min). Acceptance of alternatives to 5 μm to distinguish fine from coarse mass fractions of the dose would be more aligned with the current USP requirements.

Given these considerations, it is therefore imperative that a clear understanding of the purpose(s) for which AIM is being used, and the implications that these purpose(s) have on the use of the CI-generated data, become well understood and accepted as early as possible by stakeholders. At least one pharmaceutical manufacturer has expressed the desire (in Europe) for there to be product-specific specifications with respect to APSD metrics, when justified and approved, in particular more flexibility in the future with respect to the 5 μm upper size limit for fine-particle mass.

There are some key requirements that must be satisfied for AIM (or EDA) to become an accepted approach, irrespective of the reasons for adopting AIM that have been outlined in previous chapters. These requirements extend beyond the ordinary "method validation" requirements that would be an integral part of demonstrating the method suitability, as it is fundamental that AIM (or EDA) is verified to be "fit for purpose." Without such evidence, it is unlikely that any regulatory agency or compendial committee would accept these new approaches, regardless of the merits that have been already explained in previous chapters.

In this context, it is worth noting that "fit for purpose" does not necessarily mean that AIM (or EDA) would be deemed applicable to every inhaled dosage form or device, either now or in the future. For example, it may not be appropriate for formulations and/or devices that produce a very narrow APSD ($GSD \leq 1.2$ equivalent to near monodisperse). Hence, the applicability of either the AIM concept or an actual AIM apparatus used will have some relationship with the specification for the APSD that is specified for the product in its target product profile (TPP).

Potential strategies that will lead to demonstration of "fitness for purpose" for AIM and EDA are discussed in Sects. 11.3.2 and 11.4, respectively. Following demonstration of the "fitness for purpose," a series of activities, including publication of detailed methodology and data, comparative data analysis using various current and AIM approaches, peer-reviewed literature and conference presentations, and pharmacopeia stimuli to revise and draft monographs, will all result in an evidence-based, industry and regulatory agency-challenged approach. Under such circumstances, new robust AIM methods can eventually become approved pharmacopeial methods.

11.3.2 Strategy for AIM "Fit for Purpose"

The key parts of the strategy to show AIM "fit for purpose" are illustrated in Fig. 11.1.

(a) Calibration data for the apparatus configuration:
It is reasonable to expect that either calibration data with full traceability to the international length standard or at least justification for the claimed cut-point sizes for that configuration would be provided for each particular AIM apparatus configuration proposed for adoption. These data are likely to be provided either by the apparatus manufacturer or less probably by a pharmaceutical company, given the complexity of the process. A highly desirable approach is the so-called archival calibration, in which a reference CI is calibrated, having

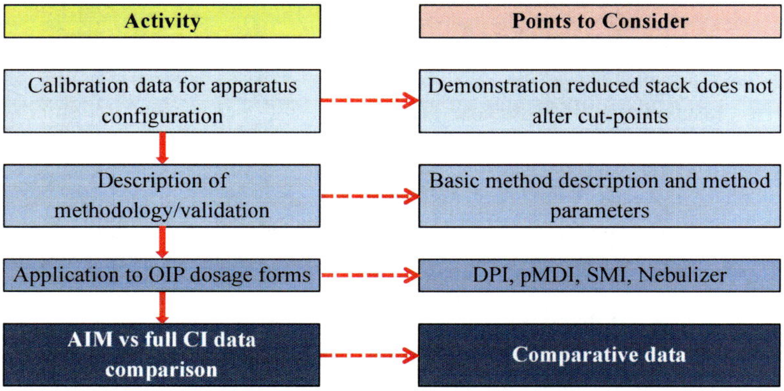

Fig. 11.1 Strategy for AIM to show that the approach is "fit for purpose"

stage nozzle diameters as close as possible to the nominal values of the manufacturing specification. The archival calibration avoids the proliferation of slightly different stage cut-point sizes arising from separate calibrations of the same CI type. Under such circumstances, it is of course highly preferable that such calibration data are published in the open literature.

(b) Description of methodology and validation:
It is critical that a full detailed description of the apparatus and methodology be provided when any data are submitted to a compendial committee or published in the open literature. This requirement is important for several reasons, principally to enable other stakeholders (regulatory agencies, National Bodies and Pharmacopieas, etc.) to be able to repeat the work when necessary. However, it is also necessary to develop an understanding whether or not the AIM apparatus configuration has any peculiar aspects that may influence data that are generated by this apparatus. Such information can only become apparent by practical experience. In addition to these AIM-specific considerations, there would be an expectation that the standard system suitability validation for CIs (as described in the pharmacopeias) would have been completed.

(c) Application to OIP dosage forms:
The applicability of AIM to the various OIP dosage forms must be established. In reality, this step may be complicated because the performance of the AIM-based apparatus may depend upon the particular aerosol and formulation characteristics, for instance, the type (presence of volatile species [17]) as well as the composition of excipients that are present [18] and/or the APSD delivered from a specific inhaler device [3]. Taking another example, passive, breath-actuated DPIs depend upon the patient-generated inhalation air flow-time profile to move the powder through the device, providing the "energy" to

de-aggregate and disperse the particles that then provide an efficacious lung dose [19]. Therefore, the air flow start-up kinetics may profoundly influence the APSD that is achieved [20]. If either the inlet and/or stage configurations of the AIM apparatus are different from those of the full-resolution CI, then it is possible that different APSD data will be obtained in comparison between the two systems. It is a concern, especially to those charged with pharmacopeial method development, that these divergences can eventually turn out to be inhaler-measurement system dependent. Sponsors therefore need to provide detailed methodology descriptions in submissions to have particular AIM-based configurations adopted into the compendia, in order to minimize the risk of inadvertently introducing such confounding effects. In particular, attention needs to be paid to the dimensions of the inlet port, internal dead space, and the stage configurations themselves. In summary, it cannot be assumed that an AIM apparatus is automatically suitable for a particular inhaled dosage form just because its acceptability has been reported elsewhere, most likely from investigations having limited scope.

(d) Comparative data for full-resolution CI and AIM apparatus:
Without doubt, the most obvious question that will be asked in connection with an application to include an AIM-based method into the compendia will be "*How does the APSD data using AIM apparatus compare to that acquired with a full stage CI?*"

The acquisition of copious data with both full-resolution and AIM CI systems will be a key requirement, as will be the subsequent demonstration of comparative performance in statistically robust terms. It is also reasonable that demonstration of system suitability through the API mass balance be provided [21]. Furthermore, cumulative undersize plots and API mass values should be demonstrated to be substantially comparable at selected cut-point sizes within the range of interest (0.5–10 μm aerodynamic diameter). Given that the basic physical aerodynamic particle sizing processes are occurring in both full-resolution and AIM CI systems (see Chap. 3), qualitatively "similar" results are to be expected as the norm. However, as already illustrated in Chap. 10, one or more of several experimental factors may require attention in order to achieve the desired goal of quantitative comparability. Therefore, it should not come as a surprise if initial data comparisons between AIM and full-resolution CIs are not always identical. The important thing is to be able to explain why such differences exist and mitigate identifiable cause(s), wherever possible. Furthermore, it may be possible to demonstrate a consistent relationship, even though the ideal correlation between full-resolution CI and AIM-based data that is directly proportional to the magnitude of API mass may not always be achieved.

It follows that the presentation to a regulatory agency as justification for using AIM as a product quality test would require comparative APSD data from the full CI together with associated metrics derived from an AIM-based apparatus of choice, in support of the proposed specification.

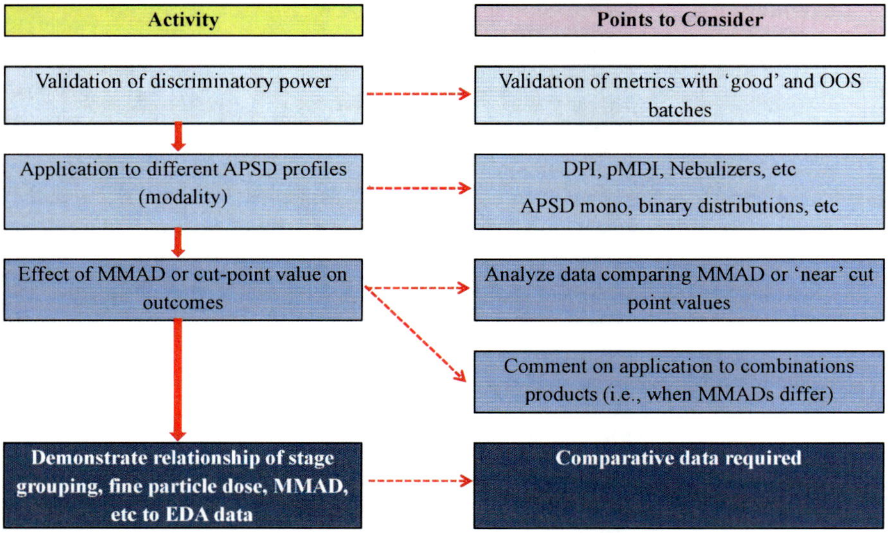

Fig. 11.2 Strategy for EDA to show that this approach is "fit for purpose"

11.3.3 Strategy for EDA "Fit for Purpose"

The key parts of the strategy to show EDA to be "fit for purpose" are illustrated in Fig. 11.2.

(a) Validation of discriminatory power:
The validation of discriminatory power may form part of the "full-resolution CI versus AIM comparison" stage of OIP development, when it is considered that EDA may be the metric of choice (Chap. 6). If an alternate AIM apparatus-derived metric is being proposed instead of *LPM/SPM* and *ISM*, attention will be needed to demonstrate discriminating power for small changes in APSD that is at least as good as current procedures, such as $FPM_{<5.0\mu m}$, in Europe or equivalent CI stage groupings in the USA.

(b) Application of EDA to different APSD Profiles arising from single component and combination products:
It should not be assumed that EDA is going to be applicable to every form of APSD profile. Although various product types with differing APSD profiles were assessed during the initial development of EDA as a concept for general applicability [22, 23], it is important that an assessment is carried out and that the requirements needed to meet EDA metrics are met for each new product under consideration.

One interesting area that has yet to be assessed is that of combination products. Clearly, using AIM to determine "fine-particle dose (mass)" would appear straightforward for both full-resolution CI and even an AIM-based apparatus having its cut-point size at 5 μm aerodynamic diameter. However, using EDA metrics with the large-to-small boundary size set at or close to the MMAD of one component may be problematic where the MMAD of the second API component is significantly different. Similar considerations would likely apply to tri-component or more complex combinations, in the event that such products are formulated for inhaled delivery. It follows that the selection of an appropriate cut point used in the AIM apparatus and proof of EDA sensitivity to all API components will be critical.

(c) Effect of OIP aerosol MMAD on large-to-small particle boundary size selection:
A key attribute of the EDA approach is the selection of the cut-point size that discriminates small from large particles. This boundary is ideally "identical" to the MMAD of the emitted aerosol. However, depending upon the skewness of the distribution, it can also be located in the range between 0.3 and 3.0 times MMAD for adequate sensitivity to distinguish small changes in APSD [23]. In the case of a new OIP coming under consideration for EDA evaluation, APSD data using a full-resolution CI method is the foundation from which the product MMAD is determined. In addition, however, there needs to be additional data justifying the selection of the large-to-small particle boundary size, if markedly different from the MMAD of the product. Such evidence may come from studies in which small perturbations in the APSD and hence MMAD of the product aerosol are intentionally introduced in order to evaluate the sensitivity of the EDA method. It follows that if an AIM-based measurement technique is being proposed instead of the full-resolution CI to obtain EDA metrics, justification will have to be provided for the selection of the cut-point size, if different from the MMAD.

(d) Demonstration of the relationship between CI stage grouping, fine-particle dose (mass), MMAD, etc., to EDA metrics:
There is already a body of evidence [24] that EDA metrics have better discriminating ability compared with current CI stage groupings/fine-particle dose assessments. Much of this material was presented in detail in Chap. 8 and will therefore not be discussed further here. It is expected that as the debate surrounding the applicability and limitations of EDA develops, more evidence supporting EDA as being less vulnerable to confounding than stage groupings will emerge, and such data will be important in securing the necessary validations to support incorporation of EDA into the compendia.

11.4 Regulatory Acceptance

In order for AIM and EDA to be developed as useful tools, it is imperative that validation studies are published, ideally in peer-reviewed journals, to make visible the strengths and weaknesses of these approaches. Most progress in the required direction can be made when the industry shares and moves forward with a set of common goals in relation to either but preferably both concepts. Fortunately, in the past 5 years, teams comprising industry experts in OIP in vitro assessment, both within the IPAC-RS and EPAG organizations, have been collaborating on many of the common practical issues associated with AIM and EDA, from which various publications have been created that are referred to in many locations within this book. Such documentation provides a sound basis from which meaningful discussions on how to develop the necessary stakeholder consensus concerning the approaches may be had with the key regulatory agencies in Europe and North America in order to gain acceptance of AIM and EDA concepts.

Initial informal discussions with individual regulators, both in Europe and the USA, have provided cautious optimism that these concepts could have a future in OIP aerosol in vitro assessments [25, 26]. However, further evidence-driven open publications together with perhaps more formal direct dialogue with the key regulatory agencies and compendial chapter committees will be critical if these concepts are to move from good ideas to practical reality as key components in the OIP life cycle.

The current thinking on AIM and EDA by the various stakeholders was outlined at an IPAC-RS-led satellite meeting held as part of the RDD Europe conference in the spring of 2011 [27]. At that meeting, the FDA representatives reiterated the position of their organization that better science should lead to better regulations. They also emphasized that they are open to discussion on alternative approaches to control strategies. They accepted that AIM would be a useful research tool and that it could save time, providing a larger throughput. AIM-based methods also have the potential to be more environmentally friendly than the current full-resolution CI approach. It appears that the following approach would be a suitable solution from the European regulatory perspective [26]:

1. Adoption of AIM methodology as option in Ph. Eur. monograph 2.9.18 (with no specific configuration). Seek harmonization with USP/JP at the outset.
2. Inclusion of EDA concept with associated *ISM* and *LPM/SPM* metrics in Ph. Eur. Chap. 2.9.18 as optional methodology.
3. Amend Ph. Eur. monograph on preparations for inhalation by deletion of fine-particle dose as mandatory test and replacement with a more general requirement to adopt limits for those parameters that are able to discriminate between acceptable and unacceptable batches.

Beyond these promising signs of eventual acceptance, some issues were raised that still need to be addressed. Most of these concerns are related to statistical aspects of EDA, such as (a) how various APSD fractions are correlated with EDA metrics, (b) whether different APSDs can share the same *LPM/SPM* ratio, and (c) how bimodal APSDs can be interpreted by an EDA-type approach.

There is also an overarching concern whether there is a loss of information potentially available from full CI stage data, by having EDA metrics alone. This consideration would be most relevant in a QC test where, in the USA, the FDA requires a specification set on several size fractions (usually at least 4), whereas EDA would provide only two metrics, the *LPM/SPM* ratio and *ISM*. Given these apprehensions, it is therefore incumbent upon the pharmaceutical industry stakeholders to provide convincing arguments in favor of both AIM and EDA through a combination of open publications and, where appropriate, direct presentation to regulators.

An important message arising from these opportunities for dialogue with regulatory agency representatives is that AIM and EDA do not present any major obstacles to becoming useable techniques. Indeed, during questions at the special symposium, the FDA representatives made it clear that to have AIM as a compendial technique would provide a useful practical approach to ensure the adoption of consistent methodology across the industry. The position of the EMA representative concerning compendial adoption was similar (see above).

Representatives from the European Pharmaceutical Aerosol Group (EPAG) provided a briefing to the UK Medicines and Healthcare Products Regulatory Agency (MHRA) on the AIM concept in June 2011, at which a cautiously favorable response was also received. The representatives from the MHRA, taking a similar position to that voiced earlier by the FDA representatives at the IPAC-RS satellite meeting entitled "*Perspectives on Efficient Data Analysis and Abbreviated Impactor Measurements as Quality Assessment Tools*" held at the Respiratory Drug Delivery Europe 2011 conference, that the development of a robust understanding of the relationship between full-resolution CI and AIM data for all OIP classes would be critical to their acceptance of the AIM approach. Given the significant differences in equipment and data assessment, AIM could be considered by the MHRA as a "new technology," and as such the justification for its adoption would have to reflect this standing through appropriate validation. Inclusion of AIM-based methodology in a Ph. Eur. monograph would go a long way towards demonstrating the approach to be valid and justifiable for use, since Ph. Eur. test methods are considered authoritative in the case of any dispute in post-marketing surveillance. These suggestions further highlight the importance to have AIM-based methods formalized by inclusion in the compendia.

Within the compendial environment, both the Aerosols Subcommittee of the USP General Chapters—Dosage Forms Expert Committee—and the Inhalanda working group of the Ph. Eur. will need to conduct their own assessments both technically and from a quality assurance standpoint as the first step towards adoption. It is anticipated that from the start, both groups will share dialogue and move, either formally or informally, towards "harmonized" AIM and EDA positions that lead ultimately to harmonized monographs. The best route for this to occur would be for AIM and later perhaps EDA to be progressed through the formal PDG harmonization process which includes all three leading pharmacopeia (USP, Ph. Eur., and JP). In Europe, pharmaceutical industry stakeholders can remain involved by lobbying their respective national pharmacopeial authorities to progress this process as well as by responding to technical articles appearing in the journals *Pharmeuropa Scientific Notes* and draft monograph texts that are published in *Pharmeuropa*. In the USA and Canada, the most effective way to remain involved will be through responses to "*Stimuli to Revision*" articles, when they appear in the journal, *Pharmacopeial Forum*, and later as draft chapters are published in this journal for public review.

11.5 Concluding Observations

There are many AIM apparatus configurations available and several reasons why one may choose to use an AIM-based technique, as have been explained in earlier chapters. It is evident that incorporation of AIM (and possibly EDA) into the pharmacopeial monographs related to OIP testing will be a key component in gaining acceptance by the regulatory agencies. Given this situation, it will be a challenge to describe each AIM apparatus configuration, even though currently several full-resolution CI apparatuses are described within the pharmacopeias. However, such a process may not be necessary, as a description of AIM and EDA as concepts associated with narrative comprising minimum apparatus requirements may be a suitable alternate approach.

In the short term, stakeholders within the inhaled pharmaceutical industry involved in OIP assessments, and desirous of seeing AIM and EDA become accepted as routine processes, need to work collaboratively to achieve the following goals:

(a) To establish precisely where AIM and EDA have utility and, perhaps more importantly, where they are not applicable
(b) To demonstrate unambiguously that AIM and, preferably, EDA as well are "fit for purpose"
(c) To assist in helping the working groups of the USP and Ph. Eur. develop appropriate methods for the compendia

Once these goals have been met, then all that remains to be achieved is that these concepts become recognized within relevant regulatory guidance as these documents are revised. Under the current mode of operation, this would give pharmaceutical industry stakeholders confidence in the application of the techniques, and it would indicate that many (all) of the general and specific concerns that the various regulatory agencies have about AIM and EDA had been addressed. Put in other words, at this mature stage the concepts would no longer be regarded as "new technology."

As a final thought, as regulatory guidance moves towards a QbD approach, it could be anticipated that specific techniques may not be mentioned in detail, as, from the user's perspective, the selection process should be based more on the suitability of the underlying science and not just because a given method is a listed technique. In other words AIM and EDA may or may not always be the most appropriate (quality) assessment tools to be used in the OIP life cycle, and the sponsoring company will need to assess the situation at each stage and justify their use to a regulatory agency with appropriate science-based arguments.

References

1. Newman SP (1998) How well do *in vitro* particle size measurements predict drug delivery *in vivo?* J Aerosol Med 11S1:S97–S104
2. Mitchell JP, Newman SP, Chan H K (2007) *In vitro* and *in vivo* aspects of cascade impactor tests and inhaler performance: a review. AAPS PharmSciTech 8(4):Article 110. http://www.aapspharmscitech.org/articles/pt0804/pt080411O/pt0804110.pdf. Accessed 20 Jan 2012

3. Newman SP, Chan H K (2008) *In vitro/In vivo* comparisons in pulmonary drug delivery. J Aerosol Med 21(1):1–8
4. International Conference on Harmonization of Technical Requirements for Registration of Pharmaceuticals for Human Use (ICH) (2009) Q8(R2): pharmaceutical development. http://www.ich.org/products/guidelines/quality/article/quality-guidelines.html. Accessed 20 Jan 2012
5. US Food and Drug Administration (FDA) (1998) CDER. Draft guidance for industry metered dose inhaler (MDI) and dry powder inhaler (DPI) drug products chemistry, manufacturing, and controls documentation, Rockville, MD, USA. http://www.fda.gov/cder/guidance/2180dft.pdf. Accessed 6 Jan 2012
6. US Food and Drug Administration (FDA) (1993) CDRH. Reviewer guidance for nebulizers, metered dose inhalers, spacers and actuators, Rockville, MD, USA. http://www.fda.gov/downloads/MedicalDevices/DeviceRegulationandGuidance/GuidanceDocuments/ucm081293.pdf. Accessed 20 Jan 2012
7. Marple VA, Roberts DL, Romay FJ, Miller NC, Truman KG, Van Oort M, Olsson B, Holroyd MJ, Mitchell JP, Hochrainer D (2003) Next generation pharmaceutical impactor – Part I: Design. J Aerosol Med 16(3):283–299
8. Adams WP, Christopher D, Lee DS, Morgan P, Pan Z, Singh GJP, Tsong Y, Lyapustina S (2007) Product quality research institute (PQRI) evaluation of cascade impactor profiles of pharmaceutical aerosols, Part 1: Background for a statistical method. AAPS PharmSciTech 8(1): article 4. http://www.aapspharmscitech.org/articles/pt0801/pt0801004/pt0801004.pdf. Accessed 20 Jan 2012
9. European Medicines Agency (EMA) (2006) Guideline on the pharmaceutical quality of inhalation and nasal products. London, UK, EMEA/CHMP/QWP/49313/2005 Final. http://www.ema.europa.eu/pdfs/human/qwp/4931305en.pdf. Accessed 20 Jan 2012
10. Health Canada (2006) Guidance for industry: pharmaceutical quality of inhalation and nasal products. File Number file number: 06-106624-547. http://www.hc-sc.gc.ca/dhp-mps/alt_formats/hpfb-dgpsa/pdf/prodpharma/inhalationnas-eng.pdf. Accessed 20 Jan 2012
11. European Directorate for the Quality of Medicines and Healthcare (EDQM) (2012) Preparations for inhalation: aerodynamic assessment of fine particles. Section 2.9.18 – European Pharmacopeia, Council of Europe, Strasbourg, France
12. European Medicines Agency (EMA) (2008) Overall summary of the scientific evaluation of Salbumalin and associated names. http://www.ema.europa.eu/docs/en_GB/document_library/Referrals_document/Sabumalin_29/WC500007491.pdf. Accessed 27 Jan 2012
13. Bonam M, Christopher D, Cipolla D, Donovan B, Goodwin D, Holmes S, Lyapustina S, Mitchell J, Nichols S, Petterson G, Quale C, Rao N, Singh D, Tougas T, Van Oort M, Walther B, Wyka B (2008) Minimizing variability of cascade impaction measurements in inhalers and nebulizers. AAPS PharmSciTech 9(2):404–413
14. Christopher D, Curry P, Doub B, Furnkranz K, Lavery M, Lin K, Lyapustina S, Mitchell J, Rogers B, Strickland H, Tougas T, Tsong Y, Wyka B (2003) Considerations for the development and practice of cascade impaction testing including a mass balance failure investigation tree. J Aerosol Med 16:235–247
15. Mitchell JP (2008) The abbreviated impactor measurement (AIM) concept for aerodynamic particle size distribution (APSD) in a quality-by-design (QbD) environment. Presentation at IPAC-RS Conference, 2008. http://www.ipacrs.com/PDFs/IPAC-RS%202008%20Conference/Day%203/5-%20IPAC-RS%20Conference%202008%20-%20Jolyon%20Mitchell.pdf. Accessed 20 Jan 2012
16. Mitchell JP, Nagel MW (2003) Cascade impactors for the size characterization of aerosols from medical inhalers: their use and limitations. J Aerosol Med 16(3):341–377
17. Stein SW (2008) Aiming for a moving target: challenges with impactor measurements of MDI aerosols. Int J Pharm 355(1–2):53–61
18. Bosquillon C, Lombry C, Préat V, Vanbever R (2001) Influence of formulation excipients and physical characteristics of inhalation dry powders on their aerosolization performance. J Contr Rel 70(3):329–339

19. Chan H-K (2006) Dry powder aerosol delivery systems: current and future research directions. J Aerosol Med 19(1):21–27
20. Blakey D, Harris D, Wilkins J (2008) The time-dependent effect of airflow profile on DPI performance. Drug Delivery to the Lungs-19, The Aerosol Society, Edinburgh, UK, pp 187–189. http://ddl-conference.org.uk/index.php?q=previous_conferences. Accessed 4 Aug 2012
21. Wyka B, Tougas T, Mitchell J, Strickland H, Christopher D, Lyapustina S (2007) Comparison of two approaches for treating cascade impaction mass balance measurements. J Aerosol Med 20(3):236–256
22. Tougas TP, Christopher D, Mitchell JP, Strickland H, Wyka B, Van Oort M, Lyapustina S (2009) Improved quality control metrics for cascade impaction measurements of orally inhaled drug products (OIPs). AAPS PharmSciTech 10(4):1276–1285
23. Tougas T (2011) Efficient data analysis in quality assessment. In: Dalby RN, Byron PR, Peart J, Suman JD, Young PM (eds) RDD Europe-2011. Davis Healthcare International, River Grove, IL, pp 209–213
24. Christopher D, Dey M (2011) Detecting differences in APSD: efficient data analysis (EDA) vs. stage grouping. In: Dalby RN, Byron PR, Peart J, Suman JD, Young PM (eds) Respiratory Drug Delivery Europe 2011. Davis Healthcare International, River Grove, IL, pp 215–224
25. Peri P (2011) Regulatory perspectives on abbreviated impactor testing (AIM). Presentations and a summary report. http://www.ipacrs.com/CI.html. Accessed 20 Jan 2012
26. Weda M (2011) Regulatory perspectives on EDA and AIM: considerations from a European perspective. Presentations and a summary report. http://www.ipacrs.com/CI.html. Accessed 20 Jan 2012
27. IPAC-RS (2011) Satellite conference at RDD Europe 2011: perspectives on efficient data analysis methods and abbreviated impactor measurements as quality assessment tools. Presentations and a summary report. http://www.ipacrs.com/CI.html. Accessed 20 Jan 2012

Chapter 12
Applying the AIM Concept in Support of Developing Improved In Vitro–In Vivo Relationships for OIPs

Jolyon P. Mitchell, Mark Copley, and Derek Solomon

Abstract The previous chapters have focused primarily on the application of AIM-EDA in the quality control part of an OIP lifecycle. An AIM-based approach would also be desirable for comparing and ideally correlating in vitro APSD-derived metrics with the likely particle deposition profile in the HRT that in turn should be linked with clinical effects. In addition to selection of appropriate size boundaries between coarse and fine particle mass fractions, there is the consideration of modifying the reduced impactor to add a third subfraction that relates to the measurement of the fraction of the dose ex-inhaler that comprises extra-fine submicron-sized particles. Furthermore, adapting the AIM concept to an alternative approach in which more clinically pertinent measures of in vitro performance are obtained raises the prospect of making the aerosol transport system more realistic in terms of human anatomy. An obvious move in this direction would be to replace the USP/Ph.Eur. induction port that was designed primarily to support OIP QC-based testing with an inlet that more appropriately models aerosol flow through the human oropharyngeal/nasopharyngeal region, depending upon patient age being studied. This chapter describes key features of how so-called AIM-pHRT systems might be constructed. The prefix "p" refers to the potential application of this alternative AIM-based approach. It will be for sponsors of this type of CI-based measurement application to undertake validation studies with their products. Such studies will likely compare

J.P. Mitchell (✉)
Trudell Medical International, 725 Third Street, London, ON N5V 5G4, Canada
e-mail: jmitchell@trudellmed.com

M. Copley
Copley Scientific Limited, Colwick Quays Business Park, Private Road No. 2, Colwick, Nottingham NG4 2JY, UK
e-mail: M.Copley@copleyscientific.co.uk

D. Solomon
Melbourn Scientific Limited, Saxon Way, Melbourn SG8 6DN, UK
e-mail: Derek.Solomon@melbournscientific.com

T.P. Tougas et al. (eds.), *Good Cascade Impactor Practices, AIM and EDA for Orally Inhaled Products*, DOI 10.1007/978-1-4614-6296-5_12,

measurements with AIM-pHRT apparatus(es) with both full-resolution CI data and particle deposition profiles utilizing imaging methods, such as gamma scintigraphy, positron emission tomography, or possibly magnetic resonance imaging. The chapter concludes with the results from the first laboratory-based evaluation of an AIM-pHRT system based on the ACI equipped with the recently commercialized "Alberta Idealized Throat" (AIT) adult upper airway geometry.

12.1 A Roadmap for Improved Comparisons of Laboratory-Generated OIP Performance Measures with Clinical Data

An AIM-based approach can potentially be used to compare and correlate in vitro APSD data with the likely particle deposition profile in the HRT that should ideally be linked with clinical effects [1, 2]. The relationship between deposition locations in the lower HRT and the clinical effect of the drug being delivered in aerosol form depends on the action of the API(s) in relation to appropriate receptors at different locations within the respiratory tract. Some drug products may be designed to penetrate deep into the periphery of the lung [3, 4], although others may be intended for deposition primarily in central or upper airways for maximum effectiveness [5, 6]. In Chap. 2, it was explained that the size selectivity of the HRT is relatively poor compared with that of a CI. It can therefore be argued that any further detail in terms of size resolution beyond the three fractions *CPM*, *FPM*, and *EPM*, traditionally reported when reducing the raw data from full-resolution CIs into stage groupings, is superfluous from the clinical perspective [7]. Interestingly, only three size fractions—oro- or nasopharyngeal, tracheobronchial, and alveolar deposition—are used in the field of occupational health to describe the inhalation of potentially toxic particles [8].

Furthermore, similar size fractions based on the mass of particulate matter <10.0, 2.5, and 1.0 μm aerodynamic diameter (PM_{10}, $PM_{2.5}$, and $PM_{1.0}$) define bounds for respiratory tract-relevant deposition fractions related to atmospheric environmental pollutants (Fig. 12.1) [8]. Given this albeit indirect evidence from related fields of study involved with inhalation of potentially harmful aerosol particles, it can be argued that any more detailed fractionation of the APSD in the context of therapeutic drug delivery to the respiratory tract both dilutes the essential information to assess clinical safety and efficacy and has the potential to magnify intrinsic data variability.

Replacement of the right-angle bend USP/Ph.Eur. induction port (Fig. 12.2) should be considered as the first step towards the development of either full-resolution or abbreviated CI measurements that provide data for comparison with lung deposition profiles.

Although the right-angle bend internal geometry suffices for measurements related to OIP quality control, where the emphasis has to be on simplicity as a key component of method robustness, this inlet pathway differs markedly from actuality [9]. Apart from considerations concerning the effect of real upper airway

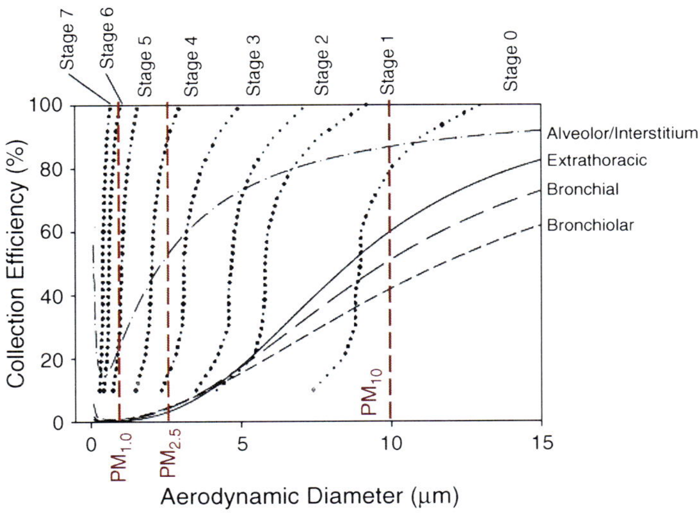

Fig. 12.1 Comparison between definitions for size-segregated pollutant aerosol (PM-series), ICRP lung deposition subfractions for a healthy male (PIFR = 28.3 L/min; V = 2 L), and cut-point sizes for the eight-stage nonviable ACI at 28.3 L/min

Fig. 12.2 Schematic of USP/Ph.Eur. induction port inlet

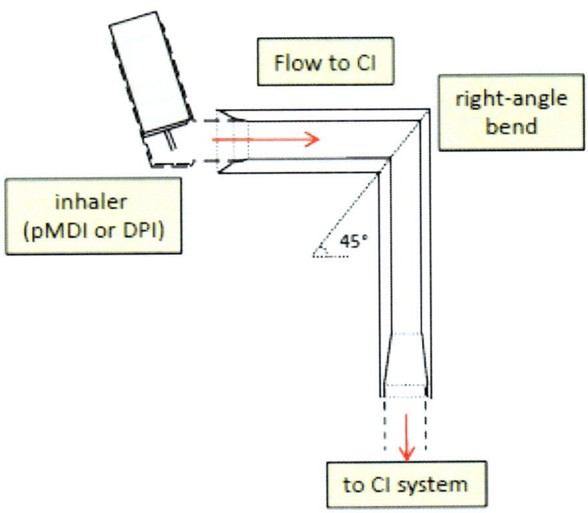

geometry on aerosol transport processes, the compendial inlet has the same internal dimensions regardless of the end user of the product (infant, small child, or adult). There is also a significant body of evidence identifying the divergence between lung deposition data and in vitro measurements of metrics derived from OIP APSDs by full-resolution systems equipped with the USP/Ph.Eur. induction port [10–12].

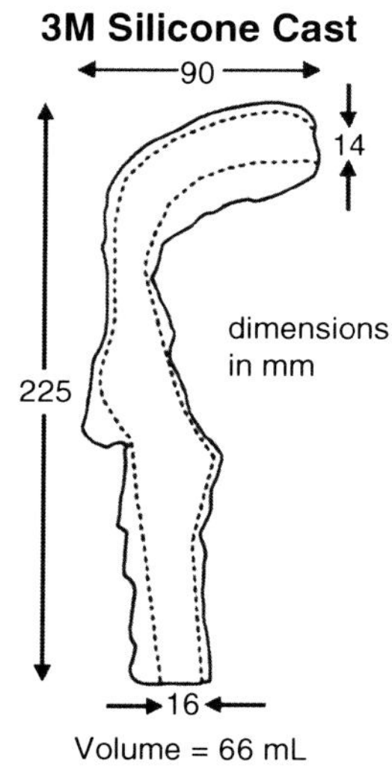

Fig. 12.3 Silicone cast of an adult upper airway developed by 3M Drug Delivery Systems (*From* [9]—*used with permission*)

As early as the mid-1990s, Berg reported [13], in a survey of OIP test methodologies, that casts of human throats of the sort described at that time by Swift [12] are better models to mimic the upper respiratory tracts of both adults and children than a standardized glass inlet having a simplified flow channel analogous to that for the USP/Ph.Eur. induction port.

In 1998, Dolovich and Rhem pointed out the wide variety of simplified and anatomically accurate inlets that are available for OIP testing, in the context of demonstrating the importance of inlet design for both the quantity (mass) and size distribution of the aerosol that eventually gets to the CI measurement system [9]. Figure 12.3 is an example of an adult upper airway cast developed around this time by 3M Drug Delivery Systems for their research. In calibration/validation experiments, Velasquez and Gabrio showed that the collection efficiency profile of this inlet was shifted significantly to finer sizes compared with the positions of other standardized inlet designs [14], including that of the USP/Ph.Eur. induction port (Fig. 12.4).

On the basis of the data from Velasquez and Gabrio [14], it might be anticipated that an AIM-based CI with an anatomically accurate upper airway would allow less coarse particulate through to the CI entry. In the case of MDI-generated aerosols, such an airway might also prevent less API mass contained in the ballistic fraction from reaching the CI system. The ballistic component of the emitted dose comprises a high-velocity droplet stream that exits from the inhaler mouthpiece as the propellant flash evaporates. This idea is taken up later in this chapter when data obtained in 2011 with the commercialized "Alberta Idealized Throat" (AIT) geometry, developed by Finlay at the University of Alberta, Canada, are discussed.

The 3M model contained other important refinements, including a realization of the upper part of the trachea-bronchial tree together with a filter to mimic

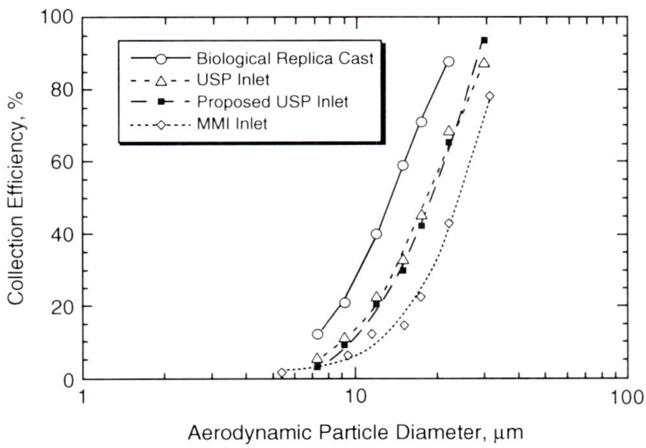

Fig. 12.4 Collection efficiency data for various inlet geometries at $Q = 30$ L/min; the biological replica cast refers to the 3 M inlet, and the MMI inlet is the induction port designed for Marple-Miller impactors. The proposed USP inlet was a new design of induction port that was not adopted (*From* [14]—*used with permission*). (**a**) CFC-albuterol. (**b**) CFC-cromoglycate

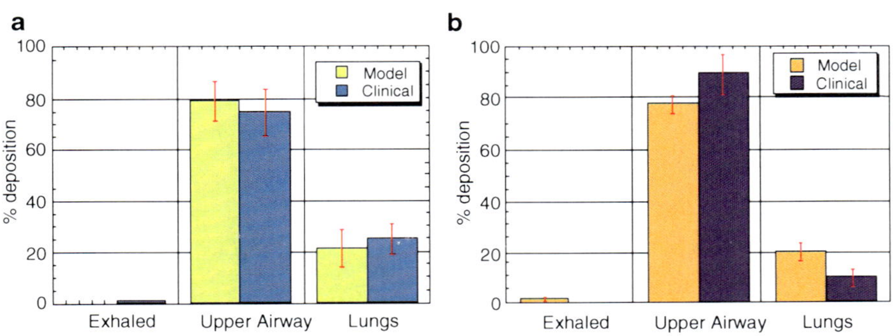

Fig. 12.5 Relative aerosol deposition patterns for the 3M physical model and from clinical HRT deposition studies (*From* [14]—*used with permission*)

deposition further into the airways that are probably impractical in an AIM-based apparatus. Importantly, Velasquez and Gabrio were able to obtain good agreement in terms of mass of API deposited in the interior of their biological replica cast as a function of the mass deposited in these upper airways, the distal airways, and exhaled mass, with regional deposition data from gamma scintigraphic studies with adults. In both instances, they used radiolabeled aerosols from CFC-salbutamol (Fig. 12.5a) and CFC-cromoglycate (Fig. 12.5b) MDI-delivered products, representing aerosols containing lower and higher mass percentages of coarse particulate, respectively.

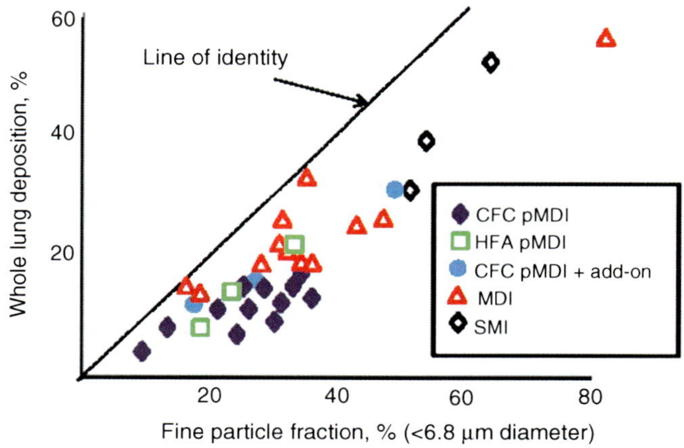

Fig. 12.6 Correlation between mean whole lung deposition (gamma scintigraphy) and mean $FPF_{<6.8\ \mu m}$ by MSLI, for 33 inhalers (*From* [2]—*used with permission*)

It is conjectured that the incorporation of more realistic models of the upper respiratory tract may offset some or all of the consistent bias observed in CI measurements comparing whole lung deposition with CI-measured $FPF_{<6.8\mu m}$ reported by Newman and Chan [2] (Fig. 12.6).

Their analysis collected data using the USP/Ph.Eur. induction port for 33 inhalers representing all classes of OIP. Importantly, Newman and Chan observed that when a glass bulb inlet to the MSLI was used, all the DPI data lay on the same regression line, but the data point for the MDI group lay away from the line [2]. After replacing the glass bulb inlet to the impinger with a nonstandard inlet made from an anatomical cast of the oropharynx, the data for all inhaler classes, including the MDIs, lay on a single regression line. In consequence, they suggested that an anatomical inlet to the MSLI would allow for more ready comparison between inhalers of different classes. It would therefore seem reasonable to extend this conclusion to the other CI types, given that the differences observed appeared to be related to the aerosol transport behavior upstream of the APSD measurement apparatus.

In this context, as a further refinement towards the realization of clinical reality, an inlet having a fixed internal geometry would not be able to reflect changes in particle deposition brought about by variations in oral cavity volume. Such effects happen with real patients if they use devices with different mouthpiece dimensions and are consequently opening their mouth wider or less wide [15]. Although anatomically correct inlets having varying oral cavity volume are the obvious solution, use of such inlets has practical complications associated with them (e.g., simulation of the mucosa, drug recovery from the complex geometries of the interior surfaces). The recent development of anatomically correct models developed from 3D MRI scans of live patients should provide at low-cost aerosol transport conditions as close to those in reality as is possible for such in vitro systems [16].

The relationships between respiratory tract internal dimensions, complexity in terms of modeling anatomic structures in the aerosol flow pathway, and their association with meaningful IVIVRs continue to be a hotly debated area [9]. It is foreseeable that as inlets representing different patient age classes (i.e., infant, small child, and adult) are developed [17–19], this debate will intensify.

There can be large inter-patient variability in clinical trials to elicit dose–response relationships with inhaled medication [20, 21], in addition to the less frequently considered added variability introduced with disease-modifying patency of airways in the respiratory tract and resulting lung function [22, 23]. Hence, small shifts in mass within the size ranges related to *CPM* and *FPM* subfractions in the event that the boundary is set at either 4.7 or 5 μm aerodynamic diameter in an AIM-pHRT system in order to match full-resolution data following North American or European practices, respectively, are unlikely to have measurable clinical consequences. This situation may be true even when a convincing IVIVR is established, as could be argued is potentially possible for some bronchodilator-based formulations [2]. To put in another way, the precision of existing CI-based methods for determining these QC metrics greatly exceeds the precision available to the clinician for the corresponding clinical metrics such as forced expiratory volume in 1 s (FEV_1), forced expiratory flow from 25 to 75 % of vital capacity ($FEF_{25-75\,\%}$), and similar indicators of airway patency obtainable from well-established spirometric measurements to assess obstructive disease [24]. The higher precision of in vitro methods is likely to become even more apparent for other therapeutic modalities such as inhaled anti-inflammatory products, where IVIVRs are not yet established with confidence [2].

12.2 Criteria for Abbreviated CI Systems Appropriate for Comparison with Clinical Data

The qualification of the clinical batches of OIP drug product together with the laboratory assessment of certain add-on devices (spacers and VHCs used with MDIs) that substantially affect the size properties of the emitted aerosol also necessitates the measurement of APSD-based data that may not always be appropriate to analyze in terms of the EDA metrics *LPM*/*SPM* and *ISM*. The main reasons for such a limitation are as follows:

1. The boundary between *LPM* and *SPM* is intended to be chosen to be as close as practical to the *MMAD* of the product in order to optimize the sensitivity of these metrics to small changes in APSD [25]. Depending upon the formulation in question, this limit may be at a significantly finer size than the traditional limit of 5 μm aerodynamic diameter used for discriminating particles that penetrate beyond the upper respiratory tract to the airways of the lungs in adults.
2. The metric *ISM* excludes the mass of API that is collected in the induction port (and pre-separator if used). These amounts are important in the assessment of add-on devices (spacers and valved holding chambers) that are widely used with MDIs to mitigate upper airway deposition [26].

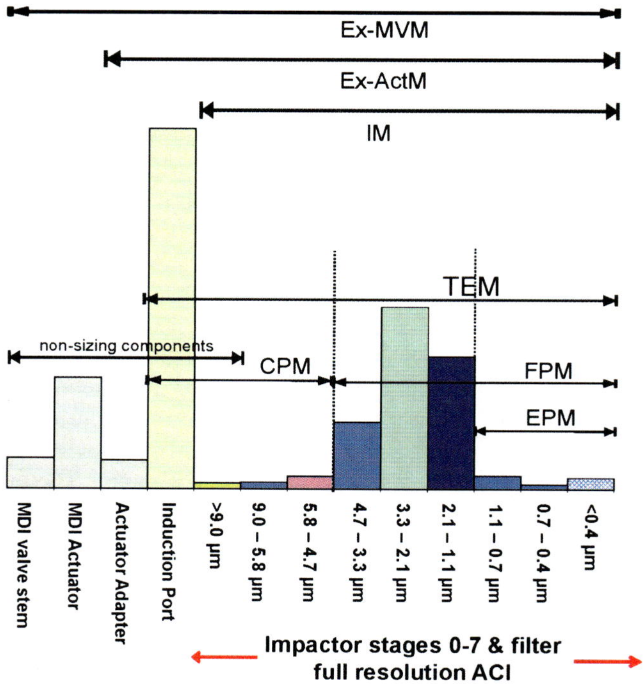

Fig. 12.7 Metrics pertinent to understanding particle deposition in a pHRT CI based on the ACI design operated at 28.3 L/min

In these instances, the abbreviated measurement concept can still apply to OIP characterization, but through measures of coarse, fine, and possibly extra-fine particle fractions (Fig. 12.7).

The metrics associated with such a so-called AIM-pHRT-based apparatus are based on the stage d_{50} values for the ACI operated at a flow rate of 28.3 L/min. The terms "*Ex-MVM*," "*Ex-ActM*," and "*IM*" refer to total mass of API ex-metering valve (in the case of an MDI), ex-inhaler mouthpiece, and mass collected in the impactor, respectively. Although some studies have demonstrated moderately good IVIVRs using anatomically correct model inlets [9, 27], more experimental work remains to be done to explore the full potential for such approaches as a prediction tool for lung dose from OIPs (Sect. 12.2). In particular, there is currently no standardized adult anatomical throat/upper airway. Those made from cadaver casts have been reported as being prone to bias caused by partial tissue collapse, but inlets reconstructed from three-dimensional imaging from living subjects using magnetic resonance imaging (MRI) show promise [28, 29].

The dimensions of the upper airways (mouth, pharynx, and larynx) vary between individuals [9], with the position of the tongue determining the velocity of inhaled air through the mouth. Such intrinsic variability will inevitably result in fluctuations

Fig. 12.8 Adult upper respiratory tract from MRI scan. Region A = oral cavity; B = pharynx; C = larynx opening into pharynx; D = larynx (*From (*[31]—*external view) and (*[32]—*MRI image)—used with permission*)

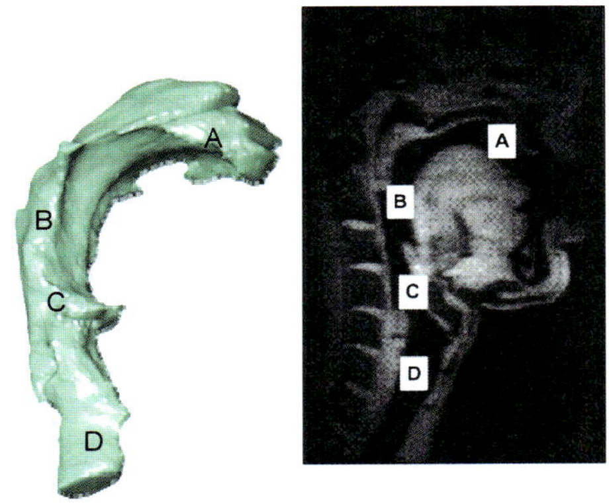

in the amount of inertial impaction that occurs in the upper airways from one patient to another, which in turn controls the mass of aerosol able to penetrate further into the respiratory tract and reach the airways of the lungs [2].

Given the importance of the upper respiratory tract geometry to the resulting aerosol deposition profile in the lungs [30], an obvious refinement to the AIM-pHRT concept would be to replace the compendial inlet with an anatomically correct throat model. One of the better examples is the model produced from MRI studies facilitated in the early 2000s by the Oropharyngeal Consortium, comprising three leading pharmaceutical companies, namely, AstraZeneca Research and Development, Aventis Pharma, and GSK plc [31]. The consortium members sought to understand the response of the human oropharynx to OIPs and the subsequent effect of that response on the lung dose of inhaled medication. As a first step to achieving these aims, structural information of the oropharyngeal region was required. The resulting model, an example of which is shown in sagittal view in Fig. 12.8, was developed with human volunteers during an appropriate inhalation maneuver [32].

Given the difficulty in agreeing upon a standard adult upper airway anatomy and commercializing such a model, it is likely to be more convenient to make use of the "Alberta" idealized adult anatomic throat (AIT) geometry that has been developed and extensively validated by Finlay and colleagues in the early 2000s [33–36] (Figs. 12.9 and 12.10).

Even making use of this model, there is still the limitation that currently there are no similar models yet commercially available to support in vitro studies to determine the performance of OIPs intended for infants and small children, in which the anatomy differs significantly from that of an adult [37], including the presence of obligate nasal breathing [38]. However, at the time of writing, this position may be about to change [Copley M (2012) Copley Scientific Ltd, personal communication], given the recent validation of an idealized child upper airway inlet [17].

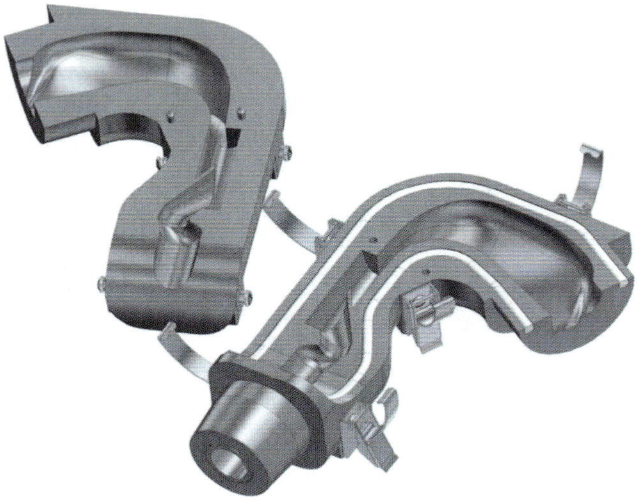

Fig. 12.9 "Alberta" idealized adult anatomic throat (AIT) commercialized by Copley Scientific Ltd, as inlet to an NGI (*Courtesy of Copley Scientific Ltd*)

Fig. 12.10 Opened "Alberta" idealized adult anatomic throat (AIT) showing flow pathway (*Courtesy of Copley Scientific Ltd*)

12.3 Validation of an AIM-pHRT System Based on the ACI

In Chap. 10, the description of the in vitro evaluation of a prototype AIM-pHRT system based on the ACI and configured to have stage d_{50} values in accordance with Fig. 12.7 was validated as one arm of a systematic investigation of precision and accuracy of this system, an AIM-QC impactor, with a full-resolution ACI as the

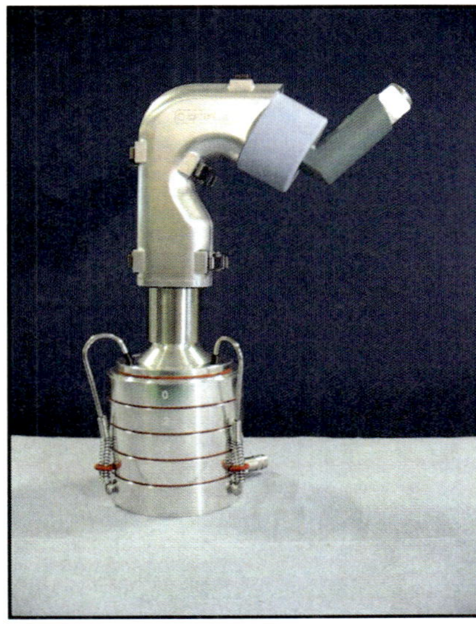

Fig. 12.11 AIM-pHRT system based on the ACI equipped with "Alberta adult throat" inlet (*Courtesy of Copley Scientific Ltd, UK*)

reference system, using MDI-delivered salbutamol as the test formulation [39]. All CIs were equipped with a USP/Ph.Eur. induction port, as the purpose of this initial validation exercise was to establish the measurement capability of the AIM-based systems as cascade impactors, rather than compare the effect of different inlets. Full details of the study are provided in Chap. 10, so the summary of the main findings provided here just relates to the AIM-pHRT system. Importantly, this study confirmed that the measurement precision and accuracy for measures of *FPM*, *CPM*, and *IM* were comparable with that using the full-resolution ACI.

The follow-on study confirmed that a filter soaked in the polyoxyethylene lauryl ether surfactant (Brij 35) used to mitigate particle bounce from the collection plates of these impactors was needed for the second stage of the AIM-pHRT impactor [40]. The surfactant-soaked filter provided additional protection from this source of bias in order to obtain measures of *EPM* that were comparable with the range of values obtained from the ACI. Although values of *FPM* and *CPM* were apparently unaffected by particle bounce arising from displacement of the surfactant layer by the incoming flow through the first impaction stage, it may be prudent for the most accurate work to consider the soaked filter option as a precaution with this type of AIM-pHRT system. Recent work by Chambers and Smurthwaite, also discussed in Chap. 9, with a fine particle dose-abbreviated impactor that is based on the parent ACVI (Westech Instruments Ltd, UK), containing surfactant-soaked filters, indicates that internal losses may be reduced with this mitigation measure for particle bounce [41].

A comparative study was undertaken in 2011, comparing an AIM-pHRT system based on the C-FSA equipped with an adult AIT idealized throat inlet with the same CI using the standard USP/Ph.Eur. induction port (Copley Scientific Ltd, UK, Fig. 12.11).

MDI-delivered salbutamol (albuterol) was chosen as the OIP for evaluation, given that the bulk of the API mass contained within the aerosol APSD is in the

Table 12.1 Mass salbutamol per actuation (μg) for the ACI-USP/Ph.Eur. induction port and AIM-pHRT with USP/Ph.Eur. and AIT inlets (*From* [42]—*used with permission*)

Apparatus	AIM-pHRT	ACI	
Location	AIT	AIT	USP/Ph.Eur.
Inhaler mouthpiece + inlet	54.7 ± 3.6	52.9 ± 2.3	55.0 ± 4.9
Stage 0		2.3 ± 0.3	5.4 ± 0.9
Stage 1		1.4 ±7 0.1	2.5 ± 0.2
Stage 2	4.5 ± 0.7	1.9 ± 0.1	2.6 ± 0.4
Stage 3		8.4 ± 0.6	9.6 ± 1.0
Stage 4		17.9 ± 1.3	15.7 ± 2.2
Stage 5	39.6 ± 4.0	10.5 ± 1.3	8.5 ± 0.8
Stage 6		1.0 ± 0.1	1.0 ± 0.1
Stage 7		0.3 ± 0.1	0.3 ± 0.1
Filter	1.7 ± 0.3	0.7 ± 0.3	0.8 ± 0.2
TM_{exMVM}	100.5 ± 7.7	97.2 ± 4.9	101.5 ± 9.5
IM	45.8 ± 4.9	44.2 ± 3.0	46.5 ± 4.6

n = 5 replicates per apparatus/inlet configuration; mean ± SD

important range from 1 to 5 μm aerodynamic diameter [42]. A detailed description of the testing and data that were obtained is provided below, as this type of information is a good guide as to the amount of detail that will be required when introducing the AIM-pHRT approach into routine use with whatever OIP class to which this apparatus is being applied.

In this investigation, the interior surfaces of the AIT were coated with a thin layer of silicone oil to simulate the mucosa, as was done by Ehtezazi et al., investigating the MDI-delivered medication to a cascade impactor via different anatomically correct oropharyngeal models [43]. The CI collection plates were also oil coated to mitigate particle bounce and re-entrainment, in accordance with the practice established in previous AIM-based studies [44]. The comparative data obtained from the full-resolution ACI as control system with either USP/Ph.Eur. induction port or with AIT and the AIM-pHRT apparatus with AIT are summarized in Table 12.1.

Individual total mass recovery values for salbutamol were within ±16 % of the label claim (100 μg/actuation), with most values lying within ±10 % label claim. The grouped mass recovery values (μg/actuation; mean ± SD) were equivalent at 100.5 ± 7.7, 97.2 ± 4.9, and 101.5 ± 9.5 for the AIM-pHRT, ACI-AIT, and ACI-Ph. Eur./USP configurations, respectively (1-way *anova*, $p = 0.64$). These system suitability data were close to expectations for the particular OIP, indicating that internal losses of API were minimal, whichever configuration was being evaluated.

Individual stage deposition data were highly reproducible, with their coefficients of variation (*C-of-V*) being comparable across configurations, and increasing in magnitude with decreasing absolute mass as would be expected as these values approached closer to the lower limit of detection of 0.1 μg salbutamol/actuation.

The corresponding APSDs, normalized either to total mass of salbutamol ex-inhaler (TM_{exMVM}, Fig. 12.12a) or just to the mass recovered from the full-resolution

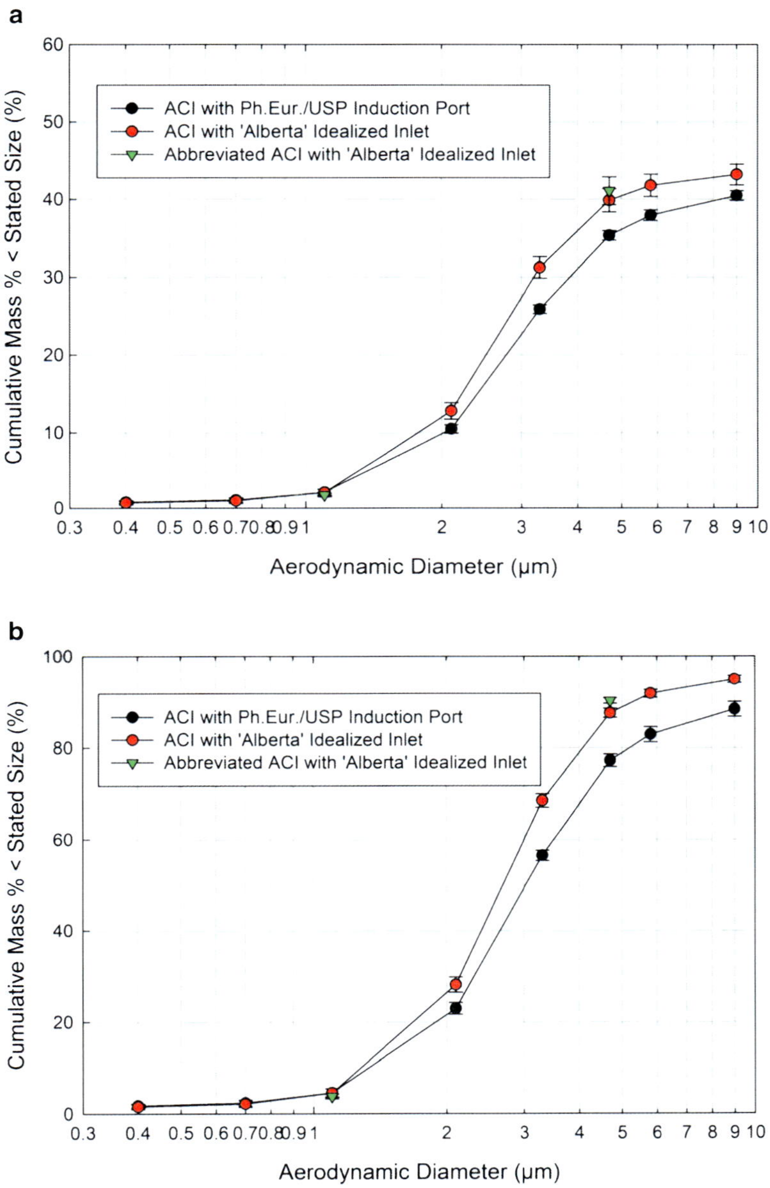

Fig. 12.12 Cumulative mass-weighted APSDs from full-resolution ACI measurements for MDI-delivered salbutamol with $Q = 28.3$ L/min compared with measures of $FPF_{<4.7\ \mu m}$ and $EPF_{<1.1\ \mu m}$ from the AIM-pHRT with AIT inlet (*From* [42]—*used with permission*). (**a**) Based on total mass ex-inhaler (TM_{exMVM}). (**b**) Based on impactor mass (IM)

Table 12.2 Descriptive statistics for the cumulative mass-weighted APSDs for MDI-delivered salbutamol (100 μg/actuation) based on *ISM* (*From* [42]—*used with permission*)

Configuration	ACI-AIT inlet	ACI-USP/Ph.Eur. inlet
MMAD (μm)	2.7 ± 0.0	3.0 ± 0.1
GSD	1.6 ± 0.0	2.1 ± 0.3

n = 5 replicates per apparatus/inlet configuration; mean ± SD

ACI (*IM* in Fig.12.12b), both indicated that replacing the USP/Ph.Eur. induction port with the AIT resulted in a slight shift to finer sizes and decreased APSD "spread," based on comparisons of *GSD*. However, although the decrease in *MMAD* determined was just below statistical significance for *MMAD*, the corresponding decrease in *GSD* was significant (Table 12.2).

This behavior is comparable with that observed in a previous investigation by the same group, using a Next Generation Pharmaceutical Impactor (NGI) with a similar MDI-delivered formulation containing salbutamol and sampling at the slightly higher flow rate of 30 L/min [45] (Fig. 12.13a).

In that previous investigation [45], a similar movement of the aerosol APSD to finer sizes was also seen evaluating a DPI (Fig. 12.13b). Taken as a whole, these measurements indicate that use of the AIT can be expected to result in impactor-sized aerosols that have higher *FPF* values than those that would be obtained using the USP/Ph.Eur induction port. Notwithstanding the fact that the upper bound size for $FPF_{6.8\,\mu m}$ in the Newman–Chan correlation [2] shown in Fig. 12.5 is at a significantly larger size than in these validation studies, such a trend, if maintained at the larger upper size limit for *FPF*, would be in keeping with a movement of the values nearer the line of identity and therefore closer to the demonstration of IVIVCs, at least with lung deposition data using gamma scintigraphy.

Grouped salbutamol mass deposition data (mean ± SD) from the study of Copley et al. [42] are summarized in Table 12.3, as these are the measures on which the comparability of AIM versus full-resolution CI measurements are perhaps best evaluated. Note that $CPM_{>4.7\,\mu m}$ included the mass retained by the inlet as well as that penetrating to stage 2 of the ACI or the upper impaction stage of the AIM-pHRT system.

Values of coarse particle mass ($CPM_{>4.7\,\mu m}$), fine particle mass ($FPM_{<4.7\,\mu m}$), and extra-fine particle mass ($EPM_{<1.1\,\mu m}$) all determined ex MDI valve, were calculated from the individual component deposition data, and are summarized in Table 12.3. The change from the USP/Ph.Eur. induction port to an AIT with the full-resolution impactor resulted in a significant decrease in $CPM_{>4.7\,\mu m}$, to values that were consistent with the same metric determined by the AIM-pHRT system with AIT. On the other hand, values of $FPM_{<4.7\,\mu m}$ and $EPM_{<1.1\,\mu m}$ were insignificantly affected. These findings demonstrated the suitability of the abbreviated system with the AIT compared with its full-resolution counterpart. Importantly, they also indicated that the change of induction port, as expected, had more influence on the ballistic fraction of the incoming dose, rather than on the aerosol that penetrated beyond the inlet into the CI.

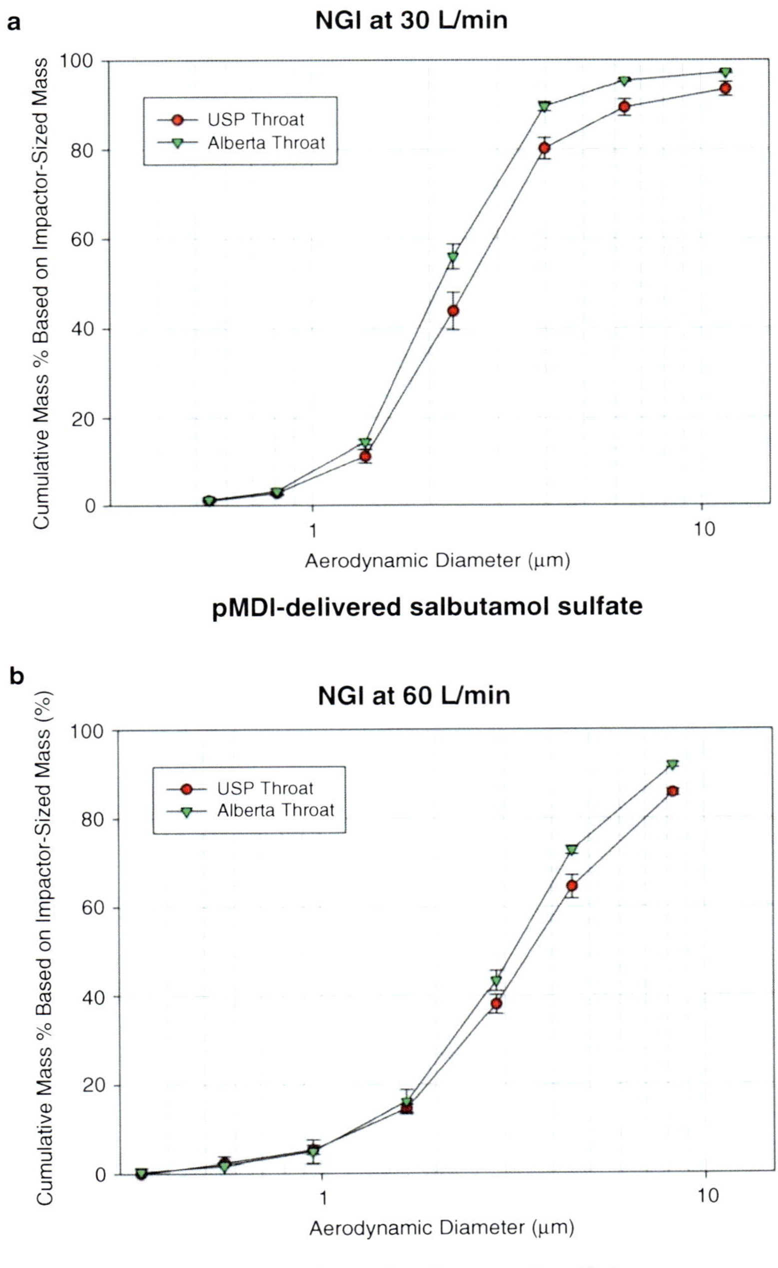

Fig. 12.13 Comparative APSDs of MDI-delivered salbutamol for an NGI equipped with either Ph.Eur./USP induction port or AIT and operated with *Q* at 30 L/min (*From* [45]—*used with permission*)

Table 12.3 Values of $CPM_{>4.7\,\mu m}$, $FPM_{<4.7\,\mu m}$, and $EPM_{<1.1\,\mu m}$ ex MDI metering valve from AIM-pHRT and full-resolution ACI configurations evaluated with MDI-delivered 100 μg/actuation salbutamol (*From* [42]*—used with permission*)

Metric (mean ± SD)	AIM-pHRT with AIT inlet (μg/actuation)	ACI with AIT inlet (μg/actuation)	ACI with USP/Ph.Eur. inlet (μg/actuation)
$CPM_{>4.7\,\mu m}$	59.2 ± 4.2	58.4 ± 2.4	65.6 ± 5.8
$FPM_{<4.7\,\mu m}$	41.3 ± 4.2	38.7 ± 3.0	35.9 ± 3.8
$EPM_{<1.1\,\mu m}$	1.7 ± 0.3	2.0 ± 0.5	2.1 ± 0.3

Table 12.4 Values of $CPF_{>4.7\,\mu m}$, $FPF_{<4.7\,\mu m}$, and $EPF_{<1.1\,\mu m}$ ex MDI metering valve from AIM-pHRT and full-resolution ACI configurations evaluated with MDI-delivered 100 μg/actuation salbutamol based on TM_{exMVM} (*From* [42]*—used with permission*)

Measure of total mass	Metric (mean ± S.D.)	AIM-pHRT with AIT inlet (%)	ACI with AIT inlet (%)	ACI with USP/Ph.Eur. inlet (%)
TM_{exMVM}	$CPF_{>4.7\,\mu m}$	58.9 ± 1.8	60.2 ± 1.5	64.7 ± 0.6
	$FPF_{<4.7\,\mu m}$	41.1 ± 1.8	39.8 ± 1.5	35.3 ± 0.6
	$EPF_{<1.1\,\mu m}$	1.7 ± 0.2	2.0 ± 0.4	2.1 ± 0.4
IM	$CPF_{>4.7\,\mu m}$	9.8 ± 0.6	12.5 ± 1.0	22.8 ± 1.4
	$FPF_{<4.7\,\mu m}$	90.2 ± 0.6	87.5 ± 1.0	77.2 ± 1.4
	$EPF_{<1.1\,\mu m}$	3.7 ± 0.5	4.5 ± 0.9	4.5 ± 0.9

Values of the various mass fractions ($CFF_{>4.7\,\mu m}$, $FPF_{<4.7\,\mu m}$, and $EPF_{<1.1\,\mu m}$) calculated based on the total mass recovered from the inhaler mouthpiece, inlet, and impactor (TM_{exMVM}) are presented in Table 12.4. The values of *C-of-V* associated with $CPF_{>4.7\,\mu m}$ and $FPF_{<4.7\,\mu m}$ were all highly reproducible at <5 %, with corresponding values in the range from 11.8 to 20.0 % for $EPF_{<1.1\,\mu m}$, reflecting less absolute mass of API associated with this subfraction. Differences in all of these measures from one configuration to another were small, as the result of the large absolute mass associated with the ballistic fraction that was collected by the appropriate inlet relative to the corresponding values of *IM* (Table 12.1).

When values of $CPF_{>4.7\,\mu m}$, $FPF_{<4.7\,\mu m}$, and $EPF_{<1.1\,\mu m}$ were determined based only on the mass captured in the impactor (*IM* in Table 12.4), the outcomes were found to be more discriminating of differences in configuration between the systems.

Thus, $CPF_{>4.7\,\mu m}$, which was now based solely on the mass collecting upon either impactor stages 0–2 for the two ACI configurations or on the first stage of the AIM-pHRT system, was substantially greater for the ACI-Ph.Eur./USP system (22.8 ± 1.4 %), compared with the same measure obtained by ACI-AIT configuration (12.5 ± 1.0 %).

The value of 9.8 ± 0.6 % for $CPF_{>4.7\,\mu m}$ obtained with the AIM-pHRT system was also slightly lower than that obtained with ACI-AIT configuration. However, such a difference could potentially be anticipated from a theoretical analysis of the effect of removing intermediate stages on performance as the result of the imperfect size selectivity of practical impactor stages, reflected by deviations in their collection efficiency-particle aerodynamic size profiles from step functions at the effective cutoff diameter appropriate to each stage. This behavior has been described in detail in Chap. 2 [46] and illustrates the importance of evaluating the effect of removing stages on CI performance for the most accurate work with abbreviated systems.

12.4 Additional Refinements for the Complete AIM-pHRT System

Two further refinements would improve realism in connection with clinically relevant aerosol size-based metrics from OIPs. The first is to operate the inhaler mimicking the process of inhalation [47]. The second is the ability to interface the inlet with anatomically appropriate representations of infant, small child, and adult faces for the testing of OIPs and add-on devices with a facemask as the patient interface [48]. Both improvements are equally applicable to full resolution as to abbreviated CI measurements.

The testing of DPIs of necessity involves the simulation of an inhalation, but the compendial methods are structured in such a way by developing a fixed pressure differential of 4 kPa and sampling a fixed volume of 4 L [49, 50], that the inhaler may not perform in the way it would in the hands of patients, whose inspiratory flow rate-time profiles, given age, obstructive lung disease severity, etc., could differ markedly from these reference conditions [51]. Although interfacing a CI to some form of breathing simulator is difficult, given the need to ensure constant flow rate at all times through the impactor, work by Daniels and Hamilton, with their electronic e-Lung™ [52], may offer promise (Fig. 12.14). Their evaluation of a FSI compared with a full-resolution NGI [53] was discussed in detail in Chap. 10. They demonstrated how a programmable bellows arrangement connected via a feedback loop to the e-Lung™ may be used to vary the pressure drop-time profile. They made the important finding that by decoupling the "inhalation" process from the DPI from the requirement of the CI to sample at a constant flow rate, bias caused by different flow ramp-up profiles between abbreviated and full-resolution CIs could be avoided.

In the context of this chapter, it is important to note that they made use of a cast, anatomically correct adult inlet to which they connected their DPI. Furthermore, they were able to replicate actual profiles of patients with asthma and COPD [53]. This arrangement is shown schematically in Fig. 12.14.

Another promising new development is a mixing inlet, originally developed by Miller [54] and now commercially available (Copley Scientific Ltd, UK, and also available through RDD Online, Virginia Commonwealth University, VA, USA), that can be interposed between the inhaler and CI, so that the former can be evaluated using breath simulation. For the purpose of discussion here, this apparatus is referred to as the "Miller" mixing inlet. In the basic setup described in the sales literature, the continuously variable flow-time profile from the inhaler (in this case with a DPI) is controlled by a breathing simulator (Fig. 12.15) that in turn is coupled to the sidearm of the mixer. A source of compressed air also provides a constant

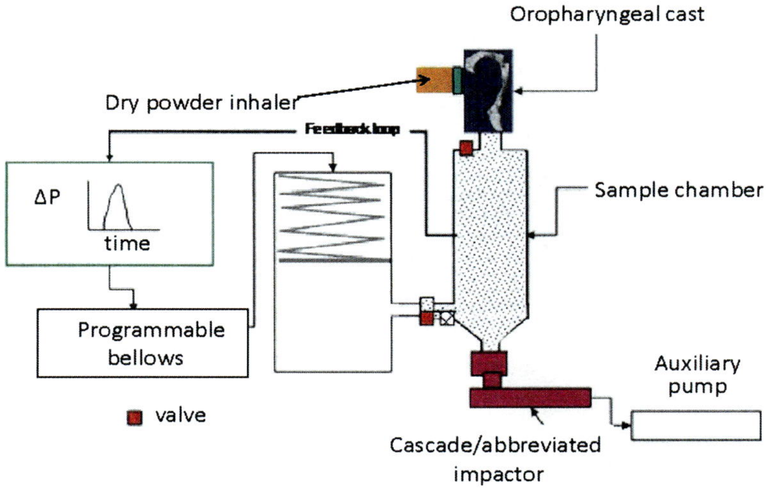

Fig. 12.14 e-Lung™ with capability to simulate patient pressure drop-inspiratory flow profiles using a programmable bellows arrangement (*From* [53]—*Courtesy G Daniels*)

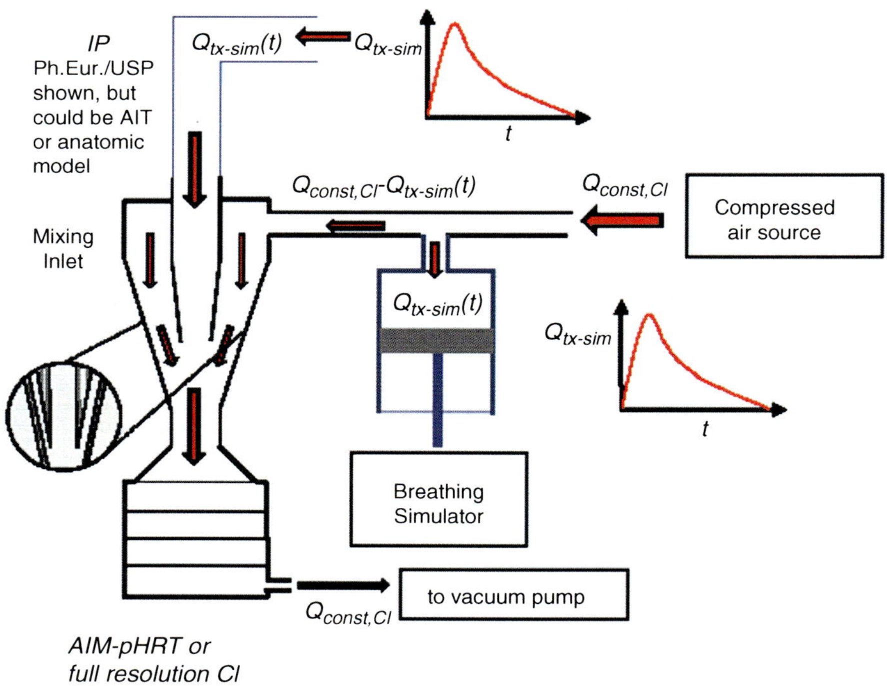

Fig. 12.15 Miller mixing inlet shown with breathing simulator and AIM-pHRT CI (Adapted from a schematic drawing supplied by Copley Scientific Ltd)

Table 12.5 Inhalation profiles used by Olson et al. [55] in studies to investigate DPI performance with mixing inlet arrangement shown in Fig. 12.15; the flow resistance of their DPI was 66 $Pa^{0.5}$/sL (*From* [55]—*used with permission*)

Profile	Peak inspiratory flow rate, PIFR (L/min)	Flow increase rate, $FIR_{10-30\ l/min}$ (L/s^2)
Weak	65	4.0
Medium	77	5.9
Strong	92	13.5

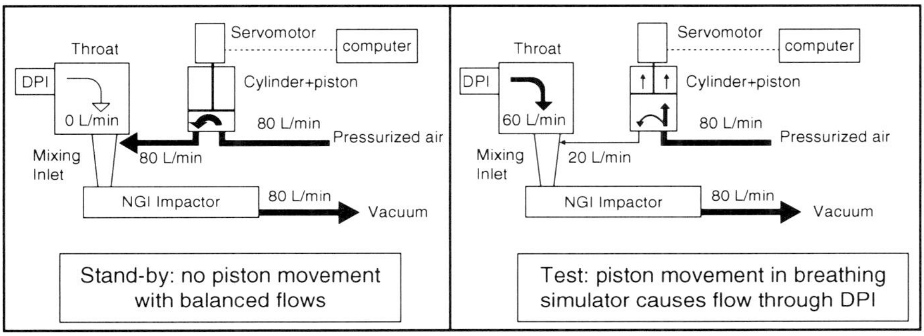

Fig. 12.16 Breathing profile simulator developed by Olson et al., using the "Miller" mixing inlet with a full-resolution NGI as CI system; an AIM-based CI could also have been used (*From* [55]—*used with permission*)

flow rate (Q_{const}) to the sidearm. This way, the flow through the impactor is kept constant also at $Q_{CI\text{-}const}$, ensuring good aerodynamic performance is maintained even when the flow rate through the device ($Q_{br\text{-}sim}(t)$) itself is low. In addition, ($Q_{br\text{-}sim}(t)$) is always maintained smaller than $Q_{CI\text{-}const}$ in this configuration.

This inlet has been applied to the assessment of DPIs by Olson et al. [55], evaluating their performance with realistic upper airway models under three different adult breathing conditions, representing weak, medium, and strong inhalations (Table 12.5). The experimental arrangement they used in conjunction with a full-resolution NGI sampling at 80 L/min is illustrated in Fig. 12.16.

Olson et al. reported excellent agreement between the target adult patient inhalation profile for which the breathing simulator was programmed (dotted line) and generated profile (solid line) for the medium flow profile for one of the DPIs evaluated (Fig. 12.17). They reported that the inlet flow rate in "standby" mode when the vacuum-generated flow from the NGI was counterbalanced by the inflow of compressed air was within ±0.05 L/min and the repeatability for a given downloaded inspiratory profile to the breathing simulator with PIF of 70 L/min and $FIR_{10-30\ L/min}$ of 8 L/s^2 was about 1 % in terms of the coefficient of variation. These findings indicate that the arrangement is sufficiently stable and repeatable for routine use. In the context of this chapter, the NGI could potentially be replaced by an AIM-based system, most conveniently the rNGI described in Chap. 10.

The "Miller" mixing inlet could also be useful for the evaluation of add-ons, such as VHCs, since these devices should ideally be evaluated while mimicking tidal

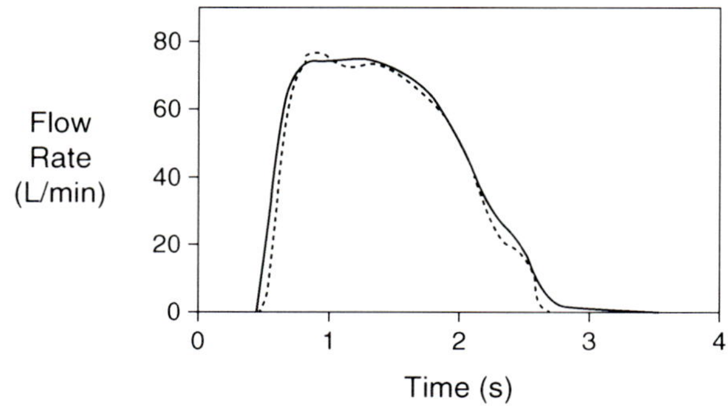

Fig. 12.17 Target (*dotted line*) and generated (*solid line*) medium inspiratory profile for a DPI to validate the breathing simulator-mixing inlet-CI measurement system described by Olson et al. (*From* [55]—*used with permission*)

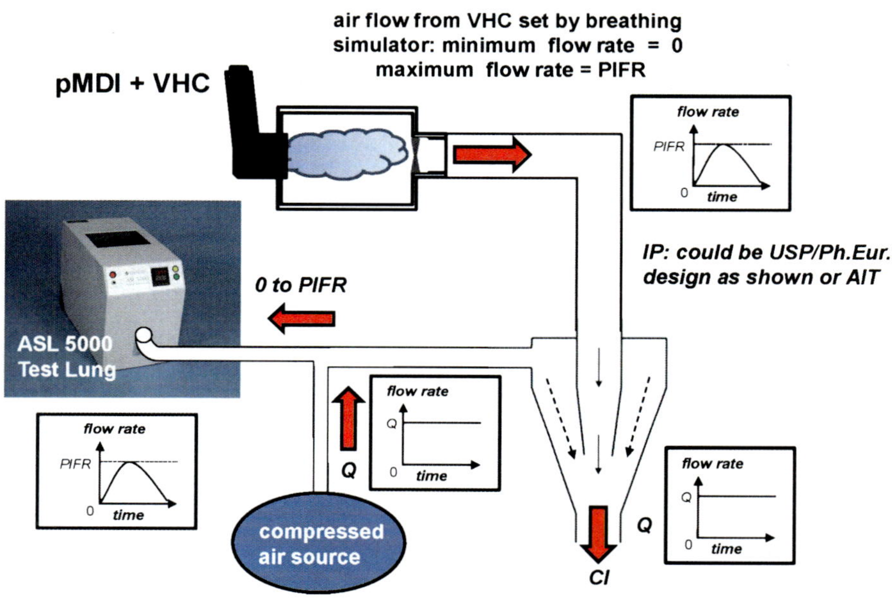

Fig. 12.18 Schematic diagram of the Miller mixing inlet interfaced with an MDI and associated VHC

breathing [56]. Such a system is illustrated in Fig. 12.18, where it should be noted that there is two-way flow into and out of the breathing simulator, reflecting the incorporation of the expiratory portion of the breathing cycle. In this setup, the selected breathing profile, $Q_{\text{br-sim}}(t)$, which could be either patient-generated or an idealized waveform, is shown with maximum inspiratory flow rate defined by the PIFR. The precise form of the exhalation profile is less important, as its function is to close the inhalation and open the exhalation valve (if equipped) of the VHC on test. However, its duration should reflect the expected duty cycle (inspiratory time as a percentage of the total time for the breathing cycle) for the patient age group being modeled (i.e., 33 % for an adult, 25 % for a child).

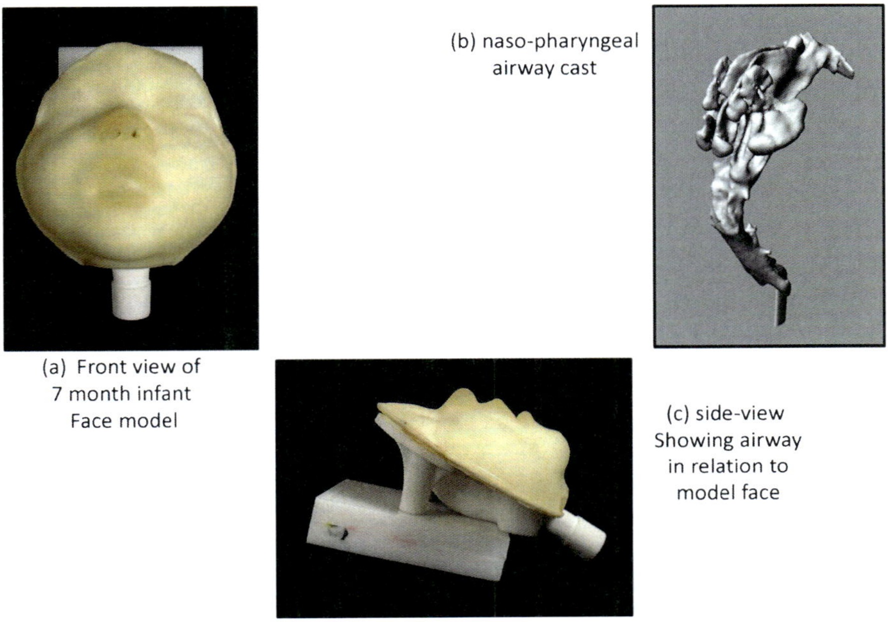

Fig. 12.19 ADAM-III infant face model with anatomically correct nasopharynx (*Facial model views from* [61]—*used with permission, nasopharyngeal model courtesy of W. Finlay*)

This arrangement makes it possible to operate for more than one breathing cycle (i.e., for the evaluation of devices intended for use by infants or small children, in which more than one breath may be needed to empty the chamber [57]). Use of such an arrangement could in principle be combined with an AIM-pHRT impactor system as well as an idealized or anatomically correct inlet. This type of setup reflects the current state of the art with regard to mimicking the patient in the laboratory setting, and as a result, significant work still needs to be done to validate these approaches and understand their limitations.

Infants and small children cannot use a mouthpiece as patient interface and so have to be prescribed inhalers with a facemask [57, 58]. Although studies have been undertaken to determine in vitro performance of VHC-facemask products with replica human faces [59], there are still no standard models that are commercially available. Such models should incorporate soft facial tissues to achieve realistic internal dead space between facemask and face when the facemask is applied with a clinically appropriate force in the range [60]. If they are combined with an anatomically correct upper airway, they are probably the closest that can be currently obtained to clinical reality in the laboratory [61, 62]. The ADAM-III infant face was developed at Trudell Medical International, using an anatomically correct nasopharyngeal upper airway, based on work by Storey-Bischoff and colleagues at the University of Alberta, Edmonton [63]. Figure 12.19 is an example of the state of the art for this type of modeling.

12.5 Guidelines for Developing AIM-pHRT Systems

If attempting to modify other full-resolution cascade impactors (not ACI or NGI) or to develop a new abbreviated apparatus, it would make sense to have guidelines for what constitutes an AIM-pHRT system. The following general considerations for the AIM-pHRT approach can be summarized, based on knowledge acquired to date (see in particular the validation studies described in this chapter and in Chap. 10):

1. An anatomically appropriate inlet should be used, based either on the Alberta idealized adult throat approach or a cast of an anatomically correct upper airway. If a cast is used, it should be preferably produced from MRI imaging rather than directly from cadaveric material to avoid possible deformation of the airway due to tissue collapse. The lining of the interior surfaces of the inlet should be thoroughly wetted with an appropriate low-volatile liquid to simulate the action of the mucosa as particle collecting media.
2. Dead space before the first impaction stage that matches as closely as possible that of the full-resolution impactor used as the reference instrument, especially if working either with MDI-based solution formulations containing low-volatile species or with DPIs, where the airflow rise time profile, should ideally mirror that achieved with the full-resolution system. In the case of DPI testing, adding a first impaction stage to match the full-resolution impactor will not by itself be sufficient to mirror the volume and flow resistance of the full impactor.

   However, in the case of the ACI, it would be a practical alternative to retain the redundant stages in the abbreviated impactor, but locate them beneath the filter stage (e.g., in the stage order: 0, 2, 5, Filter, 1, 3, 4, 6, 7) to achieve this goal.
3. The first size-fractionation stage for the separation of fine from coarse particle fractions should be chosen such that its D_{50} size is fixed as close as possible to 5 μm aerodynamic diameter at the flow rate of intended use. Note, however, that a lower size limit closer to 3 μm aerodynamic diameter may be appropriate if testing OIPs and associated add-on devices are intended for infant or small child use [64].
4. The second size-fractionation stage for the classification of extra-fine particles as a subfraction of the fine particle fraction should be chosen to have its D_{50} value close to 1.0 μm aerodynamic diameter at the flow rate of intended use, as this size represents an upper limit where significant portion of the inhaled particles may be exhaled without depositing in the lungs [65].
5. A backup filter should be used to collect the extra-fine particles. Note that the micro-orifice collector (MOC) may be appropriate in the case of modifying the NGI, except if used at flow rates <30 L/min, for instance, in the evaluation of nebulizing systems at 15 L/min [66].

On this basis, the adaptation of the FSI (MSP Corp., St Paul, MN, USA), or any similar impactor, as an AIM-pHRT system may be limited, even though it could be used with an anatomically appropriate inlet without difficulty. Its dead volume is fixed, and it lacks a second size-fractionating stage to capture the *EPF* component. Nevertheless, this abbreviated CI could still be used to determine *FPF*/*CPF*, and the importance of the fixed dead space limitation is likely to be DPI-dependent.

References

1. Mitchell JP, Newman SP, Chan H-K (2007) In vitro and in vivo aspects of cascade impactor tests and inhaler performance: a review. AAPS PharmSciTech 8(4): article 24 at: http://www.aapspharmscitech.org/view.asp?art=pt080411 0. Accessed 23 Jan 2012
2. Newman SP, Chan H-K (2008) In vitro/in vivo comparisons in pulmonary drug delivery. J Aerosol Med 21(1):1–8
3. Gonda I (1990) Aerosols for delivery of therapeutic and diagnostic agents to the respiratory tract. Crit Rev Ther Drug Carrier Syst 7:273–313
4. Patton JS, Bukar JG, Eldon MA (2004) Clinical pharmacokinetics and pharmacodynamics of inhaled insulin. Clin Pharmacokinet 43(12):781–801
5. Usmani OS, Biddiscombe MF, Nightingale JA, Underwood SR, Barnes PJ (2003) Effects of bronchodilator particle size in asthmatic patients using monodisperse aerosols. J Appl Physiol 95(5):2106–2112
6. Usmani OS, Biddiscombe MF, Barnes PJ (2005) Regional lung deposition and bronchodilator response as a Function of β_2-agonist particle size. Am J Respir Crit Care Med 172(12): 1497–1504
7. Dunbar C, Mitchell JP (2005) Analysis of cascade impactor mass distributions. J Aerosol Med 18(4):439–451
8. Vincent JH (1995) The inhalation of aerosols. In: Vincent JH (ed) Aerosol science for industrial hygienists. Pergamon Press, Oxford, pp 136–155
9. Dolovich MB, Rhem R (1998) Impact of oropharyngeal deposition on inhaled dose. J Aerosol Med 11(S1):S112–S115
10. Borgström L, Olsson B, Thorsson L (2006) Degree of throat deposition can explain the variability in lung deposition of inhaled drugs. J Aerosol Med 19(4):473–483
11. Cheng YS, Zhou Y, Chen BT (1999) Particle deposition in a cast of human oral airways. Aerosol Sci Technol 31(4):286–300
12. Swift DL (1992) Apparatus and method for measuring regional distribution of therapeutic aerosols and comparing delivery systems. J Aerosol Sci 23(S1):S495–S498
13. Berg E (1995) In vitro properties of pressurized metered dose inhalers with and without spacer devices. J Aerosol Med 8(S3):S3–S11
14. Velasquez DJ, Gabrio B (1998) Metered dose inhaler aerosol deposition in a model of the human respiratory system and a comparison with clinical deposition studies. J Aerosol Med 11(S1):S23–S28
15. Daley-Yates PT, Parkins DA, Thomas MJ, Gillett B, House KW, Ortega HG (2009) Pharmacokinetic, pharmacodynamic, efficacy, and safety data from two randomized, double-blind studies in patients with asthma and an in vitro study comparing two dry-powder inhalers delivering a combination of salmeterol 50 μg and fluticasone propionate 250 μg: implications for establishing bioequivalence of inhaled products. Clin Thera 31(2):370–385
16. Grgic B, Finlay WH, Burnell PKP, Heenan AF (2004) In vitro intersubject and intrasubject deposition measurements in realistic mouth-throat geometries. J Aerosol Sci 35(8):1025–1040
17. Finlay WH (2012) New validated extrathoracic and pulmonary deposition models for infants and children. In: Dalby RN, Byron PR, Peart J, Suman JD, Farr SJ, Young PM (eds) Respiratory drug delivery-2012. Davis HealthCare International Publishing, River Grove, IL, pp 325–336
18. Golshahi L, Noga ML, Thompson RB, Finlay WH (2011) In vitro deposition measurement of inhaled micrometer-sized particles in nasal airways of children and adolescents during nose breathing. J Aerosol Sci 42(7):474–488
19. Golshahi L, Noga ML, Finlay WH (2012) Deposition of inhaled micrometer-sized particles in oropharyngeal airway replicas of children at constant flow rates. J Aerosol Sci 49(1):21–31
20. Rennard SI (2005) Anticholinergics in combination bronchodilator therapy in COPD. In: Spector SL (ed) Anticholinergic agents in the lower and upper airways. Dekker, New York, NY, pp 97–111

21. Szefler SJ, Martin RJ, King TS, Boushey HA, Cherniack RM, Chinchilli VM, Craig TJ, Dolovich M, Drazen JM, Fagan JK, Fahy JV, Fish JE, Ford JG, Israel E, Kiley J, Kraft M, Lazarus SC, Lemanske RF, Mauger E, Peters SP, Sorkness CA (2002) Significant variability in response to inhaled steroids for persistent asthma. J Allergy Clin Immunol 109(3):S410–S418
22. Everard ML, Dolovich MB (2002) In vivo measurements of lung dose. In: Bisgaard H, O'Callaghan C, Smaldone GC (eds) Drug delivery to the lung. Dekker, New York, NY, pp 173–209
23. Dolovich MB (1993) Lung dose, distribution, and clinical response to therapeutic aerosols. Aerosol Sci Technol 18(3):230–240
24. American Thoracic Society (ATS) (1995) Standardization of spirometry—1994 update. Am J Respir Crit Care Med 152(3):1107–1136
25. Tougas TP, Christopher D, Mitchell JP, Strickland H, Wyka B, Van Oort M, Lyapustina S (2009) Improved quality control metrics for cascade impaction measurements of orally inhaled drug products (OIPs). AAPS PharmSciTech 10(4):1276–1285
26. Mitchell JP, Nagel MW (2003) Cascade impactors for the size characterization of aerosols from medical inhalers: their uses and limitations. J Aerosol Med 16(4):341–377
27. Olsson B, Borgstrom L, Asking L, Bondesson E (1996) Effect of inlet throat on the correlation between measured fine particle dose and lung deposition. In: Dalby RN, Byron PR, Farr SJ (eds) Respiratory drug delivery-V. Interpharm Press, Buffalo Grove, IL, pp 273–282
28. Ehtezazi T, Horsfield MA, Barry PW, O'Callaghan C (2004) Dynamic change of the upper airway during inhalation via aerosol devices. J Aerosol Med 17(4):325–334
29. Ehtezazi T, Southern KW, Allanson D, Jenkinson I, O'Callaghan C (2005) Suitability of the upper airway models obtained from MRI studies in simulating drug lung deposition from inhalers. Pharm Res 22(1):166–170
30. Swift DL (1994) The oral airway—a conduit or collector for pharmaceutical aerosols. In: Byron P, Dalby RN, Farr SJ (eds) Respiratory drug delivery IV. Interpharm Press, Buffalo Grove, IL, pp 187–195
31. COPHIT Consortium (2012) Anatomic oropharyngeal model, at: http://www.isam.org/. Visited 25 June 2012
32. McRobbie DW, Pritchard S, Quest RA (2003) Studies of the human oropharyngeal airspaces using magnetic resonance imaging. I. Validation of a three-dimensional MRI method for producing ex vivo virtual and physical casts of the oropharyngeal airways during inspiration. J Aerosol Med 16(4):401–415
33. Ilie M, Matida EA, Finlay WH (2008) Asymmetrical aerosol deposition in an idealized mouth with a DPI mouthpiece inlet. Aerosol Sci Technol 42(1):10–17
34. Zhang Y, Gilbertson K, Finlay WH (2007) In vivo-in vitro comparison of deposition in three mouth-throat models with Qvar and Turbuhaler inhalers. J Aerosol Med 20(3):227–235
35. Stapleton KW, Guentsch E, Hoskinson MK, Finlay WH (2000) On the suitability of k-turbulence modeling for aerosol deposition in the mouth and throat: a comparison with experiment. J Aerosol Sci 31(6):731–749
36. Grgic B, Finlay WH, Heenan AF (2004) Regional aerosol deposition and flow measurements in an idealized mouth and throat. J Aerosol Sci 35(1):21–32
37. Stocks J, Hislop AA (2002) Structure and function of the respiratory system. In: Bisgaard H, O'Callaghan C, Smaldone GC (eds) Drug delivery to the lung. Dekker, New York, NY, pp 47–104
38. Everard ML (2004) Inhaler devices in infants and children: challenges and solutions. J Aerosol Med 17(2):186–195
39. Mitchell JP, Nagel MW, Doyle C, Ali RS, Avvakoumova V, Christopher D, Quiroz J, Strickland H, Tougas T, Lyapustina S (2010) Relative precision of inhaler aerodynamic particle size distribution (APSD) metrics by full resolution and abbreviated Andersen Cascade Impactors (ACIs): Part 1. AAPS PharmSciTech 11(2):843–851
40. Mitchell JP, Nagel MW, Doyle C, Ali RS, Avvakoumova V, Christopher D, Quiroz J, Strickland H, Tougas T, Lyapustina S (2010) Relative precision of inhaler aerodynamic particle size

distribution (APSD) metrics by full resolution and abbreviated Andersen Cascade Impactors (ACIs): Part 2—Investigation of bias in extra-fine mass fraction with AIM-HRT impactor. AAPS PharmSciTech 11(3):1115–1118
41. Chambers FE, Smurthwaite M (2012) Comparative performance evaluation of the Westech Fine Particle Dose (FPD) impactor. In: Dalby RN, Byron PR, Peart J, Suman JD, Farr SJ, Young PM (eds) Respiratory drug delivery-2012. Davis HealthCare International Publishing, River Grove, IL, pp 553–558
42. Mitchell JP, Copley M, Sizer Y, Russell T, Solomon D (2012) Adapting the Abbreviated Impactor Measurement (AIM) concept to make appropriate inhaler aerosol measurements to compare with clinical data: a scoping study with the "Alberta" Idealized Throat (AIT) inlet. J Aerosol Med Pulm Drug Deliv 25(4):188–197
43. Ehtezazi T, Saleem I, Shrubb I, Allanson ID, O'Callaghan C (2010) The interaction between the oropharyngeal geometry and aerosols via pressurized metered dose inhalers. Pharm Res 27(1):175–186
44. Mitchell JP, Nagel MW, Avvakoumova V, MacKay H, Ali R (2009) The abbreviated impactor measurement (AIM) concept: part 1—influence of particle bounce and re-entrainment—evaluation with a "dry" pressurized metered dose inhaler (MDI)-based formulation. AAPS PharmSciTech 10(1):243–251
45. Copley M, Mitchell J, Solomon D (2011) Evaluating the Alberta throat: an innovation to support the acquisition of more clinically applicable aerosol aerodynamic particle size distribution (APSD) data in oral inhaled product (OIP) development. Inhalation 5(4):12–16
46. Roberts DL, Mitchell JP (2011) Influence of stage efficiency curves on interpretation of abbreviated impactor data. Drug delivery to the lungs-22, The Aerosol Society, Edinburgh, UK, pp 177–180. Available at: http://ddl-conference.org.uk/index.php?q=previous_conferences. Visited 4 Aug 2012
47. Dolovich MB, Mitchell JP (2004) Canadian Standards Association standard CAN/CSA/Z264.1-02:2002: a new voluntary standard for spacers and holding chambers used with pressurized metered-dose inhalers. Can Respir J 11(7):489–495
48. Mitchell JP, Nagel M, Finlay B (2011) Advances in models for laboratory testing of inhalers: there's more to it than meets the nose or mouth—The ADAM face models. In: Dalby RN, Byron PR, Peart J, Suman JD, Farr SJ, Young PM (eds) Respiratory drug delivery Europe-2011. Davis Healthcare International Publishing LLC, River Grove, IL, pp 457–461
49. European Directorate for the Quality of Medicines and Healthcare (EDQM). Preparations for inhalation: aerodynamic assessment of fine particles. (2012) Section 2.9.18—European Pharmacopeia, Council of Europe, 67075 Strasbourg, France
50. United States Pharmacopeial Convention (2012) USP 35-NF 30 Chapter 601: Aerosols, nasal sprays, metered-dose inhalers and dry powder inhalers. Rockville, MD
51. Feddah MR, Brown KF, Gipps EM, Davies NM (2000) *In-vitro* characterisation of metered dose inhaler versus dry powder inhaler glucocorticoid products: influence of inspiratory flow rates. J Pharm Pharm Sci 3(3):317–324
52. Burnell PKP, Malton A, Reavill K, Ball MHE (1998) Design, validation and initial testing of the Electronic Lung™ device. J Aerosol Sci 29(8):1011–1025
53. Hamilton M, Daniels G (2011) Assessment of early screening methodology using the Next Generation and Fast Screen Impactor systems. Drug Delivery to the Lungs-22, The Aerosol Society, Edinburgh, UK, 22:355–358. Available at: http://ddl-conference.org.uk/index.php?q=previous_conferences. Visited 4 Aug 2012
54. Miller NC (2002) Apparatus and process for aerosol size measurement at varying gas flow rates. US Patent 6,435,004-B1
55. Olson B, Berg E, Svensson M (2010) Comparing aerosol size distributions that penetrate mouth-throat models under realistic inhalation conditions. In: Dalby RN, Byron PR, Peart J, Suman JD, Farr SJ, Young PM (eds) Respiratory drug delivery-2010. Davis Healthcare International Publishing, River Grove, IL, pp 225–234

56. Mitchell JP, Dolovich MB (2012) Clinically relevant test methods to establish in vitro equivalence for spacers and valved holding chambers used with pressurized metered dose inhalers (MDIs). J Aerosol Med Pulm Drug Deliv 25(4):217–242
57. Dolovich MB (2004) In my opinion: interview with the expert. Pediatr Asthma Allergy Immunol 17(4):292–300
58. Morton RW, Mitchell JP (2007) Design of facemasks for delivery of aerosol-based medication via pressurized metered dose inhaler with valved holding chamber: Key issues that affect performance. J Aerosol Med 20(S1):S29–S45
59. Finlay WH, Zuberbuhler P (1999) In vitro comparison of salbutamol hydrofluoroalkane (Airomir) metered dose inhaler aerosols inhaled during pediatric tidal breathing from five valved holding chambers. J Aerosol Med 12(4):285–291
60. Shah SA, Berlinski A, Rubin BK (2006) Force-dependent static dead space of face masks used with holding chambers. Respir Care 51(2):140–144
61. Mitchell JP, Finlay JB, Nuttall JM, Limbrick MR, Nagel MW, Avvakoumova V, MacKay H, Ali RS, Doyle CC (2011) Validation of a new model infant face with nasopharynx for the testing of valved holding chambers (VHCs) with facemask as a patient interface. In: Dalby RN, Byron PR, Peart J, Suman JD, Farr SJ (eds) Respiratory drug delivery-2010. Davis Healthcare International Publishing LLC, River Grove, IL, pp 777–780
62. Mitchell JP (2008) Appropriate face models for evaluating drug delivery in the laboratory: the current situation and prospects for future advances. J Aerosol Med 21(1):1–15
63. Storey-Bischoff J, Noga M, Finlay WH (2008) Deposition of micrometer-sized aerosol particles in infant nasal airway replicas. J Aerosol Sci 39(12):1055–1065
64. Newhouse MT (1998) The current laboratory determination of "Respirable Mass" is not clinically relevant. J Aerosol Med 11(S1):S122–S132
65. Labiris NR, Dolovich MB (2003) Pulmonary drug delivery. Part I: physiological factors affecting therapeutic effectiveness of aerosolized medications. Br J Clin Pharmacol 5(12):588–599
66. Marple VA, Olson BA, Santhanakrishnan K, Mitchell JP, Murray SC, Hudson-Curtis BL (2004) Next generation pharmaceutical impactor: a new impactor for pharmaceutical inhaler testing. Part III. Extension of archival calibration to 15 L/min. J Aerosol Med 17(4):335–343

Chapter 13
Future Directions for the AIM and EDA Concepts

Terrence P. Tougas and Jolyon P. Mitchell

Abstract The abbreviated impactor measurement (AIM) and efficient data analysis (EDA) concepts have both been presented in detail in previous chapters of this book, based on the knowledge acquired by the middle of 2012. It is recognized that both issues have attracted much interest from stakeholders involved with the in vitro testing of OIPs, and so the likelihood of further significant developments in the next few years is very strong. This chapter therefore starts by looking at questions that were raised at a special symposium in the spring of 2011, at which representatives from industry, academia, compendial committees, and regulatory agencies had the opportunity to air their views on these topics, with some speculation on likely future developments. The remainder of the chapter considers some ideas about further use of both AIM and EDA concepts.

13.1 Questions Arising from the IPAC-RS-Sponsored Conference: "Perspectives on Efficient Data Analysis Methods and Abbreviated Impactor Measurements as Quality Assessment Tools"

IPAC-RS hosted a satellite conference on Friday, 6 May 2011 at *Respiratory Drug Delivery Europe 2011*, which included short presentations from representatives of regulatory agencies, pharmacopeias, and industry. These presentations were followed

T.P. Tougas (✉)
Boehringer Ingelheim Pharmaceuticals Inc., 900 Ridgebury Road, Ridgefield, CT 06877-0368, USA
e-mail: terrence.tougas@boehringer-ingelheim.com

J.P. Mitchell
Trudell Medical International, 725 Third Street, London, ON N5V 5G4, Canada
e-mail: jmitchell@trudellmed.com

T.P. Tougas et al. (eds.), *Good Cascade Impactor Practices, AIM and EDA for Orally Inhaled Products*, DOI 10.1007/978-1-4614-6296-5_13,

by interactive discussions with the audience around technical and statistical aspects of EDA and AIM and the lifecycle approach to incorporating EDA and AIM in product development, registration, and manufacturing. The presentations and a summary report are posted on the IPAC-RS public website [1]. At the outset, it is important to note that the conference mandate was on the application to OIP quality control, so issues concerning the potential application of AIM to the acquisition of more clinically appropriate data, discussed in Chap. 12, were not discussed.

Responses to the first group of questions came from the panelists covering the topic *Technical Aspects of EDA and AIM: Incorporating EDA and AIM into the Development Cycle*. Editorial comments are inserted between parentheses "[]" where needed for clarity, and some responses have added interpretation, based on information located elsewhere in this book.

1. Would we need a separate impactor for each product?
   No, we need only a couple of different cut points for the boundary between large and small particles, because the EDA method is robust with respect to shifts of the boundary (in the studied example presented by Tougas [2], it demonstrated stable performance when the *LPM/SPM* ratio was between 0.3 and 3.0). Moreover, all respiratory drugs need to get into the lung, so the sizes of interest will always be around 2–3 μm aerodynamic diameter. Setting the cut point at a different size (much below or much above 2–3 μm) could be considered if there is an interest in controlling a specific part of the distribution.
2. For solution MDIs, would you require CI data for quality control, or would delivered dose + bulk particle size testing be sufficient?
   Solutions are simpler than suspensions. Laser diffraction could be used for sizing droplets, but you still need traceability to the API to satisfy regulatory requirements. For nebulizers, it is possible to build up the traceability chain. EDA can be extended to apply to nebulizers too.
3. Can EDA detect if something is wrong with the product, or if it is a wrong product?
   The short answer is yes. EDA is an enhancement of the current CI tests. However, it should be applied only after you have defined the aerodynamic particle size distribution (APSD) of the "right" product. EDA is more sensitive to changes in APSD than current methods, so yes, you will get a signal that something is "wrong" if APSD changes, but you would need to do an investigation, potentially using a full-resolution impactor and/or other tests, to determine the cause. OOS investigations are done in the current system as well.
4. Analysts and managers need to know how to go about using EDA—not just the science and statistics but the practical how-to's and what's, so people can get on with it. We are waiting for IPAC-RS to come up with all the answers.

   We share this goal. Part of the process of getting there is obtaining feedback from users in settings like this one. We are also developing a book that will set down "ground rules" for EDA and AIM. Members of the European Pharmaceutical Aerosol Group (EPAG) are collaborating with IPAC-RS on the book writing. In addition, a series of articles has already been published [3], and more articles are in the works and will be published in due course.

5. The *LPM*/*SPM* ratio works like a charm when the cutoff (size boundary delineating *LPM* from *SPM*) is set at the *MMAD*. But if you are away from the *MMAD* value, the expected ideal ratio is not 1.0, so what happens to sensitivity? How far from *MMAD* could you go?
   EDA works as long as the size boundary defining the *LPM*/*SPM* ratio is within 0.3–3.0. Also, the goal (specification) is not to have the size boundary separating *LPM* from *SPM* set "close to 1.0" but close to whatever value you determined during development is typical for your product's APSD with the chosen boundary size.
6. Does FDA have any data on the use of EDA or AIM?
   [Mr. Bill Doub, FDA St, Louis, MO]: In my lab, we looked at four CFC MDIs and five HFA MDIs. (There was a fifth CFC MDI but it turned out to be expired.) We attempted to change their APSDs by introducing an inter-actuation delay, actuator cleaning, and looking at different life stages. We used an Anderson Cascade Impactor (ACI) and looked at the fine-particle dose as well. The experiments were conducted because patients are complaining about the CFC–HFA difference. Where we saw statistically significant effects with the full-resolution cascade impactor (FRCI), we applied EDA to see if that method showed similar effects. Although the EDA data have not yet been internally validated, qualitatively we saw the same types of APSD changes as were observed using the FRCI. Quantitatively, changes were greater with EDA although RSD was also higher with EDA. For example, the effect of cleaning was 4–11 % from FRCI data and 24 % for EDA. The effect of delay was 8 % with FRCI and 21 % with EDA. All *LPM*/*SPM* ratios were 0.2–0.3, so we were probably pushing the limits of sensitivity of EDA. *MMAD*s were 2.2–2.9 μm, and the *LPM*-to-*SPM* boundary was set at 2.1 μm.
7. If the ratio [*LPM/SPM*] is more sensitive than stage groupings, could it be used to predict *MMAD*?
   LPM/*SPM* is more sensitive to change in MMAD than stage groupings (see Chap. 8). However, if you want to predict *MMAD*, it's better to calculate *MMAD* from the cumulative mass-weighted APSD determined using a full-resolution impactor. MMAD may be estimated using an appropriate method (e.g., probit analysis). Incidentally, the USP method for *MMAD* determination assumes a log-normal APSD, which is not always valid and which can lead to a biased assessment of *MMAD* [4].
   [In addition, it may be important to look at the impact of the entire collection efficiency curve for each stage rather than just the corresponding stage d_{50} values for the most accurate *MMAD* assessment with some CIs, in particular the ACI. The NGI appears to be less prone to such effects, most probably because stage-to-stage overlap is largely absent and each collection efficiency curve is close to being symmetric about its d_{50} value.]
8. Does it matter if the impactor is based on the viable versus the nonviable type?
   Those differences can become an issue [if the chosen abbreviated CI has a parent full-resolution CI that is different in internal design, e.g., compare the FSA (which is based on the ACI and has flat collection plates) and the ACVI

that has Petri dish collection surfaces. Currently, there is very little evidence to quantify the effect of such differences on OIP aerosol APSD measurement.] Computer simulations could be used to understand the differences between specific [CI- and AIM-based] platforms.

9. As an intermediate step before going to an actual AIM apparatus, could you do a consolidated extract from a group of stages?
   Yes, you could do EDA from the full-resolution CI data, and you could pool the material from several stages by washing them together. Although, while you are still in development, you may not know which stages to group yet. Furthermore, from the time and resource perspective, it would not be much different than collecting and measuring stages individually, and then adding up the numbers.
10. The 1998 draft FDA guidance for MDIs and DPIs requires a ±15 % specification on the mass balance from CI measurements. How will it be done with EDA?
    With EDA, besides the *LPM*/*SPM* ratio, the impactor-sized mass (*ISM*) needs to be controlled. We are not discussing in detail the performance of the *ISM* portion of the EDA approach because it will be the same as with either a full-resolution CI or an AIM-based apparatus. The *ISM* includes much of the mass balance but not all (e.g., the non-sizable portions [that deposit in the induction port and pre-separator, if used] are not included). However, the delivered dose uniformity (DDU) test, which is also required for these products, does control for the total emitted dose and in a much more accurate way than current CI measurements with multistage impactors. The IPAC-RS position has always been that mass balance should be used as a "system suitability" indication, not as a specification. But your actual regulatory requirement needs to be discussed with the FDA [or appropriate regulatory agency]. We have also presented a poster at *Drug Delivery to the Lungs 21* [5] about a dual-use DDU/APSD apparatus. If such an apparatus is ever developed, it could provide a single measurement that would replace both the current DDU and CI tests.
11. The air flow [rates] recommended for ACIs are 60 and 28.3 L/min. If we want to match the full-stage ACI at 15 L/min, how can we do that? How would AIM measure at low flow rates?
    In response to the above question, a participant (at the symposium referred to at the beginning of this section) indicated that he would like to see calibration data first. He acknowledged that the full-resolution ACI can measure at 15 L/min but had not seen the calibration data. [Such data are unavailable to the best knowledge of the coeditors, although Garmise and Hickey in 2008 [6] published calibration data for stages −0, −1, and −2 (see Chap. 2) at this flow rate in the context of using the CI for sampling aerosols from nasal inhaled products.] However, the archival calibration data for the NGI at 15 L/min were published in 2004 [7]. This speaker requested that before we go forward with the ACI, we need reliable [calibration] data. Currently, the [inhaler testing community is] relying on old data from the manufacturer's operating manual for this CI. The ACI used to be made of aluminum, but it erodes. Newer ACIs are made of steel, and we now need new [archival calibration] data.

12. The purpose of using AIM is twofold: (1) development/characterization and (2) quality control. Do you plan to publish all the substantiating information, or would it be up to each company to cross-validate to existing impactors?
    Another participant indicated that stakeholders need to let things get tested by time and experience. If it takes too long to get these ideas through to USP and Ph. Eur., then each company should start cross-validating these methods themselves. [Some of the results from this type of activity are presented in Chap. 10, and earlier in 2012, the Inhalanda Committee of the Ph. Eur. started a new work initiative to develop validation data for AIM-based systems.]
13. Would AIM and EDA be mandatory or optional?
    Dr. Tougas indicated on behalf of the panel that both concepts have been thought of as *alternatives* [to current methods] in OIP QC testing, and as such they are enhancements for development. The use of these or any other methods is of course optional.
    In this second part of the conference, the questions were responded to by the panelists covering the topic "Pharmacopeial Perspectives on EDA and AIM—European and US Viewpoints."
14. We don't want to buy new equipment. We suggest that you publish the principles but not specify equipment in the Ph. Eur. and USP.
    Some [abbreviated] apparatus(es) will need to be described in the pharmacopeia, for regulatory purposes.
15. In the Ph. Eur., we have the twin impinger (apparatus A). We need to rethink the utility of this apparatus. It has a role to play, because of the bouncing issues with other cascade impactors.
    This topic deserves consideration as a future development of the AIM concept (see Chap. 10).
16. Why can't pharmacopeias write requirements for an apparatus, without specifying the apparatus? For example, if the USP is moving toward QbD approaches, why do we need to put apparatus description into USP?
    [Dr. Steve Nichols]: Someone needs to be able to go and test the product. Regulators need standard principles, stage cutoff sizes, validation information, etc. Even now, the Ph. Eur. states that you can use *ANY* method as long as it's validated. But there are legal reasons for describing a specific apparatus. To change that requirement, at least in Europe, someone would need to change the law. If there is enough interest, IPAC-RS and EPAG should develop a proposal/monograph and send it to the Ph. Eur. Inhalanda Working Party for consideration. And since the USP/EP harmonization takes 7 years on average, it would be good to submit the proposed monograph to the USP and EP simultaneously. [The peer-reviewed journal attached to the Ph. Eur.] *Pharmeuropa* can publish your proposal. If you do not come up with a monograph, the Ph. Eur. Working Party will develop one, but the outcome may not be exactly what you want to see. Someone also should compare the *FPD* (fine-particle dose <5 μm aerodynamic diameter metric in the Ph. Eur.) requirement with the EDA approach.

17. Can you comment on the pharmacopeial discussion?
[Dr. Marjolein Weda]: A full-resolution impactor is needed in the Ph. Eur. for characterization purposes. For quality control, you could remove some stages. Introducing changes to Ph. Eur. quality control methods is allowed and, for example, already happens frequently in HPLC methods (using different columns).
18. If EDA provides better discriminating power, what is the expected sample size?
[Dr. Prasad Peri]: We first need to see the data that [confirms] it indeed provides better discrimination. If this is demonstrated, and if it is validated, FDA would support the use of EDA. The data we have seen so far looks promising, and we'd like to move forward.
19. Would APSD be needed as a release test with the Quality by Design (QbD)?
[Dr. Prasad Peri]: If you can correlate APSD to some "in-process" measurement, I personally would support that approach instead of doing end-product testing. The more knowledge and process control, the better assurance that the final product will meet quality requirements.
20. On the statistical issue, and your comment that with two parameters there might be a loss of information: Since these two metrics are capable of discriminating and detecting changes while collecting fewer numbers on a routine basis—that is the whole point. If we could miraculously find a single metric that could control APSD, then that would be better still. You need all possibly relevant information in development, but once you have characterized the product and established controls, you shouldn't need to accumulate as many numbers as possible.

    [Dr. Prasad Peri]: The only two metrics in the EDA approach are *ISM* and *LPM/SPM*. If you can show that these give better assurance, discrimination, and detection that your product is changing over time, then we would consider it favorably. Right now, we do not have the complete data. Papers are being published, presentations are made at meetings, and this knowledge base should continue to build up. We want to encourage better analytical methods and better methodologies. We don't want to hamper progress and innovation.
21. Would FDA require demonstration of the same performance between EDA and the current approaches or better performance?
[Dr. Prasad Peri]: It can be the same statistically but better in terms of labor, environmental impact, etc.
22. What upcoming guidance documents should we expect to see soon?
[Dr. Marjolein Weda]: Currently, there are no concrete plans to update the EMA guidance. If the AIM and EDA topic moves forward, however, EMA/CHMP could decide to reopen the guidance or to provide clarification and additional information via a Q and A.

    [Mr. Bill Doub and Dr Prasad Peri]: Work on the new guidance is ongoing. We have another internal FDA teleconference coming up regarding the guidance. But even the current draft MDI/DPI guidance from 1998 does allow alternate approaches. So even if the guidance did not change, you could propose to use AIM and EDA in your application. We want a less prescriptive guidance, but you don't have to wait for it. We might update the guidance with alternative methods. Now the document is in the process of being updated.

There is a working group within FDA, which has been working for the past 6 months, and it will likely be a few more months. It is a slow process, with the writing and other steps. What specifications/limits would you envision for EDA? For example, would you set a range on the ratio?

[Dr. Terry Tougas—in reply to Dr. Prasad Peri]: This is a big dilemma—what should be the basis for setting specifications? Dr. Peri's slides (given at this symposium—see http://www.ipacrs.com/PDFs/CI %20Workshop/5-CI %20Workshop %20- %20Peri.pdf visited September 6, 2012) include "typical ranges" for stage groupings—we could use those as a basis and translate them into the EDA specifications. Ideally, specifications should be tied to clinical performance but in the absence of a quantitative IVIVC, or something else that links in vitro and clinical performance, we have to go with what has been required historically.

[Dr. Marjolein Weda]: Why not use development batches' data to set specifications?

[Dr. Terry Tougas]: That would be setting specifications based on capability rather than QbD. From the engineering perspective, that would be a poor way to set specifications. Today's limits are based on process capability, but there is limited experience/data at the time of registration. This results in a band that is too tight for real commercial processes. We can look at the data and use it as a rough guide regarding expected performance and standard deviation, but the modern engineering thought around "capable process" recommends a different approach (e.g., ±6 sigma, instead of ±3 sigma). Currently in the pharmaceutical industry, if a sponsor does a good job developing a product with a tight distribution, that sponsor will be penalized with very tight specifications. This is counter to the spirit of continuous improvement.

So, even though there is still an incomplete understanding of all the issues surrounding AIM and EDA, the overall outcome from the conference was one of probing the potential for both concepts to become part of the mainstream of OIP in vitro performance testing.

13.2 Future Directions: Some Further Ideas

An outline for comprehensive product lifecycle management strategy in terms of in vitro characterization of APSD has been described in Chap. 6 that is based on simpler yet more statistically powerful efficient data analysis metrics. This approach is easily combined with abbreviated impactor measurements. The EDA/AIM approach could be adopted as the norm for inhaler development and quality control, but its effective implementation will need to be undertaken on a product-by-product basis. Although it can strongly be argued that full-resolution multistage CI testing is less than ideal for QC purposes, such measurements have their place in the initial product development process, as the first resort in the event of an OOS investigation and also in OIP change management when in commercial production.

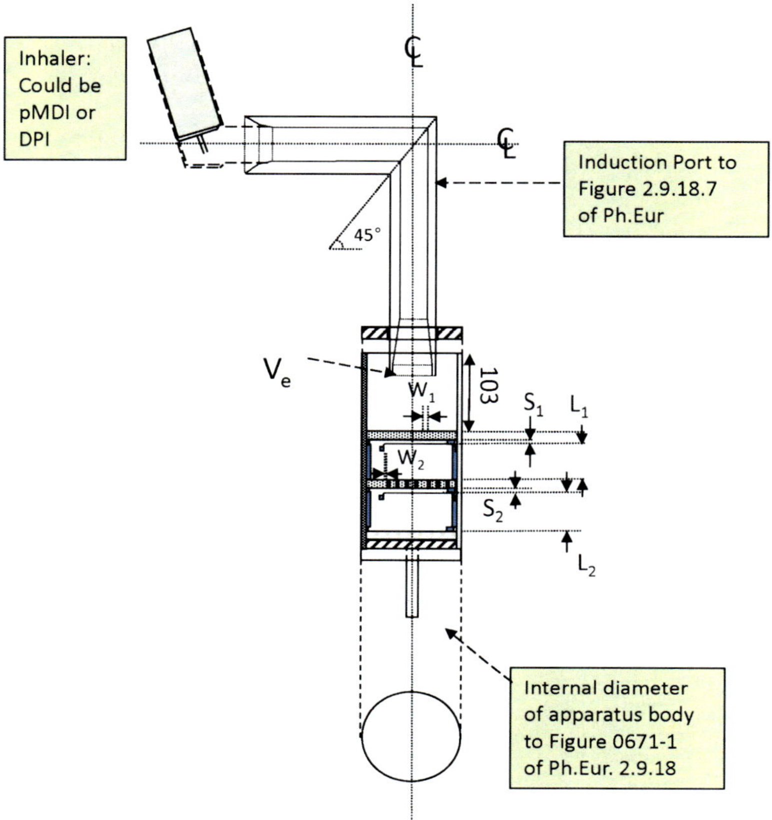

Fig. 13.1 Concept for a combined AIM-DDU apparatus

The thought process in relation to inhaler lifecycle management could potentially be extended to integrate the measurement of dose content uniformity and AIM-based APSD metrics into a single apparatus (Fig. 13.1 [5]).

This combined testing approach could thereby achieve the desirable goal from the statistical perspective of making only a single simultaneous measurement of inhaler aerosol quality.

Such an apparatus (which is in early development at the concept stage within IPAC-RS) will require experimental validation with a variety of inhaler products before gaining acceptance both from the industry and regulatory perspective. Nevertheless, by combining these measurements, intrinsic variability introduced by making them separately could be eliminated, resulting in improved overall measurement precision.

The initial designed experiment orchestrated through the CI Working Group of IPAC-RS and described in detail in Chap. 10 assessed relative precision of metrics obtained by abbreviated versus full-resolution CI, using drug product from a short period during the manufacture of a particular lot so that measurement variability was not confounded by product variability. Although helpful at identifying the capability of the abbreviated measurement methods, its chief limitation was that the findings are nor directly applicable to the assessment of commercially available product samples, subject to normal batch release testing. A further designed experiment that will take place in late 2012 or early 2013 is in the process of addressing this deficiency, again being implemented through the CI-WG of IPAC-RS, which is coordinating the activities of several independent laboratories. In this instance, the samples of drug product are coming from commercially available lots without pre-selection in terms of time of manufacture within the batch, so that the normal intra-lot variability will be present. It is anticipated that the outcome from this investigation will provide a more realistic assessment of the comparability of abbreviated with full-resolution techniques when applied in the routine OIP QC environment.

By now, it should be evident that the ability to employ both concepts to make more measurements of higher quality in terms of being able to discriminate meaningful changes in APSD-based metrics is central to both AIM and EDA concepts. Given these attributes, it is therefore anticipated that both AIM and EDA concepts will find increasing application in the QbD environment [8], in which more laboratory-based assessments in early stage product development are likely to be undertaken so that design space can be properly mapped and understood. Such an approach, rather than simply relying on end-product testing, would be in harmony with current thinking by regulatory agencies [9, 10] and add to the assurance that the final APSD specifications for the OIP in question would be appropriate.

References

1. IPAC-RS (2011) Satellite conference at RDD-Europe 2011: Perspectives on Efficient Data Analysis methods and Abbreviated Impactor Measurements as quality assessment tools. Presentations and Summary Report. Available at: http://www.ipacrs.com/CI.html. Accessed 27 Jan 2012
2. Tougas T (2011) Lifecycle aspects of incorporating AIM-EDA into development cycle: Q&A Technical aspects presented at satellite conference at RDD-Europe 2011: Perspectives on efficient data analysis methods and abbreviated impactor measurements as quality assessment tools. Presentations and Summary Report. Available at: http://www.ipacrs.com/CI.html. Accessed 27 Jan 2012
3. Efficient data analysis, abbreviated impactor measurements, aerodynamic particle size distributions: publications and presentations by members of the IPAC-RS Cascade Impaction Working Group and EPAG (2011). Available at: http://www.ipacrs.com/CI.html. Accessed 27 Jan 2012
4. Christopher D, Dey M, Lyapustina S, Mitchell J, Tougas T, Van Oort M, Strickland H, Wyka B (2010) Generalized simplified approaches For *MMAD* determination. Pharm Forum 36(2):812–823
5. Mitchell J, Tougas T, Christopher JD, Lyapustina S (2010) Extension of the Abbreviated Impactor Measurement (AIM) concept to incorporate simultaneous determination of Delivered Dose Uniformity with Efficient Data Analysis metrics pertinent to aerodynamic particle size (AIM-DDU Apparatus). Drug Delivery to the Lungs-21, The Aerosol Society, Edinburgh, 221–214. Available at: http://ddl-conference.org.uk/index.php?q=previous_conferences. Visited 4 Aug 2012

6. Garmise RJ, Hickey AJ (2008) Calibration of the Andersen cascade impactor for the characterization of nasal products. J Pharm Sci 97(8):3462–3466
7. Marple VA, Olson BA, Santhanakrishnan K, Roberts DL, Mitchell JP, Hudson-Curtis BL (2004) Next generation pharmaceutical impactor. Part III: Extension of archival calibration to 15 L/min. J Aerosol Med 17(4):335–343
8. Bowles N, Cahill E, Häberlin B, Jones C, Mett I, Mitchell J, Müller-Walz R, Musa R, Nichols S, Parkins D, Pettersson G, Preissmann A, Purewal T, Schmelzer C (2007) Application of quality by design to inhalation products. In: Dalby RN, Byron P, Peart J, Suman JD (eds) Respiratory drug delivery-Europe 2007. Davis Healthcare International Publishing LLC, River Grove, IL, pp 61–69
9. US Food and Drug Administration (FDA) (2006) Guidance for industry: quality systems approach to pharmaceutical CGMP regulations. Rockville, MD. Accessed 26 June 2012 at: http://www.fda.gov/downloads/Drugs/GuidanceComplianceRegulatoryInformation/Guidances/ucm070337.pdf. Accessed 6 Sep 2012
10. European Medicines Agency (EMA) (2011) EMA-FDA pilot program for parallel assessment of quality by design applications. EMA/172347/2011, London. Available at: http://www.ema.europa.eu/docs/en_GB/document_library/Other/2011/03/WC500103621.pdf. Accessed 26 June 2012

Chapter 14
Conclusions

Terrence P. Tougas, Svetlana A. Lyapustina, and Jolyon P. Mitchell

Abstract The concluding chapter to the book comprises a brief summing up of the overall content of the previous chapters, followed by a list of learning points to help clarify the most important aspects of good cascade impactor practice, AIM and EDA, both for the novice and for the reader with experience who is looking for "nuggets" of knowledge about these topics.

14.1 Summing Up

This book is a compilation of current knowledge concerning the use of the CI as the principal aerodynamic particle sizing technique for those wishing to characterize the in vitro performance of OIPs. It has been structured to provide initially an outline of the cascade impaction method, leading to the formalized development of a series of interconnected methods for preparation and use of these apparatuses that has been termed "good cascade impactor practice." Having established this platform of basic knowledge, the potential for rapid abbreviated impactor measurement (AIM) and effective data analysis (EDA) concepts has been explored in relation to OIP quality control. Although these approaches can readily be used together, EDA

T.P. Tougas (✉)
Boehringer Ingelheim Pharmaceuticals Inc., 900 Ridgebury Road, Ridgefield, CT 06877-0368, USA
e-mail: terrence.tougas@boehringer-ingelheim.com

S.A. Lyapustina
Drinker Biddle & Reath LLP, 1500 K Street NW, Washington, DC 20005-1209, USA
e-mail: svetlana.lyapustina@dbr.com

J.P. Mitchell
Trudell Medical International, 725 Third Street, London, ON N5V 5G4, Canada
e-mail: jmitchell@trudellmed.com

T.P. Tougas et al. (eds.), *Good Cascade Impactor Practices, AIM and EDA for Orally Inhaled Products*, DOI 10.1007/978-1-4614-6296-5_14,

is equally capable of being applied in the assessment of full-resolution CI data. Furthermore, EDA metrics have the potential for improved discriminating power in relation to batch disposition, compared with the current practices either of comparing stage groupings or the determination of $FPM_{<5.0\,\mu m}$, used as quality measures in the USA and Europe, respectively. At the time of writing, both concepts are very new to stakeholders, so that the pathway to their acceptance will require carefully crafted validation studies, as outlined in a chapter devoted to the consideration of how to go about achieving recognition as standard techniques. A chapter has been devoted to review the development of abbreviated CI-based measurement techniques for OIPs and associated add-on devices. Such methods are becoming more important in the context of providing APSD data that has been acquired by techniques that more closely mirror the use by patients that can be achieved in the necessarily simplified laboratory techniques associated with quality control. Finally, prospects for the future developments of AIM and EDA are considered, including the potential for combining AIM-based size metrics with dose content uniformity determination in a single apparatus. It is hoped that the content of the book will result in the widespread use of these new approaches to the laboratory performance assessment of OIPs.

14.2 Learning Points

The underlying purpose behind the creation of this book has been the desire by the authors to provide both the novice reader and the user familiar with the cascade impaction method, a "one-stop shop" for technical information that relates to the current knowledge of the CI for OIP testing. The intention here is to provide a series of key learning points that it is hoped the reader will have acquired by reading the book from cover to cover. However, it is also recognized that the majority of users will likely dip into only one or more chapters of specific interest at any given time, so the list below serves as an easy access to the key scientific material that has been presented.

Chapter 1: The AIM and EDA concepts are presented with diagrammatic references to generic CI configurations; the idea that both concepts can be used together or independently is also introduced.

Chapter 2: The underlying theory associated with inertial impaction is explained in sufficient detail for the reader to understand the underlying scientific basis for the AIM concept and its limitations. This chapter is also a repository of useful design information relating to the compendial CI apparatuses. The use of induction ports and pre-separators is described, as is the adaptation of the CI method using a timed-delay apparatus following inhaler actuation that is needed to evaluate OIP add-on devices, such as valved holding chambers.

Chapter 3: Changes in OIP aerosol APSD both during and following formation arise from many physical causes; the more important ones are examined in this chapter. Particle–particle agglomeration/coagulation is continuously acting on the formed aerosol cloud after actuation of the inhaler and will tend to result in a time-dependent increase in both *MMAD* and *GSD*. Agglomeration/coagulation is important when particle concentration is at its highest during the initial formation stage. Inertial impaction processes will decrease both *MMAD* and *GSD* by preferentially removing the largest particles at obstructions or bends during transport; this mechanism is important in connection with the transfer of the aerosol from the inhaler to the CI system in product performance testing or to the patient in use. Gravitational sedimentation, like agglomeration, is continuously present, selectively removing the largest particles from the formed aerosol with elapsed time. Hence, delayed aerosol inhalation/sampling by CI will result in a reduced aerosol mass concentration compared with optimal conditions in which no delay is present. Like agglomeration and gravitational sedimentation, molecular diffusion operates continuously on the formed aerosol; however, diffusion is important only for the finest particles that are in the submicron size range. This process will therefore have minimal impact on either *MMAD* or *GSD* from a typical OIP-generated aerosol. Electrostatic charge on either or both the aerosol particles and surrounding surfaces can have a major and unpredictable impact on the aerosol before it is inhaled or sampled; since the effects on APSD properties are variable, its mitigation or better avoidance altogether is therefore highly recommended when making CI measurements. Evaporative processes require the presence of a volatile component to the aerosol; the process is highly time dependent, resulting in a reduction in both *MMAD* and *GSD*, the latter by potential elimination of the finest particles that may comprise only volatile species. Finally, the condensation of ambient water vapor, also a time-dependent process, can be significant with hygroscopic particles, resulting in increases in both *MMAD* and *GSD*.

Chapter 4: The cascade impaction technique is both complex and exacting in terms of the skills required for successful measurement outcomes, leading to reproducible results. This chapter is a review of the underlying causes of variability associated with these measurements, arising from four main causes: man, machine, measurement, and material. Factors to be considered when implementing a CI-based measurement regimen are summarized, both for OIP method development and in routine (i.e., day-to-day) use, and a method failure diagnostic tree is presented, based on the good cascade impactor practice (GCIP) principle. The second part of this chapter explores the potential to simplify full-resolution CI measurements, introducing the AIM concept for the first time. Finally, consideration is also given to the ways in which data from CI-based measurements can be presented and used to derive metrics that are representative both of aerosol "quality" and likely deposition behavior in the human respiratory tract.

Chapter 5: The underlying reasons for the introduction of both AIM and EDA concepts are examined, together with the development of two different but related purposes for CI measurements, namely, OIP quality control and related regulatory activities, and advanced uses of data as aids in understanding how OIP aerosols interact with the human respiratory tract. Rationales for the application of the AIM concept to either knowledge acquisition stream are examined, and the utility of the EDA concept in the assessment of OIP aerosol quality is also introduced. This data analysis concept, though compatible with AIM-based systems, is applicable to both full-resolution and abbreviated impactor-based measurements. Guidance is also given in the selection of an appropriate AIM-based CI system, providing the foundation for the detailed results that are presented in Chap. 10, from experimental evaluations of the wide range of options that are currently available. Finally, advice is summarized in connection with the selection; appropriate approaches based on AIM and/or EDA are recommended for the particular OIP aerosol assessment task in hand.

Chapter 6: Having introduced both AIM and EDA concepts, this chapter examines their potential roles throughout the OIP life cycle from product and method development, through in vitro support to clinical trials, to development of the regulatory submission to commercial production. The foundations are laid for a CI-based measurement and data analysis management strategy that is appropriate to each stage of the life cycle, with the idea presented of selecting "the right impactor for the right purpose."

Chapter 7: The theory that underlies the EDA concept has a sound basis in measurement system analysis (MSA) theory. This chapter reviews the reasons for making measurements of CI-generated APSD as a critical quality attribute for OIPs, in a product QC environment. The various constraints that different QC metrics (e.g., stage groupings and groupings for $FPM_{<5.0\mu m}$) impose on the measurement of APSD are discussed. These ideas are then applied to demonstrate that EDA provides an approach to the assessment of APSDs that avoids confounding of the key variables, *MMAD* and *AUC*, and thereby optimizing the discriminating power of the EDA methodology.

Chapter 8: Although the EDA approach to CI-generated data analysis is simple in principle, its application in the current regulatory climate for OIPs requires a rethink concerning the appropriateness of widely practiced methods that rely on grouping of stages as the principal means of data reduction. This chapter continues the theoretical assessment of the EDA concept, examining the comparison with the stage grouping approach as recommended by the FDA. Three different assessment APSD data techniques are presented and sample data analyzed; these are measurement system analysis (MSA), operating characteristic curves (OCCs), and principal component analysis (PCA). In the case of both MSA- and OCC-based techniques, an IPAC-RS blinded database containing several thousand individual CI-generated APSDs derived from eight different types of marketed OIPs (a collection of MDI and DPI products) has been used to provide examples that demonstrate the superiority of EDA. Likewise, a smaller, but still significant, database comprising 1,738 NGI APSDs has been used with PCA, again confirming the advantage of EDA.

Chapter 9: The underlying reasons why EDA is able to detect changes to APSDs based on shifts in *MMAD* and/or *AUC* are explored, including a theoretical assessment to establish plausible scenarios in which this concept might fail. Findings from failure modes and effects analyses (FMEA) are presented. These analyses have identified the relationships for the risk dependence of the CI data analysis method for detection of important APSD shifts associated with different potential causes during the various stages in the production of both MDI and DPI forms of OIP. This chapter concludes by examining the relative performance between EDA- and FDA-type stage groupings with two case studies, both involving product previously marketed or currently available in the USA. In both instances, EDA was found to possess greater potential for discriminating APSD change in a robust manner.

Chapter 10: The focus in this chapter is the establishment of strengths and limitations of all the various AIM-based CI techniques that have been developed by many different and independent contributing groups of researchers on this topic since 2007. Every class of OIP has at some time been evaluated with one or more AIM-based apparatus, mostly with successful outcomes compared with common APSD-derived metrics determined by a reference full-resolution CI method. However, there are several practical precautions that need to be considered, in particular the following:

1. Mitigation of particle bounce and re-entrainment, a phenomenon that can be more evident with AIM-based apparatuses
2. The provision of comparable conditions for the evaporation of low-volatile species (i.e., ethanol) that are associated with certain OIPs in the AIM-based apparatus as exist within the full-resolution CI
3. The maintenance of comparable dead space within the abbreviated system to that in the full-resolution CI so that the flow rate-time curves for DPI testing are also similar

Although many abbreviated CIs have a corresponding "parent" full-resolution apparatus (e.g., the FSA and the ACI), some, in particular the FSI, do not. This restriction, however, need not be a limiting factor in the choice of equipment like the FSI, as long as the necessary due diligence is done to seek out potential sources of bias in method development. However, it is important for the ACI-based systems to use the correct abbreviated CI that relates to the parent impactor (i.e., the FSA with the nonviable ACI and the FPD with the viable ACI). This precaution is necessary, since the internal dead space of the nonviable CI is significantly smaller than its equivalent with the viable design, due to the use of Petri dish-type collection surfaces in place of near-to-flat collection plates with the nonviable design.

Chapter 11: AIM and EDA concepts have by now established a firm basis for their adoption by stakeholders involved with OIP aerosol assessments. However, consideration must now be given to the development of processes based on sound scientific principles and supported by evidence, for their adoption into the pharmacopeias and perhaps eventually into the regulatory guidance literature. This chapter provides an outline of the pathways that both concepts will likely follow in order to achieve this goal. The fact that these concepts are in harmony with the Quality-by-Design principle should be advantageous to their eventual acceptance by the regulatory agencies.

Chapter 12: The application of an AIM-based approach, combined with improvements to the architecture of the CI measurement system, in particular the induction port, has been explored with the ultimate goal of making improvements to the present less than satisfactory situation concerning in vitro–in vivo relationships that exists with OIPs. Complete resolution of the discrepancies between aerosol transport in impactors and in the HRT cannot be attained, because, among many reasons, the continuous reduction in flow velocity profile along the length of the HRT cannot be realized with current CI-based technology. However, interfacing CIs (full resolution or abbreviated) with breathing simulators is becoming more common, as more robust configurations for the aerosol transport from inlet, where the flow rate is continuously varying, to the CI, where it must be kept constant, are realized. These developments could avoid the need for parallel experiments, establishing *TM* by filter collection and breathing simulation and later mass subfractions (i.e., *CPF*, *FPF*, and *EPF*) by the CI method. The interface between OIP and the measurement apparatus has to be considered more carefully, in cases where the inhaler has a facemask instead of a mouthpiece. Novel face models that simulate the skin and sub-skin facial tissue resistance to an applied force in clinically realistic ways when the facemask is fitted are the way forward to the development of more realistic testing conditions.

Chapter 13: The future directions likely to be taken for both AIM and EDA concepts are examined with the help of feedback from stakeholders that was received at the IPAC-RS-sponsored conference in 2011 on the topic "Perspectives on Efficient Data Analysis Methods and Abbreviated Impactor Measurements as Quality Assessment Tools."

Abbreviations

Term	Description
ACI	Andersen 8-stage nonviable cascade impactor
Act	*See*: ex-Act
AVCI	Andersen 8-stage viable impactor
AIAG	Automotive industry action group
AIT	"Alberta" idealized anatomic adult throat
ANOVA	Analysis of variance
API	Active pharmaceutical ingredient
APSD	Aerodynamic particle size distribution
ASTM	American Society for Testing and Materials
AUC	Area under the curve defined by the continuous form of the differential mass-weighted APSD
CDER	Center for Drug Evaluation and Research (as of the US FDA)
CDRH	Center for Devices and Radiological Health (as of the US FDA)
CFC	Chlorofluorohydrocarbon
C-FSA	(Copley Scientific) fast-screening Andersen (abbreviated) impactor
CI	Cascade impactor
CIWG	Cascade Impactor Working Group of IPAC-RS
CMC	Chemistry, manufacturing, and controls
CR	Chapman–Richards multipoint curve fitting procedure used in context with estimation of MMAD
CU	Content uniformity
CV	Coefficient of variation
DCU	Dose content uniformity
DOE	Design of experiments
DPI	Dry powder inhaler
DUSA	Dose unit sampling apparatus

S.A. Lyapustina et al. (eds.), *Good Cascade Impactor Practices, AIM and EDA for Orally Inhaled Products*, DOI 10.1007/978-1-4614-6296-5,

ED	Emitted dose ex inhaler (used in context of clinical dosing)
EPF	Extra-fine particle fraction of the size-fractionated aerosol
EPM	Extra-fine particle mass of the size-fractionated aerosol
EDA	Efficient data analysis
e-Lung	Electronic Lung™ (GSK plc)
EMA	European Medicines Agency
EPAG	European Pharmaceutical Aerosol Group
Ex-ActM	Total mass of API ex actuator mouthpiece of the inhaler
Ex-MVM	Total mass of API ex-metering valve of an MDI
FDA	United States Federal Drug Administration
FMEA	Failure modes and effects analysis
FDF	Fine droplet fraction (for nebulizer-generated aerosols)
FEF_{25-75}	Forced expiratory flow between 25 and 75 % of vital capacity during exhalation in a forced vital capacity maneuver, associated with pulmonary performance testing by spirometry
FEV_1	Forced expiratory volume in 1 s, associated with pulmonary performance testing by spirometry
FPD	Fine particle dose (abbreviated) impactor
FPD-AVCI	Fine particle dose (Westech) impactor
FPDU	Fine particle dose uniformity
FPF	Fine particle fraction of the size-fractionated aerosol
FPM	Fine particle mass of the size-fractionated aerosol
FSA	Generic fast-screening Andersen (abbreviated) impactor
rFSA	Rapid FSA methodology (see Chap. 10)
FSI	(MSP Corporation) fast-screening (abbreviated) impactor
GageR&R	Gage repeatability and reproducibility
GCIP	Good cascade impactor practices
GSK	GlaxoSmithKline
HC	Health Canada
HFA	Hydrofluoroalkane
HPLC	High-performance liquid chromatography
HRT	Human respiratory tract
ICRP	International Commission on Radiological Protection
IE	Inspiratory effort
IM	Impactor mass including the mass of API collected on the first stage of the ACI (i.e., stage 0 of the configuration for use at 28.3 L/min), the upper-bound size of which is undefined

IP	Induction port to CI system
IPAC-RS	International Pharmaceutical Aerosol Consortium on Regulation and Science
ISM	Impactor-sized mass, defined as the total mass of API collected by the size-fractionating stages of the CI of which the upper-bound size limit is established (EDA analysis)
IVIVC	In vitro–in vivo correlation
IVIVR	In vitro–in vivo relationship
LAL	Lower acceptance limit
LC	Label claim (refers to the amount of API stated in a product's labeling)
LD	Laser diffraction, sometimes referred to as low-angle laser light scattering (LALLS)
L/min	Liters per minute
LL	Lower limit of acceptance for OCCs
LOD	Limit of detection of the API under consideration
LOQ	Limit of quantitation of the API under consideration
LPF	Large particle mass fraction (EDA analysis)
LPM	Large particle mass (EDA analysis)
MAA	Marketing authorization application (Europe)
MB	Mass balance of API within the entire cascade impactor system
MDI	(Pressurized) metered dose inhaler
MMI	Marple-Miller 5-stage impactor
MOC	Micro-orifice collector of the NGI
MMAD	Mass median aerodynamic diameter of the CI size-fractionated aerosol
MMF	Mercer–Morgan–Flodin multipoint curve fitting procedure used in context with estimation of MMAD
MPS	CI data expressed in terms of mass per stage
MRI	Magnetic resonance imaging
MSA	Measurement system analysis
MSLI	4- or 5-stage multistage liquid impinger
MT	(Inhaler) mouthpiece and (CI) throat/IP
MVM	*See*: ex-MVM
NA	Not applicable
ND	Not determined
NDA	New drug application (United States)
NGI	Next generation (pharmaceutical) impactor
rNGI	Reduced next generation impactor
NIR	Near infrared (spectrometry/spectroscopy)
NISM	Non-impactor-sized mass of the aerosol emitted by the OIP

OBIC	Open-breathe in-close
OCC	Operating characteristic curve
OINDP	Orally inhaled and nasal drug product
OIP	Orally inhaled (drug) product
OOS	Out of specification
OPC	Optical particle counter
OOT	Out of trend
PC	Principal component
PCA	Principal component analysis
PDPA	Phase Doppler particle (size) analysis
Ph.Eur.	European Pharmacopeia
pHRT	Possibly relevant for (assessing) human respiratory tract (particle deposition)
$PM_{1.0}$, $PM_{2.5}$, PM_{10}	Aerodynamic size limits used in the fields of occupational hygiene and atmospheric pollution to define suspended airborne particulate mass fractions finer than 10, 2.5, and 1.0 μm, respectively
PS	Pre-separator
PtC	Points to consider
QbD	Quality by design
QC	(Product) quality control
QTPP	Quality target product profile
RMSE	Root mean square error
rNGI	Reduced next generation (pharmaceutical) impactor
sACI	2-Stage abbreviated nonviable ACI
SEM	Scanning electron microscopy
SMI	Soft mist inhaler
SPF	Small particle mass fraction (EDA analysis)
SPM	Small particle mass (EDA analysis)
TEM	Total emitted mass from the inhaler, including all non-sizing components of the CI system such as the induction port, pre-separator (if used), and the first stage for the ACI (if used)—used in context of in vitro testing of inhaler
TM_{IP}	Total mass of API in the induction port
TM_{exMVM}	Total mass of API ex-inhaler metering valve
T-FSA/sACI	Trudell (Medical) fast-screening Andersen impactor, identical with standard reduced ACI used in AstraZeneca (UK) studies
TI	Glass "Twin Impinger"

TIR	Total recovery of API within the impactor part of the measurement system [i.e., not including induction port or pre-separator (if used)]
TM	Total mass of the aerosol emitted from an OIP per actuation
TMI	Trudell Medical International
TOF	Time-of-flight aerodynamic particle size analysis
TOI	Total mass of API in the impactor system
TPP	Target product profile
UAL	Upper acceptance limit
UL	Upper limit for acceptance in OCCs
UPLC	Ultra-high-performance liquid chromatography
USP	United States pharmacopeia
USP-NF	USP-national formulary
UV	Ultraviolet (light detection of API)
VHC	Valved holding chamber
3D	Three-dimensional

About the Editors

Terrence P. Tougas, Ph.D., is a Highly Distinguished Research Fellow in analytical development at Boehringer Ingelheim Pharmaceuticals, Ridgefield, CT, where he has over 20 years of experience and currently heads the Stability, Submission Documents and Information Systems Group. He contributed to chemistry, manufacturing, and controls sections of several New Drug Applications, including for pulmonary (metered dose inhalers, inhalation solutions, and nasal sprays) and antiviral products.

Dr. Tougas is a member and past chair of the International Pharmaceutical Aerosol Consortium on Regulation and Science (IPAC-RS) Board of Directors and has been leader or member of several IPAC-RS working groups. Dr. Tougas served on the Steering Committee and chaired the Drug Product Technical Committee of the Product Quality Research Institute (PQRI). More recently, he helped form the International Consortium on Innovation and Quality in Pharmaceutical Development (IQ), served as the first chair of its Board of Directors, and is a member of the IQ Statistics Leadership Group.

Dr. Tougas authored numerous publications related to the CMC aspects of drug development, analytical chemistry, and quality control statistics. He contributed chapters to books on chromatography and leachables/extractables testing of inhalation products. His background is in analytical chemistry; he holds a Ph.D. in chemistry from the University of Massachusetts, Amherst.

Jolyon P. Mitchell, Ph.D., is Scientific Director at Trudell Medical International, London, Canada. He is involved in several industry-wide organizations involved with inhaled medical aerosol delivery, in particular the European Pharmaceutical Aerosol Group (EPAG), as well as serving as a scientific adviser to IPAC-RS. He played major parts in the development of an international standard (ISO 20072:2009) covering the design verification of portable inhalers, as well as a Canadian Standard for Spacers and Holding Chambers (CAN/CSA Z264.1-02:2002). He is currently a Canadian delegate to ISO/TC121/SC2, involved with the ongoing development of a standard specifically for nebulizing systems (ISO 27427). In 2010, he was recently appointed to the Expert Committee: General Chapters—Dosage Forms of the

S.A. Lyapustina et al. (eds.), *Good Cascade Impactor Practices, AIM and EDA for Orally Inhaled Products*, DOI 10.1007/978-1-4614-6296-5,

United States Pharmacopeial Convention for the 2010–2015 term, where he serves as vice chair of the aerosols dosage form subcommittee.

His background is in physical chemistry. He is a fellow of the UK Royal Society of Chemistry, a Chartered Scientist, a founding member of the UK-Irish Aerosol Society, and a member of the Gesellschaft für Aerosolforschung, American Association for Aerosol Research, American Association of Pharmaceutical Scientists, and the International Society for Aerosols in Medicine. He is on the Editorial Advisory Board of Journal of Aerosol Medicine. He is also an adjunct professor at the University of Western Ontario.

Svetlana A. Lyapustina, Ph.D., is a Senior Science Advisor in the Pharmaceutical Practice Group of the Washington, DC office of Drinker Biddle & Reath LLP. As a member of that multidisciplinary team, Dr. Lyapustina has served as science advisor and secretariat for several industry consortia, including the International Pharmaceutical Aerosol Consortium on Regulation and Science (IPAC-RS), the International Consortium for Innovation and Quality in Pharmaceutical Development (IQ), an International Pharmaceutical Supply Chain Consortium Rx-360, Alliance for Biosecurity, Nanomedicines Alliance, and Allotrope Foundation. She was an active member of the Product Quality Research Institute (PQRI) and of a project team of the United States Pharmacopeia and a US national expert in the International Organization for Standardization (ISO). Dr. Lyapustina counsels clients on a wide range of topics related to drug and device product development, pharmaceutical manufacturing, regulatory compliance, quality control, bioequivalence requirements, intellectual property, and other issues. Her educational background is in physical chemistry and biophysics. She has authored or coauthored numerous articles on regulatory and scientific topics in drug product development from the US, European, and international perspectives, as well as articles and presentations about the process and value of cross-industry collaborations.

Symbols Used in Mathematical Expressions

Term	Description
A_i	Best-fit parameter for ACI collection efficiency-aerodynamic diameter profile at $Q = 28.3$ L/min to hyperbolic tangent function
A_t	Total area of the nozzle array for a multi-nozzle CI stage; the area of a single-nozzle CI stage is numerically equal to A_t
B_i	Best-fit parameter for ACI collection efficiency-aerodynamic diameter profile at $Q = 28.3$ L/min to hyperbolic tangent function
b	Slope of linear regression line
C_c	Cunningham slip correction factor; Cunningham slip correction factors related to the volume-equivalent and aerodynamic diameters respectively of a suspended particle
CMD	Count median diameter
CPF	Coarse particle fraction of the size-fractionated aerosol
CPM	Coarse particle mass of the size-fractionated aerosol
d_{10}, $d_{15.1}$, $d_{84.9}$, d_{90}	Aerodynamic diameters corresponding to the 10th, 15.1th, 84.9th, and 90th mass percentiles of the unimodal APSD
d_{ae}	Aerodynamic diameter of a suspended particle
d_{ini}	Initial droplet size in the context of evaporation of volatile components
d_p	Particle diameter expressed as volume-equivalent diameter (i.e., the size of a spherical particle occupying the same volume as the particle in question)
d_{pc}	Particle diameter corrected for slip

S.A. Lyapustina et al. (eds.), *Good Cascade Impactor Practices, AIM and EDA for Orally Inhaled Products*, DOI 10.1007/978-1-4614-6296-5,

d_{50}	Either 50th mass percentile of the unimodal APSD, in context of APSD analysis or aerodynamic diameter at which impactor-stage collection efficiency is 50% (cutoff size; effective cutoff diameter), in context of CI-stage performance
D	Characteristic size related to that of a particle in consideration of particle inertial behavior
D_d	Molecular (Brownian) diffusion coefficient of a particle
D_{eff}	Effective diameter of an array of more than one nozzle for a given CI stage; D_{eff} is numerically identical to the nozzle diameter of a stage containing a single nozzle
D_i	Initial droplet diameter in consideration of evaporative changes
D_{median}	Area-weighted median of an array of more than one nozzle for a given CI stage
D^*	Area-weighted mean diameter of an array of more than one nozzle for a given CI stage
erf(…)	Error function associated with the calculation of the mass fraction of the input aerosol that goes to a given stage of either a full-resolution or an abbreviated CI
E; E_{50}	Impactor-stage collection efficiency; stage 50% collection efficiency
E_i	Fractional collection efficiency curve of stage *i* of a CI
ED	(total) Emitted dose from the inhaler
EPF	Extra-fine particle fraction of the size-fractionated aerosol
EPM	Extra-fine particle mass of the size-fractionated aerosol
Ex-act	Total mass of API ex actuator mouthpiece of the inhaler
Ex-MVM	Total mass of API ex-metering valve of an MDI
F_d	Drag force imposed by the surrounding fluid (air) on a suspended particle
F_g	Gravitational force on a suspended particle
$F_m(d_{ae,i})$	Mass frequency of API on each stage of the size-fractionating part of the CI system
f_N	Mass fraction of the incoming aerosol that deposits on stage *N* of a CI
FDF	Fine droplet fraction (for nebulizer-generated aerosols)
FEF_{25-75}	Forced expiratory flow between 25 and 75% of vital capacity during exhalation in a forced vital capacity maneuver, associated with pulmonary performance testing by spirometry
FEV_1	Forced expiratory volume in 1 s, associated with pulmonary performance testing by spirometry
FPD	Fine particle dose (abbreviated) impactor
FPF	Fine particle fraction of the size-fractionated aerosol
FPM	Fine particle mass of the size-fractionated aerosol

g	Acceleration due to gravity
GSD, GSD_{stage}	Geometric standard deviation of the size-fractionated aerosol, size selectivity of an impaction stage represented in terms of its geometric standard deviation based on its collection efficiency-aerodynamic size profile
h	Model variable in definition of error function associated with the calculation of the mass fraction of the input aerosol that goes to a given stage of either a full-resolution or an abbreviated CI
IM	Impactor mass including the mass of API collected on the first stage of the ACI (i.e., stage 0 of the configuration for use at 28.3 L/min), whose upper bound size is undefined
ISM	Impactor-sized mass, defined as the total mass of API collected by the size-fractionating stages of the CI whose upper-bound size limit is established (EDA analysis)
K; K_{mono}	Average agglomeration coefficient for a generalized unimodal and log-normally distributed particulate system suspended in a supporting gaseous medium; average agglomeration coefficient for a monodisperse system
Kn_p	Particle Knudsen number (important for submicron-sized particles)
k	Boltzmann constant
L	Length of nozzle or nozzles in a particular impactor stage
LPF	Large particle mass fraction (EDA analysis)
LPM	Large particle mass (EDA analysis)
m	Mass of an individual particle
m_i	Mass of API on stage "i" of the CI
M_{EM}	Emitted mass of API from the inhaler that enters the CI system
MMAD	Mass median aerodynamic diameter of the CI size-fractionated aerosol
N_0; $N(t)$	Initial number concentration (density) of an aerosol in consideration of particle agglomeration; the time-dependent relationship of particle number concentration in an agglomerating aerosol
n	Number of nozzles per impactor stage
P_1; P_2; P_3	Air pressure at induction port; pressure on the impactor side of the flow-regulating valve (DPI testing); pressure on the vacuum-pump side of the flow-regulating valve (DPI testing)
p_a; p_{accept}	Probability of acceptance in context of the interpretation of data from operational characteristic curves (OCCs)

p_{diff}; p_{sed}	Root mean square (RMS) displacement of a particle in unit time due to molecular (Brownian) diffusion; RMS displacement of a particle in unit time due to gravitational sedimentation
$\Delta P_{stage(i)}$	Pressure drop (flow resistance) across a given CI stage "*i*"
Q; Q_l	Volumetric gas (air) flow rate through the CI system; volumetric liquid feed rate
Q^2	The predicted variation (in principal component analysis)
Q_{const}; $Q_{const\text{-}CI}$	Constant flow rate; constant flow rate required by cascade impactor
$Q_{br\text{-}sim}(t)$	Time-dependent flow rate profile provided by a breathing simulator
R^2, r^2	Coefficient of determination, correlation coefficient
R^2X	statistic in PCA, used to indicate how well the model explains the variability of a real data set
Re_f	Flow Reynolds number
RH_{amb}; RH_{op}	Room-ambient relative humidity; relative humidity in oropharynx
RMSE	Root mean square error
RSF	Relative span factor
S	Perpendicular distance between exit plane from nozzle plate and collection surface of impactor
S_{APSD}	Spread of a unimodal but not necessarily lognormal APSD
SD	Standard deviation
SPF	Small particle mass fraction (EDA analysis)
SPM	Small particle mass (EDA analysis)
St, St_{50}	Suspended particle Stokes number, Stokes number at which a CI stage is 50% efficient as a particle collector
t	Elapsed time in consideration of particle motion
t'	Nondimensional time in considerations of particle inertial behavior
T	Nozzle throat length for an impactor stage
T^2	Statistic used in Hotelling confidence ellipse construction in connection with PCA
T_{abs}; T_{amb}	Absolute temperature in degrees Kelvin; room-ambient temperature
TEM	Total emitted mass from the inhaler, including all non-sizing components of the CI system such as the induction port, pre-separator (if used), and the first stage for the ACI (if used)
TM_{IP}	Total mass of API in the induction port
TM_{exMVM}	Total mass of API ex-inhaler metering valve
TOI	Total mass of API in the impactor system

U; U_0	Linear gas (air) velocity associated with particle motion at a defined location within the CI; mean velocity of the gas (air) flow
ν'	Nondimensional particle velocity relative to the mean gas (air) velocity at some distance from the point of acceleration of the particle in considerations of inertial behavior
ν'**rel**	Nondimensional particle velocity relative to the gas (air) velocity during particle acceleration in considerations of inertial behavior
v_t	Suspended-particle terminal settling velocity
V_p	Particle volume
W	Impactor stage nozzle width (diameter for circular-profiled nozzles)
x	Mass of API that deposits on a collection surface of a defined stage of a CI
X_c	Impactor stage cross-flow parameter
Y_i	Best-fit parameter for generalized CI-stage collection efficiency-aerodynamic diameter profile to a hyperbolic tangent function
z	Best-fit parameter for generalized CI-stage collection efficiency-aerodynamic diameter profile to a hyperbolic tangent function
$\Delta d_{50,i}$	Size width (aerodynamic diameter scaling) associated with an individual CI stage
α	Type I statistical error (false rejection)
β	Type II statistical error (false acceptance)
κ–ω	Kappa–omega turbulence model at low flow Reynolds number
λ	Mean free-path length of gas (air) molecules supporting an aerosol
η_a; η	Air (aerosol supporting fluid) viscosity; fluid (unspecified) viscosity
ρ_0; ρ_p; ρ_g	Unit (reference) density (cgs system); particle density; supporting gas (air) density
σ_g	Geometric standard deviation of a hypothetical unimodal and log-normally distributed aerosol
τ	Particle relaxation time
χ	Suspended-particle dynamic shape factor

Index

S.A. Lyapustina et al. (eds.), *Good Cascade Impactor Practices, AIM and EDA for Orally Inhaled Products*, DOI 10.1007/978-1-4614-6296-5,

Printed by Publishers' Graphics LLC
ICISO130506.15.24.3